Ethical Issues in
Biotechnology

Ethical Issues in Biotechnology

Edited by Richard Sherlock
and John D. Morrey

ROWMAN & LITTLEFIELD PUBLISHERS, INC.
Lanham • Boulder • New York • Oxford

ROWMAN & LITTLEFIELD PUBLISHERS, INC.

Published in the United States of America
by Rowman & Littlefield Publishers, Inc.
A Member of the Rowman & Littlefield Publishing Group
4720 Boston Way, Lanham, Maryland 20706
www.rowmanlittlefield.com

12 Hid's Copse Road
Cumnor Hill, Oxford OX2 9JJ, England

British Library Cataloguing in Publication Information Available

Library of Congress Cataloging-in-Publication Data

Ethical issues in biotechnology / edited by Richard Sherlock and John D. Morrey.
    p. cm.
    Includes bibliographical references and index.
    ISBN 0-7425-1357-2 (cloth : alk. paper)—ISBN 0-7425-1377-7 (pbk. : alk. paper)
    1. Biotechnology—Moral and ethical aspects.   I. Sherlock, Richard.   II. Morrey,
John D., 1952–
    TP248.23 .E86   2002
    174′.25—dc21   2002004174

Printed in the United States of America

♾ ™ The paper used in this publication meets the minimum requirements of American National Standard for Information Sciences—Permanence of Paper for Printed Library Materials, ANSI/NISO Z39.48-1992.

Contents

Illustrations

TABLE

FIGURES

Preface

As this is being written, the president has just announced a much anticipated decision concerning federal funding for research involving human embryonic stem cells. Though heavily discussed in the press, the controversy shed only partial light on this complex topic. The same may be said of many other topics regarding biotechnology. When demonstrators complaining about genetically modified crops created havoc in downtown Seattle and Genoa, Italy, the complaint was duly noted. But the press offered little help to anyone wanting a real understanding of the relevant issues.

Our efforts in this book are one attempt to partially rectify this disconnect between interest and knowledge in the field of biotechnology. We treat biotechnology as a unit because that is what the deepest understanding of the issues requires. Other works may treat at great length human cloning or human genetic screening. For specialized research or courses they are excellent. We note, however, that the most fundamental claim about human beings playing god is made with respect to genetically modified crops and animals, as well as human cloning. Likewise the supposed dangers of permanent, life-altering changes to the DNA of living things are noted with respect to plants, animals, and human beings. The issues raised by some ecologists about genetically modified fish are mirrored in the "social ecology" concerns raised about the effect of cloning on families.

For several years we have been offering a course reflecting our perspective at Utah State University. No adequate book existed for our purposes, so we decided to create one. The readings included are ones we have used successfully. We begin with a brief review of the science of genetics and basic ethics so that students will have a general framework in which more particular discussion of issues can take place. In order to understand genetic modification, one must understand something about genes themselves. In the same way concerns about the suffering of transgenic animals are part of a much broader discussion of utilitarianism that in recent years has included the moral standing of animals.

As the contents page shows, the book is divided into six parts dealing with the major divisions of the broad field of biotechnology. In each part the most fundamental issues are examined in light of that topic, and some issues specific to that topic are noted as well. For example, general questions about benefits and harms are especially relevant to the question of genetically modified food, as are questions about food labeling that do not arise in the discussion of agriculture.

Each part is introduced by a discussion of the basic science and the basic ethics relevant to that section. This is intended to help the student place the readings in that section in the context with which we start the book. At the end of the book we have provided a number of study cases grouped according to parts of the text. These can be used by students and instructors to bring the issues discussed down to practical situations of thoughtful choice. Some of these cases provide a suggestion for their use in small group discussion. A class can be divided into small groups with the members of the group speaking up for one of a variety of positions. Groups can report to the class or write a short summary of their discussion. Our cases are designed to be balanced in their view toward biotechnology, as are our readings. They provide tools for thinking, not answers; a basis for serious reflection, not a reinforcement for prejudice. We have also appended a short set of recommendations for further research keyed to the major sections of the book. At the end there is a set of diagrams about genetics that are noted throughout the text.

The readings in this book have been chosen with two principles in mind. First, on each topic we strive for balance. Individual articles may not be balanced. They may be too optimistic or too pessimistic. But the book as a whole strives to present many sides of these difficult issues. Second, we aim at teachability. Some of the articles reflect a moderate overview of the issues. Many do not. They represent passionately held but well-expressed particular points of view. Some authors are strident but make a point of deep concern and often provoke thoughtful responses from readers. Students often read articles about topical issues such as capital punishment or abortion by passionate authors and are thereby challenged to engage the topic. Our experience is that some passionate writing helps clarify many fundamental issues and stimulate discussion. Balanced articles have a place, and we have included many to give students a solid center to their thinking. But activist writing often clarifies fundamental points of disagreement.

Almost every year we teach this course, something appears in the news each week that is relevant to the course. We frequently note news reports in class and tell students something like, "In a couple of weeks we are going to show you exactly why people might be concerned about the ecological effects of genetically modified fish" or "why the rules about animal-to-human transplants are written this way." These examples can make the materials seem more relevant to the everyday lives of the students. We encourage instructors and students to be alert to such examples, which are sure to come with regularity in the next few years.

This book would not be possible without the help, support, advice, and criticism of others. At Utah State University Brett Blanch, Nick Allan, Janeen Richins,

Mary Donohue, and Tom Sherlock helped produce the text and the graphics. Support from the Biotechnology Center, the College of Humanities, Arts, and Social Sciences, and the Department of Languages and Philosophy was also essential in producing the text. Criticism and advice from Courtney Campbell, Larry Arnhart, Paul Thompson, and reviewers for our publisher have helped make our book much better than it would have been. The weaknesses that remain are our responsibility. Eve DeVaro, our editor at Rowman & Littlefield, has been supportive of our project from the very first and has been very helpful along the crooked path from concept to product. We thank all who have helped and many who are not specifically mentioned. Finally, our students were a wonderful sounding board as we tried out ideas on them and tested and retested our readings in class. They inspired our book.

Introduction

The purpose of the scientific background presented in this book is to empower nonscience readers with the knowledge they need to evaluate the ethical and social issues of biotechnology. Without this basic knowledge, the reader is at the mercy of special interest groups who may manipulate the public and the media. Conversely, a person who has an understanding of the science of genetics and its ethical implications becomes a source of good judgments in these matters, empowered to truly affect the "ethical issues in biotechnology."

Jurassic Park the movie illustrates this concept. When Michael Crichton wrote *Jurassic Park,* there was concern in the biotechnology sector that his book would adversely affect efforts to advance the technology. The concern was amplified when the book was chosen to be made into a movie produced by Steven Spielberg. But a survey conducted after the movie was released found that the movie did not have the predicted adverse effect on public opinion toward biotechnology. The cartoon about DNA shown in the movie that prepared customers to enter the park was the most educational source of information on genetic engineering available to the public at that time. Because viewers were empowered with knowledge to feel the difference between fact and fiction, they could better appreciate the technology.

With the announcement of the cloning of Dolly the sheep, most people immediately assumed that humans could now be cloned. Yet experiments with monkeys and other animals have demonstrated that such a task may remain formidable for many years. Reports began circulating that hands, hearts, or other body parts could be cloned as donor parts, even though this has little to do with the technology that developed Dolly. More remarkable, some thought wealthy people would soon be able to clone themselves to provide some measure of immortality. Cloning from germ cells had been successful for nearly a decade previous to the cloning of Dolly. The remarkable discovery from Dolly was not the cloning of body parts or cloning for eternity, but a mammal that was produced by asexual reproduction from one adult parent, which was previously unique to nonmammalian species. By eliminating misconceptions, this book will help the reader understand the real scientific implications of Dolly and better evaluate the ethical and social implications of cloning.

Our strategy of presenting the scientific concepts of the book was summarized by Albert Einstein when he said, "We should make things as simple as possible, not simpler." We give only relevant scientific information to avoid unnecessarily burdening the reader. The science terms and figures are simplified wherever possible. Providing too much detail in an attempt to be completely accurate would be counterproductive to the ultimate goal of critically thinking through ethical issues of biotechnology, or the "new biology."

THE BASICS OF GENETICS

Since life is so complicated, early researchers predicted that proteins or other complex structures would hold the genetic blueprint for life. Today we all know that DNA holds the genetic code of life. But how can that be, since it is a relatively simple molecule compared with proteins? To understand how this works one must understand the structure of DNA.

DNA is composed of four basic building blocks—A, T, C, and G—that are held together in extraordinarily long strands. The sequence or order of these four different building blocks, or nucleotides, on a strand is the key to the complexity of life. It is not just that A's may be in greater concentration than other nucleotides, but it is the sequence order that holds the key. For example, a sequence of ATCGGACCTATA would code for a completely different characteristic than ATTCAAGCCTGGA, even though the concentration of each building block is identical between the two strands.

The different building blocks are held together by chemical linkages consisting of "phosphate bonds." These phosphate bonds exist to hold the building blocks together and are important in explaining how one strand of DNA can be linked to a different strand in the process of genetic engineering. Single strands of nucleotides, or DNA, combine into a duplex, or a double strand, to form what we identify as the "double helix." Adenine (A) and guanidine (G) are two-ringed structures (purines), whereas thymine (T) and cytosine (C) are one-ringed structures (pyrimidines) (figure 1). When nucleotides are connected together in a sequence, they form a ragged edge because the two-ringed structures take up more space than the one-ringed structures (figure 2a). In a cell, DNA usually exists as a double helix or double-stranded structure (figure 2b). For a double helix to form between two single-stranded DNA molecules, the sequences of each strand must be able to fit together along the ragged edge. The two-ringed nucleotides must be opposite the one-ringed structures, thereby forming a good fit between the two double strands of DNA. If two strands fit together well and form a duplex, they are said to be "complementary." Any mismatches between the two strands will tend to destabilize the duplex (figure 2c). Mismatches are formed when pyrimidines (one-ringed structures) are opposite pyrimidines or when purines (two-ringed structures) are opposite purines. Remember, a two-ringed structure must be opposite a single-ringed structure.

Another important feature is that there are weak bonds, called hydrogen bonds, that hold the two well-fit strands together (figure 3a). A's and T's both have two of these hydrogen bonds and G's and C's have three. To hold the duplex together, nucleotides with two hydrogen bonds must be opposite nucleotides also having two hydrogen bonds on the opposite strand. Accordingly, three-bond nucleotides need to be opposite three-bond nucleotides. Therefore, to have hydrogen bonds form that are sufficient to hold the well-fitted double strands of DNA together, the purines must not only be opposite pyrimidines, but A's must only be opposite the T's, having two bonds, and the G's must be opposite the C's, having three bonds. Consequently, any base pairing besides A's opposite T's or G's opposite C's will result in mismatches (figure 3b). If there are too many mismatches between the strands of DNA, the duplex will not form. Now you know the features of the building blocks that result in the stable formation of the "double helix."

We have talked about nucleotides (A, T, C, G) and phosphates, but as can be seen in figure 4, the DNA needs a "backbone." The sugar molecules serve that purpose, forming the joining piece between the nucleotides and the string of phosphate molecules. There are a couple of features of the sugar molecules that explain how all parts of the single-stranded DNA molecule are held together in a line. This can best be explained by numbering the carbon molecules on the sugar from one to five (figure 4). The nucleotides or bases are connected at the first carbon in the sugar molecule, denoted as 1' carbon. The phosphates are connected to the 3' and 5' carbons, thereby holding nucleotides in a line. Sugar molecules also help us understand the origin of the name "DNA." DNA stands for *deoxynucleic acid.* Notice that the 2' carbon does not have an oxygen attached to it (figure 5). The term "deoxy" refers to this absent oxygen. So the "deoxynucleotide" is a nucleotide without an oxygen at its 2' carbon. A string of nucleotides with oxygens at their 2' carbons is referred to as "ribonucleic acid," or RNA.

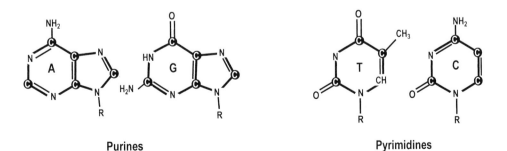

Purines   Pyrimidines

Figure 1.   Structure of Basic Building Blocks of DNA: Nucleotides

4

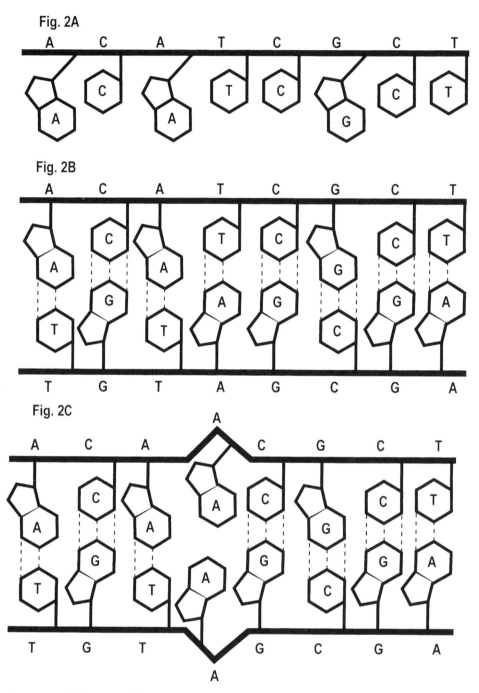

Fig. 2A

Fig. 2B

Fig. 2C

Figure 2a. Single-Stranded DNA Structure
Figure 2b. Double-Stranded DNA Structure
Figure 2c. Double-Stranded DNA Structure with One Mismatched Base Pair

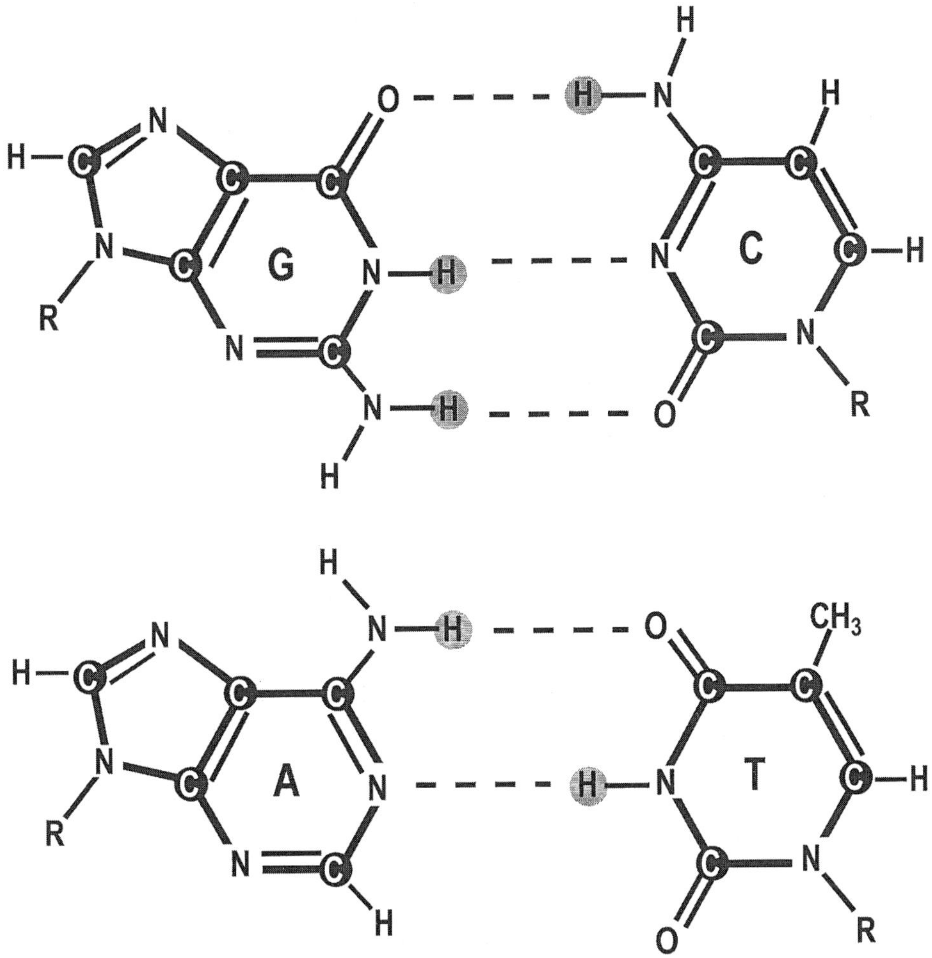

Figure 3a. **Hydrogen Bonding between Purines and Pyrimidines on Two Different DNA Strands**

Another important concept in issues of genetic testing, genetic engineering, and production of protein from DNA is that each strand of the double-stranded DNA molecule is in "antiparallel" configuration (figure 6). Notice that one terminal end of a strand of DNA is different from the other end of the same DNA molecule. The end of the DNA molecule that terminates with a phosphate connected to the 5' carbon of the sugar backbone is identified as the "5 prime end" (5' end) (figure 4). The end of the DNA molecule that terminates with an OH group on the 3' carbon of the sugar backbone is identified as the "3 prime end" (3' end). When two "complementary" DNA molecules "hybridize" or connect together to form double-stranded DNA, they are "antiparallel" with respect to each other, namely, the 5' end of one molecule lines up with the 3' end of the opposite DNA molecule

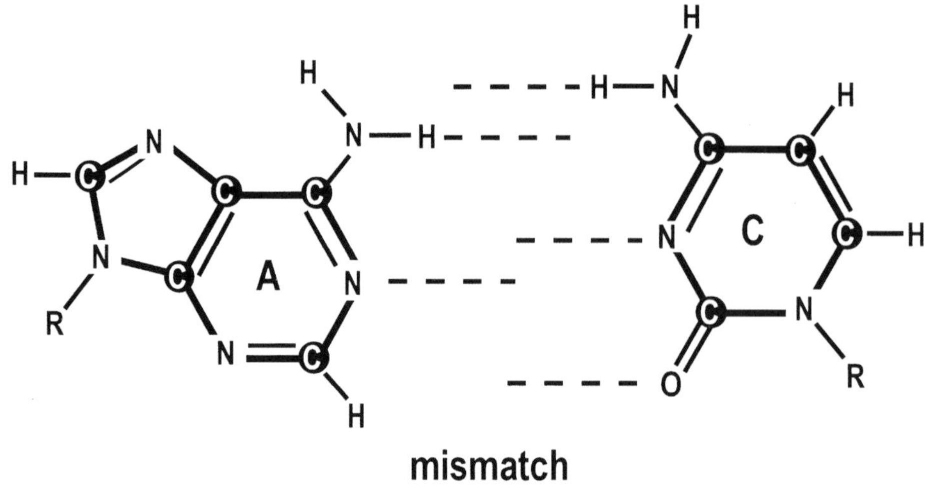

mismatch

Figure 3b. **Incompatible, Mismatched Hydrogen Bonding between Nucleotides on Two Different DNA Strands**

(figure 6). You can envision this configuration by folding your arms and observing that the hand of one arm is adjacent to the elbow of the other arm or that the arms are antiparallel with respect to each other. Hence the term "antiparallel" is used to describe this orientation of the two strands of the double-stranded DNA molecule. It is important to remember this antiparallel orientation when you read about other methods of genetic engineering described later in the book.

To summarize the structure of DNA, remember that the bases are the A's, T's, C's, or G's called nucleotides; the phosphorous is the linkage of individual building blocks of DNA; and the sugars form the backbone holding the bases to the phosphates (figure 4). The complementary base pairing of A's to T's and G's to C's is due to two features: purines are opposite pyrimidines, and nucleotides have the appropriate number of hydrogen bonds. DNA differs from RNA in that DNA does not have an OH group at the deoxy 2' carbon of the sugar and RNA does have an OH group. Finally, the two strands of a double-stranded DNA molecule are in an antiparallel orientation.

GENE STRUCTURE

The structure of DNA helps explain how a simple molecule can code for the complexity of life. One of the two single strands of DNA acts as a parent or template for new synthesis of a second strand of DNA or RNA (figures 7a and 7b, replication and transcription, respectively). The positions of the A's match with the T's

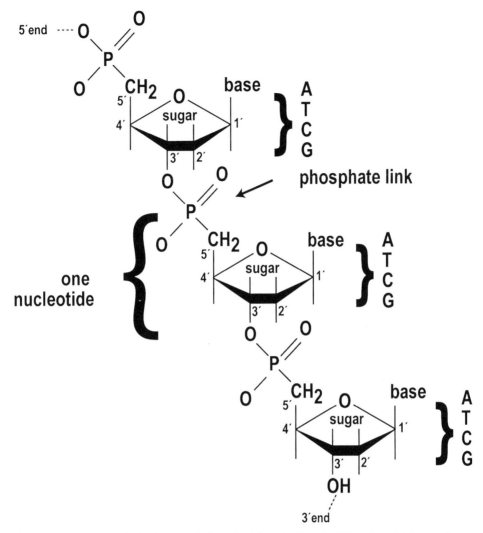

Figure 4. Structure of Single-Stranded DNA Molecule Showing Phosphate Linkages, Sugar Backbone, and Nucleotide Bases

and the G's match with the C's between the parental strand and the newly synthesized strand. Because of the complementary base pairing between the two strands of DNA, the parental template DNA can be used to synthesize a complementary strand of RNA or DNA. When DNA is produced, the process is called "replication" because old DNA is exactly replicated into new DNA (figure 7a). When RNA is produced from DNA, the process is called "transcription" (figure 7b). Knowing that the term for copying text from one page to another is called "transcription" will help you remember that the same term is used to describe the transcription of RNA, one page, from DNA, the other page. The two pages look very

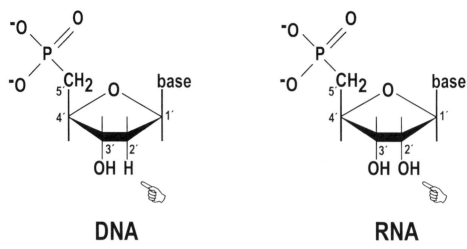

Figure 5. Structural Difference between DNA and RNA

similar except that they are duplicated on different sheets of paper, or DNA to RNA.

When cells divide, all of the DNA, or chromosomes, are also replicated; therefore, all cells have the same DNA. If all cells have the same DNA, what makes cells such as skin cells and liver cells different from each other? The answer to this question is important in understanding how to genetically engineer an organism. The differences can be accounted for through the structure of a gene.

Genes are the portions of DNA that code for individual proteins, which we call the protein coding region (figure 8). For example, there is a gene coding for a specific protein made in the liver called albumin and another gene coding for keratin protein in the skin. One gene differs from other genes in the sequence of the nucleotides, which codes for its specific protein. Knowing that different gene sequences code for different proteins is central to understanding genetics. Genes produce different proteins because of the different sequences in the protein coding region of the gene.

Each gene has other nucleotide sequences that do not code for protein, called promoter sequences. These sequences tell the cell when and in what tissues the genes are to produce their proteins (figure 8). These nonprotein-coding DNA sequences can regulate if and when the protein is made. Figure 8 shows a typical structure of a gene. Starting from the left to right is the promoter, which all genes have. Specific DNA sequences or nucleotides within this promoter are recognized and bind with cellular proteins that are involved in the process of "transcription," or copying RNA from DNA. The promoter for albumin, for example, has DNA sequences that are recognized by proteins found only in liver cells but not other cells such as skin cells (figure 9a). When the DNA sequences in this promoter bind to liver transcription proteins, RNA from the albumin gene is made, which codes for the production of albumin protein. Albumin is produced in liver cells because liver proteins recognize the albumin promoter. In contrast, keratin is a protein found only in the skin and not in the liver. Although the keratin gene is found in all cells, it is not produced in liver cells because liver transcription proteins do not recognize or bind to the keratin promoter (figure 9b).

Consequently, keratin RNA and protein are not made in the liver. Keratin RNA is made in the skin because the skin keratin protein binds to its promoter, which results in the production of keratin RNA and protein (figure 9b). One set of proteins is produced in a cell to make it a liver cell and another set of proteins is produced in the skin cell to make it a skin cell, even though both sets of cells have the same genes. This differential expression of proteins results in the development of cells having different structures and functions. As a result we are not composed of a ball of the same cells but have very different body parts, organs, and functions within the body.

By knowing this principle of "tissue-specific" promoters, genetic engineers know that a promoter specific for a certain cell type or tissue must be used if the engineered protein is to be produced in the desired specific tissue. For example, proteins that have pharmaceutical value have been produced in the milk of dairy animals. The dairy animals are used essentially as factories to produce the drug in large quantities in the milk. One of the keys to engineering these animals is to use tissue-specific promoters that are active in the specific cells found in the mammary gland. Proteins from these mammary cells that regulate transcription associate with the mammary-specific promoter to produce the gene's RNA and protein, which is then secreted into the milk. If the gene for the pharmaceutical protein were put into the mammary gland without an appropriate mammary-specific promoter, the protein would not be produced. In summary, specific DNA promoters must be used in a gene to produce a protein in its desired cell or tissue.

In the process of transcription, the transcription proteins attached to the promoter slide to the right, down the DNA of the gene (figure 8). The transcription proteins attach to the promoter and recognize the nucleotides of the DNA gene

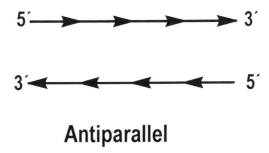

Antiparallel

Figure 6.   Antisense Orientation between Single Strands of a Double-Stranded DNA Molecule

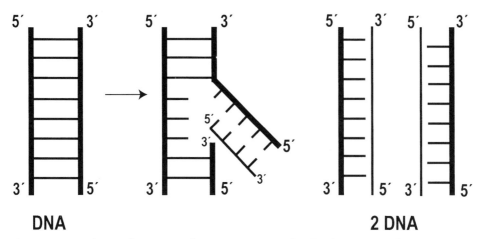

DNA   **2 DNA**

Figure 7a.   Synthesis of New Complementary DNA Molecules from Parental DNA

where the RNA synthesis begins (figure 8). At this point, called the transcription start site, the first ribonucleotide of the RNA molecule is coded by a complementary DNA nucleotide. Therefore, a transcription start site is also a necessary part of a gene.

Understanding other functions of the gene requires a knowledge of the simple structure of a cell. A living cell can be compared to an automobile manufacturing plant. The engineering department containing the master plans for the cars is analogous to the nucleus containing the genetic information. The function of the engineering department is to design the cars and then distribute copies of the plans out of the engineering office to the manufacturing facilities where the cars are actually made. The function of the nucleus is also to make copies, RNA, of the genetic information from the DNA (figure 10). The RNA is not useful until it is delivered to the manufacturing part of the cell called the cytoplasm. It is within the cyto-

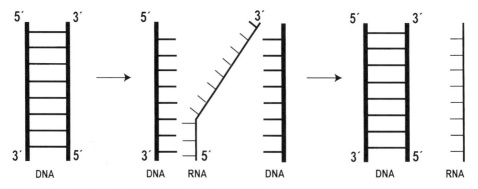

Figure 7b.   Synthesis of RNA Molecule from DNA

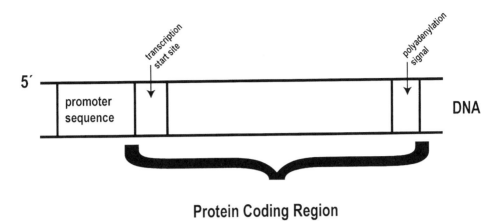

Protein Coding Region

Figure 8. Typical Structure of a Gene with a Promoter and Protein Coding Region

plasm that the manufacturing takes place—the proteins and other components are made to allow the cell to function properly and to build new cells.

At the right-hand side of the gene in figure 8 is the DNA sequence called the polyadenylation (PA) signal. When DNA is transcribed into complementary RNA within the nucleus, the PA signal causes many A's to be added at the end of the RNA molecule. This tract of poly A's is important in protecting the RNA from being broken or destroyed by enzymes in the nucleus and also in transporting the RNA from the nucleus into the cytoplasm where the protein manufacturing can begin (figure 10). The PA signal should, therefore, be included in the genetic engineering of a gene if the RNA is to be transported to the cytoplasm where the protein is made.

When the RNA in the cytoplasm of cells is compared with the DNA from which it was derived, the RNA is shorter. While the RNA is in the nucleus, portions within the RNA are precisely cut out or spliced by nuclear enzymes (figure 11). These removed sequences are called "introns" and the sequences that are preserved in the intact RNA are called "exons." The exons have the ribonucleotides that code for the protein. Once the introns are spliced out from the RNA, the coding sequences for the protein are contiguous so that there are no breaks in the final protein. The RNA introns must be removed for the RNA to be translated into the correct functional protein. Scientists have learned that it is valuable to have these introns and exons in the DNA gene because without the introns in a genetically engineered gene, much lower amounts of proteins are produced from that gene. This is because some intron sequences are involved in regulation of gene expression.

The process of making proteins from RNA is called translation (figure 11), since it is analogous to translating a message from one language to another. The RNA appears chemically quite different from the protein just as Spanish appears

12

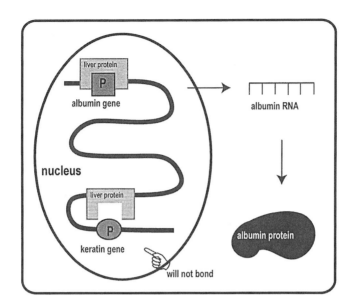

liver cell with albumin transcription

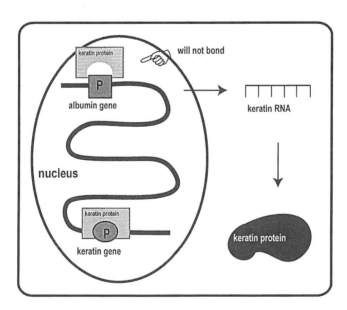

skin cell with keratin transcription

Figure 9. How Different Types of Proteins Are Produced in Different Tissues Using Tissue-Specific Promoters (P)

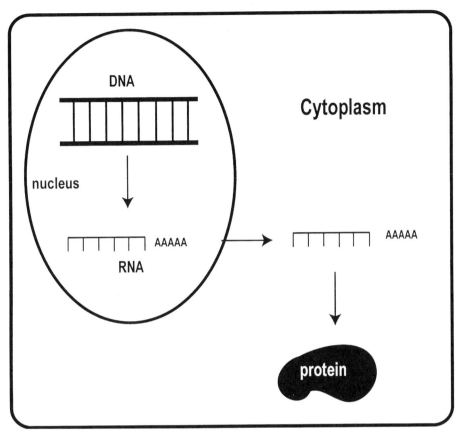

Figure 10. Importance of Polyadenylation (adding A's to RNA) for Transport of RNA out of the Nucleus of a Cell into the Cytoplasm for Synthesis of Protein

different from English. Nevertheless, the message is contained within both languages. Just as there is a start site of transcription to make the RNA, there is also a start site of translation to make the protein from the RNA. This translation start site must also be included in gene construction for proper protein synthesis to occur. A complex of proteins and some small RNA molecules, called a ribosome complex, bind precisely at this start site to add the first amino acid of the protein (figure 12). Three nucleotide sequences, called codons, code for one amino acid. For example, a GGA RNA sequence codes for the amino acid glycine. Proteins are made up of a combination of twenty-one different amino acid combination sequences. Amino acid building blocks are to proteins what nucleotide building blocks are to DNA or RNA. As the ribosome transcription complex moves down the linear RNA molecule, it adds more amino acids to the growing protein (figure 12). The specific amino acids added to the protein are coded by the sequence of the RNA codons. Therefore, the specific protein produced is linked with a specific RNA; the specific RNA is complementary to its DNA gene.

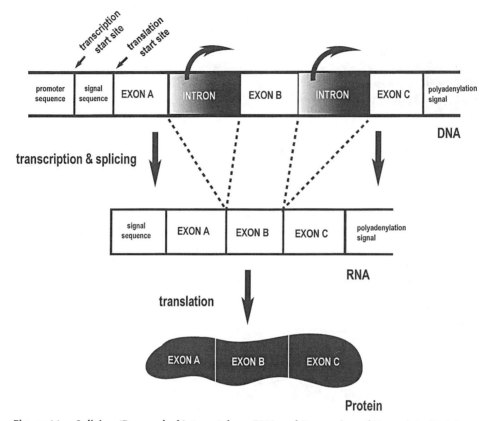

Figure 11. Splicing (Removal of Introns) from RNA and Expression of Exons into Proteins

This linkage of information from protein to RNA to DNA is the way DNA acts as the genetic code of complex life. Simple linear sequences of DNA result in proteins that fold into three-dimensional molecules with vastly different functions. Some proteins provide structure or architecture to cells and tissues. Some proteins are enzymes, which facilitate the synthesis of other types of molecules, such as fats, lipids, and carbohydrates. Vast numbers of functions are provided by these molecules (table 1). The difference between amino acid chains allows for considerably different structures and interactions between other molecules. It is at the protein level, not the nucleic acid level, that the diversity of life has its basis.

To summarize the structure of a gene, certain components of a gene must be present for the production of a functioning protein. A tissue-specific promoter helps express the gene in the proper tissue. The transcription signal for the start site, introns and exons, and the polyadenylation signal in the DNA gene result in production of complementary RNA that is spliced and transported to the cytoplasm. Once in the cytoplasm, the ribosome complex initiates translation of the RNA at the start site to add specific amino acids of a protein. These proteins then

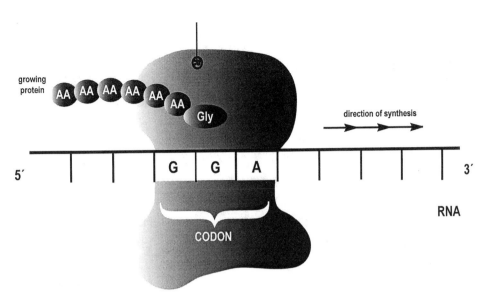

growing protein

direction of synthesis

5′   3′

RNA

CODON

amino acids - building blocks of protein

Figure 12.   Synthesis of Protein from RNA Using a Ribosomal Complex

lead to the structure of cells and tissues, as well as synthesis of other cellular components, to ultimately create the living organism.

CHROMOSOME STRUCTURE

The long strands of DNA are tightly packaged in the nucleus of a cell. A person going on a trip cannot just gather clothes, shoes, and belongings into her arms and go to the airport. Nothing would be compartmentalized and items would be lost along the way. A suitcase is the answer, and so it is with DNA in the cell. The human mind cannot comprehend the enormous length of a DNA strand. We can visualize thirty people in a room, maybe hundreds as a discrete number of individuals, but large numbers of individuals, such as a couple hundred thousand, are best understood as a numerical representation (e.g., 210,246). A number like 3,200,000,000 is too large for our minds to really comprehend, yet it is the number of nucleotides within a human genome. Suffice it to say that it is enormous! Consider, then, that this enormous stretch of DNA needs to fit into the nucleus of a cell. Obviously "suitcases" are needed for packing the DNA.

This is accomplished by winding DNA strands around specialized proteins and essentially twisting the proteins/DNA strands into knots. Remarkably, this is an orderly process that allows DNA to replicate when the time comes in the replica-

Table 1.   Protein Functions

Type	Example	Function
Enzymes	glucuronidase	help chemical synthesis and reactions
Hormones	insulin	signal to cells to perform a function
Antibodies	IgG	part of the immune response
Transporters	ion channels	allow movement of molecules into and out of cells
Structural Proteins	muscle, hair, eye lens, nails, skin	provide support and protection

tion cycle of the cell and to selectively "unravel" portions of DNA for the transcription of genes and production of proteins. The complexity is amplified when you realize that there are twenty-three duplicated chromosomes in each individual human cell.

Because of the complexity of DNA in the cell, genetic engineering, or describing life, is not a simple process. We try to describe it with mental pictures or language, but we should use caution in accepting scientific explanations in casual conversations or in the news as fact without qualification. The methods used to create dinosaurs in *Jurassic Park* are simply explained, but such explanations do not constitute complete fact. The identification of the breast cancer gene does not mean that it is the only determining factor for breast cancer and that anyone who does not have the gene will not get breast cancer, or vice versa. People need some understanding of such developments to evaluate their merit. By recognizing the complexity of life, scientists have devised tools to genetically engineer life.

THE BASIS OF ETHICS

Naturalism

Some fields of study introduce students to areas of inquiry that they have never thought about before. Particle physics is likely to be such a field for most of us. Another example might be the Sanskrit language and the history of India two thousand years ago. Many fields of philosophy, such as advanced logic or philosophy of language, do not come readily to most students' minds. If I ask a room full of people including philosophers whether Gödel's theorem should be taken as a decisive refutation of computational models of the brain, I could start a discussion among the philosophers that would go on for hours. Everyone else would head for the door.

But if I asked whether it is reasonable to believe in an all-knowing, all-powerful God, few would leave. Philosophers might have some interesting things to say and they might use fancier language in making their points. But everyone would be engaged. If I further asked whether God can be certain about what you are going to do before you do it, I would really start the group in a debate. There would be few wallflowers in this discussion and few sleepers.

What is the difference between the two topics? Some areas of philosophy address topics that we all think about. I know many students and others who do not believe in any god and many who do not believe in the all-knowing, all-power-ful God we are familiar with in Western civilization. But I do not know anyone who has not thought about the question of whether God exists. I know many people who believe in such a God. But I do not know any believer who has never thought about the problem of Job: If such a God exists, why is there so much suffering in the world?

It seems that all students have done philosophy. They have thought about some of the great questions and tried to find answers that satisfy them intellectually. The same thing is true about the philosophical field of ethics. Students may not have thought about every small topic in ethics. They have, however, thought about some of the great questions. They have all wondered whether ethics is just a personal opinion. They have all wondered about the basis of right and wrong, and most have wondered if any principles of morality are universal. Should we always tell the truth or always respect the lives of others? Is it ever right to lie or intentionally kill someone?

When students consider these questions, they find that they have always done "moral philosophy." They may not have done it with the rigor and precision of professionals, but they have considered the questions of ethics and often deeply.

Many readers will remember what they first thought in February 1997 when they saw the headline with a photo in *USA Today*, "Hello Dolly," announcing the first mammal cloned from an adult cell. Did you think, "Wow, what a great achieve-ment for humanity!" or, "Wait a minute, aren't we playing God? Are we doing something wrong?" When you read in a newsmagazine that people with a genetically based disease can't get insurance, do you say to yourself, "That's not fair"? These are moral reactions to developments in the study of genetics. Ten or fifteen years ago no one thought about them specifically but by now all have thought about the general principles of fairness or the possible moral limits to science.

What you will find in this book is a discussion of a variety of topics that you have probably thought about, discussed by writers with a wide variety of view-points. This introduction is designed to give you some basic knowledge of genet-ics and biotechnology and some basic knowledge of ethical theories so that you can more intelligently consider the questions of biotechnology. Each part in the book has an introduction that reviews the basic scientific and ethical issues related to that topic (e.g., genetically modified food). At the end of the book there are discussion cases that will help you understand and apply what you have learned and considered.

In the long history of serious thinking about ethics two different sorts of ques-tions have emerged as crucial. Almost everyone who has ever thought about ethics has wrestled with these questions and offered many complex and highly developed answers. What follows is a brief overview of these questions and the most important types of thought about them. This discussion is keyed to the specific context of the issues considered in this volume.

What is the basis of morality? When most of us are asked about whether some act is wrong or right (e.g., lying to a girlfriend about where you were), we frequently respond by referring to a rule or principle. In this case we might refer to a rule we accept such as, "It is wrong to tell a lie." In many simple and common situations appeal to a shared principle is sufficient. Some principles or rules such as "lying is wrong" are so widely accepted that we need only agree that the rule applies to a particular case for us to agree that the act (in this case, lying to a friend) is wrong.

Suppose, however, on some highly complex moral problem you appealed to a rule like "keep your promises," and someone asked, "Why is this a sound rule to follow?" or "How do you justify such a rule as this?" We are not now asking what the best rule is but what is the ground or basis of any moral rules at all. We want to know why lying is wrong or keeping promises is right as a general rule.

In the long history of considering questions of ethics two different and sometimes overlapping answers have been given to this question. The first and perhaps the oldest answer is the appeal to nature in general or human nature in particular. We might appeal to some requirement of survival such as our need to live together in groups. In this way one might argue that there are certain rules that are important for group cooperation. Trust among members of a group might be regarded as essential if the group is to remain together. This trust might require members of a group to follow certain basic rules such as promise keeping or truth telling. For example, without being able to rely on drivers to keep their promise to obey traffic rules, our mobile society would come to a halt. Cooperation among drivers is essential, and this cooperation is nurtured by and developed in certain rules such as promise keeping that we all follow. Such rules are ultimately justified by an appeal to requirements of human nature for group survival and cooperation.

We might also appeal to human nature to justify a moral rule by appealing to certain regularities of human life or behavior. We might appeal to our pursuit of pleasure and avoidance of pain. From such a feature of human nature we might develop a moral rule such as "Promote pleasure in yourself and others and try to reduce and avoid pain in yourself and others." One of the great works of moral philosophy, Jeremy Bentham's *Principles of Morals and Legislation*, begins with just such an appeal to human nature. "Mankind," wrote Bentham in 1789, "is born under the service of two masters, pleasure and pain. It is for them to decide all that we shall do and all that we should do." From this observation about human nature as a regular pattern of behavior, Bentham develops a moral rule: Always promote the greatest amount of pleasure for the greatest number of people. In the rest of his book Bentham extends and qualifies his basic principle in a number of ways. He asks us to consider the sheer amount of pleasure our acts will produce. He asks us to examine the duration of the pleasure and the number of persons affected over long periods of time. But the starting point for such a moral principle is the initial claim about human nature and the belief that we should adopt those moral principles that are consistent with human nature.

Finally, we might view human nature not from the vantage point of survival or the observations of how we act, but in the context of what we can become if we act in certain ways. Consider not human beings but trees. You can examine what the minimum requirements in soil nutrients and water are for the survival of the seedling. You can consider how this particular species of tree usually grows in a particular soil and climate. By contrast, you can consider what the tree can become if it has an optimum amount of water and fertilizer and the perfect soil conditions: what the nature of the tree is if given the chance to flourish. Hunting dogs like Labradors can survive in cramped pens. But they can only thrive and flourish in a much more open setting. In their nature as large hunting dogs, they require a certain environment to be all they can be.

Moral principles may express and encourage human thriving or flourishing. Love or nurturing may not be all that we need, but our expressions of love and our nurturing of others may be examples of the best that human beings can become. Moral rules may develop as principles that encourage us to develop these qualities of love and nurturing.

The idea of deriving practical rules from an analysis of human nature or from nature in general considered in the three ways just noted is a pervasive theme in the discussion of biotechnology. First, many persons claim that living in harmony with nature around us, limiting technology, and using traditional methods of breeding and cultivation are necessary for our survival. Technological diminution of nature is a dangerous use of science to supplant or upset a balance of nature. This argument might be that living in the way nature "intends" is necessary for human beings to continue to survive on the planet.

Second, nature is understood in terms of the pretechnological rhythms of human life. Certain plants and animals are regular inhabitants of certain areas of earth and not others. Nature supposedly shows us that rice is not to be grown in Alberta nor are salmon to be fished in the Indian Ocean. We learn what the rules are by studying the regularities and patterns of biological life, diversity, and ecology of particular places. In this sense of observed regularities, agricultural biotechnology or animal transgenics are thought to be wrong because they violate rules derived from the observation of natural patterns of existence.

Finally, human beings are said to flourish only to the extent that they live in harmony (notice the morally positive word) with nature. Human beings can become whole only as a part of a larger natural world. We are a part of, not apart from, nature. Organic food is thought to be better because it is "more natural," less technological, less prone to chemical pollution, or less likely to trigger allergic reactions. So too human cloning represents a threat to a natural symbiotic relationship between sexuality and parenthood, a connection that must be maintained for our ultimate happiness.

In these three ways appeals to human nature or to nature in general are pervasive in biotechnology as the basis of rules or principles that ought to govern it.

Moral Sense

Suppose you read in a campus newspaper about a graduate student in another country who put kittens in ice water to study the freezing point of warm-blooded mammals like human beings. The student said that this did not violate nature because kittens freeze all the time. The study would be valuable to human rescue personnel and doctors. Animals have no rights anyway. So this study cannot be wrong.

We suspect that your first response would not be "That is unnatural and is therefore wrong" or "It will benefit mankind and we should applaud it." We also suspect that you would not say "That is just the way they do things in that country so we should not criticize them." We suspect your first response would be "That is repugnant or repulsive" or, more mildly, "That is wrong." It seems to be just an immediate sense of "wrong" or in reverse "right" if the student was expelled or even charged with a crime.

This example displays vividly the other great tradition of justifying moral principles: an appeal to moral sense or moral intuition, which we employ constantly. For example, we cheer the person who saves lives while risking his own. We feel disgusted by the father who deserts his family but not by the one who divorces his wife. There is a difference. Finally, we feel revulsion with regard to justly convicted child abusers or pedophiles.

In the topics treated in this volume we see this sort of claim loom large, especially on core issues regarding the treatment of animals and human beings. For example, many persons seem to have very strong moral sentiments in regard to deliberately causing suffering for other sentient creatures, such as the cat experiment noted above. When people hear or read about Nazi experiments of a similar nature on prisoners, they often have the same sense of moral rejection. In the same way many persons have a sense that using prisoners or homeless persons to test an AIDS vaccine without their consent would be wrong. This sense of wrongness would remain, we suspect, even if a person were told that using human beings for an initial test would be a faster, more cost-effective means of developing a vaccine. Finally, the appeal to a moral sense of wrongness is often used in the debate over human cloning. Many writers, like Leon Kass in this volume, seem to initially appeal to a sense of what Kass calls "repugnance" at the thought of human cloning. We will see this appeal to moral sense as a grounding for specific moral principles throughout the readings in our book.

Moral sense theory has a long and honorable tradition in ethics. Many nuanced versions of it have been developed by first-class thinkers such as David Hume and Adam Smith. Most thinkers who employed "moral sense" as a foundation for ethics also believed that if the supposed data of our moral sense were systematized and structured we would find very broad agreement on a set of simple rules such as those against lying and physical assault of human beings. These are principles that can also be developed and grounded in several ways, different systems of ethics offering different grounds for such principles. We might (and often do) start with the basic sense of right with respect to a principle such as "lying is wrong" and then work out a more elaborate argument based on the requirements of survival or flourishing to complement the basic feeling of rightness or wrongness.

The view is often expressed that since human beings have these "moral emotions," as Darwin called them, they must play some role in the maintenance and thriving of the human species as found in communities. For example, we may ground a rule such as "keep your promises" in an immediate moral intuition or sense of rightness that we as human beings broadly share. Further, we may explore what features of human existence would lead to the development and persistence of any moral sense concerning such a rule. In other words, why would such a specific "moral emotion" survive in the human species over time? This is what Kass does with his discussion of human cloning. He believes that the general sense of rejection of the prospect of human cloning is the beginning of moral wisdom on the part of ordinary persons. This intuition can be buttressed with arguments drawn from human nature to develop and amplify these data gathered from our moral sense.

PRINCIPLES OF RIGHT AND WRONG

When we first think of a practical moral problem such as when or whether to require the labeling of genetically modified food, we most often initially appeal to a moral rule or principle that we accept and hope others will accept. In the case of genetically modified food, we might appeal to a principle such as "people have a right to know what they are buying." In such a way we would analogize the case of genetically modified food to other cases involving a buyer and a seller. If I buy a used car, for example, few would dispute the idea that I have a claim on the seller to be told of the characteristics of the car, such as age, miles driven, repairs, gas mileage, and so on. The seller should tell me what he is selling and we frequently say that we have "a right to know." In the case of genetically modified foods many people assert the same principle. Food ingredients and additives are required to be labeled. I have a right to know what salt, sugar, and preservatives are in the food I buy. Why not genetic modification as well?

In the long history of thinking about moral practice among human beings and even between human beings and other parts of nature three different sorts of principles have been widely developed.

Utilitarianism

The first sort of theory about good moral rules is the theory known as utilitarianism. This theory holds that there is only one fundamental principle of ethics, though there may be useful rules of thumb that embody it. The principle is: Always do that act which produces the greatest good for the greatest number of people. "Good" in this view is generally equated with pleasure and the absence of pain.

According to utilitarian theory, when we make a moral decision we must look at how many persons are benefited or harmed by each of the possible courses of action. How intense are the effects for each person? If five people are only mildly benefited and one person is drastically harmed, we probably should not do it because even though the ratio is 5 to 1, the actual amount of harm is very substantial. How long will the effects last? A very short pain, such as that experienced in a dental procedure, may be outweighed by a long-term benefit. We add up the costs/harms in terms of numbers, duration, and intensity and compare the harms with the benefits in the same way.

Utilitarianism is a forward-looking, consequentialist moral theory. To the utilitarian the outcome of an action is a key to any moral evaluation. A classical example of this sort of moral reasoning is in the discussion of capital punishment. Those in favor of capital punishment frequently argue that capital punishment saves lives by deterring crime, especially murder. "It sends a message to others about what will happen" if they kill anyone. Hence there are fewer murderers because no one wants to die. The claim is that capital punishment is morally justified because of its beneficial consequences (i.e., the greatest good for the greatest number in reducing the murder rate).

Many critics oppose capital punishment because they believe that no good effects are yet demonstrated while there are many bad effects. They deny that capital punishment deters murders and further note that with so many appeals, putting them to death actually costs more than incarcerating them for life. Finally they note the possibility, for some the probability, that innocent persons will be put to death. This is a great harm. Critics, however, still seem to be looking at the problem in a utilitarian way. They are still considering the greatest good for the greatest number. They are disagreeing about how the facts are to be understood in a specific way in the context of capital punishment.

Utilitarianism is one of the great theories of practical moral rules. Its influence is enormous and it is widely applied in the field of biotechnology. Anyone who attacks genetically modified food as "unsafe" or "dangerous," thus requiring a ban, is engaging in utility analysis. References to bad results for farmers when they use Roundup Ready soybeans (e.g., genetic migration causing the development of "super weeds" or weeds becoming resistant to Roundup so that over time using Roundup is a waste of money for farmers) are a utilitarian form of reasoning. Food that has not been properly tested may cause more harm to health than other kinds of food. Hence a utilitarian may argue that all genetically modified food must be tested for safety so that we can avoid harm to our health. When supporters of genetically modified food argue that increased nutrition at lower cost will result from such food, they are utilitarians.

Utility theory, broadly conceived, is a widely employed form of moral reasoning and policy analysis regarding practical problems. Whatever other considerations people wish to bring to practical problems, no one denies at least the relevance of good and bad consequences to the decision of what we should do. One can claim that cloning is wrong because it is unnatural. This is a much debated question. But few deny that the possible benefits and harms of human cloning are important factors in deciding whether or not to proceed with experiments that will lead to the development of human cloning. It is easy to see how utilitarianism became a position that is most often appealed to in public debate. Everyone on each side of a position agrees that it is important.

Utility theory is highly significant for its claim that all important features of a moral decision can be reduced to a calculation of pleasure and pain resulting from a decision. Certainly these calculations need to be made carefully. As noted above, attention must be paid to the distant effects—the number of individuals affected and the intensity of various effects. These, however, are just ways of ensuring that all of the pleasures and pains have been accounted for in making a decision. None of these subsidiary rules require us to consider any fundamental good other than mental or physical pleasure or any other evil than mental and physical pain. If this concept is plausible, it suggests a way of resolving otherwise apparently intractable moral problems. We can reduce all the outcomes of any action to a single common denominator. Then the outcomes of various activities can be compared in terms of this single scale and we can judge what is the best course of action.

To see the point, consider an example from nutrition. If we want to make a decision about what is the most effective way to get vitamin C in our diet, we can compare units of vitamin C obtained per dollar of fruit eaten (e.g., apples, oranges, grapes) and even compare fruit with tablets. The calculation is not contested because we know what the scale of judgment is: units of vitamin C per unit of money. Once the scale or rule of judgment is agreed to, the decision is just a matter of time and calculation.

In the case of biotechnology, once we accept that the fundamental concern with genetically modified food is the amount of nutrient per dollar, without allergic reactions, we can calculate the cost of production, analyze the size of the harvest, examine the possibility of allergic reactions, and make an informed judgment about the relevant choices before us. Similarly, if we want to decide about a DNA database of persons convicted of a crime, we can calculate the cost of the testing and storage of the results, determine how many guilty people will be caught or innocent persons released, and see how much pain will be produced by false results.

This example, however, suggests the problem that critics see in utility theory. They believe that it too easily reduces all moral concerns to a single scale of pleasure and pain. But aren't there other concerns that might be relevant to deciding about such a database? Think of the privacy of those tested. DNA carries much more information than a fingerprint, and some of it could be used to discriminate against people who have served time in jail. It could be used to identify predispositions to illness or specific genetic links to a fatal condition. This might make such persons unable to get health or life insurance. It would also reveal private information about an individual to the government permanently.

Critics contend that the efficient or cost-effective administration of criminal justice is one matter and the right to privacy of individuals is another. One cannot be sacrificed for the other. In the case of transgenic animals, one way to make judgments is to see what the specific transgenic application will be, the cost of creating the transgenic animal, the possible suffering or loss of life of animals in the process, and the benefits derived from the transgenic animal. But we would not consider such a series of questions sufficient to determine whether to create transgenic human beings. We would almost certainly reject an experiment on human beings that would require some loss of life to develop transgenic human beings without regard to the potential benefits to humanity at large from the results of such a development. Many persons concerned about animal rights believe that the same consideration applies to animals. They argue that we should respect their right to exist and thrive independently of our manipulations. Animals, it is argued, want and feel and believe. They have life plans too. If it constitutes a moral wrong to sacrifice a human being's life plan or hopes to someone else's plans or hopes, then it is also wrong to do the same thing for animals.

Whether or not this is a sound conclusion with regard to higher sentient animals, it seems to reflect a moral truth with respect to human beings. Some things, like liberty or fairness, we do not want to sacrifice to the principle of the greatest good for the greatest number.

This criticism of utility theory has been worked out in two different ways. One way has been to reformulate utilitarianism not as a theory of how we should make specific decisions but as a theory of how we should decide between sets of moral rules that guide our decisions. On this view we should make our specific decisions about what to do in situations in light of rules that we have previously adopted, for example, "always tell the truth" or, in the case of biotechnology, "respect the genetic integrity of species." But when we ask what sorts of rules we should adopt in the first place, we can refer to a broad utility principle to justify a specific rule or set of rules. In a specific situation I should tell the truth because that is a rule I have agreed to. But if I ask why I should agree to such a rule, then I refer to a principle of utility. This theory, known as "rule utilitarianism," has much to commend it and has been widely discussed in the last few decades. Its strengths are obvious. It seems to provide a place for rules such as human autonomy and fair play that we do not want to give up, while preserving the strength of utilitarianism as a procedure for justifying rules.

The problems with the theory are also well-known. If we always follow the rule irrespective of the consequences in a specific set of circumstances, we have not preserved much of utility theory. It may be that we need another theory entirely. On the other hand, if we sacrifice the specific rules in any case where they do not seem to lead to good outcomes as judged by utility theory, then why have any rules except "the greatest good for the greatest number"?

Formalism

This concern for rules or principles of justice or autonomy that are not exactly synonymous with utilitarianism gives rise to a different practical theory. This theory stresses fairness and equal respect for other human beings. It holds that we must consider carefully the liberty and equal standing of other persons along with benefiting them in the way that utilitarianism advises when we make moral choices. On this view we cannot sacrifice individual liberty or equal rights just because the greater good for the greater number can be achieved this way.

Consider, for example, the problem of abortion, a subject that intersects with human cloning and stem cell research as they are treated in the last part of this book. One way to look at abortion is to ask whether any given case of abortion satisfies the utility principle. We would ask whether the unborn child would be well cared for after birth. We would want to know if the mother would get prenatal care. Could she keep the baby, or would adoption be a realistic option? In some cases our best judgment would be that the future prospects for the child and the burdens placed on society would make abortion the preferable option. In some cases we would judge that the utility principle was satisfied best by keeping the child or having it adopted. For example, one highly relevant factor would be whether the child would be born with handicaps, most frequently caused by genetic abnormalities. Such handicaps would profoundly influence the utilitarian calculation because they would impact the future happiness of the child and the prospects for adoption.

A fundamentally different way of looking at the problem of abortion would be to consider the rights of the child or the woman as being of primary importance. If we thought primarily of the liberty of those persons who unquestionably have rights, then we would conclude that the choice of the woman is decisive, whatever the general utility calculation might lead us to conclude. On the other hand, if we were convinced that the unborn fetus had equal standing, then we might conclude that abortion is wrong and that adoption into a good home is the morally correct option. In the same way, human cloning might be thought of as an immoral experiment on a human being (the unborn) without his or her consent and without any benefit for the unborn.

Using genetic technologies to increase agricultural yields can also be examined from either the point of view of utility theory or its alternative. In terms of utility theory we might ask how much yield per acre will be increased at what cost. From the point of view of justice or fairness, we might want to know whether transgenic agriculture will increase the disparity between rich and poor. Will it provide benefits for all farmers and all countries or will the benefits go to wealthy regions and farmers who can afford it?

As we will see, these concerns for autonomy and fairness are of crucial importance in the discussion of biotechnology. Inequitable social impacts of biotechnology are of crucial concern to many critics of agricultural and food biotechnology. Concern for justice also is a great concern in the discussion of human genetic testing and gene therapy. Genetic testing can be used to discriminate against people in access to health insurance. Individuals may not be insurable for a genetic disability that becomes known. If they test positive for a genetic sequence that is highly correlated with a health problem, then they may not be able to obtain insurance for that problem. Insurance is based on shared risk; a known risk that is close to certain may mean that the affected person must pay the full cost of treatment. In many cases this will increase the cost of insurance beyond what a person can pay or it will mean that the insurance will specifically not cover the problem linked to genetics. An individual woman might be able to get insurance for every medical condition except breast cancer if she tests positive for the BRCA1 "breast cancer" gene.

Justice concerns have also been raised about gene therapy, especially if the technology is used to "enhance" offspring. Only the wealthy, it is argued, will be able to afford this technology. This will further exacerbate the division between the well "endowed" and the less well off.

In each of these cases utilitarian considerations may lead us to conclude one way, and justice or fairness considerations may lead us to conclude another. Thus, according to this alternative principle of moral decision making, we must ask in any situation of moral choice what the effects of our actions will be on a just distribution of the burdens and benefits of life together and on individuals' liberty to lead a life of their own choosing.

Virtue Theory

The two approaches to practical moral judgment that we have just considered are theories that focus on human actions. In each case we will want to know the impact of our actions on ourselves and others. Will an action benefit others? Will it increase human suffering? Will it increase an inequitable distribution of resources or a divide between rich and poor?

Suppose, however, that we are fundamentally interested most of all in our character, not our actions. Not what we do but what we become through what we do. This is the moral theory that focuses on virtue, not just benefits. This concern for character goes back to Aristotle in the fourth century before the common era. His *Nicomachean Ethics* is still considered by many to be the greatest single work on ethics ever written.

In recent years this theory has been redeveloped by many writers concerned to develop an ethics of care as distinct from an ethics based on rules of justice. Such an ethical emphasis might focus on the virtue of compassion or on the character trait of always giving more of yourself than is required by the rules of justice.

Virtue theory is almost always grounded in an understanding of "human nature." It is typically argued that human nature is such that certain character traits enrich or perfect it. Human beings are said to thrive when they adopt virtues such as compassion, courage, or self-control. They do not thrive when they adopt practices such as immoderation and extreme selfishness.

The exact nature of virtue is a matter of debate, but we can best understand it as a character trait or habit of action. In situations of choice we are said to habitually act in a certain way such as thinking of others first. To see the difference between the theory of virtue and theories of justice and utility, we might consider again the problem of abortion. Utilitarian theorists would ask about the consequences of various courses of action in any case where abortion was under consideration. Justice theorists would ask about whether the right of the woman or the unborn takes precedence. Virtue theorists would ask about how the practice of easy access to abortion affects the character of those involved (e.g., the woman, her partner, the doctors, etc.). Virtue theorists would be less concerned about how abortion might be right in some special cases and more about how the practice of abortion affects us.

Though not as obvious as the other two practical theories of ethics we have examined, the idea behind virtue theory is seen in debates on biotechnology. Obviously the discussion of stem cell research, which is connected to the question of abortion, may involve questions of virtue. However, agricultural biotechnology, as well as animal and human cloning, is connected to questions of character. Many critics of biotechnology see in it a process of human domination of the world—a character trait they do not like. They believe that permanent engineered genetic change manifests a will to power on the part of human beings toward other humans and nature. Genetic enhancement of humanity is also a dangerous problem of power and domination. In these ways a virtue ethical theory is relevant to the questions of biotechnology.

ANALYZING PRACTICAL MORAL PROBLEMS

This volume focuses on practical issues in a specific area of human life. It will require you to think about these practical questions and to analyze the arguments and disagreements against them. To analyze disagreement or potential agreement on a practical moral issue, you need to consider carefully the source of the disagreement. One disagreement may be over what is the best moral rule or theory to apply to a given case. One writer may be a utilitarian and another writer may be a human rights or human dignity theorist.

But the disagreement does not need to be about moral theory. It may be a disagreement about the facts of the case or problem under discussion. Two persons might have a strong disagreement about what to do in a specific situation, though they may both accept utilitarianism. They may disagree about the set of facts to which utilitarianism is applied in this case. They may disagree about the intensity of suffering that may be caused by a course of action or the duration of the suffering. For example, the use of animals in medical research is a topic we address in later parts of this text. One might believe that animals suffer in anticipation of what will happen to them and as a regret for what they have lost. Human beings typically suffer from fear in unfamiliar and cramped circumstances and from regret over life plans that have been thwarted or delayed in some circumstances. But fear in unfamiliar circumstances and regret over lost plans seem to be a complex cognitive and affective response to one's situation. These responses are far more complex than a mere physical response to pain. Regret seems to presuppose the concept of a "life plan" that is subjectively valued even if it is not widely shared and even if a considerable effort is required to pursue it.

Someone who accepts utilitarianism fully might disagree with others over whether animals have the cognitive capacity to suffer in this way. One might easily agree that higher-capacity animals can feel pain and thus experience some pleasure. Another might strongly disagree over whether animals can have the joy of achievement or the sadness of lost love. Thus two writers might strongly disagree about the relative weight of human suffering relieved and animal suffering caused in a medical experiment. The animal suffering would be a result of pain but not mental anguish or loss. Some other writers might disagree and cite evidence of complex consciousness in higher animals. A third person might agree that primates have such a consciousness, so using them in any experiment would be wrong. This third person might hold that mice may not have such a neurological structure so that using them is acceptable on utility grounds. This disagreement would not be about the moral theory to be followed but about one's assessment of the world in which this theory is to be used.

Finally, a source of disagreement about practical cases or problems may lie in what we shall call, for want of a better term, quasi-religious beliefs, which many people hold as a result of sincere membership in a religious tradition. Everyone has them, and it is not even necessary to derive them from a religious tradition. For example, I can ask whether a specific measure of health or well-being suggests that children from two-parent households seem to be better off than children from single-parent households. To find an answer, I would study data from a variety of empirical sources. Suppose, however, we said that women should be wives, mothers, and homemakers and leave everything else to the husband. This assertion does not debate a factual objection or appeal to a moral principle. We are not claiming that women would be happier or children would have fewer problems if women did not work outside the home. Nor are we claiming that a moral principle requires a course of action. We are not saying that children have a right not to have working mothers. We are saying that a certain way of living is preferred on cosmic grounds. Another example might be a belief in reincarnation. Someone who believed in reincarnation might oppose cloning on the basis that since it is impossible to clone a soul, the clone of a being would be inherently imperfect and limited. Reincarnation is not a moral principle, but believing in it or not affects whether one approves of cloning.

In general we might say that these quasi-religious beliefs give us a cosmic non-empirical view of what constitutes nature and thus what a harm might be. The grounding of moral principles in nature will entail some view about what is nature. In the typical case nature is all that is observed empirically about the world around us and about ourselves. But it is often bound up with transempirical beliefs that connect empirical nature with a cosmic whole that is more than we can observe. These three elements of any moral decision will help you identify the real debate or disagreement in any topic.

Consider, for example, the debate about using genetically modified organisms in agriculture. To make a judgment about this topic in general or in any specific case, we need to understand the facts of genetics and of a specific organism as it will be used in a specific case. We also need to adopt a specific moral theory such as utility theory or a theory about the good inherent in nature. But we will also presuppose in any practical judgment a quasi-religious view about the ultimate character of nature and human beings.

One specific example might be this: To increase the growth of plants in a specific situation, we plan to introduce genetically modified bacteria into the soil, which will increase nitrogen fixing and promote plant growth. To make a decision we will need first an assessment of the facts about this particular bacterium. We will also need facts about how it functions in soil and some data about how it seems to function in laboratory tests of the soil in which it will be used and the plants that will be affected. Second, we will need to apply some moral theory. Perhaps in this case utility theory is the most obvious theory to apply. In that case, we will need to know about the short- and long-term effects of the organism in the soil. How will it affect plant growth, animal feed, farmers, ranchers in the area, long-term and short-term? Will it cause problems for growers who have not used it? Will it increase yields and will that require more water, increased use of herbicides and pesticides? These sorts of data will be important in any utility calculation about agricultural use of biotechnology.

Third, we will need to be clear about background beliefs regarding nature and harms to nature. Is nature static? Is it moving? Suppose someone believed that the nature around us is God or at least an abode of God. We might very well agree with someone else about the facts of a particular use of biotechnology. We might also agree about the moral theory to employ. But if we believe that living nature is God, then changing it deliberately through biotechnology might seem incorrect or foolish and impossible. You can't change God. Hence on nonmoral grounds one might oppose agricultural biology.

You may find it helpful to keep these three elements of practical decision making in mind as you evaluate the readings in this volume. Where does the debate lie? What is the source of disagreement on various topics? Is it factual, moral, or metaphysical? Can the debate be resolved in practice by greater clarity in metaphysics or by more and better information about a problem?

I

FUNDAMENTAL ISSUES OF ETHICS AND BIOTECHNOLOGY

BENEFITS AND HARMS OF BIOTECHNOLOGY

As we noted in the introduction, one question that always seems relevant to ask of any human activity is simple: what are the consequences? What do we believe will happen if we do or do not do something and how confident are we of our prediction of the future in this regard? This is the consequentialist, mostly utilitarian approach to ethics that we reviewed earlier. With regard to new and unfamiliar technologies like biotechnology the evaluation of positive and negative consequences is often heavily dependent on other more general beliefs we hold about the world. No belief just appears with a green flag saying "accept me." All of our typical beliefs about the world are dependent on other beliefs. We do not usually ask the question, Am I the only person in the world? Why isn't the world around us only a dream? Our beliefs about specific other people are dependent on our more general beliefs about the existence of some other people besides us. It is that general sort of belief and some others—for example, what counts as the most basic sort of harm that can befall a person and how threatening the world is—that allow us to evaluate new activities of human beings like biotechnology in general or a specific part of biotechnology like genetically modified food (GMF).

The discussion of the possible consequences of many new technologies like biotechnology has been strongly influenced in recent years not only by the common utilitarian approach to technology assessment (what activity will produce the greatest good for the greatest number) but also by a new twist on considering the consequences of actions known as "the precautionary principle." Since this principle has been widely applied to biotechnology, even in international treaties, we will briefly take note of it at the end of this introduction.

31

Background Beliefs

No one's assessment of the consequences of an activity like biotechnology occurs in a vacuum. We all carry with us beliefs about human nature, about the world in general, and about the place of human beings in a cosmic scheme. We may be skeptical atheists but that is still a set of beliefs about the place of humankind in the wider universe. For purposes of considering the public assessment of biotechnology, beliefs about human beings in relation to the world of other living things, the natural world if you will, seem most relevant.

Several years ago the late political scientist Aaron Wildavsky developed a helpful typology of views about human beings and the natural world. This typology can be of use in sorting out beliefs that will powerfully influence our views of biotechnology. Broadly, these background beliefs can be seen as clustered in four groups.

Cornucopian

People who hold this view "see the glass half or more full." They see nature as plentiful and this bounty as continuing into the future. If we run low on oil, clean coal technologies will develop. If coal becomes scarce, expensive, or unsafe, solar or other technologies will develop. These persons tend to believe in individual ingenuity as the means by which human beings will thrive. With respect to biotechnology, they tend to be strong optimists about scientific creativity as making for a better world.

Tolerant

These people tend to be moderate optimists about nature and technology. They see nature as tolerant of the human effort to prosper. We can thrive but only with care and foresight. People like this do not gloss over problems with promises of bounty. But neither do they avoid new technology. They like to study a problem carefully and rely on experts to judge what sorts of technology ought to be promoted. These are people who want individual creativity with regard to biotechnology to be carefully monitored by publicly accountable experts.

Fragile

These persons are moderate pessimists. They see the world as a fragile balance of human civilizations and ecosystems. They hold that technical safety review is not all that one needs with respect to new technology. We need to see how technology alters the human world: how it changes patterns of life, how it disrupts communities and creates and sustains unequal access to the world's bounty. For example, serious inequities created by technology may be a key factor in population growth, which puts more pressure on an ecosystem, causing more poverty, and the system repeats itself. With respect to biotechnology, these are persons strongly concerned about the social disruptions of native agriculture and production from vast corporate structures over which local communities have little control. They want to know not just that biotechnology is physically safe, but that it will lead to more equitable distribution and local control of food and agriculture.

Capricious

These people are mostly hostile to biotechnology. They are often fatalists about the impact of new technology. This stems from a more general view about a deep natural balance between the inanimate, animate, and human worlds. This "deep ecology" of the physical, social, and even spiritual is a fragile balance that we only partially understand. Permanent life-altering technologies like biotechnology are dangerous to this balance.

We shall see these four positions reflected in many of the readings below. The two readings in this section are very different in this regard. Hayry is quite clearly a moderate optimist. She believes that with care and expertise biotechnology can benefit humanity. Epstein is much more hostile. In some passages he seems to be a moderate pessimist. In others he is much more pessimistic, writing of the "lethal threat of genetic engineering to life on the planet."

In our analysis, these two views—moderate optimism and serious hostility—seem to dominate the attitudes of most writers toward biotechnology. One way in which serious concerns about biotechnology have been expressed in the critical literature is through a revised version of consequentialism that has come to be known as "the precautionary principle." The precautionary principle has been widely employed in a number of international environmental declarations from the United Nations and other bodies such as the European Community. Specifically with respect to biotechnology it has been written into the international treaty known as the Cartagena Biosafety Protocol in the following language: "Lack of scientific certainty due to insufficient relevant scientific information and knowledge regarding the extent of potential adverse effects of a living modified organism on the conservation and sustainable use of biological diversity in the party of import, taking also into effect risks to human health, shall not prevent that party from taking a position as appropriate with regard to the import of the living modified organism . . . in order to avoid or minimize such potential adverse effects." In other words, even if a government only thinks that there is some risk from a genetically modified organism, it may ban the import of such an organism (e.g., a plant) without violating general free trade rules. Advocates of this principle see it as a more formal statement of the common sense of "look before you leap." They believe that because of the potentially adverse effects of many technologies, "precautionary measures" are warranted even when a full understanding of cause-and-effect relationships between technologies and harms is present. One prominent formulation that would be included but not limited to biotechnology is from the Wingspread Statement of 1998:

> When an activity raises threats of harm to human health or the environment precautionary measures should be taken even if some cause and effect relationships are not fully established scientifically.

For the writers of the Wingspread Statement, precautionary measures should include requiring those proposing a potentially dangerous activity to demonstrate that it is actually safe or to prove that no safer alternative exists before engaging in the activity. This can be regarded as a risk-averse consequentialism: one considers potential harms more serious than potential benefits. In the case of biotechnology, that could, in some cases, be because changes to living systems will be permanent and cannot be recalled. This might be especially true with respect to agricultural and food biotechnology, where changes are made to open living systems. As we shall see, it might also be the case in germ-line gene therapy and in the case of animal-to-human transplants where new viruses may be introduced into human beings.

In general, defenders of the precautionary principle think of examples of long-term nonobvious harms such as global warming, ozone depletion, or the once touted fire retardant insulation, asbestos. They want us to examine technology much more carefully before adopting it. As some have argued, precaution or "risk avoidance" should be the "default" mode of approaching new technology in the face of incomplete information.

Is this actually different than standard utilitarian calculation? The idea that it is may be doubted. If the harms are of sufficient magnitude and permanent it seems that utilitarian analysis will account for it. In the case of a harm that is serious and irreversible why shouldn't we avoid it or mitigate it? Why shouldn't this be a part of any sensible utilitarian calculation of new technologies? It may be that in the past we have not carefully enough considered the potential harms from technology. But is the answer adopting a principle that requires an impossible proof of safety or that seems to consider the prospect of harm worse than the prospect of benefit? We should consider harm carefully in making evaluations of technology. But it does not seem reasonable to bias our evaluation by considering negative effects as more serious than beneficial ones.

Advocates of the precautionary principle often overlook the negative effects of delaying or avoiding the use of a new technology. If transgenic crops will increase yields per acre, then delay in using them will have a negative effect on solutions to hunger and malnutrition. If transgenic pig organs can be used in human beings, then delay in using them will have serious negative effects for those on waiting lists for transplants. These negative effects should be balanced against the risks of harm from new technology. This point has been made best in the authoritative *Communication* on the precautionary principle by the European community: "A comparison must be made between the most likely positive or negative consequences of the envisaged action and those of inaction in terms of the overall cost to the community both in the long and the short term." We find this to be an eminently sensible position but one that is hard to distinguish from classical utility theory.

1

How to Assess the Consequences
of Genetic Engineering

Heta Hayry

Assuming that there are no intrinsic ethical grounds for condemning biotechnology and genetic engineering,[1] the moral status of these practices can only be determined by an evaluation and assessment of their expected consequences. In theory, such an assessment should not be too difficult to carry out, since the advantages and disadvantages of gene-splicing techniques are fairly well publicized.[2] In practice, however, the situation is different. The actual consequences of applying advanced molecular biology depend upon the social and political setting in which the application takes place. As different groups of people hold different views concerning the structure and dynamics of social and political life, these groups inevitably also disagree upon the consequences of the development of recombinant DNA techniques.

My aim in this paper is to sort out and analyze, first, the major consequences of genetic engineering, and second, the most important ideological factors which cause dispute concerning their weight. I shall begin by reviewing the advantages and disadvantages of various forms of biotechnology as they have been presented by the proponents and opponents of these practices. I shall then go on to consider the different ways in which the effects of genetic engineering can be evaluated, depending on the degree of optimism or pessimism inherent in the referee's judgment.

THE ADVANTAGES OF GENETIC ENGINEERING

The advantages of biotechnology, as seen by its proponents, include many actual and potential contributions to medicine, pharmacy, agriculture, the food industry, and the preservation of our natural environment. Within medicine and health care, the most far-reaching consequences are supposed to follow from the mapping of the human genome, an enterprise which is still in its infancy but which holds the key to many further developments.[3] Accurate genetic knowledge is a precondition for many foreseeable improvements in diagnoses and therapies, as well as an important factor in the prevention of hereditary diseases, and possibly in the general genetic improvement of humankind. Provided that such knowledge becomes available, potential parents can in the future be thoroughly screened for defective genes and, depending on the results, they can be advised against having their own genetic offspring, or they can be informed about the benefits of pre-natal diagnosis. One of these benefits is that an adequate diagnosis may indicate a simple monogenic disease in the foetus which can be cured by somatic cell therapy at any time during the individual's life. Another possibility is that an early diagnosis reveals a more complex disorder which can be cured by subjecting the embryo to germ-line gene therapy—this form of pre-natal treatment also means that the disorder will not be passed down to the offspring of the individual. Even in the case of incorrigible genetic defects the knowledge benefits the potential parents in that they can form an informed choice between selective abortion and deliberately bringing the defective child into existence.

Pre-natal check-ups and therapies do not by any means exhaust the medical applications of future genetic engineering. Somatic cell therapies are expected to help adult patients who suffer from monogenic hereditary diseases, and the increased risk of certain polygenic diseases can be counteracted by providing health education to those individuals who are in a high-risk bracket. In addition, the purely medical benefits of genetic engineering are extended by the fact that gene mapping will probably prove to be useful to employers and insurance companies as well as to individual citizens. Costly mistakes in employment and insurance policies could be avoided by carefully examining applicants' tendencies towards illness and premature death prior to making the final decisions.

Gene therapies and genetic counseling are practices which require advanced knowledge concerning the human gene structure. But the genetic manipulation of non-human organisms can also be employed to benefit humankind. The applications of gene splicing to pharmacy, for instance, may in the future produce new diagnosing methods, vaccines and drugs for diseases which have to date been incurable, such as cancer and AIDS. The agricultural uses of biotechnology include the development of plants which contain their own pesticides. In dairy farming, genetically engineered cows give more milk than ordinary ones, and with the right kind of manipulation the proteins of the milk can be made to agree with the digestive system of those suffering from lactose intolerance. As for other food products, biotechnology can be applied to manufacture substances like vanilla, cocoa, coconut oil, palm oil and sugar substitutes. And biotechnology can even provide an answer to the growing environmental problems: genetically engineered bacteria can be used to neutralize toxic chemicals and other kinds of industrial and urban waste.

THE DISADVANTAGES OF GENETIC ENGINEERING

The disadvantages of biotechnology, as seen by its opponents, are in many cases closely connected with the alleged benefits. An efficient strategy in opposing genetic engineering is to draw attention to the cost and risk factors which are attached to almost all inventions and developments in the field.

One general critique can be launched by noting that the applications of recombinant DNA techniques are enormously expensive. Millions of pounds are spent every year by governments and multinational corporations in biotechnological research and development. These resources, opponents of the techniques argue, would do more good to humankind if they were allocated, for instance, to international aid to the Third World. Another problem is that, despite the undoubtedly good intentions of the scientists, the actual applications of genetic engineering are often positively dangerous. Consider the case of plants which are inherently resistant to diseases, or which contain their own pesticides. Although there are no theoretical obstacles to the production of such highly desirable entities, corporations who also sell chemical pesticides might prefer to market another type of genetically manipulated plant, which is unprotected against pests but highly tolerant to toxic chemicals. The result of this policy would be an increase in the use of dangerous chemicals in agriculture, particularly in the Third World, which is to say that the outcome is exactly opposite to the one predicted by the proponents of biotechnology. Besides, it is quite possible that genetically engineered grains are less nourishing than the grains which are presently grown. If this turns out to be the case, then the employment of biotechnology will intensify instead of alleviating famine in the Third World. And to top it all, genetically manipulated plants may contain carcinogenic agents, and thus contribute to the cancer rates of the developing countries.

An oft used criticism against agricultural biotechnology is that the introduction of altered organisms into the natural environment can lead to ecological catastrophes. Scientists working in the field of applied biology have themselves noticed this danger, and set for themselves ethical guidelines which are designed, among other things, to minimize this particular risk. But as the opponents of genetic engineering have repeatedly pointed out, not all research teams follow ethical guidelines if the alternative is considerable financial profit.

Apart from the excessive expense and the increased risk of physical danger caused by biotechnology, the opponents of gene splicing can appeal to yet another disadvantage, which is related to widely shared moral ideals rather than to straightforward estimates concerning efficiency. Genetic engineering, it can be argued, is conducive to economic inequity and social injustice, both nationally and globally. Even in the most affluent Western societies, gene therapies are too expensive to be extended to members of all classes and age groups. Subsequently, these therapies are likely to become the privilege of an elite, and they will drain scarce resources from the more basic areas of health-care provision. In the developing countries, the situation is even more absurd. Medical problems which originally stem from lack of democracy, lack of education, shortages of fresh water, population explosion, archaic land-ownership arrangements, and the like cannot possibly be solved by high-tech western innovations which can barely be made to work in the most affluent and democratic of countries.

Another type of injustice emerges from the fact that the natural national products of many developing countries are superseded in the market by the biotechnological products of multinational corporations. Genetically engineered substitutes for sugar, for instance, could adversely affect the lives of nearly fifty million sugar workers in the Third World. Biotechnological vanilla could increase the unemployment figures by thousands in Madagascar, Reunion, the Comoro Islands, and Indonesia. And plans to produce cocoa by genetically manipulating palm oil threaten the current export market of three poverty-stricken African countries, namely Ghana, Cameroon and the Ivory Coast. In all these cases, the profits of multinational Western corporations are clearly and directly drawn from the national income of the developing countries. Finally, as regards medical biotechnology, there are those who believe that advanced knowledge concerning the human genome will inevitably become an instrument of genetic programming, which in its turn leads to subtle forms of genocide and general injustice. The opponents of gene splicing argue that the development begins inconspicuously with attempts to eliminate hereditary diseases. This practice of what is called "negative" eugenics will, however, soon be followed by more "positive" efforts towards altering the human genome: the inborn qualities of future individuals will first be improved in their own alleged interest, and then, later on, in the interest of society at large. When this development has gone far enough, scientists will also be asked to design special classes of sub-human beings who can do all those occupations which are too dangerous or too tedious for ordinary people. The outcome, according to the opponents of biotechnology, will be something like Aldous Huxley's *Brave New World.*[4]

TECHNOLOGICAL VOLUNTARISM VERSUS
TECHNOLOGICAL DETERMINISM

When one compares the views for and against biotechnology, one can easily see that the prevailing differences of opinion are not purely factual in nature. It is an undeniable fact that genetic engineering can be employed to eliminate diseases, but it is also an undeniable fact that genetic engineering can create bizarre and dangerous life forms. Likewise, it is possible that agricultural biotechnology will relieve famine in the Third World, but it is also possible that it will intensify it.

One ideological difference between those who defend recombinant DNA techniques and those who oppose them is based on their attitudes towards what have been called in the literature "technological imperatives."[5] The proponents of genetic engineering tacitly assume that the development and implementation of new techniques can always in the end be controlled and steered by human decision-making. If this were the case, then there would be no binding imperatives dictated to us by technological processes. This view has been labeled in the literature technological voluntarism. On the other hand, however, the opponents of genetic engineering have argued that technological development in fact has its own internal laws and logic, which cannot be altered or checked by human choices or human action. This view, which can be called technological determinism, implies, among other things, that all new techniques, however dangerous or morally offensive they may be, will ultimately be implemented.[6]

Those who believe in technological determinism can support their views by referring to real-life examples drawn from the history of science and technology. The splitting of the atom, for instance, made it possible to devise nuclear explosives, and despite the expected evil consequences, the atom bomb was constructed and used as soon as the technical difficulties had been solved.[7] In medicine, the drastic improvement of life-sustaining therapies since the 1950s has led to a situation where irreversibly unconscious and intolerably painful human lives are prolonged beyond all reasonable limits, the only boundaries being set by medical technology. These and many other examples seem to show that every technological possibility will ultimately be exploited whatever the effects on human wellbeing. Those who place their trust in technological voluntarism like to point out, however, that the "imperatives" created by scientific know-how can always be rejected. Atom bombs were not launched by abstract technological systems, but by political decision-makers. Hospital patients are not kept alive against their own wishes and beyond any hope of recovery by medical technologies as such, but by doctors and medical teams whose actions are sanctioned by democratically chosen legislators. Even if it is true that technical solutions often "offer themselves" to those making important decisions, these "offers" can be ignored or rejected if the consequences of accepting them seem to be harmful or otherwise undesirable. The fact that technological imperatives can be disobeyed makes it possible, according to the voluntarist, for us to reject those forms of technology which are harmful while preserving those forms which are useful to humankind.

Both technological determinism and technological voluntarism appear to have some truth in them. There are cases in which humankind seems to be guided, to a certain extent at least, by scientific and technical inventions. But there are also cases in which scientific knowledge and its applications are more or less firmly in development and use of poison gases was foregone by the political and military decision-makers of the period. These and many other real-life examples indicate that the dispute between determinism and voluntarism cannot be solved at a general level. But they also indicate that technologies can be divided into those which support the determinist interpretation and those which uphold the voluntarist view.

Given that technologies differ from each other in this respect, the important question in our present context concerns the category into which genetic engineering can be placed. Is genetic engineering one of those technological advances which control our actions, or does it belong to the class of techniques that we can control for our own benefit?

The empirical evidence which is presently available is mostly reassuring, and allows for moderate optimism concerning the controllability of biotechnology. Although there have been a number of unauthorized and potentially dangerous experiments with recombinant DNA, the guidelines set first by the scientists themselves and later on by governmental bodies have been rather leniently followed by the majority of research teams.[8] During the last two decades, gene splicing has been transformed from invention to industry, but up to this time human decisions have definitely played an important role in the development and application of the technique.

THE THREE STAGES OF
TECHNOLOGICAL DEVELOPMENT

Granted that biotechnology has not yet taken total control over human affairs, the pessimist can nevertheless argue that the situation will in the near future be drastically different. The pessimist's view can be based on the idea that new techniques develop gradually, each stage provoking its own peculiar responses in the business world and among the general public.[9]

The first stage of technological development is the theoretical invention of a new technical tool or process. Depending on the nature of the invention, this stage can either go unobserved by all but a few experts in the field, or it can become widely publicized and commented upon. The latter is often the case when the new technique in question is connected, however remotely, to human sexuality and reproduction. At the second stage of technological development the newly invented tools and processes are transformed by practical innovations into industrial products and means of production. Since this phase usually extends over a relatively long period of time, public attention tends to fade away, along with the fears which may initially have been caused by the novel device. At the third and final stage the completed products are marketed and distributed among the consumers. Assuming that the preceding innovation phase has been long enough, the products often enter the market without protestation, regardless of the reactions which may have accompanied the introduction of the new technique.

The three-stage model of technological development can be employed by the opponents of genetic engineering to construct an argument against scientific and industrial gene splicing. When recombinant DNA techniques were first invented in the 1970s, their potential applications were publicly discussed and widely condemned. The reaction was for the most part based on religious taboos and on horror images created by twentieth-century science fiction, not on any real-life facts. The practical outcome of the protest was, however, that scientists between themselves decided to proceed cautiously, and to avoid arousing further popular outrage. Under the cover of professional ethical codes, biotechnology was then gradually advanced to its present innovational stage, at which the actual and potential industrial applications have become numerous and diverse. Attitudes towards gene splicing have, in the course of this development, inconspicuously become more permissive, largely because of the fact that there have been no major catastrophes.[10]

According to technological pessimists, the relaxation of popular attitudes marks the beginning of a new and increasingly dangerous era in genetic engineering. When scientists are no longer censoriously controlled by the general public, their readiness to follow voluntarily assumed guidelines will decrease, and their codes will be gradually slackened. At first the reasserted freedom of the scientists will be qualified by assessments of the expected harms and benefits. Although the majority of gene-splicing activities will be sanctioned, those forms of genetic engineering which are considered to be particularly hazardous will be kept under the closest surveillance. After a while, however, the qualifications will be removed. By the time biotechnological development fully reaches its third stage— the stage at which completed products are marketed and distributed—consumers will have become accustomed to whatever forms of genetic engineering scientists care to pursue. And at that point, so the opponents of biotechnology claim, humankind will only be one step away from the Brave New World.

There are two reasons to doubt the validity of the pessimistic view. First, as regards empirical facts, most western countries will during the 1990s enter a phase where genetic engineering becomes regulated by law. The emergence of legal regulations undermines the immediate importance of public attitudes and voluntary guidelines, and the progress from qualified to unqualified licence predicted by the pessimist cannot take place. Second, even if the pessimist's prediction turned out to be correct, it remains unclear whether the unrestricted practice of gene splicing would in fact lead to the kind of genetic totalitarianism that Huxley described. One cannot reasonably assume that biotechnology could by itself destroy the prevailing democratic institutions and replace them by an authoritarian system. It is true that decision-making concerning the development of genetic engineering is at present controlled by multinational corporations rather than democratically elected governments. But it does not follow from this fact that humankind would be on the path to genocide and totalitarianism.

It seems to me that there are two types of pessimism involved in the issue, one of which is more readily justifiable than the other. The first, legitimate type of pessimism concerns the power of biotechnology to solve global problems like famine and pollution. Since it is not in the interest of multinational corporations to organize international aid to Third World countries, it is indeed unlikely that suffering would in the near future be effectively alleviated by implementing gene-splicing techniques. But this is not to say that genetic engineering ought to be banned, or that the situation could not be different sometime in the future. The second, illegitimate type of pessimism cynically assumes that nothing can be done to change the situation, and that the total prohibition of gene-splicing activities is the only way to save humankind from the slippery slope to which mad scientists and big corporations are leading us. This attitude is groundless and, moreover, positively dangerous in that the prophecies of the pessimist of this second type can easily become self-fulfilling. If ordinary citizens do not even attempt to control the corporation giants to make the benefits of biotechnology available to everybody, it is obvious that they will not attain such control. Pessimism in this sense is a form of ideological defeatism rather than considered intellectual scepticism.

WHAT ARE THE CONSEQUENCES?

To return to the consequences of genetic engineering and their assessment, it is now easy to see why competing views on biotechnology are so often so uncompromisingly antagonistic to each other. Factual disputes, which could be reconciled by empirical observation, play a relatively minor part in the controversy, and the real disagreement can be traced to commitments to optimism or pessimism, voluntarism or determinism. It seems, moreover, that these commitments are unavoidable, i.e., that it is impossible to produce a genuinely impartial assessment of the consequences of genetic engineering. It is therefore necessary to assess the validity of the underlying views before an evaluation of biotechnology can be completed.

As regards technological voluntarism and technological determinism, the empirical evidence which is available at present seems to indicate that genetic engineering is, by and large, controllable by human decision-making. The temptations created by technological possibilities should not be ignored or mitigated, but they should not be exaggerated either. Like all advanced technological innovations, gene splicing offers many possibilities, some of which will appeal to certain unscrupulous research groups despite the inherent hazards and despite the voluntary codes agreed upon by more responsible scientists. But the legal regulation of biotechnology, which in most western countries will take effect during the 1990s, provides an adequate device against such unethical conduct.

The issue of optimism and pessimism is slightly more complex, since different variations of both views can be either acceptable or unacceptable in the context of biotechnology. To begin with the gloomier views, epistemic pessimism is justifiable when it prevents people from believing uncritically what genetic engineers tell them about the advantages of the new techniques. Graver types of ideological pessimism cannot, however, be condoned, since they are prone to be self-fulfilling: widespread resignation and defeatism is likely to produce consequences which would never have occurred if people had been more hopeful. Similar observations can be made about the variants of optimism. It would be rather naive to believe that the multinational corporations which presently control biotechnological development would voluntarily undertake to further general welfare and global development. Epistemic optimism in this sense is therefore unwarranted. But it is not necessarily naive to believe that changes are possible in economic and political life, and that genetic engineering can in the future serve wider interests than it does at the present time. Ideological optimism is in fact supported by the feature which can be used to disqualify its pessimistic counterpart: if the belief in a better world is in any way self-fulfilling, it is obviously better to hope and prosper than to despair and perish. Assuming that democratic decision-making can be gradually extended to biotechnology, and assuming that genetic engineers continue to be controlled by voluntary agreements or legal regulations, the consequences of gene splicing can be expected to be prima facie good and desirable. Democracy both within and outside the world of biotechnology guarantees that no Brave New Worlds will be created as a result of misguided social experiments or malevolent dictatorial schemes. Surveillance and regulation, in their turn, provide a safeguard against reckless and inhumane research on areas such as biological warfare, the manipulation of our natural environment, and designs for new breeds or hybrids of animals. Granted that these possibilities are excluded, the remaining forms of genetic engineering are without doubt prima facie acceptable.

There are, however, certain embarrassing points which ought to be accounted for before recombinant DNA techniques can be wholeheartedly condoned. The moral problem at the core of the issue is that whatever advantages may ensue from the employment of genetic engineering, these advantages will not be allocated to those who need them most. The medical advances "will almost exclusively benefit the wealthy and the privileged of the affluent western societies, and the industrial applications will mainly profit the big corporations, sometimes at the expense of destitute Third World countries. The prima facie desirable consequences of biotechnology are not genuinely worth pursuing, one might argue, unless they can be equitably distributed among nations and among individuals.

But although it is easy to point out instances of injustice in the application of gene splicing, none of these inequities is primarily caused by genetic engineering. The allocation of scarce resources is a growing problem in medicine and health care, but it has not been either originated or, to any significant degree, aggravated by the present-day societies. And although it is true that certain Third World countries will be even worse off than they are at present after multinational corporations have taken over their export markets, the roots of their suffering surely lie deeper. The invention of recombinant DNA techniques is, after all, a relatively minor evil compared to the centuries of Western colonialism and imperialism which have left the national economies of these countries exceedingly vulnerable to any sudden changes in the world market.

What these remarks amount to is that genetic engineering should not be blamed for the injustice related to its implementation. If humankind can find a way to solve the problems of just distribution at a general level, then the benefits of biotechnology will also be distributed justly among individuals and nations. According to the moderately optimistic view that I have put forward and defended here, our best chance is to believe that this is possible, and then proceed to make it possible by our own actions. If this can be done, and if the requirements of democracy and adequate control are fulfilled, then the conclusion can be drawn that, in the last analysis, the expected consequences of genetic engineering favour its acceptance.

NOTES

Reprinted by permission from John Harris and Anthony Dyson, *Ethics and Biotechnology* (London: Routledge, 1994).

I should like to thank Mark Shackleton, lecturer in English, University of Helsinki, for improving the style of this paper.

1. See Matti Hayry, "Categorical Objections to Genetic Engineering: A Critique," in John Harris and Anthony Dyson, eds., *Ethics and Biotechnology* (London: Routledge, 1994).

2. For a balanced account of the advantages and disadvantages, see, e.g., P. Wheale and R. McNally, *Genetic Engineering: Catastrophe or Utopia?* (Hemel Hempstead: Wheatsheaf, 1988); P. Wheale and R. McNally, eds., *The Bio-Revolution: Cornucopia or Pandora's Box?* (London: Pluto, 1990).

3. See, e.g., S. Kingman, "Buried Treasure in Human Genes," *New Scientist*, July 8, 1989; reprinted in *Bioethics News* 9 (1990): 10–15.

4. Aldous Huxley, *Brave New World* (Harlow, Essex, 1983; first edition 1932).

5. I. Niiniluoto, "Should Technological Imperatives Be Obeyed?" *International Studies in the Philosophy of Science 4* (1990): 181–89.

6. Niiniluoto, "Technological Imperatives," 182–84, makes an additional distinction between pro- and anti-technological variants of technological determinism and voluntarism. In what follows I shall confine my comments to pessimistic anti-technological determinism and optimistic pro-technological voluntarism.

7. J. Ellul, *The Technological Society* (New York: Alfred A. Knopf, 1964).

8. The situation in the United States up to the end of the 1980s is described in D. M. Koenig, *The International Association of Penal Law: United States Report on Topic II: Modern Biomedical Techniques* (Lansing, Mich.: Thomas M. Cooley Law School, n.d.). Many European countries are now preparing legal regulations on gene splicing.

9. The following is mostly based on Niiniluoto, "Technological Imperatives," 185, who is broadly citing the ideas of Joseph Schumpeter.

10. In the words of Daniel Callahan: "It's very hard to sustain a great deal of worry about these things when, after ten years of pretty constant interest and attention, there have been no untoward events." Cited by Koenig, *International Association,* p. 29.

2

Redesigning the World: Ethical Questions about Genetic Engineering

Ron Epstein

Until the demise of the Soviet Union, we lived under the daily threat of nuclear holocaust extinguishing human life and the entire biosphere.[1] Now it looks more likely that total destruction will be averted, and that widespread, but not universally fatal, damage will continue to occur from radiation accidents from power plants, aging nuclear submarines, and perhaps the limited use of tactical nuclear weapons by governments or terrorists.

What has gone largely unnoticed is the unprecedented lethal threat of genetic engineering to life on the planet. It now seems likely, unless a major shift in international policy occurs quickly, that the major ecosystems that support the biosphere are going to be irreversibly disrupted, and that genetically engineered viruses may very well lead to the eventual demise of almost all human life. In the course of the major transformations that are on the way, human beings will be transformed, both intentionally and unintentionally, in ways that will make us something different than what we now consider human.

Heedless of the dangers, we are rushing full speed ahead on almost all fronts. Some of the most powerful multinational chemical, pharmaceutical and agricultural corporations have staked their financial futures on genetic engineering. Enormous amounts of money are already involved, and the United States government is currently bullying the rest of the world into rapid acceptance of corporate demands concerning genetic engineering research and marketing.

WHAT IS GENETIC ENGINEERING?

What Are Genes?

Genes are often described as "blueprints" or "computer programs" for our bodies and all living organisms. Although it is true that genes are specific sequences of

DNA (deoxyribonucleic acid) that are central to the production of proteins, contrary to popular belief and the now outmoded standard genetic model, genes do not directly determine the "traits" of an organism.[2] They are a single factor among many. They provide the "list of ingredients" which is then organized by the "dynamical system" of the organism. That "dynamical system" determines how the organism is going to develop. In other words, a single gene does not, in most cases, exclusively determine either a single feature of our bodies or a single aspect of our behavior. A recipe of ingredients alone does not create a dish of food. A chef must take those ingredients and subject them to complex processes which will determine whether the outcome is mediocre or of gourmet quality. So too the genes are processed through the self-organizing ("dynamical") system of the organism, so that the combination of a complex combination of genes is subjected to a variety of environmental factors which lead to the final results, whether somatic or behavioral.[3]

> . . . a gene is not an easily identifiable and tangible object. It is not only the DNA sequence which determines its functions in the organisms, but also its location in a specific chromosomal, cellular, physiological and evolutionary context. It is therefore difficult to predict the impact of genetic material transfer on the functioning of the extremely tightly controlled, integrated and balanced functioning of all the tens of thousands of structures and processes that make up the body of any complex organism.[4]

Genetic engineering refers to the artificial modification of the genetic code of a living organism. Genetic engineering changes the fundamental physical nature of the organism, sometimes in ways that would never occur in nature. Genes from one organism are inserted in another organism, most often across natural species boundaries. Some of the effects become known, but most do not. The effects of genetic engineering which we know are usually short-term, specific and physical. The effects we do not know are often long-term, general, and also mental. Long-term effects may be either specific[5] or general.

Differences between Bioengineering and Breeding

The breeding of animals and plants speeds up the natural processes of gene selection and mutation that occur in nature to select new species that have specific use to humans. Although the selecting of those species interferes with the natural selection process that would otherwise occur, the processes utilized are found in nature. For example, horses are bred to run fast without regard for how those thoroughbreds would be able to survive in the wild. There are problems with stocking streams with farmed fish because they tend to crowd out natural species, be less resistant to disease, and spread disease to wild fish.[6]

The breeding work of people like Luther Burbank led to the introduction of a whole range of tasty new fruits. At the University of California at Davis square tomatoes with tough skins were developed for better packing and shipping. Sometimes breeding goes awry. Killer bees are an example. Another example is the 1973 corn blight that killed a third of the crop that year. It was caused by a newly bred corn cultivar that was highly susceptible to a rare variant of a common leaf fungus.[7]

Bioengineers often claim that they are just speeding up the processes of natural selection and making the age-old practices of breeding more efficient. In some cases that may be true, but in most instances the gene changes that are engineered would never occur in nature, because they cross natural species barriers.

HOW GENETIC ENGINEERING IS CURRENTLY USED

Here is a brief summary of some of the more important, recent developments in genetic engineering.[8]

(1) Most of the genetic engineering now being used commercially is in the agricultural sector. Plants are genetically engineered to be resistant to herbicides, to have built in pesticide resistance, and to convert nitrogen directly from the soil. Insects are being genetically engineered to attack crop predators. Research is ongoing in growing agricultural products directly in the laboratory using genetically engineered bacteria. Also envisioned is a major commercial role for genetically engineered plants as chemical factories. For example, organic plastics are already being produced in this manner.[9]

(2) Genetically engineered animals are being developed as living factories for the production of pharmaceuticals and as sources of organs for transplantation into humans. (New animals created through the process of cross-species gene transfer are called xenografts. The transplanting of organs across species is called xenotransplantation.) A combination of genetic engineering and cloning is leading to the development of animals for meat with less fat, etc. Fish are being genetically engineered to grow larger and more rapidly.

(3) Many pharmaceutical drugs, including insulin, are already genetically engineered in the laboratory. Many enzymes used in the food industry, including rennet used in cheese production, are also available in genetically engineered form and are in widespread use.

(4) Medical researchers are genetically engineering disease carrying insects so that their disease potency is destroyed. They are genetically engineering human skin[10] and soon hope to do the same with entire organs and other body parts.

(5) Genetic screening is already used to screen for some hereditary conditions. Research is ongoing in the use of gene therapy in the attempt to correct some of these conditions. Other research is focusing on techniques to make genetic changes directly in human embryos. Most recently research has also been focused on combining cloning with genetic engineering. In so-called germline therapy, the genetic changes are passed on from generation to generation and are permanent.

(6) In mining, genetically engineered organisms are being developed to extract gold, copper, etc. from the substances in which it is embedded. Other organisms may someday live on the methane gas that is a lethal danger to miners. Still others have been genetically engineered to clean up oil spills, to neutralize dangerous pollutants, and to absorb radioactivity. Genetically engineered bacteria are being developed to transform waste products into ethanol for fuel.

SOME SPECIFIC DIFFICULTIES WITH GENETIC ENGINEERING

Here are a few examples of current efforts in genetic engineering that may cause us to think twice about its rosy benefits.

The Potential of Genetic Engineering for Disrupting the Natural Ecosystems of the Biosphere

At a time when an estimated 50,000 species are already expected to become extinct every year, any further interference with the natural balance of ecosystems could cause havoc. Genetically engineered organisms, with their completely new and unnatural combinations of genes, have a unique power to disrupt our environment. Since they are living, they are capable of reproducing, mutating and moving within the environment. As these new life forms move into existing habitats they could destroy nature as we know it, causing long term and irreversible changes to our natural world.[11]

Any child who has had an aquarium knows that the fish, plants, snails, and food have to be kept in balance to keep the water clear and the fish healthy. Natural ecosystems are more complex but operate in a similar manner. Nature, whether we consider it to be conscious or without consciousness, is a self-organizing system with its own mechanisms.[12] In order to guarantee the long-term viability of the system, those mechanisms insure that important equilibria are maintained. Lately the extremes of human environmental pollution and other human activities have been putting deep strains on those mechanisms. Nonetheless, just as we can clearly see when the aquarium is out of kilter, we can learn to sensitize ourselves to Nature's warnings and know when we are endangering Nature's mechanisms for maintaining equilibria. We can see an aquarium clearly. Unfortunately, because of the limitations of our senses in detecting unnatural and often invisible change, we may not become aware of serious dangers to the environment until widespread damage has already been done.

Deep ecology[13] and Gaia theory have brought to general awareness the interactive and interdependent quality of environmental systems.[14] No longer do we believe that isolated events occur in nature. Each event is part of a vast web of inter-causality, and as such has widespread consequences within that ecosystem.

If we accept the notion that the biosphere has its own corrective mechanisms, then we have to look at how they work and the limitations of their design. The more extreme the disruption to the self-organizing systems of the biosphere, the stronger the corrective measures are necessary. The notion that the systems can ultimately deal with any threat, however extreme, is without scientific basis. No evidence exists that the life and welfare of human beings have priority in those self-organizing systems. Nor does any evidence exist that anything in those systems is equipped to deal with all the threats that genetically engineered organisms may pose. Why? The organisms are not in the experience of the systems, because they could never occur naturally as a threat. The basic problem is a denial on the part of many geneticists that genetically engineered organisms are radical, new, and unnatural forms of life, which, as such, have no place in the evolutionarily balanced biosphere.

Viruses

Plant, animal, and human viruses play a major role in the ecosystems that comprise the biosphere. They are thought by some to be one of the primary factors in evolutionary change. Viruses have the ability to enter the genetic material of their hosts, to break apart, and then to recombine with the genetic material of the host to create new viruses. Those new viruses then infect new hosts, and, in the process, transfer new genetic material to the new host. When the host reproduces, genetic change has occurred.

If cells are genetically engineered, when viruses enter the cells, whether human, animal, or plant, then some of the genetically engineered material can be transferred to the newly created viruses and spread to the viruses' new hosts. We can assume that ordinary viruses, no matter how deadly, if naturally produced, have a role to play in an ecosystem and are regulated by that ecosystem. Difficulties can occur when humans carry them out of their natural ecosystems; nonetheless, all ecosystems in the biosphere may presumably share certain defense characteristics. Since viruses that contain genetically engineered material could never naturally arise in an ecosystem, there is no guarantee of natural defenses against them. They then can lead to widespread death of humans, animals or plants, thereby temporarily or even permanently damaging the ecosystem. Widespread die-off of a plant species is not an isolated event but can affect its whole ecosystem. For many, this may be a rather theoretical concern. The distinct possibility of the widespread die-off of human beings from genetically engineered viruses may command more attention.[15]

Biowarfare

Secret work is going forward in many countries to develop genetically engineered bacteria and viruses for biological warfare. International terrorists have already begun seriously considering their use. They are almost impossible to regulate, because the same equipment and technology that are used commercially can easily and quickly be transferred to military application.

The former Soviet Union had 32,000 scientists working on biowarfare, including military applications of genetic engineering. No one knows where most of them have gone, or what they have taken with them. Among the more interesting probable developments of their research were smallpox viruses engineered either with equine encephalitis or with Ebola virus. In one laboratory, despite the most stringent containment standards, a virulent strain of pneumonia, which had been stolen from the United States military, infected wild rats living in the building, which then escaped into the wild.[16]

There is also suggestive evidence that much of the so-called Gulf War Syndrome may have been caused by a genetically engineered biowarfare agent which is contagious after a relatively long incubation period. Fortunately that particular organism seems to respond to antibiotic treatment.[17] What is going to happen when the organisms are purposely engineered to resist all known treatment?

Nobel laureate in genetics and president emeritus of Rockefeller University Joshua Lederberg has been in the forefront of those concerned about international control of biological weapons. Yet when I wrote Dr. Lederberg for information about ethical problems in the use of genetic engineering in biowarfare, he replied, "I don't see how we'd be talking about the ethics of genetic engineering, any more than that of iron smelting—which can be used to build bridges or guns."[18] Like most scientists, Lederberg fails to acknowledge that scientific researchers have a responsibility for the use to which their discoveries are put. Thus he also fails to recognize that once the genie is out of the bottle, you cannot coax it back in. In other words, research in genetic engineering naturally leads to its employment for biowarfare, so that before any research in genetic engineering is undertaken, its potential use in biowarfare should be clearly evaluated. After they became aware of the horrors of nuclear war, many of the scientists who worked in The Manhattan Project, which developed the first atomic bomb, underwent terrible anguish and soul-searching. It is surprising that more geneticists do not see the parallels.

After reading about the dangers of genetic engineering in biowarfare, the president of the United States, Bill Clinton, became extremely concerned, and, in the spring of 1998, made civil defense countermeasures a priority. Yet, his administration has systematically opposed all but the most rudimentary safety regulations and restrictions for the biotech industry. By doing so, Clinton has unwittingly created a climate in which the production of the weapons he is trying to defend against has become very easy for both governments and terrorists.[19]

Plants

New crops may breed with wild relatives or cross breed with related species. The "foreign" genes could spread throughout the environment causing unpredicted changes which will be unstoppable once they have begun. Entirely new diseases may develop in crops or wild plants. Foreign genes are designed to be carried into other organisms by viruses which can break through species barriers, and overcome an organism's natural defenses. This makes them more infectious than naturally existing parasites, so any new viruses could be even more potent than those already known.

Ordinary weeds could become "Super-weeds": Plants engineered to be herbicide resistant could become so invasive they are a weed problem themselves, or they could spread their resistance to wild weeds making them more invasive. Fragile plants may be driven to extinction, reducing nature's precious biodiversity. Insects could be impossible to control. Making plants resistant to chemical poisons could lead to a crisis of "super pests" if they also take on the resistance to pesticides.[20]

The countryside may suffer even greater use of herbicides and pesticides: Because farmers will be able to use these toxic chemicals with impunity their use may increase threatening more pollution of water supplies and degradation of soils.

Plants developed to produce their own pesticide could harm non-target species such as birds, moths and butterflies. No one—including the genetic scientists—knows for sure the effect releasing new life forms will have on the environment. They do know that all of the above are possible and irreversible, but they still want to carry out their experiment. THEY get giant profits. All WE get is a new and uncertain environment—an end to the world as we know it.[21]

When genetically engineered crops are grown for a specific purpose, they cannot be easily isolated both from spreading into the wild and from cross-pollinating with wild relatives. It has already been shown[22] that cross-pollination can take place almost a mile away from the genetically engineered plantings. As has already occurred with noxious weeds and exotics, human beings, animals and birds may accidentally carry the genetically engineered seeds far vaster distances. Spillage in transport and at processing factories is also inevitable. The genetically engineered plants can then force out plant competitors and thus radically change the balance of ecosystems or even destroy them.

Under current United States government regulations, companies that are doing field-testing of genetically engineered organisms need not inform the public of what genes have been added to the organisms they are testing. They can be declared trade secrets, so that the public safety is left to the judgment of corporate scientists and government regulators many of whom switch back and forth between working for the government and working for the corporations they supposedly regulate.[23] Those who come from academic positions often have large financial stakes in biotech companies,[24] and major universities are making agreements with biotech corporations that compromise academic freedom and give patent rights to the corporations. As universities become increasingly dependent on major corporations for funding, the majority of university scientists will no longer be able to function as independent, objective experts in matters concerning genetic engineering and public safety.[25]

Scientists have already demonstrated the transfer of transgenes and marker genes to both bacterial pathogens and to soil fungi. That means genetically engineered organisms are going to enter the soil and spread to whatever grows in it. Genetically engineered material can migrate from the roots of plants into soil bacteria, in at least one case radically inhibiting the ability of the soil to grow plants.[26] Once the bacteria are free in the soil, no natural barriers inhibit their spread. With ordinary soil pollution, the pollution can be confined and removed (unless it reaches the ground-water). If genetically engineered soil bacteria spreads into the wild, the ability of the soil to support plant life may seriously diminish.[27] It does not take much imagination to see what the disastrous consequences might be. Water and air are also subject to poisoning by genetically engineered viruses and bacteria.

The development of new genetically engineered crops with herbicide resistance will affect the environment through the increased use of chemical herbicides. Monsanto and other major international chemical, pharmaceutical, and agricultural corporations have staked their financial futures on genetically engineered herbicide-resistant plants.[28]

Recently scientists have found a way to genetically engineer plants so that their seeds lose their viability unless sprayed with patented formulae, most of which turn out to have antibiotics as their primary ingredient. The idea is to keep farmers from collecting genetically engineered seed, thus forcing them to buy it every year. The corporations involved are unconcerned about the gene escaping into the wild, with obvious disastrous results, even though that is a clear scientific possibility.[29]

So that we would not have to be dependent on petroleum-based plastics, some scientists have genetically engineered plants that produce plastic within their stem structures. They claim that it biodegrades in about six months.[30] If the genes escape into the wild, through cross-pollination with wild relatives or by other means, then we face the prospect of natural areas littered with the plastic spines of decayed leaves. However aesthetically repugnant that may seem, the plastic also poses a real danger. It has the potential for disrupting entire food-chains. It can be eaten by invertebrates, which are in turn eaten, and so forth. If primary foods are inedible or poisonous, then whole food-chains can die off.[31]

Another bright idea was to genetically engineer plants with scorpion toxin, so that insects feeding on the plants would be killed. Even though a prominent geneticist warned that the genes could be horizontally transferred to the insects themselves, so that they might be able to inject the toxin into humans, the research and field testing is continuing.[32]

Animals

The genetic engineering of new types of insects, fish, birds, and animals has the potential of upsetting natural ecosystems. They can displace natural species and upset the balance of other species through behavior patterns that are a result of their genetic transformation.

One of the more problematic ethical uses of animals is the creation of xenografts, already mentioned above, which often involve the insertion of human genes. (See the section immediately below.) Whether or not the genes inserted to create new animals are human ones, the xenografts are created for human use and patented for corporate profit with little or no regard for the suffering of the animals, their feelings and thoughts, or their natural life-patterns.

Use of Human Genes

As more and more human genes are being inserted into non-human organisms to create new forms of life that are genetically partly human, new ethical questions arise. What percent of human genes does an organism have to contain before it is considered human? For instance, how many human genes would a green pepper[33] have to contain before one would have qualms about eating it? For meat-eaters, the same question could be posed about eating pork. If human beings have special ethical status, does the presence of human genes in an organism change its ethical status? What about a mouse genetically engineered to produce human sperm[34] that is then used in the conception of a human child?

Several companies are working on developing pigs that have organs containing human genes in order to facilitate the use of the organs in humans. The basic idea is something like this. You can have your own personal organ donor pig with your genes implanted. When one of your organs gives out, you can use the pig's.

> The U.S. Food and Drug Administration (FDA) issued a set of xenotransplant guidelines in September of 1996 that allows animal to human transplants, and puts the responsibility for health and safety at the level of local hospitals and medical review boards. A group of 44 top virologists, primate researchers, and AIDS specialists have attacked the FDA guidelines, saying, "based on knowledge of past cross-species transmissions, including AIDS, Herpes B virus, Ebola, and other viruses, the use of animals has not been adequately justified for use in a handful of patients when the potential costs could be in the hundreds, thousands or millions of human lives should a new infectious agent be transmitted."[35]

England has outlawed such transplants as too dangerous.[36]

Humans

Genetically engineered material can enter the body through food or bacteria or viruses. The dangers of lethal viruses containing genetically engineered material and created by natural processes have been mentioned above.

> The dangers of generating pathogens by vector mobilization and recombination are real. Over a period of ten years, 6 scientists working with the genetic engineering

of cancer-related oncogenes at the Pasteur Institutes in France have contracted cancer.[37]

Non-human engineered genes can also be introduced into the body through the use of genetically engineered vaccines and other medicines, and through the use of animal parts genetically engineered with human genes to combat rejection problems.

Gene therapy, for the correction of defective human genes that cause certain genetic diseases, involves the intentional introduction of new genes into the body in an attempt to modify the genetic structure of the body. It is based on a simplistic and flawed model of gene function which assumes a one-to-one correspondence between individual gene and individual function.

Since horizontal interaction[38] among genes has been demonstrated, introduction of a new gene can have unforeseen effects. Another problem, already mentioned, is the slippery slope that leads to the notion of designer genes. We are already on that slope with the experimental administration of genetically engineered growth hormone to healthy children, simply because they are shorter than average and their parents would like them to be taller.[39]

A few years ago a biotech corporation applied to the European Patent Office for a patent on a so-called pharm-woman, the idea being to genetically engineer human females so that their breast-milk would contain specialized pharmaceuticals.[40] Work is also ongoing to use genetic engineering to grow human breasts in the laboratory. It doesn't take much imagination to realize that not only would they be used for breast replacement needed due to cancer surgery, but also to foster a vigorous commercial demand by women in search of the "perfect" breasts.[41] A geneticist has recently proposed genetically engineering headless humans to be used for body parts. Some prominent geneticists have supported his idea.[42]

Genetically Engineered Food

Many scientists have claimed that the ingestion of genetically engineered food is harmless because the genetically engineered materials are destroyed by stomach acids. Recent research[43] suggests that genetically engineered materials are not completely destroyed by stomach acids and that significant portions reach the bloodstream and also the brain-cells. Furthermore, it has been shown that the natural defense mechanisms of body cells are not entirely effective in keeping the genetically engineered substances out of the cells.[44]

Some dangers of eating genetically engineered foods are already documented. Risks to human health include the probable increase in the level of toxins in foods and in the number of disease-causing organisms that are resistant to antibiotics.[45] The purposeful increase in toxins in foods to make them insect-resistant is the reversal of thousands of years of selective breeding of food-plants. For example when plants are genetically engineered to resist predators, often the plant defense systems involve the synthesis of natural carcinogens.[46]

Industrial mistakes or carelessness in production of genetically engineered food ingredients can also cause serious problems. The l-tryptophan food supplement, an amino acid that was marketed as a natural tranquilizer and sleeping pill, was genetically engineered. It killed thirty-seven people and permanently disabled 1,500 others with an incurable nervous system condition known as eosinophilia myalgia syndrome (EMS).[47]

Dr. John Fagan has summarized some major risks of eating genetically engineered food as follows:

> . . . the new proteins produced in genetically engineered foods could: a) themselves, act as allergens or toxins, b) alter the metabolism of the food producing organism, causing it to produce new allergens or toxins, or c) causing it to be reduced in nutritional value. . . . a) Mutations can damage genes naturally present in the DNA of an organism, leading to altered metabolism and to the production of toxins, and to reduced nutritional value of the food. b) Mutations can alter the expression of normal genes, leading to the production of allergens and toxins, and to reduced nutritional value of the food. c) Mutations can interfere with other essential, but yet unknown, functions of an organism's DNA.[48]

Basically what we have at present is a situation in which genetically engineered foods are beginning to flood the market, and no one knows what all their effects on humans will be. We are all becoming guinea pigs. Because genetically engineered food remains unlabeled, should serious problems arise, it will be extremely difficult to trace them to their source. Lack of labeling will also help to shield the corporations that are responsible from liability.

MORE BASIC ETHICAL PROBLEMS

Junk DNA

The 100,000 or more genes found in the human genome constitute perhaps 5% of the approximately 3.5 billion base pairs of DNA sequence in the haploid human genome. Most of this noncoding DNA lies between genes and has been called spacer, or even "junk" DNA.[49]

The overwhelming proportion of the DNA—perhaps up to 99% in some genomes—appears to have no known function. It has been described as "junk DNA" or "selfish DNA"—selfish because it serves no purpose except to get itself replicated along with the rest of the genome.[50]

An interesting parallel in thinking exists between junk DNA theory and earlier theories of the prefrontal lobes of the brain as being non-essential to human well-being. In 1955 Egas Moniz was awarded the Nobel Prize for Physiology or Medicine for the treatment of schizophrenia by prefrontal lobotomy. In this procedure a long wire is inserted through the side of the skull and the prefrontal lobes are then stirred like scrambled eggs. In the United States the foremost practitioner of prefrontal lobotomy, Dr. Walter Freeman, a professor at George Washington University Medical School and a president of the American Board of Psychiatry and Neurology, inserted an ice-pick through the tear ducts to sever the prefrontal lobe connections. The procedure, initially used on mental hospital patients, became popular as a way of dealing with mental problems.

> A sanitized version of the operation and its consequences was invariably given, and never more so than in an influential article entitled "Turning the Mind Inside Out," published in the *Saturday Evening Post* in 1941. The writer, the science editor of the *New York Times,* began in dramatic fashion by stating that there must be at least 200 men and women in the United States who had had worries, persecution complexes, suicidal intentions, obsession, and nervous tensions literally cut out of their minds with a knife.[51]

About 20,000 people eventually received the operation. Scrambling the frontal lobes of the brain is, of course, an irreversible process.

The "junk DNA" concept is a lot like the attitude toward the functioning of the prefrontal lobes of the brain that led to the travesty of prefrontal lobotomies. The attitude in both cases was and is that the latest scientific research does not show that there is any useful function going on there so the prefrontal lobes or the junk DNA must not have any important function and we can, therefore, remove or ignore them. Just as the performers of lobotomies operated "blind"—they could not see what they were doing, so too researchers who insert genes into new organisms operate "blind," with a scatter-gun approach, not knowing where the gene is going to end up in the new DNA or what effects it is going to have apart from the most crude measures. As with performing lobotomies, creating genetically engineered organisms is an irreversible process. The lobotomies cannot be undone and the organisms, once released, cannot be recalled. In both cases science and the "responsible" popular press laud the great benefits for humankind of the procedures.

Life as a Genetic Commodity

In 1971 the United States government issued the first patent on a living organism, a genetically engineered bacterium for cleaning up oil spills. That slippery slope has led not only to the patenting of genetically engineered plants and animals, but also to the patenting of human genes, often without either the consent of the people from whom they are taken or any benefit to them.[52]

A proprietary attitude toward living organisms is based on philosophies of instrumental values, in which intrinsic value is disregarded. In other words all life is evaluated only in terms of its specific use for the individual. Absent is any sense of respect for life and the right of other living beings to work out their own destiny.

Given the historical role of the United States in championing the notions of equality and individual rights, the legalization of instrumental values with regard to human genes is somewhat surprising. If "a man's home is his castle," how much the more so our bodies and genetic makeup. One would think that people would have legal control over their own genes; however, that does not seem to be the case.[53]

ASSESSING THE PRICE

For all the advantages claimed for genetic engineering, in the overwhelming number of cases the price seems too high to pay. In order to insure megaprofits for multinational corporations well into the next century, we will have to mortgage the biosphere, seriously compromise life on the planet, and even risk losing what it means to be a human being. We have seen that genetic engineering poses serious risks to human health and to the environment. It raises serious ethical questions about the right of human beings to alter life on the planet, both sentient and non-sentient, for the benefit of a few.

If there are some areas of genetic engineering that can safely benefit humanity while respecting other forms of life, then efforts need to be redoubled not only in the area of scientific risk assessment, but also in developing broad ethical guidelines. If experts in both scientific and ethical areas are to be trusted and respected, they must be free from the taint of personal monetary gain and other forms of self-aggrandizement. The public's right to know and assess potential dangers and ethical problems must have priority over both corporate secrecy and naïve views of academic freedom that accord scientists the right to experiment with whatever strikes their fancy without regard for the consequences. Decisions should not be left solely to the so-called experts, whatever their value. Ordinary citizens need to inform themselves, insist upon a mandate, and take responsibility for the grave decisions that must be made. The public welfare must be restored as the primary consideration, and the unrestrained amoral greed of multinational corporations somehow curtailed.

Is such a program of action possible? Certainly even slowing the inexorable progress of the current trends will be extremely difficult. Yet there is hope. In Europe, for example, heightened public awareness of the dangers of genetically engineered foods has significantly affected corporate plans for their widespread introduction there. Fortunately there also continue to be a vocal minority of well-trained scientists in the field, who see clearly the dangers of what is occurring, and who are brave enough to voice their consciences, despite very real personal and professional risks.[54] Clearly the key is educating the public about what is happening. We need to have confidence that ordinary citizens working together can build a foundation of integrity from which can arise a collective wisdom that can show us the way through the incredibly complicated maze of issues surrounding genetic engineering.

NOTES

Published with the permission of the author.

1. Further information related to the present article is available on his Web site: Genetic Engineering and Its Dangers <http://online.sfsu.edu/~rone/gedanger.htm>.

2. Ho et al. summarize the old and new genetic models as follows. In the old standard or "central dogma" model:

- Genes determine characters in linear causal chains, one gene giving rise to one character.
- Genes are not subject to influence from the environment.
- Genes remain stable and constant.
- Genes remain in organisms and stay where they are put.

According to the new genetics:

- No gene ever works in isolation, but rather in an extremely complicated genetic network. The function of each gene is dependent on the context of all the other genes in the genome. So, the same gene will have very different effects from individual to individual because other genes are different. There is so much genetic diversity within the human population that each individual is genetically unique. And, especially if the gene is transferred to another species, it is most likely to have new and unpredictable effects.
- The genetic network, in turn, is subject to layers of feedback regulation from the physiology of the organism and its relationship to the external environment.
- These layers of feedback regulation not only change the function of genes but can rearrange them, multiply copies of them, mutate them to order, or make them move around.
- And genes can even travel outside the original organism to infect another—this is called horizontal gene transfer.

Mae-Wan Ho, Hartmut Meyer, and Joe Cummins, "The Biotechnology Bubble" *Ecologist* 28, no. 3 (1998): 148.

3. Stewart A. Newman, "Genetic Engineering as Metaphysics and Menace," *Science and Nature* 9–10 (1989): esp. 114–18. See also Richard C. Strohman, "Epigenesis and Complexity: The Coming Kuhnian Revolution in Biology," *Nature Biotechnology* 15 (1997): 194–200; and Mae-Wan Ho, *Genetic Engineering: Dream or Nightmare. The Brave New World of Bad Science and Big Business* (Bath, U.K.: Gateway, 1998).

4. Vanaja Ramprasad, "Genetic Engineering and the Myth of Feeding the World," *Biotechnology and Development Monitor* 35 (1998): 24.

5. Examples of specific long-term effects are the death of a particular species or the introduction of a new disease organism.

6. After a major pesticide spill in Dunsmuir, California, in 1991, there was much

opposition to stocking the river because it would strongly discourage the revival of the native fish populations.

7. See Vandana Shiva, *Monocultures of the Mind* (London: Zed, 1993); and Shiva, *Biopiracy: The Plunder of Nature and Knowledge* (Boston: South End, 1997), 87–90. The use of genetic engineering in agriculture is a radical new extension of the monoculture paradigm. See also the discussion of the so-called Terminator genes below.

8. Most of this section is based on Jeremy Rifkin, *Biotech Century: Harnessing the Gene and Remaking the World* (J. P. Tarcher, 1998), 15–32.

9. See the section on plants entitled "Some Specific Difficulties with Genetic Engineering."

10. "The FDA cleared Organogenesis (Canton, MA) Apligraf® (graft-skin) for marketing. The product is the only living, bilayered skin construct approved for marketing in the U.S., according to Novartis Pharmaceuticals (E. Hanover, NJ), which will market Apligraf worldwide. Like human skin, Apligraf has two primary layers, including an outer epidermal layer made of living Keratinocytes. The dermal layer of Apligraf consists of living human fibroblasts. The human Keratinocytes and fibroblasts utilized to manufacture Apligraf are derived from donor tissue that is thoroughly screened for a wide range of infectious pathogens, notes a Novartis spokesperson. Apligraf is applied by a physician in a hospital outpatient facility or a wound care center" (*Genetic Engineering News,* June 15, 1998).

11. "The End of the World as We Know It: The Environmental Costs of Genetic Engineering" <www.greenpeace.org/~comms/cbio/brief2.html>.

12. This idea is central to Gaian theory, the theory that the entire planetary biosphere constitutes a single living organism or self-regulating system. Gaian theorists are divided into two main camps, those who believe that the earth is a mindless self-organizing physical system, and those who believe that the earth has its own consciousness or awareness.

13. The term deep ecology was coined by Arne Naess in his 1973 article, "The Shallow and the Deep, Long-Range Ecology Movements." "Naess was attempting to describe the deeper, more spiritual approach to nature. . . . He thought that this deeper approach resulted from a more sensitive openness to ourselves and nonhuman life around us." Bill Devall and George Sessions, *Deep Ecology: Living as If Nature Mattered* (Salt Lake City: Peregrine Smith, 1985), p. 65.

14. Although deep ecology and Gaia theory are still somewhat controversial, the existence of such legislation as the Endangered Species Act and the National Environmental Policy Act in the United States is indication that there is an increasing awareness of the importance of environmental systems.

15. The first genetic engineers called for a moratorium in the Asilomar Declaration of 1975, precisely because they were afraid of inadvertently creating new viral and bacterial pathogens. The worst-case scenario they envisaged may be taking shape. Commercial pressures led to regulatory guidelines based largely on untested assumptions, all of which have been invalidated by recent scientific findings. For example, biologically "crippled" laboratory strains of bacteria can often survive in the environment to exchange genes with other organisms. Genetic material (DNA) released from dead and living cells, far from being rapidly broken down, actually persists in the environment and transfers to other organisms. Naked viral DNA may be more infectious, and have a wider host

range than the virus. Viral DNA resists digestion in the gut of mice, enters the blood stream to infect white blood cells, spleen and liver cells, and may even integrate into the mouse cell genome. ("Scientists Link Gene Technology to Resurgence of Infectious Diseases. Call for Independent Enquiry," Press Release 6.4.98 from Professor Mae-Wan Ho: <http://home1.swipnet.se/~w-18472/prhortra.htm>)

See also Mae-Wan Ho et al., "Gene Technology and Gene Ecology of Infectious Diseases," *Microbial Ecology in Health and Disease* 10, no. 1 (1998): 33–59.

16. See Richard Preston, "Annals of Warfare: The Bioweaponeers," *New Yorker*, March 9, 1998, 52–65. See also Judith Miller and William T. Broad, "Iranians, Bioweapons in Mind, Lure Ex-Soviet Scientists," *New York Times,* December 8, 1998 <http://online.sfsu.edu/~rone/GE%20Essays/Iranians%20Bioweapons%20ExSoviet%20Scientists.htm>; and *Frontline,* "Plague War" <www.pbs.org/wgbh/pages/frontline/shows/plague>.

17. See Dr. Garth and Dr. Nancy Nicolson's Institute for Molecular Medicine Web site for details: <www.trufax.org/gulfwar2/newsrel.html>.

18. Quoted from e-mail I received from Dr. Lederberg in spring 1998.

19. William J. Broad and Judith Miller, "Germ Defense Plan in Peril as Its Flaws Are Revealed," *New York Times,* August 7, 1998. See also Wendy Barnaby, "Biological Weapons and Genetic Engineering," *GenEthics News,* June-July 1997.

20. The original says "resistance to herbicides" but "resistance to pesticides" is clearly meant.

21. "The End of the World as We Know It: The Environmental Costs of Genetic Engineering" <www.greenpeace.org/~comms/cbio/brief2.html>.

22. "The most immediate and easily observable impacts of transgenic plants on the ecological environment are due to cross-pollination between transgenic crop-plants and their wild relatives to generate super-weeds. Field trails have shown that cross-hybridization has occurred between herbicide resistant transgenic Brassica napa and its wild relatives." Ho, *Genetic Engineering,* p. 133. Some evidence indicates that transgenic plants have a greater ability to pollinate other plants than ordinary plants, which increases the danger of the emergence of superweeds. See J. Bergelson, C. B. Purrington, and G. Wichmann, "Promiscuity Increase in Transgenic Plants," *Nature,* September 3, 1998, 25; and "Genetically Engineered Plant Raises Fears of 'Superweeds,'" *Los Angeles Times,* September 3, 1998.

23. For example, it is public knowledge that there is a revolving door between Monsanto Corporation, one of the major transnational forces in genetic engineering, and the White House:

... Mr. Mickey Kantor, former United States Trade Representative and, until January 21 of last year, the Secretary of Commerce for the United States, has been made a member of the board of directors of Monsanto Corporation, a leading transnational biotechnology firm. The appointment was confirmed by a staff member of Mr. Kantor's Washington, DC law firm. Mr. Kantor joins others in the US who recently changed job assignments from service in government to positions in the biotechnology industry. Marcia Hale, earlier this month, moved from assistant to the President of the United States for intergovernmental affairs to senior official with Monsanto to coordinate public affairs and corporate strategy in the United Kingdom and Ireland. Also putting on the industry hat this month was L. Val Giddings. Less than two weeks ago, Giddings went from being a biotechnology regulator at the United States Department of Agriculture (USDA/APHIS) to being the vice president for food and agriculture at the Biotechnology Industry Organization (BIO). Giddings, who had been a member of the U.S. delegation at the first meeting of the Open Ended Ad Hoc Working Group on a Biosafety Protocol, attended last week's second meeting on the protocol as the representative of BIO. (E-mail news release from Beth Burrows of the Edmonds Institute entitled "Government Workers Go Biotech," May 22, 1997. Also posted on the Internet at <www.corpwatch.org/trac/corner/worldnews/other/other53.html> and <www.geocities.com/Athens/1527/appt.html>.)

Burrows distributed the following addendum at the biosafety negotiation meetings in Montreal, August 17, 1998:

Once again, this time in alphabetical order, we note with interest the following changes in job assignments:

David W. Beier . . . former head of Government Affairs for Genentech, Inc., now chief domestic policy advisor to Al Gore, Vice-President of the United States.

Linda J. Fisher . . . former Assistant Administrator of the United States Environmental Protection Agency's Office of Pollution Prevention, Pesticides, and Toxic Substances, now Vice President of Government and Public Affairs for Monsanto Corporation.

L. Val Giddings . . . former biotechnology regulator and (biosafety) negotiator at the United States Department of Agriculture (USDA/APHIS), now Vice President for Food & Agriculture of the Biotechnology Industry Organization (BIO).

Marcia Hale . . . former assistant to the President of the United States and director for intergovernmental affairs, now Director of International Government Affairs for Monsanto Corporation.

Michael (Mickey) Kantor . . . former Secretary of the United States Department of Commerce and former Trade Representative of the United States, now member of the board of directors of Monsanto Corporation.

Josh King . . . former director of production for White House events, now director of global communication in the Washington, D.C. office of Monsanto Corporation.

Terry Medley . . . former administrator of the Animal and Plant Health Inspection Service (APHIS) of the United States Department of Agriculture, former chair and vice-chair of the United States Department of Agriculture Biotechnology Council, former member of the U.S. Food and Drug Administration (FDA) food advisory committee, and now Director of Regulatory and External Affairs of Dupont Corporation's Agricultural Enterprise.

Margaret Miller . . . former chemical laboratory supervisor for Monsanto, now Deputy Director of Human Food Safety and Consultative Services, New Animal Drug Evaluation Office, Center for Veterinary Medicine in the United States Food and Drug Administration (FDA)*.

William D. Ruckelshaus . . . former chief administrator of the United States Environmental Protection Agency (USEPA), now (and for the past 12 years) a member of the board of directors of Monsanto Corporation.

Michael Taylor . . . former legal advisor to the United States Food and Drug Administration (FDA)'s Bureau of Medical Devices and Bureau of Foods, later executive assistant to the Commissioner of the FDA, still later a partner at the law firm of

King & Spaulding where he supervised a nine-lawyer group whose clients included Monsanto Agricultural Company, still later Deputy Commissioner for Policy at the United States Food and Drug Administration, and now again with the law firm of King & Spaulding.

Lidia Watrud . . . former microbial biotechnology researcher at Monsanto Corporation in St. Louis, Missouri, now with the United States Environmental Protection Agency Environmental Effects Laboratory, Western Ecology Division.

Clayton K. Yeutter . . . former Secretary of the U.S. Department of Agriculture, former U.S. Trade Representative (who led the U.S. team in negotiating the U.S. Canada Free Trade Agreement and helped launch the Uruguay Round of the GATT negotiations), now a member of the board of directors of Mycogen Corporation, whose majority owner is Dow AgroSciences, a wholly owned subsidiary of The Dow Chemical Company.

Such problems are by no means restricted to the United States:

The government's advisers on genetically engineered crops should be sacked because too many have close links to the biotech industry, environmentalists said yesterday. Friends of the Earth (FoE) said that 8 of the 13-strong Advisory Committee on Releases to the Environment (ACRE) have ties with companies or organisations involved in carrying out crop trials or other genetic engineering research. Members of ACRE are the Government's statutory advisers on allowing genetically modified crops to be planted in the countryside. They have so far passed more than 150 applications without any refusals. Although panel members do not vote on any application in which they have a personal interest, Adrian Bebb, FoE's food campaigner, said the process was flawed. "How can people have confidence in the government advisory panel when so many members have close financial links to the biotech industry?" (London *Independent,* July 8, 1998)

24. Russell Mokhiber, "'Objective' Science at Auction," *The Ecologist,* March-April 1998. See also Dan Fagin, Marianne Lavelle, and the Center for Public Integrity, *Toxic Deception: How the Chemical Industry Manipulates Science, Bends the Law, and Threatens Your Health* (Birch Lane Press, 1997).

25. According to a 1991 study of universities in the United States, MIT, Stanford, and Harvard respectively had the highest rates of commercial penetration into their biotechnology-related departments. (Sheldon Krinsky et al., "Academic-Corporate Ties in Biotechnology: A Quantitative Study," *Science, Technology, and Human Values* 16, no. 3 (1991): 275–87). The College of Natural Resources, University of California at Berkeley recently signed a $50 million agreement with a subsidiary of Novartis Corporation to give the latter exclusive access to patent rights on genetic engineering research done at the college. Novartis also has the right to appoint some faculty members. See Carl T. Hall, "Research Deal Evolving between UC, Biotech Firm: Berkeley Campus Could Get $50 Million," *San Francisco Chronicle,* October 9, 1998; Peter Rosset and Monica Moore, "Research Alliance Debated," *San Francisco Chronicle,* November 16, 1998; Charles Burress, "UC Finalizes Pioneering Research Deal with Biotech Firm: Pie Tossers Leave Taste of Protest," *San Francisco Chronicle,* November 24, 1998.

26. Ho, *Genetic Engineering,* 133. See also M. T. Holmes and E. R. Ingham, "The

Effects of Genetically Engineered Micro-organisms on Soil Food-webs," *Bulletin of the Ecological Society of America,* supplement, 75 (1994): 97.

27. The conscious choice of a few genes for mobilization and widespread replication substitutes human judgment for natural selection. From a theological viewpoint, it is questionable that the agribusiness scientific staff have the collective wisdom to determine what constitutes the "good" when it comes to desirable genes. The fact that their choice could become self-sustaining (e.g., if the gene escaped into the wild) is cause for further concern. Initially, this and other adverse impacts potentially resulting from mass scale transgenic operations are likely to be invisible. One potentially insidious effect of reliance on genetically engineered herbicide-resistant technology is the repeated use of single-herbicide preparations. The repeated applications of a controlled sequence of four or more different herbicides typical of transgenic farming could be expected to transiently affect soil microorganisms. But the sustained reliance on a single herbicide such as glyphosate [Roundup] or bromoxynil would predictably shift the soil microflora for longer periods, perhaps changing the overall composition of the soil's living matter irrevocably. Such an effect, should it occur, could affect soil quality for future plantings, particularly since germination in some herbicide-treated soils has been reported to be impaired. Here the ethical concern is responsibility for future generations. . . . (Marc Lappe and Britt Bailey, *Against the Grain: Biotechnology and the Corporate Takeover of Your Food* [Monroe, Me.: Common Courage, 1998], 114).

28. Lappe and Bailey, *Against the Grain,* esp. 50–62.

29. On March 3, 1998, the US Department of Agriculture (USDA) and an American cotton seed company, Delta & Pine Land Co., received a US patent on a technique that genetically alters seed so that it will not germinate if re-planted a second time. The technology aims to prevent farmers from saving seed from their harvest to re-plant the following season. Because it is a potentially "lethal" technology, Rural Advancement Foundation International (RAFI) has dubbed it the "Terminator technology." . . . If commercially viable, the Terminator technology will have profound implications for agriculture. It is a global threat to farmers, biodiversity and food security. The seed-sterilizing technology threatens to eliminate the age-old right of farmers to save seed from their harvest and it jeopardizes the food security of 1.4 billion people—resource poor farmers in the South—who depend on farm-saved seed. The developers of the technology say that it will be targeted for use primarily in the South as a means of preventing farmers from saving proprietary seeds marketed by American seed corporations. Delta & Pine Land Co. and USDA have applied for patents on the Terminator technology in at least 78 countries. If the Terminator technology is widely utilized, it will give the multinational seed and agrochemical industry an unprecedented and extremely dangerous capacity to control the world's food supply. (*RAFI Communique,* March-April 1998 <http://www.rafi.ca/communique/ 19982.html>).

Geneticist Joseph Cummins has commented:

> Pollen escaping from the terminator crop is sterile and cannot spread to weeds or other crops. Pollen escaping from the tetracycline treated seed producing crop can spread the terminator blocking genes. When a weed is fertilized, for example, with the terminator pollen, the new generation of seeds will bear plants with fertile pollen. In the next generation, 25% of the terminator plants will produce sterile pollen. Since the sterile pollen cannot spread the terminator genes, the spread of terminator genes by normal sexual means is limited, but the terminator genes will always be in the population. The situation is similar to lethal genetic diseases in humans. Terminator doesn't threaten plant populations if it is spread only by normal sexual processes. However, spread of terminator by other means is more intimidating. . . . Spreading terminator genes by virus could easily cause a wide array of weeds and crops to be rendered sterile, and genetic recombination could easily eliminate the reversing action of tetracycline. The terminator virus could have a profound influence on crop production. . . . [Such genes] are potentially able to create chromosome mutations leading to genetic erosion and untoward changes in gene regulation and expression. They are very highly mobile, and, once introduced into higher plants and animals, are likely to spread and not want to leave ever! ("Genetics of Terminator," e-mail from Prof. Cummins, Wednesday, June 17, 1998. Some punctuation added).

See also Volker Lehmann, "Patent on Seed Sterility Threatens Seed Saving," *Biotechnology and Development Monitor,* June 1998, 6–8; Rural Advancement Foundation Internationals (RAFI) news releases on "Terminator Technology" <www.rafi.org/misc/terminator.html>; and Martha L. Crouch, *How the Terminator Terminates: An Explanation for the Non-Scientist of a Remarkable Patent for Killing Second Generation Seeds of Crop Plants,* rev. ed. (Edmonds, Wash.: Edmonds Institute, 1998) <www.bio.indiana.edu/people/terminator.html>. Zeneca BioSciences (UK) has recently applied for a patent of their own version of the Terminator. See Rural Advancement Foundation Internationals (RAFI) news release "And Now, the 'Verminator': Fat Cat Corp. with Fat Rat Gene Can Kill Crops," August 24, 1998 <www.rafi.org/pr/release19.html>.

30. "The first samples of transgenically grown biodegradable plastics, or polyhydroxy-alkanoates (PHAs), have been shipped to plastics companies in the United States and Europe. Metabolix Inc., of Cambridge, MA, has patented transgenic technology in the U.S. that inserts genes into transgenic highly efficient fermentation systems and field crops to produce plastics, which will eventually be cost competitive with petroleum-based plastics used in packaging, diapers, containers, bottles, and garbage bags. A number of companies and research institutions in the U.S. and Europe are working to patent transgenic plants for plastic production. . . . University of Warwick scientists in England have made breakthroughs in transgenic plastic production, while researchers in Canada are close to making plastics from similar transgenic plants" (*BIOINFO: An Agricultural Biotechnology Monitor* 4, no. 3 [1996]).

31. Dr. John Fagan, a molecular biologist, has warned that the new constituents that are used in these plastics are oils that are probably toxic to animals and humans. Thus when cross-pollination occurs with wild brassica, the wild plants produce things that are toxic to the deer, rabbits, and other wildlife, as well as humans (personal communication).

32. Geneticist Joseph Cummins warned: "The questionable experiment is to insert a

gene for scorpion toxin into an insect virus then to spray the tinkered virus onto produce crops in the field. The tinkered virus is now highly potent in destroying insects, both pests and their natural predators and the pollinators. The scorpion toxin may not be threatening to humans as a toxin when it is eaten but its impact on cuts and open sores is a concern. Such toxins are frequently allergens as well as nerve toxins. Food allergy causes effects ranging from migraine headache to death. The danger from a small field test is tangible provided the experiment is not well thought out and controlled. Genetic recombination is a significant concern in such experiments. The scorpion toxin gene can be spread by recombination to insects that suck blood as well as insects that suck plant juice. The virus that acquires a toxin gene will achieve a new ecological niche and is likely to be a formidable parasite" (*Gene Tinkering Blues,* August 1996).

33. In the People's Republic of China, Professor Chen Zhang-Liang, director of the National Laboratory of Protein Engineering and Plant Genetic Engineering and vice president, Peking University, directs a laboratory "that is transplanting human protein genes into tomatoes and sweet peppers to control ripening" (Arthur Fisher, "A Long Haul for Chinese Science," *Popular Science,* "Chinese Science and Technology," special issue, August 1996, 42).

34. See "Surrogate Fathers," *New Scientist,* January 31, 1998.

35. *IP/Biodiv News,* January 24, 1997. For further information on the dangers of xenotransplants, see Alix Fano et al., *Of Pigs, Primates, and Plagues: A Layperson's Guide to the Problems with Animal-to-Human Organ Transplants* (New York: Medical Research Modernization Committee, n.d.).

36. "LONDON (AP)—Britain barred the transplant of animal organs to humans on Thursday, saying the risk of disease transmission must be better understood. The decision came after a report by a government-appointed panel of experts, which was chiefly concerned that animal viruses in the transplanted organs could introduce new diseases into humans. . . . Pigs are known to harbor several retroviruses—the family of virus that includes HIV—one of which was only discovered in the last two years. Research has shown that some of them can infect animal cells in the laboratory. . . . But Britain has made it clear that emergency legislation will be brought in if necessary to halt human trials. . . . The pigs are bred with human genes that cause their organs to be coated with human molecules, which is meant to prevent the most severe form of rejection in a human patient" (Associated Press, January 16, 1997).

37. Ho (1996) reporting on information from *New Scientist,* June 18, 1987, 29.

38. Horizontal gene transfer refers to "the transfer of genes to unrelated species by infection through viruses, through pieces of genetic material, DNA, by being taken up into the cells from the environment, or by unusual mating taking place between unrelated species" (Mae-Wan Ho, *Genetic Engineering,* 13).

39. Andrew Kimbrell, *The Human Body Shop: The Engineering and Marketing of Life* (New York: HarperCollins, 1994), 142–57.

40. Kimbrell, *Human Body Shop,* 191.

41. See Rifkin, *Biotech Century,* 24–25.

42. "Jonathan Slack, professor of developmental biology at Bath University and a leading embryologist, says he can now create headless frog embryos relatively easily by manipulating certain genes. . . . He said the breakthrough could be applied to human embryos because the same genes perform similar functions in both frogs and humans.

Using intact cloned human embryos to grow organs would be out of the question because they would have to be killed and this would be equivalent to murder, Slack said. . . . Slack's ideas have angered some academics. Professor Andrew Linzey, an animal ethicist at Oxford University, denounced his research. 'This sort of thinking beggars belief. It's scientific fascism because we would be creating other beings whose very existence would be to serve the dominant group. It is morally regressive to create a mutant form of life,' Linzey said. Other scientists, however, support Slack in raising the profile of such controversial research. . . . Lewis Wolpert, professor of biology as applied to medicine at University College London, said Slack's suggestions were perfectly sensible and could in principle be possible. 'There are no ethical issues because you are not doing any harm to anyone. It is a question of whether it is acceptable or not to the public and that depends on the 'yuk' factor" ("Headless Frog Opens Way for Human Organ Factory," *London Sunday Times,* October 19, 1997).

43. Textbooks say that DNA in food should be digested and destroyed. But Dorfler and his student Rainer Schubbert found that when they fed a bacterial virus called [1]M13 to a mouse, sections of its genetic material about 700 DNA "letters" long—large enough to contain a gene—survived to emerge in faeces. The researchers wondered whether a few of these genetic snippets had managed to penetrate the mouse's cells. They took cells from the mice and probed them with a dye molecule that lights up when it binds to the M13 DNA. The probe lit up inside cells not only from the intestine, but the spleen, white blood cells and liver. "They weren't hard to find," says Dorfler. "In some cases as much as one cell in a thousand had viral DNA." Usually the DNA does not stay long inside the cells. After 18 hours, most cells had somehow ejected the viral intruders. But Dorfler speculates that occasionally some foreign DNA may remain. (*New Scientist,* January 4, 1997)

Geneticist Joseph Cummins has commented:

The matter of DNA being incorporated into human chromosome from food is still alive and active. The fact is clear that such incorporation takes place. . . . Uptake into cells may trigger apoptosis (cell suicide) in some cases but Doerfler showed that mature healthy cells do incorporate foreign DNA into chromosomes. It is too early to make conclusions about the consequences of such incorporation. Since food is implicated in a variety of immune and autoimmune diseases as well as mental illness (Celiacs have very high incidence of mental illness) the role of incorporated DNA disease should be investigated and blind conclusions should be avoided. Certainly food DNA has been incorporated into somatic cell chromosomes throughout evolutionary history. The danger of genetic engineering is related to the greatly amplified incorporation of viral promoters, bacterial genes and synthetic genes (widely employed by Monsanto) causes new risks. (personal e-mail communication, August 18, 1998)

See also R. Schubbert, C. Lettman, and W. Doerfler, "Ingested Foreign (Phage M13) DNA Survives Transiently in the Gastrointestinal Tract and Enters the Bloodstream of Mice," *Mol. Genet.* 242 (1994): 495–504.

44. "It has long been assumed that our gut is full of enzymes which can rapidly digest

DNA. In a study designed to test the survival of viral DNA in the gut, mice were fed DNA from a bacterial virus, and large fragments were found to survive passage through the gut and to enter the blood stream. This research group has now shown that ingested DNA end up, not only in the gut cells of the mice, but also in spleen and liver cells as well as white blood cells. 'In some cases, as much as one cell in a thousand had viral DNA'" (Ho, "Gene Technology," 141).

45. The spread of antibiotic-resistant organisms is a dangerous by-product of genetic engineering. Antibiotic resistant genes are often used as marker genes in the process of gene-splicing. They can then be spread by horizontal transfer. As mentioned above, antibiotics are also sprayed on crops in large quantities to unlock the effect of the so-called Terminator technology.

46. *Science,* June 16, 1989, 1233.

47. John B. Fagan, Ph.D., "Tryptophan Summary," <http://home1.swipnet.se/~w-18472/jftrypt.htm>. Apparently genetically engineered tryptophan is already back on the market with the same dangerous characteristics that caused the initial problems:

A reformulated food supplement could contain harmful contaminants similar to those found in an earlier banned version of the product, according to US researchers. L-trytophan, a naturally occurring amino acid, was banned in 1990 after the product was linked to a Japanese outbreak of a rare blood disease called eosinophilia myalgia syndrome (EMS). The outbreak affected 1,500 people and killed 30. Studies of the product, which was promoted as a sleep and diet aid, revealed it contained an unidentified contaminant nicknamed 'peak X'. Scientists could not determine whether the contaminant, or L-trytophan, or a combination of the two caused the disease. Several manufacturers have since reformulated the product into brands containing 5-hydroxy-L-trytophan. The product is widely available over-the-counter in the US. Stephen Naylor and Gerald Gleich at the Mayo Clinic in Minnesota examined six brands for traces of the original contaminant. All six showed the 'peak X' signature, they said on Monday 31 August. The levels varied between 3–15% of those observed in a test on the original product. Gleich said they were not aware of the new formulations being associated with any outbreaks of EMS, but said the 'potential was there.' The Food and Drug Administration said it had confirmed Naylor and Gleich's findings, which are published in the September issue of the journal Nature Medicine. ("Remade Food Supplement as Bad as Banned Original" [posted by Shane Morris on the GENTECH@ping.de listserv on September 2, 1998]

Several articles on the subject, including those reporting the research mentioned above, were published in *The Lancet* (August 29, 1998).

48. John Fagan, "Assessing the Safety and Nutritional Quality of Genetically Engineered Foods" <http://home1.swipnet.se/~w-18472/jfassess.htm>.

49. Mark V. Bloom, Ph.D., "Polymerase Chain Reaction," <www.gene.com/ae/RC/CT/polymerase_chain_reaction.html>.

50. Mae-Wan Ho, *Genetic Engineering: Dream or Nightmare,* 110.

51. "Adventures with an Ice Pick: A Short History of the Lobotomy," adapted from Robert Youngson and Ian Schott, *Medical Blunders* (London: Robinson, 1996). Adaption copyrighted by The Independent on Sunday, March 3, 1996.

52. See Vandana Shiva, *Biopiracy,* 19 ff., for a summary. See also Vandana Shiva, *Biotechnology and the Environment* (Malaysia: Third World Network, n.d.).

53. For example, see the interesting case of Seattle businessman John Moore, discussed by Phillip L. Bereano in "Body and Soul: the Price of Biotech" (*Seattle Times,* August 20, 1995), B5. <http://online.sfsu.edu/~rone/GE%20Essays/Biotech%20Price.htm>. The case is also discussed in Monte Paulsen, "Biotech Buccaneers" (*Fairfield County Weekly,* August 29, 1998).

54. For example, several years ago I talked with a distinguished professor in the field who is a department chair at a well-known university. After publicly writing about the theoretical possibility of some serious dangers with genetic engineering, he was publicly reprimanded by colleagues for needlessly scaring the public and blacklisted, so that he was denied government funding, even after subsequent experiments proved him to be correct.

3

"Playing God" and Invoking a Perspective

Allen Verhey

One of the most commonly used phrases in the debate over transgenic uses of biotechnology is the claim that human beings are "playing God." Though this phrase obviously depends for much of its popularity on the Western religious traditions in which God is said to be Creator and Lord of creation, it does not need to be developed in this specific theological context. It can also be developed within the framework of Western theism, within the context of various eastern religious traditions, and within nontheistic and nonreligious naturalism.

Broadly the idea is that fixed genetic barriers divide species or at most genuses. Crossing these lines is wrong in much the same way that crossing property lines is wrong. As chapter 4 by Mira Foung shows, we can call this either "playing God" in the theistic context explored by Verhey or in Foung's phrase "genetic trespassing" (i.e., crossing natural barriers that we should not cross by means of transgenic technologies).

Actually this claim about trespassing is found in three different versions in the literature, depending on what sort of wrong or harm is done by such an activity. The first version is what we might call the "hard trespassing" version. In this view God or nature has established species boundaries much like property lines. These lines should be respected. Nature is said to have in a sense "decreed it," as one prominent writer has said:

Natural genetic change is extremely restricted in scope and rate. Mutations and rearrangements of DNA can occur and genetic information can be transferred from closely related species under special circumstances. They occur in minute steps that are strictly circumscribed by a host of biological constraints. In contrast genetic engineers are far less restricted. In principle and in practice they are limited only by their own imaginations. They can splice together genetic information from any

71

two or more organisms on the planet and can introduce that DNA into virtually any organism that they choose. . . . Practically speaking such transformations would never take place in nature.

Unfortunately this view is open to two serious objections. First, it offers no reason why these lines should not be crossed: we should not cross them because we should not cross them. This is a tautology, not a reason. Second, it adopts a much too static view of nature. Over the long course of evolution virtually all genetic lines have been crossed and mixed. Thus it seems odd to say that if human beings do what nature itself has done, it is wrong because it is unnatural.

A second version of this claim seeks to supply the missing argument for why it is wrong to cross the lines of genetic species. This has to do with the concept of genetic isolation. All species that survive as part of the ecological fabric that sustains life at any point in time represent successful adaptations to the changing conditions in that fabric. If these species are to persist, they must be genetically isolated from one another for long periods of evolutionary time. As Ernst Mayr, one of the greatest biologists of the twentieth century, has written, "The reproductive isolation of a species is a protective device that guards against the breaking up of its well integrated, co-adapted gene system." If successful adaptations are to remain and sustain the ecological web of life they must maintain isolation. Deliberate gene transfer breaks this isolation and thus threatens life on earth. It is unnatural because it breaks this isolation, but it is altogether wrong because of the extreme danger that results.

This view might be likened to going the wrong way on a freeway. There is no safe speed for doing so. It is not that freeway lines are natural or God given and should not be crossed because God so decreed it. Rather, they should not be crossed because of the extreme physical danger involved in doing so. Ron Epstein's argument in chapter 2 can be seen as an example of this sort of claim about the severe danger of transgenic technology.

Though this does give a reason for complete opposition to transgenic technology, it is faced with essentially the same problem as the harder version of the "unnaturalness" claim. It focuses too exclusively on static species and not enough on change and adaptation. Species emerge as well as persist. Their coming into existence will require the crossing of genetic barriers, i.e. breakdown of species isolation. Failure to adapt to changing environments and competition from other species is just as much of a threat to life as the rapid breakdown of genetic isolation. Hence we cannot say that all crossing of species lines is so dangerous that it is morally forbidden to human beings.

The two views we have just noted seem to analyze the idea of "playing God" as a matter of ontology—human beings are assuming a power once reserved to God and, for most writers using the phrase, a power that should remain only with God. Suppose, however, we revised the claim as one concerning epistemology, not ontology. In this view transgenics would be problematic because we do not know how to do it without serious risk of harm. God's perfect knowledge enables God to act without harm. But our crossing genetic barriers is an attempt to go much faster than typical evolutionary change and in an epistemic position is far from God's.

This might be like speeding. It is not wrong per se or so dangerous as to be completely forbidden. In general in the transgenic case we would be going too fast, drastically speeding up processes of genetic change when we do not know well the context in which these changes must fit. We may be playing God by using God's power without God's knowledge.—Eds.

Should human beings play God? It is a question frequently raised in discussions of bioethics and of genetics. The question is sometimes asked rhetorically, as though the answer is obvious: "Human beings should not play God!"[1] Sometimes the question is set aside as if it were not a serious question, as though human beings have no choice but to play God, as though it is what human beings do. "The question is not whether we will play God or not, but whether [we will play God responsibly or not]" (Augenstein, 1969, p. 145).

We are sometimes invited to play God, and we are sometimes warned against it, but before we decide whether to accept the invitation or to heed the warning, it would be good to know what it means to "play God."

When my daughter, Kate, was very young, she once invited the rest of the family to play "52-semi." She was holding a deck of cards, obviously eager to play. But when we asked for an explanation of this game, she would give none, only repeating her invitation to play "52-semi." Finally we said, "OK, Katie, let's play '52-semi.'" She threw the cards up into the air and when they had fallen back to the floor, commanded triumphantly, "Now pick 'em up." She had gotten her trucks mixed up, confusing "52-semi" with "52 pick-up" but suddenly—too late—we knew what she meant. Should human beings "play God"? It depends, you see, on what it means to "play God."

Unfortunately, the phrase does not mean just one thing; it means different things to different people in different contexts. That is hardly surprising, I suppose, given the fact that neither "play" nor "God" is a simple term. Moreover, sometimes the phrase is used in ways that have nothing to do with either "play" or "God."

In one recent survey of the uses of the phrase, Edmund Erde decided that the phrase is meaningless. Using the phrase as though it meant something, he said, ". . . is muddle-headed" (Erde 1989, p. 594); moreover, he regarded the phrase not only as "non-sensical," but also as "unconstitutional or blasphemous" (Erde 1989, p. 599), even "immoral" (Erde 1989, p. 594). Erde demanded that, for the phrase to be meaningful, it must mean a single moral principle, and as a universal moral principle at that. That seems a bit much to ask.

This article undertakes to sort through at least some of the uses of this phrase. I hope to indicate that the phrase does not so much state a principle as invoke a perspective on the world; a perspective from which other things, including scientific and technological innovations in genetics—and the phrase itself—are meaningful. I hope to indicate that we must be attentive not only to particular moral problems raised by genetic engineering but also to the perspective from which we examine and evaluate these new powers and problems. And I hope to suggest, finally, the relevance of a perspective in which "God" is taken seriously and "play" playfully.

The President's Commission report on Splicing Life (1982) would seem a good place to begin. The commission noted the concerns voiced about "playing God" in genetics and, to their credit, undertook to make some sense of the phrase. It even invited theologians to comment on the phrase and its relevance to genetic engineering. The "view of the theologians" is summarized in a single paragraph.

> [C]ontemporary developments in molecular biology raise issues of responsibility rather than being matters to be prohibited because they usurp powers that human beings should not possess. . . . Endorsement of genetic engineering, which is praised for its potential to improve the human estate, is linked with the recognition that the misuse of human freedom creates evil and that human knowledge and power can result in harm. (President's Commission 1982, pp. 53–54)

There is much here that could reward a closer analysis. It is clear that the theologians rejected the warnings against "playing God" when those warnings were understood as warnings against usurping powers that are properly God's—but how else might they be understood? The theologians evidently thought that the notion of "responsibility" might be suggestive. It is indeed suggestive, and I will return to it. The President's Commission, however, decided to leave the notion of responsibility to God aside. It decided that the phrase "playing God" does not have "a specific religious meaning" (President's Commission 1982, p. 54).

If, in stating this, the Commission had meant simply that the phrase does not mean one thing, and that the meaning of the phrase varies with the particular religious tradition and perspective within which it might be used, then one could hardly object. However, the Commission proceeded to assert that "at its heart" the phrase was "an expression of a sense of awe [in response to extraordinary human powers]—and concern [about the possible consequences of these vast new powers]" (President's Commission 1982, p. 54). The Commission simply translated the warnings against "playing God" into a concern about the consequences of exercising great human powers (Lebacqz 1984, p. 33).

The Commission reduced the meaning of the phrase to secular terms and made "God" superfluous. "At its heart," according to the Commission, the phrase "playing God" has nothing to do with "God." Moreover, there is nothing very playful about "playing God" either. The human powers in genetics and their possible consequences are too serious for playfulness.

"Playing God" might mean what the Commission interpreted it to mean, something like, "Wow! Human powers are awesome. Let's not play around!" It evidently does mean something like that to many who use the phrase. Such an interpretation of the phrase is hardly trivial, but it is also not very useful to guide or limit human powers. Moreover, it is worth pointing out that the President's Commission invoked a particular perspective in interpreting the phrase, and it then used the phrase as shorthand to invoke that perspective when interpreting developments in genetics.

The President's Commission highlighted one very important feature of contemporary culture, the hegemony of scientific knowledge. "Since the Enlightenment," it said, "Western societies have exalted the search for greater knowledge" (President's Commission 1982, p. 54). Scientific knowledge, beginning with Copernicus, has both "dethrone[d] human beings as the unique center of the world" and delivered "vast powers for action" into their hands (President's Commission 1982, pp. 54–55).

Leroy Augenstein had made the same point in *Come, Let Us Play God*. Science has taught us the hard lesson that human beings and their earth are not "the center of the universe" (Augenstein 1969, p. 11), but it is now putting into human hands powers and responsibilities "to make decisions which we formerly left to God" (Augenstein 1969, p. 142). Borrowing the phrase of Dietrich Bonhoeffer, Augenstein described this situation as humanity's "coming of age" (Augenstein 1969, p. 143).

Where this is the context for talk of "playing God," it is not surprising that "God" is superfluous, that "God" is not taken seriously when we try to make sense of the phrase. Bonhoeffer, after all, described humanity's "coming of age" as an effort to think the world *etsi deus non daretur* ("as though God were not a given") (Bonhoeffer 1953, p. 218). Science has no need of God "as a working hypothesis" (Bonhoeffer 1953, p. 218); in fact, it is not even permitted for science *qua* science to make use of "God." There are assumptions operative in this perspective, however, not only about "God" but about humanity, knowledge, and nature as well. With respect to humanity, science has taught us that we are not "the center of the universe." However, science has not taught us where we do belong. As Nietzsche aptly put it, "since Copernicus man has been rolling from the center into x" (cited in Jungel 1983, p. 15). Once human beings and their earth were at the center. They did not put themselves there; God put them there, and it was simply accepted as a matter of course that they were there. After Copernicus had shown that they were not at the center, humanity was left to fend for itself (or simply to continue "rolling"). This positionlessness was the new assumption, and it entailed that humanity had to attempt to secure (if somewhat anxiously) a place for itself—and what better place than at the center. After Copernicus, humanity was not simply at the center, it had to put itself at the center, make itself into the center. Fortunately, the very science that destroyed the illusion that humanity was at the center gave to humanity power in the world and over the world. Such mastery, however, has not eliminated human insecurity and anxiety; in fact, the new powers and their unintended consequences evoke new anxieties.

In this context "playing God" *etsi deus non daretur* might well be interpreted as "an expression of a sense of awe [before human powers]—and concern [about unanticipated consequences]" (President's Commission 1982, p. 54).

There are assumptions concerning knowledge, too. The comment of the President's Commission that "[s]ince the Enlightenment, Western societies have exalted the search for greater knowledge" (President's Commission 1982, p. 54) requires a gloss. They have exalted a particular kind of knowledge, the knowledge for which they reserve the honorific term "science."

It is simply not the case that the search for knowledge only began to be exalted with the Enlightenment. Thomas Aquinas, for example, had exalted the search for knowledge long before the Enlightenment, affirming "all knowledge" as "good." He distinguished, however, "practical" from "speculative" (or theoretical) sciences, the difference being that the practical sciences are for the sake of some work to be done, while the speculative sciences are for their own sake (Aquinas, Commentary on Aristotle's *On the Soul*, 1,3; cited in Jonas 1966, p. 188).

That classical account (and celebration) of knowledge must be contrasted with the modern account epitomized in Francis Bacon's *The Great Insaturation* and "exalted" in Western societies. In Bacon all knowledge is sought for its utility, "for the benefit and use of life" (Bacon [1620] 1960, p. 15). The knowledge to be sought is "no mere felicity of speculation" (Bacon [1620] 1960, p. 29), which is but the "boyhood of knowledge" and "barren of works" (Bacon [1620] 1960, p. 8). The knowledge to be sought is the practical knowledge that will make humanity "capable of overcoming the difficulties and obscurities of nature" (Bacon [1620] 1960, p. 19), able to subdue and overcome its vexations and miseries. "And so those twin objects, human knowledge and human power, do really meet in one" (Bacon [1620] 1960, p. 29). The knowledge "exalted" in Western societies is this power over nature which presumably brings human well-being in its train.

In the classical account, theory (or the speculative sciences) provided the wisdom to use the practical sciences appropriately. The modern account may admit, as Bacon did, that for knowledge to be beneficial humanity must "perfect and govern it in charity" (Bacon [1620] 1960, p. 15), but science is "not self-sufficiently the source of that human quality that makes it beneficial" (Jonas 1966, p. 195). Moreover, the compassion (or "charity") that responds viscerally to the vexations and miseries of humanity will urge us to *do something* to relieve those miseries, but it will not tell us *what thing* to do. Bacon's account of knowledge simply arms compassion with artifice, not with wisdom (O'Donovan 1984, pp. 10–12).

For the charity to "perfect and govern" human powers and for the wisdom to guide charity, science must call upon something else. But upon what? And how can humanity have "knowledge" of it? Knowledge of that which transcends "use"—and transcends the "nature" known scientifically, even the "human nature" known scientifically—has no place in Bacon's theory.[2]

Knowledge of that which might guide and limit the human use of human powers was the subject of classical theory, but not of the Enlightenment "search for greater knowledge." In this context there is no place for either "play" (because play is not "useful") or "God" (because God is transcendent and will not be used).[3]

With the different assumptions concerning knowledge come different assumptions concerning nature, too. The Baconian project sets humanity not only over nature but against it. The natural order and natural processes have no dignity of their own; their value is reduced to their utility to humanity—and nature does not serve humanity "naturally." Nature threatens to rule and to ruin humanity. Against the powers of nature, knowledge promises the power to relieve humanity's miseries and "to endow the human family with new mercies" (Bacon [1620] 1960, p. 29). The fault that runs through our world and through our lives must finally be located in nature. Nature may be—and must be—mastered (Jonas 1966, p. 192).

This is the perspective invoked by the President's Commission. From this perspective "playing God" has nothing to do with either "play" or "God." Rather it is concerned with human scientific knowledge and power over nature and it raises doubts about the taken-for-granted assumption that human well-being will come in the train of such knowledge and power.

Religious people have sometimes celebrated this Baconian perspective and its quest for scientific knowledge and technical power—and have sometimes lamented it. Some who have lamented it have raised their voices in protest against almost every new scientific hypothesis (witness Galileo and Darwin) and against almost all technological developments (for example, anaesthesia during childbirth). These evidently regard scientific inquiry as a threat to faith in God and technical innovation as an offense to God. These lament a "humanity come of age" and long to go back to a former time, a time of our childhood (if only we knew the way!). They regret a world *etsi deus non daretur* and wish to preserve the necessity of "God" in human ignorance and powerlessness. But such a "God" can only ever be a "God of the Gaps" and can only ever be in retreat to the margins.

It is an old and unhappy story in Christian apologetics that locates God's presence and power where human knowledge and strength have reached their (temporary) limit. Newton, for example, saw certain irregularities in the motion of the planets, movements which he could not explain by his theory of gravity, and in those irregularities he saw, he said, the direct intervention of God. When later astronomers and physicists provided a natural explanation for what had puzzled Newton, "God" was no longer necessary. And there is the old joke of the patient who, when told that the only thing left to do was to pray, said, "Oh, my! And I didn't even think it was serious." The God of the Gaps is only invoked, after all, where doctors are powerless.

In the context of such a piety, when there is a defensive faith in the God of the Gaps, "playing God" means to encroach on those areas of human life where human beings have been ignorant or powerless, for there God rules, there only God has the authority to act. In this context "playing God" means to seize God's place at the boundaries of human knowledge and power, to usurp God's authority and dominion. In this context it is understandable that humanity should be warned, "Thou shall not play God."

Once again the phrase is used not so much to state a principle as to invoke a perspective. To be sure, such warnings serve to remind humanity of its fallibility and finitude, and such warnings are salutary. There are, however, at least two problems with this perspective and with such warnings against "playing God."[4]

The first and fundamental problem with this perspective is that the God of the Gaps is not the God who is made known in creation and in scripture. The God of creation and scripture made and sustains the order we observe and rely upon. To describe that order in terms of scientific understanding does not explain God away; it is to give an account of the way God orders God's world. The order of the world comes to us no less from the gracious hand of God than the extraordinary events humans call "miracles." "Nature" is no less the work of God than "grace." The world and its order are not God, but they are God's. They are the work of God. And, to understand the world and its order as God's is not to understand it in a way that prohibits "natural scientific" explanations. It is to be called to serve God's cause, to be responsible to God in the midst of it.

The second problem with this perspective and with such warnings against "playing God" is that they are indiscriminate; they do not permit discriminating judgments. There are some things which we already know how to do (and so can hardly be said to trespass the boundaries of human ignorance and powerlessness), but which we surely ought never to do. And there are some things (including some things in genetics) which we cannot yet do, but which we must make an effort to learn to do if God is God and we are called to "follow" one who heals the sick and feeds the hungry. The warning against "playing God" in this perspective reduces to the slogan "It's not nice to fool with Mother Nature (at least not any more than we are currently comfortable with)." Ironically, then, the warning enthrones "nature" as god rather than the One who transcends it and our knowledge of it.

Some other religious people celebrate the advances of science and the innovations of technology, urging humanity bravely to go forward, uttering a priestly benediction over the Baconian project. These sometimes use the phrase "playing God," too, usually in inviting humanity to "play God." Joseph Fletcher, for example, responded provocatively to the charge that his enthusiasm for genetic technology amounted to a license to "play God" by admitting the charge (Fletcher 1970,

p. 131) and by making the invitation explicit; "Let's play God," he said (Fletcher 1974, p. 126).

The "God" Fletcher invited us to "play" was still the God of the gaps (Fletcher 1970, p. 132), the God at the edges of human knowledge and power. For Fletcher, however, "that old, primitive God is dead" (Fletcher 1970, p. 132; 1974, p. 200). Dead also are the "taboos" which prohibited trespass on the territory of that God's rule (Fletcher 1974, p. 127), the "fatalism" that passively accepted the will of that God (Fletcher 1974, p. 128), and the "obsolete theodicy" (Fletcher 1970, p. 132) that attempted to defend that God. "What we need," he said, "is a new God" (Fletcher 1970, p. 132), but Fletcher's "new God" bore a striking resemblance to the God of the eighteenth-century deist, and indifference to a God so conceived is inevitable; life may proceed—and "playing God" may proceed—*etsi deus non daretur.*

Although Fletcher said little more about this "new God," he did say that "any God worth believing in wills the best possible well-being for human beings" (Fletcher 1974, p. xix). Fletcher's "new God" turns out to be a heavenly utilitarian, and this God, too, humanity must "play."

So, the invitation to "play God" comes to this: humanity should use its new powers to achieve the greatest good of the greatest number of people (not intimidated by "taboos"), to take control over "nature" (not enervated by "fatalism"), to take responsibility, to design and make a new and better world, to substitute for an absent God. "It was easier in the old days," Fletcher said (1974, p. 200), to attribute at least some of what happened to God's will—we could say with a moral shrug that we weren't responsible. Now we have to shoulder it all. The moral tab is ours and we have to pick it up. The excuses of ignorance and helplessness are growing thin.

Notice what has happened to responsibility. Fletcher underscores human responsibility, but we are responsible not so much to God as instead of God.[5] That shift puts an enormous (and messianic) burden on genetics, a burden which leaves little time for "play." The phrase "playing God" here does state a principle, namely, utility, but it also does more than that—it invokes a perspective, a perspective in which the God of the Gaps is superfluous, in which humanity is maker and designer, in which knowledge is power, and in which nature must be mastered to maximize human well-being. Such a perspective makes the invitation to "play God"—and much else in Fletcher's discussion of genetics—meaningful.

Christians may welcome Fletcher's burial of the God of the Gaps, but they still wait and watch and pray not for the invention of some "new God" but for the appearance of the one God who continues to create, preserve, and redeem humanity and its world. Moreover, Fletcher's invitation to "play God" need not seem blasphemous to those trained to "imitate God," to "follow" God, to be disciples of one who made God present among us. But, to map the path of discipleship and imitation as "the utilitarian way" must seem strange to those who know the law and the prophets, and the gospels.

It seemed strange, at least, to Paul Ramsey. In Ramsey's usage, although we are usually warned against "playing God," we are sometimes encouraged to "'play God' in the correct way" (Ramsey 1970a, p. 256) or to "play God as God plays God" (Ramsey 1978, p. 203)—and God is no utilitarian.[6] "God," Ramsey said (1978, p. 205), is not a rationalist whose care is a function of indicators of our personhood, or of our achievement within those capacities. He makes his rain to fall upon the just and the unjust alike, and his sun to rise on the abnormal as well as the normal. Indeed, he has special care for the weak and the vulnerable among us earth people. He cares according to need, not capacity or merit. These divine patterns and images are, according to Ramsey, at "the foundation of Western medical care" (Ramsey 1978, p. 205).

One might expect Ramsey, then, simply to echo Fletcher's invitation to "play God" while engaging him and others in conversation concerning who this God is whom we are invited to "play." However, he also (and more frequently) warned against "playing God." The phrase itself, he admitted, is "not [a] very helpful characterization" (Ramsey 1970b, p. 90), but he used it to name—and to warn against—an "attitude," an "outlook," certain "operating, unspoken premises" at work in western scientific culture (Ramsey 1970b, p. 91), and to invite a different perspective on the world.

The fundamental premise of the perspective Ramsey warns against is that "God" is superfluous. "Where there is no God . . ." he said (Ramsey 1970b, p. 93), there humanity is creator, maker, the engineer of the future (Ramsey 1970b, pp. 91–92), and there nature, even human nature, may be and must be controlled and managed with messianic ambition (Ramsey 1970b, pp. 92–96). Where "God" is superfluous and human beings are cast in this role of "the maker," there morality is reduced to the consideration of consequences, knowledge is construed simply as power, and nature—including the human nature given to humanity as embodied and communal—is left with no dignity of its own.

Ramsey's warnings against "playing God" are not immediately identified with a particular moral rule or principle; rather, they challenge the wisdom and the sufficiency of the assumptions too much at work in Western culture. It is not that some "God of the Gaps" is threatened. It is not simply that human powers are awesome or that the consequences of "interfering with nature" are worrisome, as the President's Commission suggested. It is rather that the fundamental perspective from which we interpret our responsibilities is critically important to seeing what those responsibilities are (Ramsey 1970b, pp. 28, 143).

The fundamental perspective which Ramsey recommends and to which he contrasts "playing God" is "to intend the world as a Christian or as a Jew" (Ramsey 1970b, p. 22), i.e., *etsi deus daretur*—and not just any old *deus* (nor Fletcher's "new God") but the God who creates and keeps a world and a covenant. That means, among other things, that the end of all things may be left to God. Where God is God and not us, there can be a certain eschatological nonchalance. From this perspective, our responsibilities, while great, will not be regarded as being of messianic proportion. There will be some room, then, for an ethic of means as well as the consideration of consequences (Ramsey 1970b, pp. 23–32), for reflection about the kind of behavior which is worthy of human nature as created by God, as embodied and interdependent, for example.

When joined with such reflection, Ramsey's warnings that we should not play God do provide some prohibitions. When joined with an interpretation of human procreation, for example, the warning against "playing God" bears the prohibition against putting "entirely asunder what God joined together," against separating *"in principle"* the unitive and procreative goods of human sexuality, against reducing procreation either to biology or to contract (Ramsey 1970b, pp. 32–33), and that prohibition supports in turn a series of more particular prohibitions, for example, a prohibition against artificial insemination using the sperm of a donor (Ramsey 1970b, pp. 47–52).

When joined with an interpretation of the patient as *"a sacredness in the natural, biological order"* (Ramsey 1970a, xiii), the "edification" drawn from the warning against "playing God" includes prohibitions against deliberately killing patients, including very little patients, for the sake of relieving their (or another's) suffering, against using one without consent, even a very little one, even one created in a petri dish, to learn to help others.

Ramsey warns against "playing God," against trying to substitute for an absent God, against trying to "be" God, but there remains room for "playing God" *etsi deus daretur*. Indeed, as we have seen, Ramsey can invite people to "'play God' in the correct way" (Ramsey 1970a, p. 256). Such "playing" is not to substitute for an absent God, not to "be" God, but to "imitate" God (Ramsey 1970a, p. 259), to follow in God's way like a child "playing" a parent.

In both the warning and the invitation a perspective is invoked, an outlook which assumes that God is God and not us, that humanity is called to honor and to nurture the nature God gave, that knowledge of that which transcends use is possible, and that the fault that runs through our lives and our world is not simply located in nature but in human pride or sloth.

One who—like me—shares this perspective will make sense of the phrase "playing God" in the light of this perspective. Sometimes it will be appropriate to sound a warning against "playing God," and sometimes it will be appropriate to issue an invitation to "play God" in imitation of God's care and grace. Permit me to focus on the invitation to "play God"—and first to underscore the invitation to "play." Many have complained that "playing God" is serious stuff and regretted the implication of "playfulness" in the phrase (e.g., Lebacqz 1984, p. 40, n. 19). Some "play," however, can be very serious indeed—as anyone who plays noon-hour basketball knows quite well. "Playfulness" is quite capable of being serious, but it is not capable of being purely instrumental.

When Teilhard de Chardin said that "in the great game that is being played, we are the players as well as . . . the stakes" (1961, p. 230), he created a powerful image to call attention both to the extraordinary powers of human beings and to the awesome consequences of exercising those powers. No wonder playfulness seems inappropriate. Precisely because the stakes are high, however, it may be apt to set alongside de Chardin's image a Dutch proverb: "It is not the marbles that matter but the game" (quoted in Huizinga 1950, p. 49). When the stakes are high, or even when the stakes alone are taken seriously, then one is tempted to cheat in order to win. And when one cheats, then one only pretends to play; the cheat plays neither fair nor seriously.

Play, even marbles, can be serious, but it cannot be purely instrumental; it cannot allow attention to be monopolized by the stakes, by the consequences of winning or losing. When our attention is riveted by de Chardin's image that we are "the stakes," it may well be important to allow our imagination to be captured by his image that we are "the players," too. Then we may be able to avoid reducing the moral life to a concern about consequences, even where the stakes are high. We may be able to avoid reducing ourselves to makers and designers and our existence to joyless and incessant work. We may see that we are at stake, not just in the sense of some plastic destiny our powers may make but already in the imagination, in the image of ourselves with which human creativity begins (Hartt 1975, pp. 117–134).

The invitation is an invitation to "play," but it is more specifically an invitation to "play God," and that invitation requires attention to the God whom we are invited to play. In the foreword to *Should Doctors Play God?* Mrs. Billy Graham wrote (1971, p. vii),

> [i]f I were an actress who was going to play, let's say, Joan of Arc, I would learn all there is to learn about Joan of Arc. And, if I were a doctor or anyone else trying to play God, I would learn all I could learn about God.

That seems a prudent strategy for an actress—and good advice for people called to imitate God. The invitation to "play God," to cast ourselves playfully in the role of God, invites theological reflection; it invites reflection about "God."

The invitation goes out to all, not just to Christians. When ancient Greek physicians swore the Hippocratic Oath by Apollos, Aesclepius, Hygiea, Panacea and all the gods and goddesses, they invoked a story. Healing had its beginnings among the gods, and the Hippocratic physicians swore to make that story their own. And when the temple to Aesclepius in the Areopagus was inscribed with the message that, like a god, Aesclepius healed both rich and poor without discrimination, a path was laid out for physicians to follow.

The invitation goes out to all, but reflection about God is always formed and informed by the particular stories and communities within which it is undertaken, and Christians will heed this invitation in the light of their own tradition and its talk of God. We play God in response to God, imitating God's ways and providing human service to God's cause. Our responsibility to God limits and shapes an account of what we are responsible for in God's good world—and its genetics.

Permit me, then, simply to select a few images of God in the Jewish and Christian tradition and to suggest something of their relevance to "playing God" in genetics. Two of these images are regularly invoked in these discussions: creator and healer—and the third is often overlooked: God is the one who takes the side of the poor.

First, then, what might it mean playfully to cast ourselves in the role of the creator? This, of course, has been the topic of much discussion. If I read the story right, however, to cast ourselves in the role of the creator might mean something too much overlooked. It might mean that we look at the creation and at its genetics and say to ourselves, "God, that's good." It might mean, that is, first of all, to wonder, to stand in awe, to delight in the elegant structure of the creation and its DNA. It would mean a celebration of knowledge which was not simply mastery. It would mean an appreciation of nature—and of human nature—as given, rather than a suspicion of it as threatening and requiring human mastery.

And if I read the story right, it might mean a second thing too much overlooked. It might mean to take a day off, to rest, to play. But we have already talked of that.

It also means, of course, a third thing, a thing seldom overlooked in these discussions—that human creativity is given with the creation. Human beings are created and called to exercise dominion in the world—and I see no reason to suppose that such creativity and control does not extend to genetics. It is not "Mother Nature" who is God, after all, in the Christian story. Human creativity and control, however, are to be exercised in response to God, in imitation of God's ways, and in service to God's cause. That's a part of the Christian story, too, a part of the story usually captured in describing ourselves as stewards and our responsibility as stewardship.

We can discover something of God's cause, the cause stewards serve, in a second feature of the story. God is the healer. Jesus, the one in whom God and the cause of God were made known, was a healer. We discover there that the cause of God is life, not death; the cause of God is human flourishing, including the human flourishing we call health, not disease. What does it mean to cast ourselves playfully in the role of God the healer? It means to intend life and its flourishing, not death or human suffering. Therefore, genetic therapy, like other therapeutic interventions which aim at health, may be celebrated. Healing is "playing God" the way God plays God. Genetic therapies, however, are still mostly (but not completely) a distant hope. The more immediate contributions of genetics to medicine are in genetic diagnosis. And where there are therapeutic options, these too may be celebrated. However, genetic diagnoses without therapeutic options are sometimes deeply ambiguous.

Prenatal diagnoses, for example, are frequently ambiguous. Already we can diagnose a number of genetic conditions in a fetus, and the number is constantly growing. For most of these there is no therapy. The tests allow parents to make a decision about whether to give birth or to abort. How shall we "play God" here in ways responsible to God? If God's cause is life rather than death, then those who would "play God" in imitation of God will not be disposed to abort; they will not celebrate abortion as a "therapeutic option."

There are, I think, genetic conditions which justify abortion. There are conditions like Tay-Sachs which consign a child not only to an abbreviated life but to a life subjectively indistinguishable from torture. And there are conditions like Trisomy 18 which are inconsistent not only with life but with the minimal conditions for human communication. Prenatal diagnosis—and abortion—can be used responsibly. However, when some children with Down's Syndrome are aborted because they have Down's, there seems to exist a reasonable possibility that prenatal diagnoses have been—and will be—used irresponsibly. And when some girls are aborted because they are girls, it seems obvious that the tests have been—and will be—used irresponsibly. When the slogan about "preventing birth defects" is taken to justify preventing the birth of "defectives," those who do not measure up to the standards or match the preferences of parents, then there are reasons to worry a little, to worry that the disposition of a good "parent" will change from the sort of uncalculating nurturance that can evoke and sustain trust from children to the sort of calculating nurturance that is prepared to abandon or abort the offspring who do not match specifications. "Playing God" the way God plays God—or, if you will, the way God plays "parent"—would sustain care for the weak and the helpless, and for the little ones who do not measure up.

Genetic therapy, I said, may be celebrated as service to God's cause of health. It is to "play God" as God plays God. However, to use this knowledge and technology responsibly it must be aimed at "health," not genetic enhancement. The distinction between intervening for health and intervening for genetic enhancement may be a slippery one, but casting ourselves playfully in the role of God the healer will encourage us to make such a distinction and to abide by it. Eugenics is not the way to "play God" the way God plays God.

Consider, finally, this third image: God is one who takes the side of the poor. What would it mean to cast ourselves in the role of one who takes the side of the poor? It would mean, at the very least, I think, a concern for social justice. It would mean, for example, to ask about the allocation of resources to the human genome project. When cities are crumbling, when schools are deteriorating, when we complain about not having sufficient resources to help the poor or the homeless, when we do not have the resources to provide care for all the sick, is this a just and fair use of our society's resources? Is it an allocation of social resources that can claim to imitate God's care and concern for the poor?

Having raised that question, let me focus instead on the sharing of the burdens and benefits of the human genome project itself. Who bears the burdens? Who will benefit? And is the distribution fair? Does it fit the story of one who takes the side of the poor and powerless?

If we cast ourselves in this role, if we attempt to mirror God's justice and care for the poor and powerless, we will not be eager to create human life in order to learn from it with the intention of destroying it after we have learned what we can from it. We will not be eager to use the unborn for experiments to learn some things that would benefit others, even if it were a great benefit, even if it would benefit a great number of others. And we would be cautious about stigmatizing some as diseased and others as carriers.

But consider also the sharing of benefits. Who stands to benefit from the human genome initiative? Will genetic powers be marketed? Presumably they will, given the patenting of micro-organisms. And so the rich may get richer while the poor still watch and pray. Will the poor have access to health-care benefits that their taxes helped develop? Since health care reform has died in Congress again, can we have any confidence that genetic technology will be available to the uninsured or to those with public insurance? Or will insurance companies use genetic information to screen candidates for insurance? Will the category of "preexisting condition" be redefined to make it easier for insurance companies to make a still larger profit? Will genetic information be included in actuarial tables? Will corporations use genetic information to screen applicants in order to hire those with greatest promise of long-term productivity? The point of these questions is not simply to lament our failure to accomplish health care reform. It is to suggest that "playing God" as God plays God will be attentive not only to intriguing questions about the frontiers of technology and science but also to mundane questions about fairness, about the effect of such innovations on the poor. If we are to "play God" as God plays God, then we have a pattern for imitation in God's hospitality to the poor and to the stranger, to the powerless and to the voiceless, to one who is different from both us and the norm, including some genetic norm. If we are to "play God" as God plays God, then we will work for a society where human beings—each of them, even the least of them—are treated as worthy of God's care and affection.

This has been just a selection of images of God, and I admit that the moves to claims about genetic interventions were made far too quickly. But enough has been said, I hope, to suggest the importance of the invitation to play God as God plays God. Enough has been said, I hope, to suggest the importance of the perspective in terms of which we think about genetics and in terms of which we make sense not only of our powers but of the phrase "playing God."

NOTES

Reprinted by permission from *Journal of Medicine and Philosophy* 20 (1995): 347–65.

1. Ted Howard and Jeremy Rifkin, for example, ask their readers *Who Should Play God?* in the title of their book (1977), but a reader who expects an extended discussion of the question or a reasoned defence of an answer will be disappointed. The question is evidently rhetorical and the answer is "no one."

2. To be sure, Bacon recommended his "great instauration" as a form of obedience to God, as a restoration to humanity of the power over nature which was given at creation

but lost through the fall. Indeed, he prays "that things human may not interfere with things divine, and that there may arise in our minds no incredulity or darkness with regard to the divine mysteries" (Bacon [1620] 1960, pp. 14–15). Even so, such mysteries have no theoretical place in Bacon's account of knowledge.

3. Jonas (1966, p. 194) contrasts the relations of leisure to theory in the classical and modern traditions. In the classical account leisure was an antecedent condition for speculative knowledge, for contemplation; in modern theory leisure is an effect of knowledge (as power), one of the benefits of that knowledge that provides relief from the miseries of humanity, including toil. "Wherefore," Bacon says ([1620] 1960, p. 29), "if we labor in thy works with the sweat of our brows, thou wilt make us partakers of . . . thy sabbath."

4. This account of "playing God" was the one rejected by the theologians consulted by the President's Commission (1982, p. 53).

5. On the shift from theodicy to "anthropodicy," see Becker (1968, p. 18) and Hauerwas (1990, pp. 59–64).

6. A delightful essay by Jan van Eys (1982) also underscores the invitation to "play" in the phrase "play God"; unfortunately, van Eys treats "play" as a kind of psychological therapy and so renders it instrumental finally.

REFERENCES

Augenstein, L. 1969. *Come: Let Us Play God.* New York: Harper & Row.

Bacon, F. [1620] 1960. *The New Organon and Related Writings.* Edited by F. H. Andersen. Indianapolis: Liberal Arts Press, Bobbs-Merrill.

Becker, E. 1968. *The Structure of Evil.* New York: George Braziller.

Bonhoeffer, D. 1953. *Letters and Papers from Prison.* Edited by E. Bethge. Translated by R. H. Fuller. New York: Macmillan.

Chardin, T. de. 1961. *The Phenomenon of Man.* Translated by B. Wall. New York: Harper & Row.

Erde, E. 1989. "Studies in the Explanation of Issues in Biomedical Ethics: (11) On 'On Playing God, Etc.'" *Journal of Medicine and Philosophy,* 14: 593–615.

Fletcher, J. 1970. "Technological Devices in Medical Care." In *Who Shall Live? Medicine, Technology, Ethics,* edited by K. Vaux, 115–42. Philadelphia: Fortress.

Fletcher, J. 1974. *The Ethics of Genetic Control: Ending Reproductive Roulette.* Garden City, N.Y.: Anchor.

Graham, R. 1971. Foreword to *Should Doctors Play God?* Edited by C. A. Frazier. Nashville: Broadman.

Hartt, J. 1975. *The Restless Quest.* Philadelphia: United Church Press.

Hauerwas, S. 1990. *Naming the Silences: God, Medicine, and the Problem of Suffering.* Grand Rapids, Mich.: Eerdmans.

Howard, T., and J. Rifkin. 1977. *Who Should Play God?* New York: Dell.

Huizinga, J. 1950. Homo Ludens: *A Study of the Play-Element in Culture.* Boston: Beacon.

Jonas, H. 1966. *The Phenomenon of Lift: Toward a Philosophical Biology.* New York: Dell.

Jungel, E. 1983. *God as the Mystery of the World.* Translated by D. Guder. Grand Rapids, Mich.: Eerdmans.

Lebacqz, K. 1984. "The Ghosts Are on the Wall: A Parable for Manipulating Life." In *The Manipulation of Life*, edited by R. Esbjornson, 22–41. San Francisco: Harper & Row.

O'Donovan, O. 1984. *Begotten or Made?* Oxford: Oxford University Press.

President's Commission for the Study of Ethical Problems in Medicine and Biomedical and Behavioral Research. 1982. *Splicing Life: A Report on the Social and Ethical Issues of Genetic Engineering with Human Beings*. Washington, D.C.: U.S. Government Printing Office.

Ramsey, P. 1970a. *The Patient as Person: Explorations in Medical Ethics*. Yale University Press, New Haven.

Ramsey, P. 1970b. *Fabricated Man: The Ethics of Genetic Control*. New Haven: Yale University Press.

Ramsey, P. 1978. *Ethics at the Edges of Life: Medical and Legal Intersections*. New Haven: Yale University Press.

Van Eys, J. 1982. "Should Doctors Play God?" *Perspectives in Biology and Medicine* 25: 481–85.

Genetic Trespassing and Environmental Ethics

Mira Foung

Faces of hope, voices of space
None for us to seek and prey
They are children from the sun
Innocent in their mortal wound
Singing earth songs

THE METAMORPHOSIS

Will fish mate with tomatoes or soybeans crossbreed with petunias? Will pigs mate with humans or rabbits with mice? Of course not, but some scientists are combining the genes of these diverse creatures against the laws of natural selection. This unsacred liaison invented by chemical corporate giants is called genetic engineering. This high tech species metamorphosis makes the Existential writer Franz Kafka, with his man Gregor who woke up one morning and found that he had become a giant bug, a prophet of our time. Millions of cows imprisoned inside factory farms suddenly wake up to find their udders engorged to an enormous size. Instead of carrying twelve pounds of milk to feed their calves, the cows are forced to pump out fifty to sixty pounds of milk just for human consumption, not knowing they've been injected with a genetically engineered growth hormone.

Genes are blueprints composed of thousands of genetic codes. They carry information for the proteins that make up the structure, function and outward traits that constitute the individual organism. DNA ultimately dictates the distinctive qualities of a species, from microorganism to insect, plant, animal and human being. The genetic codes in DNA determine physical forms, skin color, size of fruits, sensory structures of animals, types of trees, specific times for flowers to blossom, and billions of other features and functions.

Genetic engineering (or bioengineering) is a technique to splice, delete, add, isolate, recombine or transfer genes from one organism to another that may be totally unrelated. Alteration in genes and chromosomes causes disruption and disturbance in the biochemical structure of species and can result in species mutation. It is a kind of artificially programmed evolution (or devolution) changing the individual organism as its starting point, in contrast to natural evolution in which changes occur among diverse populations through natural selection.

Since the early nineteen fifties, biologists began to turn their attention to the mysterious double helix called DNA. Within twenty years scientists were already mixing DNA extracted from different species. The quantum leap of this new technology allowed the human creature to become the new creator of life on earth, creating a variety of plants and animals. Now natural evolution can be halted at our fingertips, forever altering the meaning of life and forcing us to redefine religion, nature and individuality.

Cellular dynamics in all living systems requires mutual acknowledgment and interdependence, a constant cooperation between the individual life and the entire biosphere to maintain the stability and equilibrium suitable for species survival. The holistic concept of the Gaia Hypothesis proposes a subtle mutual participation between organic life (the moving part) and the geological environment (the unmoving part) as an integral whole in the evolutionary journey. Bioengineering disregards this fundamental intricacy by disrupting species integrity, a gesture in contempt of nature's wisdom. Science can alter other creatures' very genetic structure to suit our desires and the market value. Do animals, plants, forests, mountains, and oceans exist only for human benefit?

THE SILENCED PLEA

Among the many victims of these artificial mutations, farm animals suffer the most. Their entire lives are locked inside factory warehouses, manipulated by machines as if their sole purpose to be born was to be harvested by man. They never have a chance to see the sky or smell the earth. They can never experience the pleasure or the freedom of living beings like our pets, the wildlife, or ourselves. Farm animals are subjected to lifelong abuse by the most atrocious, appalling manipulation invented by agribusiness. Their utter misfortune is caused by being labeled as food animals, but they are still sentient beings not so different than we are.

The super pig, a product of genetic engineering, is a sick animal, fattened artificially by human growth hormone. This super pig must endure side effects including crippling arthritis and distorted vision caused by the human growth gene that makes them cross-eyed. Pigs are being modified with human genes so that the organs of their offspring can be transplanted into humans. Soon, in addition to factory pig farms, there will be pig organ farms. A new creature called a GEEP is part goat and part sheep. In nature, the two species never mate, but our modern alchemists have already perfected such a new species that never existed before.

And then there is the ordinary chicken. The modern bird has been bred to grow at twice its normal rate. Its legs can no longer carry its massive body weight, and the animal suffers leg pain and deformities as well as an enormous strain on its heart and lungs. Often these chickens experience heart failure before the age of six weeks. Many others die due to rampant infectious diseases caused by intensive breeding. A transgenic chicken is engineered with a cow's growth hormone gene, which imbalances its entire metabolism. One cannot imagine the intensity of suffering caused by such mutations.

Someday, chickens might be engineered with genes from centipedes, giving the birds more than two legs, so that we can have more drumsticks for our dinner table. Or the chicken may be further modified into a kind of tube, without head, wings or tail, but with many legs, so it will produce more meat for us and be easier to manage for commercial exploitation. No one will know how to take care of this new breed of animal; in fact there will be no need for veterinarians. The new food machine, no longer a real creature by definition, can put an end to hundreds of years of debate on animal rights.

Are farm animals not part of the animal kingdom sanctified by nature? Are they not "the breathing shapes, many voiced landscape," a phrase borrowed from David Abram's book, *The Spell of the Sensuous?* They also have their special journey on Earth and deserve equal compassion and protection. The primary reason that they are excluded from ethical considerations, and even from the nature programs on public television, is because of their innocence and gentleness that allows them to easily be raised and turned into our food. We would be very outraged if wild animals like elephants and dolphins were subjected to such conditions.

THE BRAVE NEW WORLD

Over the last three million years, human beings developed slowly from a species that was mostly vegetarian like other primates, living in harmony with other animals. Then we gradually developed agriculture, languages and weapons. During the last two hundred years since the industrial revolution our power has soared and the development of our techno-culture has escalated at an alarming rate. The human population will soon reach six billion and will double its number again in the next thirty years. Daniel Quinn describes the scenario of population explosion in his book *The Story of B*. He demonstrates that caged mice continue to multiply as long as their food supply is unlimited. Quinn's conclusion is based upon a fundamental law of ecology: An increase in food availability for a species results in population growth for that species. Genetic engineering aims at unnaturally increased food production to fuel the already excessive human population explosion that is burdening the planet and its resources.

Presently we are breeding 1.28 billion cattle, which further deplete the Earth's resources. In America, one hundred thousand cows are slaughtered every day to satisfy us human carnivores. Eventually, the planet will be crowded with human species along with our billions of food animals. As for the rest of the species, they will go extinct by losing their natural habitat. There is a strong connection between diet and behavior, with the global emphasis on meat-eating reinforcing aggression in society. As a result, we humans have become the deadliest predators on the planet. The fast growing new industry of biotechnology will eventually usher us into a brave new world beyond imagination. Not even Plato, Darwin or contemporary evolutionists and ethicists can provide meaning for such a strange world.

Biotech companies also profit from patenting new species, genetically engineered bacteria, seeds, primates, pigs, cows, chickens, dogs, rabbits, and mice and owning the new species under patent rights. The first ever patented animal was the Oncomouse in 1992, a mouse genetically engineered for cancer research, and many other patented species are soon to follow. Patenting lab created animals is not only religiously and ethically offensive, it opens endless possibilities for humans to exploit other living beings.

CONFUSION IN THE AIR

Mono-agriculture, the production of a few selected crops for mass production, itself is an artificial manipulation of nature. Along with the heavy spray of pesticide and herbicide it is abusive to the soil and threatens biodiversity. Planting bioengineered herbicide-resistant crops, which is one of the main projects of genetic engineering, will only allow farmers to spray higher levels of herbicides without damaging crops. A vicious cycle will be created that will seriously contaminate our environment and poison animals.

Another danger is that biotechnology promises us a new variety of disease resistant crops. Transgenic crops contain genes from viruses, bacteria, animals and other plants. For example, transgenic tomatoes and strawberries contain the anti-freeze gene from Arctic fish so they are more frost resistant. Such bizarre, surreal combinations not only can disrupt the host genetic functions but also can cause confused, chaotic biochemical mutations in the plants.

When transgenic crops cross pollinate with wild plants, it can cause migration of their gene traits, including making them resistant to antibiotics. In time this migration will lead to new mutations and the fields will be eventually taken over by the super grass created by our genetic indiscretion.

The production of new lab crops in developed countries poses a threat to the livelihood of millions of farmers in undeveloped countries. For example, the lab product of cocoa butter and a new sugar substitute could put ten million farmers in poor countries out of work. The new product will not help farmers in poor countries who cannot afford such technology. The increased crops mainly benefit the countries already living in abundance, and the profit primarily goes to the transnational industries that are forging new global commercial monopolies in the name of scientific advancement.

Transgenic salmon contains genes from Arctic sea flounder, which enables them to grow six times larger and faster. Yet eventually these salmon can escape into the wild and cause unpredictable ecological disruption. The DNA of a virus can pass through even the gut of mice and find its way into every kind of cell, creating genetic disturbances including cancer, a disease that more than thirty years of medical research has been unable to find the cure.

A gene can replicate indefinitely, spread and combine. We have no means to stop this process but must let it pass on in its invisible ways. When a massive load of virus genes combines with wild relatives it can result in creating super viruses that can lead to deadly diseases. Dr. Mae-Wan Ho from the Open University biology department in the United Kingdom believes that "a vector currently used in fish has a framework from marine leukemic virus, which causes leukemia in mice, but can infect all mammalian cells. Vectors used in genetic engineering can infect a wide range of species. It's a bad science and a bad business making dangerous alliance."

PATHOGENS: THE ULTIMATE PREDATOR

Along with ecological disasters, the year 1997 can be rightly named a year of the holocaust of farm animals. Outbreaks of infectious diseases among farm animals all over the world have caused researchers to worry that we are due for another global epidemic, primarily owing to the overuse of antibiotics. Two strains of E. coli as well as Staphylococcus bacteria now contaminate meat, poultry and dairy products. Genetic engineering can greatly compound this problem. Laboratory contained transgenic organisms when released into the environment are capable of spreading across species barriers and creating new diseases. A new danger is that they can easily develop multiple antibiotic resistance. In recent years old diseases like cholera, malaria and tuberculosis are coming back in new strains resistant to treatment. At the same time new pathogens are arising. To cope with this, medical laboratories will have to sacrifice billions more animals for medical experimentation. What befalls other creatures also befalls ourselves.

In the future, allergy specialists will have to study gene behavior in order to treat new allergies, because genetic engineering involves adding new proteins to artificially altered food products. This can aggravate allergies since proteins cause most allergies. We are becoming guinea pigs and without our consent being herded into the giant laboratory of biotechnology.

PLANETARY ENCROACHMENT

Can we entrust our food supply and the future of the Earth to those who have no respect or ethical consideration for the living planet, who are motivated mainly by short term profits? Bioengineering is promoted by a multibillion-dollar agribusiness, which controls large segments of the world food supply. It is spearheaded by scientists whose strange alchemical adventure recognizes no species boundary; not even God can predict the consequences. The potentials of bioengineering can become the most dangerous device to destroy nature ever invented, worse in the long run than nuclear weapons.

Why is our government so complacent about this important issue and not keeping us properly informed? Because biotechnology promises the security and abundance of our food supplies, therefore more population growth. For the time being, we are comforted by the deceptive appearance of affluence and continuous economic growth. We can continue cluttering our environment, encroaching into the wilderness, and trespassing territory that naturally belongs to other creatures.

New technologies are erasing the most vital processes that human beings need to form direct maternal bonding with nature. Human beings, like other animals, need physical contact with nature, to live and play with curiosity and humility, sharing nature's offering and wisdom with other Earth residents. Modern men are obsessed with power, possession, production, technological efficiency and speed. Unlike ants and bees, which are at least capable of living in altruistic cooperative societies, we continue to operate out of our own self-interest only. As Daniel Quinn wrote in his book, *Ishmael,* instead of being a "leaver" on Earth, we have become the only "taker." The deterioration of our sense of moral responsibility will only accelerate the current ecological crisis.

All species and habitats are members of the bio-community. From the daisies of the field to great whales in the ocean, from the desert to the rainforest, each has its own intelligence, personality, and consciousness, to evolve creatively with mutual consent. Each has the right to be protected. Human beings as one of these species are indeed out of control. This is evidenced by the growing population explosion, the outbreak of new infectious diseases, the accelerated crime rate, our exploitative economic policies and the way each of us is destroying the planet as a wasteful consumer. Millions of years of planetary evolutionary efforts can become obsolete in a few decades. Sensitive species, such as frogs in some areas in the United States, are already displaying deformities owing to mutation caused by environmental pollution. Though we have not figured out the mystery of the Big Bang behind the existence of the universe, human history is already entering into the second big bang, an explosion through genetic engineering that promises to radically alter everything that we know. The current measures we use to secure our own species mean the end of nature. Without biodiversity the Earth cannot evolve and is doomed to decay.

HOPE AGAINST HOPE

In this time of global crisis, each one of us needs to awaken a new ethical vitality and put forth the energy and moral responsibility that our planet desperately needs to resist the forces of commercial exploitation. We need this for our own sanity and meaningfulness, sacredness of the living Earth. The survival of their future is our own survival. Without collective effort, we will not be able to remedy the ill fate of the planet. Instead of self-gratification, we all need to make some sacrifices in order to give hope to other creatures. Bill McKibben made a deeply moving and refreshing non-anthropocentric statement in his book, *The End Of Nature*: "So I hope against hope, though not in our time, and not in the time of our children, or their children, if we now, TODAY, limit our numbers, our desires and our ambitions, perhaps nature could someday resume its independent working."

Since there is no regulation in labeling genetically engineered products, we have no way to avoid them. Hence we must demand that our government enforce strict regulation in labeling all transgenic products. We can boycott processed food made by genetic engineering, and begin to educate our communities about this important issue. We should support local farmers by purchasing locally grown, organic produce, and switch to an ECO-VEGETARIAN DIET. As long as we breathe fresh air, eat food, and enjoy the beauty of nature, we owe it to mother Earth and her billions of years of sustainability.

NOTE

Published with the permission of the author.

REFERENCES

Ehrlich, Paul R. *The Population Explosion.*
Epstein, Ron. *Why You Should Be Concerned with Genetically Engineered Food.*
Ho, Mae-Wan. *Transgenic Transgression of Species Integrity and Species Boundaries.*
————. *The Unholy Alliance.*
McKibben, Bill. *The End Of Nature.*
Regan, Tom. *The Unnatural Order.*
Rifkin, Jeremy. *Algeny.*
————. *Beyond Beef.*
Singer, Peter. *How Are We To Live.*
Stone, Christopher D. *Earth and Other Ethics.*

II

AGRICULTURAL BIOTECHNOLOGY

AGRICULTURAL BIOTECHNOLOGY:
RISKS AND BENEFITS

To genetically engineer an organism is to transfer DNA originating in one organism into another organism. Recombinant or spliced DNA is constructed so that the correct promoter produces the protein in the anticipated tissue. How DNA is put into organisms will be described in later chapters, but you need to understand aspects of the process at this stage to best understand the benefits or risks of such technologies in agriculture. When DNA is put into the nucleus, it might be "free floating" in the nucleus or it might be integrated into the chromosomes of the cell (figure 13). The free-floating DNA might produce protein for a time, but as the cell metabolizes and divides, the foreign DNA is lost or degraded. This is referred to as transient expression because both the DNA and the protein production are eventually lost. If a scientist only requires transient expression in cells to study the function of the foreign protein, such temporary functionality of the gene is sufficient. However, if a long-lasting production of the foreign protein is needed, the foreign DNA must integrate into the cellular DNA. This integration occurs when cellular enzymes recognize and attach to the foreign DNA and act as though it is DNA that needs to be "repaired." Cellular enzymes whose function is to repair DNA cut the cellular DNA and insert the foreign DNA into the chromosome. Except under certain circumstances, the site of integration is essentially random. The foreign DNA might integrate outside of any existing genes where it has little physiological consequence to the cell or organism. Alternatively, the foreign DNA might integrate in a cellular gene so that its function is destroyed. Obviously, unplanned destruction of cellular functions is not beneficial. If the foreign DNA is not integrated in the genome, it will be degraded and used by the cell for nutrients. Remarkably, genetically engineered organisms are often completely normal except for the presence of foreign DNA or protein. Because there is so much "space" in DNA or because the cell has backup functions to override the inactivated genes, there are often no abnormalities of the cell or organism.

Another feature of chromosomes can also affect the expression of transgenes. At particular stages of the life cycle of the cell, certain large regions of the chromosome are tightly wound in a knot and other regions are unraveled so that transcription proteins can access the DNA and make new proteins. Consequently, some genes with similar types of functions are clustered in the same region of the chromosome so that they can be expressed when that part of the chromosome is unraveled or loosened. Therefore, the region of the chromosome in which the transgene integrates will affect its expression, depending on whether that region of the chromosome is unraveled or not. It is said, then, that expression of a transgene is dependent on the microenvironment of the chromosome in which it is integrated. If the transgene integrates in a portion of the chromosome that rarely unwinds, the transgene will not readily make protein. Because of this random integration, genetically engineering an organism is an uncertain process, especially in attempts to engineer an organism as genetically complex as a person.

Some specific techniques have been developed that cause foreign DNA to be inserted or integrated at a precise location into the cellular chromosome. This process is called "targeted" integration because it is not random. Obviously, this control of the integration site is highly desirable compared to random integration, but the technique to date is restricted to a few organisms such as mice, yeast, and certain livestock animals. Scientists are making considerable efforts to expand targeted integration technology to more organisms. Understanding targeted or random integration of foreign DNA assists the reader in evaluating the benefits and risks of genetic engineering. If targeted integration can be employed, there is more control of the process and fewer mistakes are likely to occur. With targeted integration, there is less chance of destroying important genes in an organism.

If stable integration occurs in a cell (figure 13), progeny of that cell will "inherit" the foreign DNA. When the cell divides, it will replicate the integrated, foreign DNA along with the rest of the chromosome. Because the integrated DNA is inherited from one cell generation to the next, the process of putting the DNA into the nucleus does not have to be repeated when new cells with the foreign DNA are needed.

If DNA is put into cells of an animal, the foreign DNA is referred to as transgenic DNA and the animal is referred to as a transgenic animal. Examining the roots of the word is helpful. "Trans" implies moving across, beyond, or through, as in the word "transcontinental." The root "genic" implies gene, so transgenic DNA is DNA moved from one organism to another. This term needs to be clearly understood because genetically engineered organisms are often referred to as transgenic organisms, even in the popular press. If the transgenic DNA is integrated in germ cells, sperm, egg, or early embryos, the transgenic embryo cell can replicate into every cell of the body. Consequently, every cell in the body would contain the transgene. Remember, however, that every cell of the body would

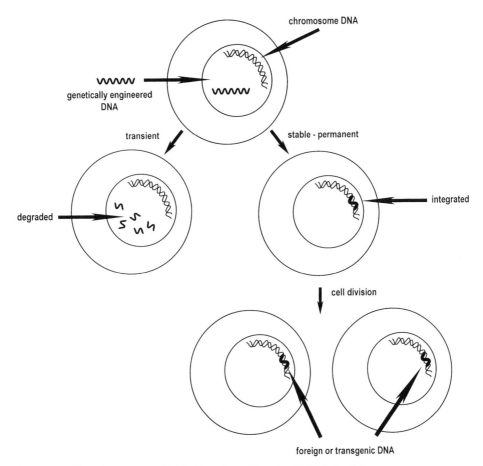

Figure 13.   Transient versus Stable Transfer of Foreign DNA into Cells

probably not produce the protein because of differential expression from the promoter. Therefore, when the adult organism replicates, the new germ cells will contain the transgene (figure 13), and the transgene will be bequeathed indefinitely to the progeny. For example, if new genetically engineered cows are needed, people simply breed genetically engineered cows to produce genetically engineered progeny. The process of manually putting the foreign DNA into a cell is not necessary every time a new transgenic animal is produced.

The ethical and social issues of genetically engineered organisms depend to some extent on the type of organism used, whether they are microscopic or visible, able to adapt to new environments, pathogenic or harmless, and cross-hybridizing or restricted in transfer of genetic material. Consider a genetically engineered cow, for example. It is certainly visible and does not reproduce uncontrollably. In most countries, cattle are domestic animals, so they do not uncontrollably breed with wild animals or other nonrelated animals. Moreover, they are certainly not pathogenic. Consequently, it would be difficult for the transgene within the cow to be "lost" to some other organism or to the environment. Since cattle are considered higher-order organisms (i.e., they are phylogenically higher than a beetle, for example), some people have ethical concerns in altering a cow as compared to a beetle. What about lower-order mammals, such as mice? Domestic laboratory mice are marginally survivable in the environment and might cross-breed with wild mice to possibly spread the transgene outside of its intended host. However, they are visible animals that can be traced and probably controlled with enough effort. Insects, however, may be more difficult to keep contained in captivity or restricted to a designated area. Conceivably, they can undetectably spread and reproduce. Like other animals, they do not breed outside well-defined limits. Plants, however, can cross-pollinate or otherwise spread their genetic material. Seeds or pollen from plants can spread beyond their vegetative boundaries. Clearly, plants can present special problems to accidental spread of genetic material.

Organisms that are microscopic present a whole new set of issues. They cannot be seen without the aid of a microscope and might move about undetected. Some types of microorganisms have the potential of being pathogenic by infecting their hosts in harmful ways. Certain categories of bacteria have the ability of spreading their genetic material, DNA, to other bacteria. They can adapt to pressures of their environment by exchanging genes, such as antibiotic-resistant genes. Huge populations of individual bacteria can die and still have successful propagation of the species if at least a few individuals genetically mutate to overcome selective pressures in their environment. Mutations help bacteria change to the environment for their own survival.

It is important to bear in mind that some bacteria have been genetically engineered to cripple their ability to spread into the environment, where they could inadvertently transfer genes to other bacteria. These bacteria have been used as tools of recombinant DNA for many years and have been monitored to make sure that there has been no inadvertent release or harm done by these bacteria; they have proven to be safe for other organisms and the environment.

Genetically engineered products for use in agriculture may be simply a protein chemical or a living organism. Chymosin is a protein chemical that is added to milk to assist cheese manufacturing. It is obtained from calf intestines, since the enzyme is naturally present to help the calf digest milk. For economic and production reasons, the ability to produce chymosin by some other method would be beneficial to the dairy industry. So microorganims were genetically engineered to produce chymosin safely within the containment of a laboratory. The chymosin protein is extracted from the microorganism as a chemical and is not a living organism. The chymosin dairy product is only a chemical, and some of the ethical and social issues involved with genetically modified organisms do not apply to chemicals derived from genetically engineered organisms in the factory or laboratory. Chymosin was approved by the USDA for use in cheese making in 1990 and is presently used in 70 percent of all cheeses bought in the United States. The safety of genetically engineered proteins (recombinant proteins) used as agricultural products still needs to be proven, but when the product is not a replicating organism or genetically modified organism (GMO), the issues are not as complicated.

To better understand the differences between recombinant proteins and genetically modified organisms, you need to understand how the recombinant proteins are produced. We will use the example of a protein being produced in bacteria. Plasmids are naturally occurring, double-stranded circular DNAs in some bacteria. They have DNA sequences that interact with bacterial enzymes, which cause the circular DNA to replicate exact copies of each plasmid (figure 14). This is an important concept to understand because it is a natural form of "cloning." To clone means to make copies of one thing. In this sense plasmids are clones of each other because the DNA of the plasmid replicates exactly. As the plasmids grow in the bacteria, many copies of the original plasmid are produced. The strategy is to put DNA of particular value in the plasmid by recombinant DNA methods. When the plasmid grows, exact copies of the valuable DNA also grow in number (figure 15). This is the principle of DNA cloning! Why is cloning DNA important? Let's assume that you put DNA that codes for chymosin into one plasmid molecule. Such a small amount of chymosin would be produced from only one plasmid. The protein would be undetectable and would have no value. If, however, the chymosin gene were put into a plasmid and the plasmid replicated to extremely large numbers of cloned plasmids, then larger amounts of chymosin could be produced. If enough protein is produced and purified from the bacteria, then it can have value as an agricultural product.

Let's think of the value of DNA cloning in a different way. Suppose that you wanted to study the sequence of the gene that produces beta-casein found in milk. The beta-casein gene is about 23,000 nucleotide bases long. It is found in mammalian chromosomes that are about 10^9 nucleotide bases long. Even if you knew approximately where the gene was located in the chromosomes, the DNA of the chromosomes would mask or confuse the chemical analysis of this relatively small sequence of 23,000 bases. Some way is needed to expand the numbers of this short beta-casein gene to far greater numbers than the chromosome sequences. Maybe if 1 trillion more molecules of this short sequence could be obtained, it could be chemically purified from the genomic DNA of the organism and studied away from the rest of the genomic DNA sequences. This is done using recombinant DNA techniques—by putting the valuable DNA into plasmids and cloning the DNA. Here are the itemized steps to be explained (figures 14 and 15).

Cut genomic DNA containing the beta-casein gene.
Put the beta-casein gene into the plasmid.
Put the plasmid into bacteria.

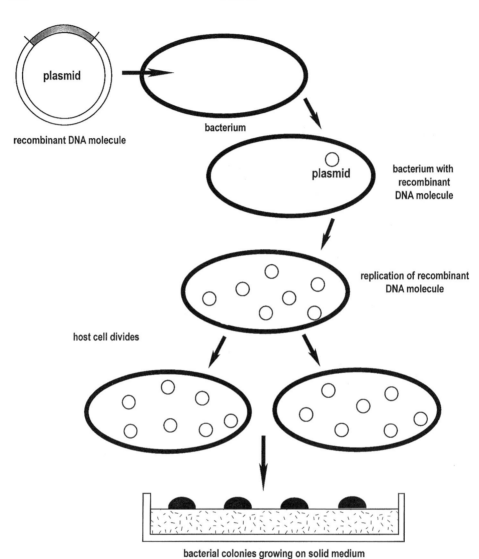

Figure 14.   Cloning DNA: Transfer and Replication of Self-Replicating Plasmid DNA in Bacteria

Allow the plasmid and bacteria to grow and make exact copies of the beta-casein gene in large numbers, or clone the gene.

It is hard for readers not having a chemistry background to picture how molecules are changed or manipulated. Obviously, we cannot cut and glue bonds using scissors and paste: we have to use chemical reactions in test tubes on molecules that are invisible to the naked eye. But using the terms "scissors" and "glue"

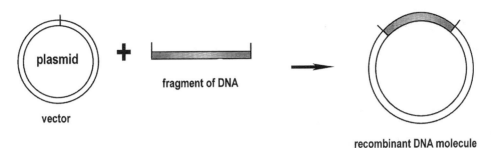

Figure 15.   Insertion of DNA of Interest into a Plasmid DNA for the Eventual Purpose of Cloning the DNA of Interest

helps explain the events in the chemical reactions. The protein enzymes that break DNA bonds in preparation for splicing different DNAs together are called restriction endonucleases, or restriction enzymes, which you can think of as the scissors. The root "ase" in chemical terms means enzyme. The root "nucle" refers to nucleic acid in DNA molecules. The root "endo" means that the enzyme does not act on nucleotide bases at the terminal ends of the DNA, but on nucleotides inside the DNA molecules. Some "nucleases" can only act on nucleotides at the terminal ends of the DNA, so the distinction is made that restriction endonucleases act within the DNA. The word "restriction" implies that the enzyme is limited as to what nucleotide bases it will cut, and that it will not cut all bases. This restriction is good for our purposes because we would not want the enzyme to simply cut the DNA into unrecognizable pieces. So restriction endonucleases are protein enzymes that cut DNA only at very specific nucleotide sequences, such as at the sequence GATATC. The restriction endonuclease that recognizes this particular sequence of DNA is called EcoRI. It has a unique name to distinguish it from all other restriction endonucleases such as Hind III, Bam HI, Sal I, or Pst I. It is important to make this distinction between the different enzymes because they all cut different sequences of DNA.

Another helpful feature of restriction endonucleases is that they can cut double-stranded DNA so that the strands are staggered with overhanging ends (figure 16). The phosphate bonds of the backbone of the DNA are cut by the enzyme (figure 4). The number of hydrogen bonds now connecting the two DNA molecules together is so few in that region that the hydrogen bonds cannot hold the two strands of DNA together. These overhanging ends of DNA are complementary, or match with other pieces of overhanging ends of DNA that have been cut with the same restriction endonuclease, such as with EcoRI. Look at the diagram (figure 16) carefully by matching the ends of the DNA to see that they are complementary. The A's match with the T's. All ends of DNA cut with the same restriction enzyme, such as EcoRI, have exactly the same matching ends.

In cloning a DNA of interest, both a plasmid and the valuable DNA are cut with

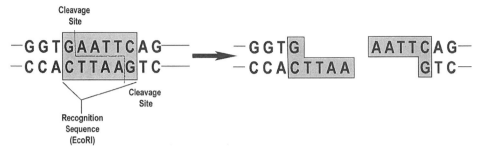

Figure 16. Cutting DNA Precisely with Restriction Enzyme

the same restriction enzyme, such as EcoRI. When the two DNAs are cut with the same enzyme, the ends of DNA from the plasmid can potentially connect with the DNA of interest. Even though the plasmid and the DNA connect by the overhanging ends, the phosphate bonds need to be rebuilt; otherwise the hydrogen bonds will just come apart again. Figuratively speaking, some glue is needed to reconnect the phosphate bonds. The gluing enzyme is DNA ligase. The word "ligate" means to connect and that is just what the enzyme does: it reconnects the phosphate bonds of the backbone. When this backbone is connected, the hydrogen bonds stay together and the double-stranded DNA is completely repaired. The ligase reaction is a reversal of the enzyme cutting, as shown in figure 16. Chemically speaking, this repaired DNA is indistinguishable from any other DNA. It functions in the same way and replicates in the same way, except that one part of the DNA is the plasmid and the other is the newly inserted DNA.

The next step is to put the DNA into bacteria so that the plasmid can replicate. The bacteria used for this purpose have had certain functions removed to prevent them from escaping into the environment, exchanging DNA with bacteria in the environment, or recombining their DNA with other DNA. The bacteria are specifically made for genetic engineering purposes and have proven safe during the last twenty-five years. There are a number of different methods for putting plasmids into bacteria, but these methods will not be discussed except to say that only one plasmid molecule usually enters a bacterial cell. This is an important concept to remember—all the plasmid molecules growing out of one bacteria were derived from one original plasmid molecule; therefore, they are identical clones of each other.

Not only do the plasmids amplify inside the bacteria, but the bacteria divide and grow to very large numbers. A single bacterial cell can divide until it becomes a visible group of bacteria, which is referred to as a colony. They appear as a dome of billions of cells that are identical to each other, since they are derived from a single bacterial cell. When the bacterial cells are dividing and replicating, the plasmids are dividing and replicating so large numbers of cloned plasmids are produced! Do these plasmids contain the gene for the cloned gene such as beta-casein? Yes. There are many billions of identical copies of beta-casein gene that had been inserted, or ligated, into the plasmid. With this number, the beta-casein DNA can be chemically analyzed or sequenced. Also, if a promoter that works with the bacterial transcription proteins is connected to the beta-casein gene, then large amounts of beta-casein protein can be produced in the bacterial cells. None of the other genes from the cow DNA would be expressed because they are not present in the bacteria; only the cloned plasmid having the beta-casein gene is present. By comparing the levels of protein production from a single beta-casein gene with the many cloned beta-casein genes in the plasmid, we can appreciate the value of the methods for cloning DNA.

ETHICAL ISSUES: RISK AND REWARD

During the mid-1980s a fierce debate began about the agricultural uses of biotechnology. The specific starting point in the discussion was a proposal by scientists at the University of California to begin testing on strawberry fields a genetically altered bacterium known as ice-minus *Pseudomonas syringae*. This bacterium typically helps ice crystals form on fruit. The genetically modified variety would not do this. Sprayed on sensitive early-season crops like strawberries, the GM variety would replace the natural variety and thus significantly reduce losses to strawberry growers from late frost.

Scientists who worked on *Pseudomonas syringae* in the laboratory assured the public that the ice-minus genetically modified organism, used in this fashion, was safe. They had worked on it for years and they believed that they knew its properties in enormous detail. After all, it was only altered by the deletion of one gene. Critics of the deliberate release were not as sure. Most of them were not "bench scientists" who worked on laboratory genetics. Some of the critics were activists concerned about the social impacts of technology in general and genetic technology in particular. Scientific critics tended to come from "whole systems" fields like ecology and population biology. Ecologists in particular argued that what happens in a laboratory does not correlate well with ecosystems, which are extremely complex and little understood, especially at the microlevel. They claimed that this complexity meant that ecosystem disruption can occur in ways that are poorly understood and cannot be better understood through work done in a closed system like a lab.

The debate over *Pseudomonas syringae* is an example of the debate over the benefits and risks of agricultural biotechnology in general. Partially this is a debate between scientific disciplines. Microbiologists and molecular biologists are used to working in environments where input factors such as temperature and chemicals can be carefully controlled and the results measured. This is classic experimental science.

Ecologists, by contrast, start by describing very complex systems that are not well understood and cannot be tested in the manner of experimental science. They see complexity where laboratory scientists seek simplicity. They try to describe the complexity and see a whole system striving for balance. Molecular biologists seek to understand genotypes as a prelude to precise human-originated change. Evolutionary biology and ecological study may one day show that human life evolved on the African continent in a symbiotic relationship with malaria. But the molecular biologists try to develop a vaccine that would dramatically alter this supposed balance.

Scientific critics of open system agricultural biotechnology are worried about three specific features that they believe distinguish agricultural biotechnology from laboratory work. First, they are worried about what is called horizontal or lateral gene transfer, especially at the microlevel in open environments. Basically this can be understood as a sort of "genetic infection" among species that do not interbreed in nature. Sometimes viruses provide a "vector," or transfer mechanism, but often genetic elements simply migrate from one bacterium to another. Under environmental stresses the genetic material from one microlevel species can be "taken up" by another. According to the critics, deliberate release of genetically modified organisms thus might have systemic effects such as altering nitrogen-fixing bacteria or creating new environmental pathogens.

Second, critics are worried about the possibility of a decline in the genetic diversity of useful species. Take fish, for example. Suppose salmon are genetically altered to grow larger than wild varieties. Their size and food consumption will tend to crowd out the wild, native types of salmon; as a result much valuable genetic diversity will be lost. Genetic diversity among the members of a species is widely thought to reflect the "health" of a species in evolutionary and ecological terms. Hence the argument is that transgenic technologies will narrow the gene pool of a species and threaten its long-term health.

Finally, critics are worried about the "permanent" effects of agricultural biotechnology. This seems to mean that mistakes are difficult if not impossible to correct. Take the salmon example again. If, despite our best efforts, genetically altered salmon escape from a containment area, they will tend to survive. They may be impossible to recall or contain apart from simply destroying all life in the area. Consider what is often done when nonnative species are found in some lakes to be crowding out the native species: all the fish are killed with chemicals. In the case of ocean species like salmon this is impossible. Even in the case of nonnative species in lakes and rivers, should we be using a technology like transgenics that may require us to kill thousands of fish with chemicals if a mistake is made?

This debate, especially between laboratory scientists and ecologists, is nicely summarized in the Hilleman chapter. These questions are addressed in a moderately optimistic fashion by Fincham and Ravetz. They point out that the risk of harm occurring from the deliberate release of GMOs is actually very low. By analogy with chemical discharges, we have a sound body of background knowledge in the context of which we can evaluate the possible harms and benefits of any use of GMOs. We can use the information we have about the organism and the environment to evaluate the harms and benefits and make an informed decision about the deliberate release of any GMO, just as we do in many other cases such as new drugs.

Alternatively, since we cannot be certain of positive or at least neutral effects we should do nothing. But this position is hardly adequate and has never proven effective in other areas of science or technology. The dynamism inherent in technology seems to push toward the use of a technology, either with acceptance and regulation or without. If we use the knowledge we have and proceed with public involvement and appropriate caution, we can carefully evaluate the effects as we go forward and realize whatever benefits there may be from agricultural biotechnology. Will we make mistakes? Yes. Are there risks? Certainly. Any technology has some risks. A no-risk society may be the most harmful because technology that is widely beneficial such as vaccines would not be developed because some people have allergic reactions. This cannot be a welcome result.

AGRICULTURAL BIOTECHNOLOGY: VIOLATING NATURE

In a well-known article biologist Martha Crouch writes about her personal alternative to biotechnology in agriculture:

> The subsistence tomato I grow at home has a very different network of connections. First I grow about a dozen different kinds each with unique characteristics. I save seed from year to year. They are grown in a polyculture with other different species. Predator insects, birds, earthworms, nitrogen fixing legumes, trap plants and many unknown interactions make up the system. I do not use purchased chemical inputs nor is irrigation required. . . . Seasonality is savoured. We have a contest to see who can coax the first ripe tomato and the earliest variety usually has ripe fruit by July. The last green tomato is fried up in November. Surplus fruit is dried for use in the winter and I don't eat salads out of season. ("Biotechnology Is Not Compatible with Sustainable Agriculture," *Journal of Agricultural and Environmental Ethics,* 8 [1995]: 98–111.)

Crouch is obviously concerned about the social impacts of biotechnology that are treated in this section. This passage also implies that there is a natural rhythm or process of growing tomatoes and an unnatural, technologically dominated form of growing and using tomatoes.

This dichotomy of the natural and the unnatural is a persistent theme of the critical literature regarding agricultural biotechnology. "Natural," "ecologically sound," and "respecting nature" are phrases that persist in this literature. Deeply indebted to the language of "playing God," blended with respect for nature, buttressed by a radical, static view of ecology, these critics believe that nature is a guide to practice. They are critical of technical agriculture as a domination of the world that will ultimately come to dominate humanity with technology. Generally they see nature as a fragile work of art easily destroyed by technology. They note that global warming is enhanced by the very technology that was once thought of as progress. Chlorofluorocarbons, which cause ozone depletion, were thought to be good for human welfare when they were introduced.

This view is represented in chapter 4 by Mira Foung and in chapter 7 by Miriam MacGillis. They see the world, including evolution, as a slow process of development, representing the "natural," which sustains all life. We are part of that life. Trying to dominate the world with technology will, in her view, only end in disaster. It will destroy the very life we seek to redesign. Rather, we must reject biotechnology in all of its forms and "live within the limitations of the unity of the whole."

MacGillis's use of evolution to make her point provides an opening for her critics, as do some of the examples regularly cited. Evolution shows that nature is not static and cannot be compared to a work of art, where change would mean diminution or even destruction. But MacGillis's reading of evolution seems one-sided. Students of nature know that other species, especially primates, use tools and craft nests and share means of food preparation. If so, how can it be wrong for human beings to survive by doing the same thing? Are human beings forbidden from using technology while other species are not?

MacGillis does not use the concept of "speeding," but she might usefully employ it. What she might say is that some modern technologies, especially biotechnology, promote too vast a change in too short a time for us to learn from and correct our mistakes the way evolutionary feedback mechanisms do. Evolution seeks to find natural balances through a slow process of change. Information is fed back in a loop, allowing for change that preserves the process. Are we going too fast to slow down if one of our supposed improvements goes awry? If a drug proves harmful, it can be recalled before a wide harm results. But transgenic plants are living organisms released in open evolving systems where recall may be impossible.

Biotechnology may be thought of as unnatural in a sense because of the speed at which it promotes change in living systems. Alternatively we may be "playing God" not by making such changes (which happen in nature anyway), but by making them on a vast scale without being in God's epistemic position with respect to the whole of the system.

AGRICULTURAL BIOTECHNOLOGY: SOCIAL EFFECTS

The current debate about GMOs in agriculture is focused most strongly on the social effects of such technological changes. On one side are optimists, like McGloughlin in chapter 9, who believe that genetic technologies offer many advantages over traditional agriculture, advantages that will benefit all countries, especially less-developed countries (LDCs). For the most part they are moderate optimists who, while recognizing that political, economic, and natural forces may alter their projections, believe that the benefits of biotechnology in agriculture are potentially highly significant. Like McGloughlin, these writers generally focus on three supposed benefits of biotechnology for agriculture.

First is both the need for and the possibility of increased crop yields. Even with the projected serious drop in family size worldwide, the planet is growing more populous. Within the next fifty years world population is expected to reach 9–10 billion. More food per acre will be needed in what is historically a very rapid time. Yields must increase faster and in more complex ways than traditional breeding produces. Hence transgenic technology must be used to increase gross production and reduce loss to pests (as in the case of Bt corn, which McGloughlin notes). These writers do not believe that "living close to nature" or small farming will feed the urban masses of the next century. We will need the productivity increases brought on by technology, not nature, and the economies of scale brought on by large farms to feed 10 billion people.

Second is increased nutritional quality of food. Genetic technology has already been used to increase the iron content and add beta-carotene (precursor to vitamin A) to rice. While so-called golden rice is several years away from wide use, such examples suggest to optimists the possibility of using transgenic technologies to add nutrition to crops to combat problems such as anemia and blindness from lack of vitamin A, especially in LDCs.

Third is fewer harms to the environment through reduced pesticide and herbicide use. Crops such as Bt corn that create natural pesticides do not require the level of chemical pest control nor do herbicide-resistant crops like Roundup Ready soybeans require multiple sprayings of different herbicides. Supposedly a good herbicide like Roundup will kill all weeds and leave the crop. Fewer chemicals mean less harm to the environment and safer foods.

Those who disagree with these assessments tend in varying degrees to be pessimists about transgenic agriculture, like Altieri and Rosset (chapter 10). They note biological problems like gene transfer in GMOs in agriculture and they dispute some of the specific claims that optimists make about greater productivity and reduced chemical use. But they also focus on social and political structures of international agriculture that optimists often ignore.

First, they argue that current malnutrition and hunger are the result of political and ethnic strife and profit-driven corporations. Conflict prevents food from getting where it is needed, and poor people cannot be robust consumers for corporations. They believe that the world as a whole is producing more food than ever before. It is social, economic, and political structures that undermine food security.

Second, they point to possible problems with transgenic agriculture. They believe that herbicide-resistant crops may encourage less careful use of chemicals because the crop cannot be harmed. They do not believe that yields per acre have been shown to be significantly greater or that GM foods are always safe. Finally, they note the problems of gene transfer and increased pest resistance, such as we noted already regarding the issue of so-called super weeds. Further, if crops produce their own pest toxins, pests will develop a resistance and nothing will be gained.

Third and most significant is the critics' view of the structure of agriculture internationally. Multinational agricultural companies want marketable commodities. They allegedly encourage cash crop mono-agriculture in LDCs. This will increase the power of corporations, decrease the diversity and nutritional quality of local agriculture taken as a whole, and do little for the poverty of growers in much of the world. Corporate agriculture will displace farmers and provide little safety or security in return.

The critics' preferred solution is what they variously refer to as "sustainable agriculture" or "agroecological farming," which supposedly works in harmony with local practices, local people, and local ecology. Native growers can and will produce enough food for their people in a balanced way that can be sustained over generations.

As chapters 9 and 10 show, the debate is less about biology or even moral goods. Both sides of the debate want enough nutritious food for the world now and in the future. They differ about the influence of globalization on this process and about the need to use extensive technologies to achieve these goals. The debate is not fundamentally about biotechnology but about the impact of the twin forces of technology and global capitalism on local communities, and impacts that will only grow in the decades ahead.

5

Differing Views of the Benefits and Risks of Agricultural Biotechnology

Bette Hilleman

As the 21st century approaches, the world is facing problems many experts see as intractable. All but four of the world's 17 major fisheries are seriously depleted. Irrigation water is in short supply in many parts of the globe, and erosion threatens the productivity of much farmland. While the amount of land under cultivation cannot be expanded greatly, almost 100 million people are expected to be added to the world's population each year for the next 30 years. As a result, per capita food production is likely to continue to decline in Sub-Saharan Africa and barely rise in South Asia.

Biotechnology—specifically that aspect involved in transferring genes from one species into the genome of another—has the potential to alleviate many of these problems. Genetically engineered (transgenic) fish that grow much faster than wild and traditional aquaculture varieties could relieve some of the pressure on the world's fisheries. Crops bioengineered for pest resistance could increase yield, eliminate the use of several insecticides now derived from fossil fuel and reduce health risks and groundwater contamination from pesticides. In some areas, crops genetically modified for herbicide tolerance could decrease the amount of herbicide used and allow for no-till agriculture, which can minimize erosion. Crops engineered to produce oil-derived chemicals could relieve pressure on oil supplies. Recombinant bovine growth hormone already enables cows to use feed more efficiently and produce more milk.

Although each of these applications holds great promise, each also may cause harm to ecological systems, human health, or economic and social structures. Proponents of biotechnology claim that most transgenic agricultural products pose no unique hazards to health or to the environment. Critics say the advantages of

111

bioengineered products have been exaggerated and that they involve potential dangers that have not been fully investigated. In general, it is molecular biologists, most of whom have a background in biochemistry, who see little reason to believe that transgenic organisms could injure ecological systems, while ecologists and experts in fisheries and marine biology think great caution should be used when releasing such organisms.

In the U.S., the Administration generally is confident that most uses of biotechnology will not cause undue harm and that, on an international basis, it can be used safely with voluntary and cooperative oversight. The U.S. takes the position that a legally binding international protocol governing uses and releases of genetically modified organisms—a biosafety protocol—would interfere with research and the development of the biotechnology industry.

In contrast, some European nations are quite leery about uses of biotechnology, especially in developing countries. Last month, the European Parliament passed a resolution saying that "a legally binding international biosafety protocol is necessary and a matter of urgency and must be immediately negotiated by the states party to the [United Nations] Convention on Biological Diversity," which was agreed to at the Earth Summit in Rio de Janeiro in 1992. To justify the need, the resolution states: "Deliberate releases of genetically modified organisms are being carried out in many developing countries, which have no legislation or infrastructure to ensure their safe use . . . [and] this situation is putting the entire biosphere of the planet at risk."

Delegations from 80 countries met at a UN experts meeting in Madrid, July 24 to 28, to decide on the need for, and the issues to be addressed in, a biosafety protocol under the Convention on Biological Diversity. Even though the U.S. is not among the 120 nations that ratified the convention, it plays a strong role in the conferences of the parties and experts meetings.

At the Madrid meeting, the delegations reached consensus on a document calling on the parties to the convention to begin drafting a legally binding biosafety protocol. This document was approved, even though the U.S., Germany, Japan, and Australia officially took the position that a protocol is not needed and pushed instead for voluntary guidelines. The protocol issue will be dealt with further when the second conference of the parties to the convention meets in Jakarta, Indonesia, in November.

In its strong support for a binding protocol, the Madrid meeting represents an about-face from a report prepared by 15 UN experts at a meeting in Cairo last spring. The Cairo statement roughly equates genetically engineered organisms with organisms produced by traditional technologies—such as crop hybridization—implying that, therefore, no international control over the products of biotechnology is needed. The Madrid statement not only calls for a binding protocol, it also says the protocol should be based on the precautionary principle—the idea that a transgenic organism should not be released if there is significant uncertainty about its risk.

If a binding protocol is eventually adopted, international trade and corporate profits could be affected. The Biotechnology Industry Organization (BIO), a Washington, D.C.–based association of biotechnology companies, and other biotechnology industry groups lobbied strongly at the Madrid meeting to keep controls at the national, rather than the international, level.

After many years of development, agricultural bioengineered products are rapidly entering the U.S. marketplace. The first major product, Monsanto's recombinant bovine growth hormone, was approved for sale in February 1994, and since then nine other products have garnered regulatory approval. In addition, more than seven products are going through the approval process and likely will be on the market within two or three years.

Although no variety of transgenic fish has been commercialized anywhere in the world, some companies are developing brood stocks of these fish and expect to begin marketing them within a few years. China plans to start selling its genetically modified fish as soon as it has produced stocks that grow fast enough to make the effort worthwhile.

Because the development costs have been so high, no company has yet made a profit on a transgenic agricultural product, despite the upsurge in commercial approvals; however, some firms expect to do so within a few years. BIO predicts that annual sales of such products will reach several billion dollars by 2000.

With the exception of recombinant bovine growth hormone, agricultural products with current or pending regulatory approval in the U.S. fall into six categories: plants engineered for herbicide resistance, plants designed for insect resistance, tomato plants engineered for delayed tomato ripening, plants modified to produce products now made from other crops, plants engineered so the crop can be processed more easily, and bacteria designed to enhance nitrogen fixation in alfalfa or to control insects. (The plants and animals that are genetically modified to produce pharmaceuticals are not discussed in this article.)

PATCHWORK OF STATUTES

In the U.S., three different agencies regulate agricultural transgenic organisms under a patchwork of statutes. So far, Congress has not enacted a single new law to govern biotechnology research and commercial applications. Instead, the agencies have adapted the existing body of legislation to regulate recombinant DNA research and products. However, they have written some new guidelines for the conduct of research and for regulatory approval processes.

Currently, for product research and commercialization, companies must seek approval from one to three agencies, depending on the nature of the transgenic organism. Under the Food & Drug Administration's (FDA) transgenic plant food policy, few products will require agency approval under the Food, Drug, and Cosmetic Act. However, most companies thus far have voluntarily consulted with

FDA to gain its stamp of approval. The act has very few data requirements and specifies that consultations should go on behind closed doors, unless the company involved requests open meetings. Under the Plant Pest Act, the U.S. Department of Agriculture (USDA) regulates transgenic plants to be grown on a large scale. If the gene-modified organism expresses a pesticide or functions as a pesticide, the Environmental Protection Agency (EPA) regulates it under the Federal Insecticide, Fungicide & Rodenticide Act (FIFRA). Under the Toxic Substances Control Act (TSCA), EPA also controls genetically engineered microorganisms that have no pesticidal attributes. An example would be a bacterium engineered to produce ethanol from agricultural residue.

Corn that is genetically engineered to express an insect toxin from the soil bacterium *Bacillus thuringiensis* in its cells is an example of a crop that is regulated by all three agencies. Its insecticidal properties must be approved by EPA, its large-scale growing by USDA, and the corn as a food product by FDA.

Under the existing legal framework, environmental releases of most genetically engineered animals are essentially unregulated. Some varieties of transgenic fish have extra copies of a fish growth hormone gene, and FDA may decide to regulate them as new animal drugs. If FDA decides it does not have authority to regulate transgenic fish, no statute covers the environmental impacts of commercializing them, says James Maryanski of FDA's Center for Food Safety & Applied Nutrition. But he says FDA would be in charge of the quality of the fish in commerce.

Despite the frequent need for multiple approvals, industry is generally satisfied with the current regulatory structure. "The system is working pretty well," says BIO President Richard Godown. "Obviously, there are difficulties, but we're getting our products to market."

Agency officials also consider the current system effective. Janet L. Andersen, acting director of EPA's biopesticides and pollution prevention division, says, "The agencies have coordinated their efforts together well so they appropriately cover their mandate to regulate transgenic organisms."

However, the Union of Concerned Scientists (UCS) in Cambridge, Mass., and a number of environmental groups such as the Environmental Defense Fund have criticized the patchwork approach for years. Such groups prefer a new statute that would establish a comprehensive program administered by EPA to regulate the release of engineered organisms into the environment.

Jane Rissler, a UCS senior scientist based in Washington, D.C., says the use of USDA's Plant Pest Act is an especially weak part of the current system. This statute is vulnerable to legal challenge because it applies to plant pests, not plants, she says. "Furthermore, it is not a strong enough statute," she notes. "Unlike FIFRA, it is not a registration statute, which would allow USDA greater control over a product, and it does not require labeling. The product is simply approved or not approved."

TRANSGENIC FISH

Nearly all researchers in the biotechnology area agree that the techniques of recombinant DNA are a wonderful research tool for understanding plant and animal physiology. They say that with biotechnology the desired results in traditional breeding can be reached much more quickly and with less trial and error.

But scientists' views are sharply divided over transgenic fish. Some experts regard such fish as the potential answer to the depletion of the world's fisheries and a practical way to increase protein intake for many malnourished people. They believe that even if some fertile transgenic fish manage to escape to oceans or freshwater bodies, the fish will not pose a severe ecological hazard. Others see gene-modified fish as a potential nightmare that could permanently eliminate many species of wild fish.

Currently, about 40 or 50 labs around the world are working on transgenic fish. About a dozen of them are in the U.S., another dozen in China, and the rest in Canada, Australia, New Zealand, Israel, Brazil, Cuba, Japan, Singapore, Malaysia, and several other countries. Some of these labs are associated with companies that expect to commercialize their fish in a few more years. Many of the fish under development are being modified to grow faster than their wild or traditionally bred aquaculture siblings. Faster growth is usually accomplished by transferring a fish growth hormone gene from one species of fish into another. The faster growing fish not only reach market size in a shorter time, they also use feed more efficiently, researchers say.

For example, Thomas T. Chen, director of the Biotechnology Center at the University of Connecticut, Storrs, transferred into common carp the growth hormone DNA from rainbow trout fused to a sequence from an avian sarcoma virus. The genetic material was injected into fertile carp eggs with microinjection. The offspring of the first generation of transgenic fish grew 20 to 40% faster than their unmodified siblings. Chen is also developing transgenic catfish, tilapia, striped bass, trout, and flounder.

Another example is work by Robert H. Devlin, a research scientist with Fisheries & Oceans, Canada, in West Vancouver, British Columbia. He has modified the growth hormone gene in coho salmon by developing a gene construct in which all the genetic elements are derived from sockeye salmon (*Nature*, 371, 209 [1994]). The transgenic coho grew on average 11 times faster than unmodified fish and the largest fish grew 37 times faster. The growth hormone levels in the transgenic fish are high year-round, rather than falling off in the winter as occurs in ordinary salmon, Devlin says. The modified salmon are large enough to be marketed after one year, in contrast to standard, farmed salmon that do not reach market size for at least three years.

Research associate Amy J. Nichols and professor Rex Dunham in the department of fisheries and allied aquaculture at Auburn University, Auburn, Ala., have developed transgenic carp and catfish that grow 20 to 60% faster than standard farmed varieties. They use microinjection and electroporation to inject another copy of a fish growth hormone gene into fertile fish eggs. The growth of the resulting modified carp and catfish is stimulated by extra fish growth hormone.

Transgenic fish are wild types or nearly so, often created from eggs hatched from gametes collected in the wild, so they are fully capable of mating with wild fish. Consequently, one important risk associated with transgenic fish is that, if they escape to fresh water or to the ocean and mate with wild fish, they could destroy the diversity of the wild population gene pool.

Such an event occurred in Norway with farmed salmon. Seals occasionally broke the net cages where the salmon were being raised in fjords and some of the salmon escaped and mated with Norway's wild salmon. Because the numbers of wild salmon had already been depleted as a result of acid rain on freshwater spawning grounds, the wild salmon were easily overwhelmed by aquaculture salmon. As a result, says Anne R. Kapuscinski, professor of fisheries and Sea Grant extension specialist at the University of Minnesota, the genes of the wild salmon were homogenized, and Norway lost one of its most important resources, a tremendous amount of genetic diversity in its wild salmon, and the associated commercial and sport fishing industries.

In addition, transgenic fish could also eliminate whole aquatic ecological systems by preying on and outcompeting native species, as many introduced exotic (nonindigenous) fish have done. In the U.S., the introduction of exotic species is responsible for 28 of the 86 endangered and threatened fish species and subspecies.

At U.S. and Canadian research facilities, elaborate precautions are being taken to prevent the release of transgenic fish into the environment. The fish are often raised in ponds covered with nets to keep birds out; enclosed by electric fences to keep muskrats, raccoons, and humans out; and the outlets are fitted with screened drains to prevent the loss of small fish or eggs.

USDA's Agricultural Biotechnology Research Advisory Committee recently developed voluntary performance standards for safely conducting research with genetically modified fish and shellfish. The committee wrote the standards in consultation with fish and shellfish researchers around the world and will publish them in a few weeks. Many researchers in the U.S. have expressed an interest in using the guidelines. The standards may also fulfill a real need in developing nations that lack expertise or resources to develop such guidelines. Eric M. Hallerman, associate professor of fisheries and wildlife sciences at Virginia Polytechnic Institute & State University, Blacksburg, surveyed countries around the world and found that only 12 have specific national policies or regulations on transgenic aquatic organisms.

Commercialization of transgenic fish will occasionally involve some escapees, no matter how carefully pools or net pens are designed. So researchers are planning strategies to prevent the reproduction of transgenic fish. The main strategy is to raise triploid transgenic fish. Instead of having the usual two sets of chromosomes, triploid fish have three sets and are sterile. To produce triploid fish, eggs from diploid fish are given a heat or pressure shock at the appropriate time of development.

But this method has potential drawbacks. Using heat or pressure shock makes only 90 to 98% of the eggs triploid. This problem can probably be solved, however, by using an alternate method to produce triploid fish. The eggs are made tetraploid by using heat or pressure and the resulting tetraploid fish are then mated with diploid fish. The offspring are 100% triploid.

Even if 100% of the transgenic fish are triploid, however, there is another difficulty that has only recently come to light—the possibility that some of them will revert to diploid and thus become fertile. Standish K. Allen, associate professor of marine science at Rutgers University, New Brunswick, N.J., found that when individually tested triploid Pacific oysters were placed in the Chesapeake Bay, many of them reverted to diploid. Each oyster was tested twice—before it was put in the bay and shortly after—and all were triploid. But when tested several months later, 14% had reverted to diploid at one site and 20% at another.

Not enough testing has been done to rule out the possibility that some species of triploid fish might also revert, Allen says. "It's a little difficult to know whether fish will revert because the subject is not well documented in the literature," he says. One of the reasons might be that people have not looked carefully for reversion in many artificially induced triploid fish, he explains.

So much is unknown about transgenic fish that it is difficult to assess what risks their release might pose to the environment. Researchers do not even have a definitive answer to the simple question: Will transgenic fish ultimately grow larger than unmodified varieties? Chen believes that, in general, fish keep growing throughout their lives, so faster growing transgenic fish will end up larger than their standard counterparts. In contrast, Kapuscinski cautions that not enough has been published about the growth of modified fish to provide any general answers to this question.

Another unknown is how the fitness of fertile genetically modified fish would change through the generations. Researchers suspect that some species would become more fit as the generations proceed, while others would become less well adapted. But there are almost no experimental data. Some experts believe that nearly all varieties of transgenic fish would have a hard time surviving and reproducing in the wild.

William M. Muir, a population geneticist at Purdue University, is studying transgenic medaka, small fish with a 10-week generation time, to see how their fitness changes over generations. Muir does not yet have results on his medaka but he believes that some transgenic fish if allowed to reproduce will get more fit through the generations. "I suspect that when we first put the transgene in the fish that we are not seeing its worst face," Muir says. "Originally, the fish may be out of sync with its physiology, but natural selection may help the organism cope with this new functionality, and probably over time it will get more aggressive or more fit."

In China, fertile transgenic fish are being raised in large open ponds without nets or fences to keep out birds, or humans who might want to steal the fish and grow them in their own ponds. Marc Welt, assistant professor of molecular biology at Xavier University, New Orleans, who spent a month at Chinese facilities that develop transgenic fish, says this lack of security docs not present an ecological risk because China's environment is so degraded that it supports few wild fish.

Alvin L. Young, director of the Office of Agricultural Biotechnology at USDA, agrees. "The Chinese have had a very disrupted ecosystem for 2,000 years," he says. "So how does one track the adverse effects of transgenic fish from the normal things that have gone on there for centuries?"

Hallerman admits that China no longer has commercial quantities of wild fish. For centuries, however, the Chinese have employed the genetic diversity in wild fish to replenish and revitalize the stocks used in aquaculture. If transgenic fish homogenize the wild fish in China, the nation will have lost a valuable source of genes, he says.

"There is a lot of talk now about how aquaculture, especially with transgenic fish, is a great answer because most commercial fisheries and wild populations are declining and increasingly are going extinct," Kapuscinski says. "But it may not be in the nation's best interests to totally turn our backs on wild populations and even on commercial fishing, and rely on aquaculture, because then we may be tempted to allow the genetic diversity of the wild populations to erode even further. And these are the populations we need to go to periodically to get new genes to infuse into our aquaculture stocks."

GENETICALLY MODIFIED PLANTS

Scientists' views of transgenic plants, though not as polarized as their views about modified fish, vary greatly. USDA officials essentially consider transgenic plants as no different from standard varieties. Therefore, the plants require no more care in research and commercialization than standard varieties. A number of other scientists believe that genetically modified plants may pose more risks than traditionally bred crops.

So far, transgenic crop plants with final regulatory approval include two varieties of tomato with delayed ripening, cotton resistant to the herbicide bromoxyril, insect-resistant potato, insect-resistant corn, herbicide-resistant soybean, virus-resistant squash, tomato engineered to have a higher solids content for easier processing into sauces, and canola with the oil composition altered to be high in lauric acid.

Monsanto developed the potato that resists the Colorado potato beetle. Monsanto spokesmen and most officials at USDA and EPA believe this crop is almost entirely risk free and, in fact, very beneficial to the environment. The crop will increase yield and protect workers, consumers, and the groundwater by eliminating or cutting back on the use of toxic insecticides.

The insect-resistant potato is created by using a modified virus as a vector to insert into the plant a truncated gene for the production of a potent insect toxin from *Bacillus thuringiensis* (Bt). The truncated gene is from the variety of Bt that produces a toxin for the Colorado potato beetle.

Organic farmers and other growers have employed bacterially produced insect toxins derived from Bt and the spores of Bt itself for 30 years as a tool in integrated pest management. Used in this manner, Bt stays on the plants for only a few days at a time before it is broken down by sunlight. Nevertheless, some insect resistance to Bt has already begun to show up in grain silos and in Hawaii where farmers have used it heavily. Unless precautions are taken, insects might also become resistant to the Bt plants because the Bt toxin, which is very similar to that produced by the bacteria, is present in the plants throughout the growing season and can't be broken down in sunlight.

Because of this potential problem, Monsanto worked with EPA to develop a strategy that farmers can use voluntarily to ward off resistance. The plan involves making refugia available, that is, growing in the same field with the transgenic plants or nearby fields some unmodified plants that insects can eat. Thus, some nonresistant insects will be preserved to mate with resistant ones. The strategy also involves monitoring the crop so that if resistance develops in one area, action can be taken to keep it from spreading.

An EPA science advisory panel of top entomologists approved the plan, says M. Lisa Watson, director of public affairs at Monsanto. However, UCS's Rissler says the resistance plan bears more resemblance to a work-in-progress with many unanswered scientific questions than to a concrete, well-worked-out strategy.

Early this month, EPA gave separate approvals to Mycogen and Ciba Seeds, a division of Ciba-Geigy, to market corn hybrid seeds, genetically engineered with a gene from Bt, to fight damage from the European corn borer. The seeds were developed in part by a collaboration between Mycogen and Ciba Seeds. Both companies have developed grower education materials to help farmers ward off insect resistance. To further reduce the likelihood that the corn borer will develop tolerance to a single control mechanism—the Bt toxin—the seeds Mycogen will be selling have both the Bt-based gene and a natural resistance gene acquired from a wild relative of corn and incorporated into the seed with traditional breeding.

UNAVOIDABLE RESISTANCE

Some scientists doubt that resistance can be avoided for more than a few years, even with carefully designed strategies, when thousands of acres of commercially grown crops are producing the Bt insect toxins throughout the growing season and when each crop has only one type of Bt. "Agricultural entomologists are well aware that the value of a 'single-bullet' approach to insect control is likely to be short-lived," says Allison A. Snow, associate professor of plant biology at Ohio State University.

Charles M. Benbrook, a private agricultural consultant based in Washington, D.C., points out that if insects become resistant to Bt plants, they will also be resistant to the Bt used by organic farmers and this Bt will become useless.

Nearly all researchers agree that growing herbicide-tolerant plants in the U.S. will probably result in the use of less herbicide on some crops, at least in the short run, and, in most cases, the use of more environmentally friendly herbicides. And they do not expect herbicide-resistant crops with no wild relatives to pose a major risk to the environment.

The herbicide-resistant soybean developed by Monsanto will allow greater use of the herbicide glyphosate (Roundup), which breaks down quickly in soil, and will enable farmers to use fewer herbicides. In some areas, this soybean will allow elimination of all herbicides with detrimental properties, says Frank S. Serdy, director of regulatory affairs at Monsanto.

Critics do worry that growing Calgene's bromoxynil-tolerant cotton on U.S. cropland will lead to the use of more bromoxynil—a suspected cause of birth defects. On the other hand, according to Carolyn E. Hayworth, manager of public relations at Calgene, bromoxynil-tolerant cotton reduces by half the number of applications of herbicide required and decreases by about 40% the total amount needed.

Researchers are also concerned about the long-term effects of herbicide-resistant crops. "Evidence suggests the effects on the environment of herbicide-tolerant crops are not clear-cut," says Roger P. Wrubel, assistant professor in the department of urban and environmental policy at Tufts University, Medford, Mass. In the aggregate, the crops could help phase out environmentally damaging herbicides and reduce overall herbicide use, he says. Or they could promote reliance on chemicals that lead to weed resistance because they will encourage the rise of a few broad-spectrum herbicides.

Some experts warn that herbicide-resistance or insect-resistance genes could spread from the transgenic crop to its wild relatives and create new weeds that are especially difficult to control, or that the crop itself could become a weed. Corn, potato, and soybean do not have wild relatives in the U.S., but canola—the rapeseed plant—has six wild relatives in the mustard family (Brassica), some of which are already rather troublesome weeds in regions where canola is grown. Monsanto, Hoechst, and Rhone-Poulenc are developing herbicide-resistant canolas, which will probably be commercialized in a few years.

Jack Brown, a plant breeder geneticist at the University of Idaho, has been studying the spread of herbicide-resistant genes from transgenic canola to its relatives in the mustard family. He has found that the resistance gene moved through pollination from the canola to a small fraction of one type of Brassica weed (birds-crape mustard), and that gene then moved to wild mustard, a weed that is much more troublesome. Even though the fraction of weeds affected was low, so much herbicide-resistant canola will soon be growing in the U.S. that a large number of herbicide-resistant weeds could potentially be created in a few years, he says. "We need to know much more about gene movement in nature and what steps to take if gene introduction into weeds occurs," Brown says. "Modern agriculture has happened at a price," he warns. "We should learn from our experiences what disasters could befall us before we jump into large-scale production of gene-modified plants."

USDA's Young admits that herbicide-resistance genes move from canola, but he does not think this will have "great ecological consequence. We just asked Brown to come up with a worst case scenario," Young says. "But that doesn't say it would be a bad case in reality."

BIOENGINEERED TOMATOES

The delayed-ripening tomato developed by Calgene, called either Flavr Savr or MacGregor, is already in supermarkets, and the one designed by DNA Plant Technology, Endless Summer, has been test marketed in New York State. Except for those researchers who are opposed to all bioengineered foods, few expect these tomatoes to have a harmful effect on the environment or on the development of weeds.

Most tomatoes are picked in the green hard-ball state so that they can be shipped long distances without rotting. Both of these transgenic tomatoes are designed so that they ripen more slowly and can be picked later in the ripening process. Calgene's tomato has an extra gene—a reverse copy of the gene responsible for an enzyme, polygalacturonase, that breaks down cell walls. As a result, Calgene's tomato softens more slowly. DNA Plant Technology's tomato has a gene that controls the enzyme 1-amino-cyclopropane-1-carboxylic acid (ACC) oxidase, which is necessary for the production of ethylene, one factor that makes a tomato soft.

A crookneck squash with virus resistance developed by Asgrow Seed Co., Kalamazoo, Mich., is now being grown and will be sold as frozen squash. It is modified by inserting into the genome genes coding for viral coat proteins of two viruses that often infect squash. For reasons not fully understood, the expression of low levels of viral coat proteins in the plant prevents infection by the original virus. The yield of the Asgrow squash is about five times greater than it is with standard seed because the virus that reduces yields cannot infect the crop, BIO's Godown says.

Despite its advantageous yields, the squash is a subject of scientific controversy. Some experts worry that when the squash is infected with other viruses, recombinational events could occur that would generate new viral strains. Plant biologists Anne Green and Richard Allison of Michigan State University found that recombinational events occurred in transgenic cow pea plants modified with a virus coat protein. When the plants were inoculated with a different virus, viral RNA or DNA recombined with genetic material from the invading virus to form a new more virulent strain (*Science*, 263, 1423 [1994]). However, it is not known whether this would happen with squash plants. Nor is it known whether the widespread use of transgenic crops would increase the rate at which pathogens evolve, Snow says.

Calgene's genetically engineered canola, which will produce large amounts of lauric acid, was modified with one gene from the California bay tree. The gene shuts off fatty acid synthesis at 12 carbons instead of the usual 18-carbon length normal for the plant. The oil from some of the plants is more than 40% laurate. Currently, laurate oils are obtained from imported coconut and palm kernel oils produced in Southeast Asia.

One risk some see in high-laurate canola is that after harvest, the seeds will get mixed up with the seeds from the unmodified edible crop, since they are identical. As a result, some people might unknowingly eat high-laurate canola, which in large quantities is hard to digest. The problem will be further exacerbated when other types of transgenic canola, such as one now under development producing the industrial oil erucic acid—a feedstock for making nylon 13-13—are also grown commercially.

Another risk is an economic and social one. High-laurate canola could displace some of the coconut and palm kernel oils used in coffee whiteners and hair products, and as a result it could displace some exports from Southeast Asia. So rather than feeding or helping economically developing nations, high-laurate canola may simply cause more poverty among farmers who now grow or harvest coconuts and palm trees, say some experts.

Hundreds of field trials of transgenic plants are now being carried out each year, and for most of them, USDA requires only that the researcher give notification. It does not usually require a permit, nor does it request data showing a probable lack of harmful environmental effects. "Using the data from all the field studies done in the U.S., we've concluded that transgenic plants represent minimal risk," says USDA's Young. "They are no different from traditionally bred plants." Such plants do present some risk "but we have in place a very sophisticated agricultural system that monitors new varieties. Every time a new variety is introduced, researchers from the Agricultural Research Service and companies track those varieties to see if there are any unique problems," he adds. Transgenic plants will be monitored under this system as well.

UCS's Rissler disagrees. USDA's field tests of transgenic plants have not been conducted in a way that could assess environmental risks, she says. "We did an analysis of the field test reports about a year ago," she explains. "They showed very little attention to the experimental analysis of risks. USDA relies too much on intellectual arguments and not enough on field data."

Sheldon Krimsky, professor of urban and environmental studies at Tufts University, also analyzed USDA's field trials. He says USDA does not require sufficient precautions to prevent the spread of genes from transgenic plants to wild relatives in its field trials. USDA accepts the level of cross pollination acceptable to plant breeders, which provides only about 95 to 98% pure seed, he explains. This means that the department is allowing in its field trials about as much pollen dispersal from transgenic plants as plant breeders would permit in producing seed (*BioScience*, 42, 280 [1992]).

Those who see important risks in agricultural biotechnology say USDA should spend less money on biotechnology research and more on sustainable agriculture research. In the past five years, USDA has spent a total of about $28.5 million for sustainable agriculture research and about $640 million for biotechnology research. And for risk assessment of biotech research, USDA has spent only 1%, or $6.4 million, of the biotechnology budget. "We think that biotechnology is not a very good long-term solution to problems in agriculture—that, in fact, sustainable agriculture methods are a better long-term investment," says UCS's Rissler.

Many biotechnology companies have branches or joint ventures around the world and are prepared to introduce transgenic plants globally. For example. Pioneer Hi-Bred International, Des Moines, Iowa, which has large biotechnology investments, has branches in 31 countries.

Transgenic plants can be a mixed blessing for developing nations. Plants that are more nutritious or productive than standard varieties could be one factor that helps feed the world: China, for example, is now cultivating thousands of acres of Bt rice, which may increase its rice yields if insects do not become resistant to the crop.

On the other hand, Rissler explains, if genetically modified crops are grown in areas where they have not been adequately field tested, they could have unexpected harmful effects, such as the spread of genes to wild relatives or the creation of new weeds or new viruses from virus-resistant plants.

For those countries that are so-called centers of diversity—areas that have high concentrations of traditional crop varieties and their wild relatives—transgenic plants pose special risks. Crop breeders from around the world often go to these areas to find new genes for disease or insect resistance. Crop varieties are already disappearing rapidly from these areas because of the green revolution and population pressure. For example, according to the Ottawa-based Rural Advancement Foundation International, about 30,000 rice varieties have been lost in India, largely because farmers have abandoned traditional varieties in favor of the more productive seeds introduced by the green revolution.

Transgenic crops may accelerate this loss of species by gene spread to wild relatives. If a transgene gives one wild relative an advantage, such as insect resistance, this relative may out-compete the others or itself become a weed.

Furthermore, some bioengineered products could wipe out the major exports of some developing nations. For example, a genetically modified bacterium is under development that produces vanilla flavoring. If this bacterium is commercialized, say many researchers, it could eliminate markets for vanilla beans, one of Madagascar's major agricultural products.

RECOMBINANT BOVINE GROWTH HORMONE

After more than a decade of controversy, recombinant bovine growth hormone (rBGH) was first marketed in February 1994 amid threats of boycotts by many consumer groups. At the time, a number of supermarket chains and milk wholesalers said they would not accept milk from cows injected with rBGH, and opponents predicted that milk consumption would go down.

Since then, the controversy has subsided in most regions, and some businesses that said they would not accept rBGH milk now do so. Total milk consumption increased 4% in the first 10 months of 1994, compared with the same period the previous year, and fluid milk consumption rose 1%.

Farmer acceptance is also fairly high. Monsanto estimates that farmers who purchase rBGH—about 11% of U.S. dairy farmers—own about 30% of total US. dairy cow herds and use it on some of the cows in their herds. Cows injected with the recombinant hormone produce about 10% more milk per day than cows without the hormone.

Thomas Lyson, professor of rural sociology, and coworkers at Cornell University surveyed all the dairy farmers in one New York county and found that 40% use rBGH. The users tend to be the larger farms with other advanced technologies, such as computers. However, 28 of the 46 respondents said that rBGH is "not a good thing for the (dairy) industry in New York or in the U.S."

One of the opponents' concerns is that rBGH will cause increased mastitis in cows and lead to greater use of antibiotics and higher levels of antibiotic residues in the milk supply. Critics are also concerned that treated cows will experience reproductive problems. But based on the 806 complaints FDA received about the hormone as of March 1995, the agency said in a press release, "FDA does not find any cause for concern." However, the agency is continuing to monitor herds treated under field conditions.

Despite U.S. commercialization of rBGH, opposition remains fairly strong in Europe. The European Union (EU) has in place a moratorium on the use of rBGH. It contends that the hormone may put small dairy farmers out of business, that it may have adverse effects on cow health, and that there are unresolved questions about human health, such as the possible adverse effects of increased levels of insulin-like growth factor-I (IGF-1) present in the milk from treated cows. And last month, the Codex Alimentarius Commission, the World Trade Organization's intergovernmental panel that develops global standards to protect health and the environment, voted to delay action on rBGH until it meets again in 1997. Technically, the panel could overrule EU, so this decision to delay action means that EU's ban on rBGH remains in place.

LABELING PROS AND CONS

Agency scientists and observers also have widely divergent opinions about whether genetically engineered food should be labeled. FDA has decided not to require labeling unless the food contains a known allergen or its composition is substantially different from their standard food. For example, high-laurate canola is labeled because it has a high concentration of lauric acid. But under FDA's policy, most genetically engineered food will not be labeled.

Philip Bereano, a professor of engineering technology and public policy at the University of Washington, Seattle, strongly disagrees with FDA's labeling policy. "Consumers have an absolute right to know the processes by which their foods are made," he says. And there are precedents for that, he adds. Even though many kosher products are chemically the same as non-kosher products, the process by which they are made is identified, he says. Dolphin-free tuna is a similar example.

FDA's Maryanski counters: "FDA is not saying that people don't have a right to know how their food is produced. But the food label is not always the most appropriate method for conveying that information. The Food, Drug & Cosmetic Act is not a comprehensive right-to-know act."

Another argument is one over allergens. Not all allergens are known, and a genetically modified food could contain an unknown allergen, Krimsky says. Therefore, if a problem arises, he adds, it would be much easier to track epidemiologically if the food is labeled.

Maryanski admits the question of allergens in transgenic food is a difficult one. FDA convened a conference on the issue, he explains, and decided that it was very unlikely that people would become allergic to the possible allergens in genetically modified food, largely because they are expressed at such low levels, and because the vast majority of allergic people are sensitive to a very limited number of proteins.

Rita R. Colwell, a marine biologist who is president of the Maryland Biotechnology Institute at the University of Maryland, Baltimore, sees a real urgency to use agricultural biotechnology. For example, she says, the world's oceans can produce at most 100 million metric tons of fish annually, and the harvest has dropped to 80 million metric tons. Soon, the world's population will require 135 million metric tons of fish. "It is no longer a question of should or shouldn't we raise transgenic fish," she says. "It is a necessity."

Simon G. Best, chief executive officer and managing director of Zeneca Seeds, Wilmington, Del., has a similar view of transgenic crops: "Without biotechnology, we will simply fail to meet the challenge" of dramatically increasing the availability of affordable basic food.

In contrast, David R. MacKenzie, who until last month was head of USDA's biotechnology risk assessment program and is now executive director of the Northeastern Regional Association of State Agricultural Experiment Station Directors, urges much more restraint before transgenic organisms are commercialized on a large scale. He is convinced that transgenic organisms may present some risks. "We're still unsure of what is going to happen with transgene spread into wild species," he says. "Some people say it doesn't matter because they are wild species and who cares. But people who do care about the ecology of natural systems get very upset at the idea that we would be spreading these genes.

"The chorus of people, especially at USDA," he continues, "who say there is no risk in biotechnology subscribe to what I think is the wrong notion that DNA is DNA. I think there are barriers between species. I think there are aspects of these revolutionary [transgenic] designs that we don't understand yet. Before we start messing around too much, we ought to understand what happens when we jump these barriers, not just put things out there and see what happens."

NOTE

Reprinted by permission from *Chemical and Engineering News*, August 24, 1995.

6

Risk and Risk Management

J. R. S. Fincham and J. R. Ravetz,
in collaboration with a working party of the
Council for Science and Society, Canada

RISK ASSESSMENT

The previous chapters may have conveyed the impression that the hazards of genetically engineered organisms as now handled are not such as to cause serious immediate concern. This is generally correct; by itself genetic manipulation does not normally confer qualities that will make an organism more harmful to mankind or to its environment. When genetic manipulation is being done under controlled laboratory conditions, the hazards of such procedures, while naturally requiring care, have not proved to be a significant addition to those already presented by the organisms being used. Of course, when the organisms are designed to be robust in order to survive in the wild, or when they are used on a very large scale in agricultural practice, then new problems may arise. Our experience in coping with these is limited; and so when we consider the problems of regulating this new sort of scientific work, we become aware of the depth of our uncertainties about the processes and their containment. It is in such cases that the skills of risk assessment will be deployed to their fullest extent.

The Precedent from Genetic Manipulation

In considering how to assess the risks associated with the release of genetically engineered organisms or viruses, it may be useful to recall the precedents set by the Genetic Manipulation Advisory Group (GMAG) and their successors, the Advisory Committee on Genetic Manipulation, which is now an agency of the Health and Safety Executive. The system in current use has been unchanged in principle since 1979, when the GMAG adopted a method of risk assessment that had been devised by the eminent molecular biologist Sydney Brenner. Broadly speaking, the effect was to place less emphasis on the supposed inherent risk of the DNA itself and more on the scenario through which it could become an actual threat to health.

Attention was focused on three considerations labeled access, expression, and damage. The first of these headings covered the likelihood that the recombinant DNA would escape from the system of containment and gain access to the human body. The access factor was rated most highly when the DNA was cloned into a bacterium, such as wild-type *Escherichia coli,* that could readily colonize humans, lower when the cloning vehicle had no such natural ability and lowest when it had been "crippled" by genetic modification. Expression referred to the ability of the DNA to get itself transcribed and translated into some product. The expression factor was highest when the cloning vector had been deliberately designed to promote expression of genes inserted into it and lowest when the vector was designed to preclude expression. Finally, the damage factor was a measure of the ill effect of the DNA on human health if it both gained access and was expressed. It was obviously highest where the product was known to be a toxin and lowest where it was known to be harmless.

The problem with this very logical scheme was the great difficulty of quantification of any of the three factors in any particular case. This difficulty was dealt with in a very bold and arbitrary way. Instead of attempting to make precise estimates of risk, which would certainly have been spurious in view of the great difficulty or impossibility of empirical tests, each experiment was assigned, with respect to each of the three criteria, to a very broad and approximate category. Virtual certainty that the DNA could gain access, or be expressed, or pose a danger was in each case represented by a factor of 1. Below this there were different levels of supposed improbability represented by the factors 10–3 (one in a thousand), 10–6 (one in a million) or 10–9 (one in a thousand million). Then these factors were multiplied together to give an overall measure of the chance of the DNA actually damaging human health. If the product came to 10–15 (that is, one chance in one million billion) or less, the proposed experiment was judged safe under conditions of good microbiological practice—which, incidentally, is defined as something a good deal more careful than most molecular biologists had been accustomed to before genetic engineering came on the scene.

It is quite easy to scoff at this system. The thousandfold difference in hypothetical risk between experiments placed in adjacent categories could not possibly be taken literally, and the idea that one needed to take any precautions at all when faced with estimated odds against danger of one million billion to one might seem absurd—anyone who took this order of risk seriously would hardly dare to draw a single breath. But the GMAG system did embody two important principles that it would be well to keep in mind in the new context of environmental release. The first is that the probability of an outcome that depends on a sequence of events is the product of the individual probabilities of all the steps in the sequence. The second is that large uncertainties in calculation need to be compensated for by large safety margins. The system provided a fairly rational basis for deciding that some procedures were likely to be far more, or far less, risky than others, and the built-in safety margins appeared sufficiently broad to accommodate the large uncertainties in determining the absolute magnitudes of any hazards. With hindsight, we can say that, however arbitrary it may have seemed (and still seems), it was a social and political success. It provided a formula for compromise between alarmism and complacency—a formula that allowed work to proceed safely and for experience to be gained. It was a system that led to decisions that nearly everyone involved felt were adequately cautious.

Extension to Released Organisms

However well it worked for laboratory experiments, the GMAG formula can hardly be applied unchanged to environmental release. The very term "release" implies access to the environment and although farm animals can he effectively fenced in and thus not really released, it will hardly be practicable to apply effective containment to crops growing in large open fields of bacteria inoculated into the soil of such fields. One can, however, distinguish between release of the engineered organism as such and spread of modified genetic material through hybridization with wild species. An informed estimate can be made of the chance of this spread of genes, taking into account the availability of related wild species or strains capable of receiving genes from the engineered variety by hybridization, the viability and fertility of the hybrids where these have been tested and the likelihood that the transferred novelty would confer a competitive advantage. In the case of an engineered bacterium, one would want to know whether there was any known way, for example through plasmids transmissible between different strains and species, that could effect transfer of novel genetic material. These complex considerations could all contribute to a rough estimate of a secondary access factor.

The expression factor is hardly relevant in the case of an organism engineered for environmental release; if its genetic novelty were not expressed there would be no point in releasing it. Thus, if we are considering the primary product of genetic engineering designed for environmental release, everything rests on the danger factor; in the case of hypothetical products of secondary spread the overall risk will be the product of the danger and secondary access factors.

It is hardly possible to propose even a three-orders-of-magnitude estimate of danger unless one has some hypothesis as to how danger might arise. The obvious examples of projects involving foreseeable risks are those for making new kinds of live vaccines. Assessments of risk in this area will carry more weight when the roles of particular bacterial or viral genes in determining pathogenicity are better understood. Meanwhile, regulatory agencies will have to make cautious and approximate assessments on the basis of the best evidence available. Their task will be the more difficult in that the hypothetical risks of new vaccines will have to be balanced against the prospect of meeting particularly urgent human needs.

The genetically modified farm animals or crop plants at present under consideration present a different kind of problem. It is difficult to see how any of them could be dangerous in themselves unless they have the potential to release pathogenic viruses, a possibility that can and should be avoided. The main perceived risk, which applies to crop plants and hardly at all to animals (except perhaps fish), is that some of the new types will turn out to be invasive, swamping important crops or natural habitats. What has to be assessed here is the likelihood of invasiveness, and the ecological or agronomic damage that might result from it. To some extent this question can be approached experimentally through the use of contained "microcosms," but these can never simulate the enormous variety of the outdoor environment. Furthermore, to the extent that the danger is thought to reside in transfer of genes to wild species, it may not be easy to guess the kind of hybrid that it would be important to test.

Coping with these problems requires knowledge rather remote from the molecular biology that has dominated discussion hitherto. The scientists best qualified to judge these questions are biologists who have made a close study of the particular species groups to which the organisms under consideration belong. Knowledge of comparative morphology, taxonomic relationships, life histories and environmental preferences must be the best basis for judging how a particular genetic modification is likely to affect competitive ability or invasiveness, either of the organism primarily modified or of possible secondary recipients of engineered DNA. In other words, we would look to taxonomists and students of genetic variation in relation to the fine detail of ecology. This is the kind of knowledge that tends to be included in the category of "natural history," a term used nowadays, more often than not, as a term of disparagement. It attracts very little research funding or student interest, and many university biological departments seem virtually to have given up the struggle to keep it going. It seems important to reverse this trend if we are seriously concerned about environmental hazards of engineered organisms—or indeed about the many other kinds of environmental degradation that are arguably more immediate and serious.

In our view, the discussion of the risks of genetic engineering has been too much dominated by an over-dramatic view, verging at times on the apocalyptic. A comparatively moderate opinion sometimes expressed is that we are faced with a very small chance of an extremely large catastrophe. This "zero-infinity" postulate has a certain ironic intellectual appeal, the point being that the product of the two terms can take any value that one might like to imagine. Reality, we believe, is likely to be altogether more mundane. It seems more realistic to think in terms of fairly small chances (perhaps roughly calculable) of less-than-catastrophic problems that one could overcome or live with. The risks in prospect are likely to be no different in kind from those that have always been associated with the introduction of the novel products of plant and animals breeding, or new vaccines. That is not to say, of course, that they do not need to be anticipated and guarded against.

What May the Unforeseen Provide?

It is sometimes said that predictions of the probable harmlessness of genetically manipulated organisms are cast in doubt by the inevitability of new scientific discoveries. Of course new discoveries will be made in, say, the ecological genetics of microorganisms (a subject almost called into being by this very consideration). How can we guess what effect they might have? We can perhaps gain some clues by looking backwards, at findings in the not too distant past which were unforeseen and which expanded our ideas of what is possible in the field of genetics, either molecular, microbial or ecological.

We can divide these findings into two types. The first has to do with the mechanisms that exist for the propagation and expression of the genetic material, while the second does not upset our ideas about basic mechanisms but has to do with the contingent facts of what actually happens in the real world.

The first class, of "fundamental surprises," includes the following:

1. The existence of intervening sequences ("introns") in the genes mainly of higher eukaryotes. These are transcribed along with the coding portions ("exons") of the gene, but are spliced out of the messenger RNA before it assumes its mature, translatable form.
2. The capacity of RNA to adopt a catalytic role in biological reactions; this was once regarded as the special province of proteinaceous enzymes.
3. The role of "reverse transcriptase," the RNA-dependent DNA polymerase that copies RNA into DNA in the life cycle of certain viruses.
4. The non-universality of the genetic code, especially in the mitochondria of eukaryotes.

With hindsight, we can see that all of these extend the known characteristics of certain classes of macromolecules, rather than providing something wholly new. The special properties of RNA are inherent in the "adaptor" role of a certain type of RNA, the amino-acyl transfer RNA, predicted in the late 1950s. The ability of RNA to be copied into DNA and the (rather limited) non-universality of the code are not incompatible with the complexities of the biochemical events in base pairing. The likely impact of such findings on our view of the proclivities of DNA in the natural environment would seem to be limited.

The second class, of "contingent surprises," is very varied; a few examples are as follows:

1. The existence of "transposable elements," blocks of DNA that change location in the genetic material. That their discovery is not new (for plants it dates from the 1950s, for bacteria from the late 1960s to early 1970s) should not blind us to their originally apparently revolutionary quality.
2. The ability of the Ti plasmid DNA of the bacterium *Agrobacterium tumefaciens* to transfer itself into plant cells, followed by the insertion of a segment ("T-DNA") of this plasmid to enter plant chromosomal DNA.
3. More recently discovered is the ability of some bacterial plasmids to move between widely diverse bacteria (the so-called Gram-positive and Gram-negative types) and even to yeast.[1]
4. Similarly, we have recently become aware of the presence of unexpectedly large numbers of bacterial viruses in natural bodies of water. These must exert strong selective pressure on the bacteria, and may in addition mediate DNA transfer between them (by the process called "transduction").[2]

Unlike the class of "fundamental surprises," the "contingent surprises" certainly have implications for our theme. But these implications can work two ways. We can point to the fact that since these observations involve the remarkable natural movement of genetic material, within cells and between them, they suggest that the diversity of natural genetic transfer is so vast that nothing we can do is likely to make much difference. Or we can look at the awesome amplifying power of such systems to spread throughout the natural world any construct made by our in vitro methodology that is allowed to escape. The dichotomy in viewpoints seems inescapable. Thus the problems of the risks of released genetically manipulated organisms cannot be definitely solved along the same lines as those of the classic experimental and field sciences.

MANAGING THE UNCERTAINTIES
OF RISK ASSESSMENT

Special Problems of Biohazards

Biohazards present significantly greater difficulties of assessment than hazards from physical plant or even natural accidents. We may appreciate the reason for this difference in terms of the idea of information. Man-made machines and living organisms both carry information about their own functions. The unique feature of organisms, however, is that their information is self-propagating and so can produce more organisms of the same kind. Moreover, as we now know, it cannot be assumed that the information will always remain within the bounds of a single kind of organism. Especially in bacteria, informational pieces of DNA can be transferred between species at low but significant frequencies through the agency of plasmids or viruses, or even without the help of such vectors. Some of the species in receipt of information could be harmful to us. The microbial flora of natural and agricultural environments is so diverse that there is little hope of enumerating and quantifying all the possible pathways of information transfer; and it is even quite difficult to monitor empirically any particular DNA fragment after it has been released. Most probably it disappears, but who can be sure that it is not hiding out somewhere ready for a resurgence when opportunity arises?

Here we are very far from the controlled laboratory experiment upon which "hard" science is traditionally based. In an inevitably poorly defined situation it is very easy to be guided by ill-founded or even self-interested perceptions. If debate over the hazards of genetic engineering, and over planned release in particular, is to have any scientific content, we must be clear about the crucial differences between the style of laboratory science and that involved in controlling hazards.

Almost by definition, a laboratory experiment cannot be conducted in an ill-defined situation. Even if it is exploratory, or motivated by speculative theories and ideas, in its specification and techniques it must be precise and controlled, or else it is valueless. Normally, there will be a hypothesis being tested; and that must be stated as clearly as possible. At the end of a successful experiment there will be a few well-defined alternative interpretations and conclusions, which can be presented for review and criticism.

All this is possible when the substances or organisms in the system under investigation are well defined and accurately reproducible, and can be subjected to controlled variation taking one variable at a time. In the case of environmental hazards, very few of these conditions will hold. Although there have been few cases of damage to the general public arising from biological technology, the problem is a familiar one with chemical and radioactive agents. Sometimes there is a situation approaching a local epidemic, and it is only selfishness and stubbornness of the offending institutions that prevents a quick and clear identification of the cause; such was the case with the incident at Mishimata in Japan, where a discharge of mercury into the sea resulted in the poisoning of people who ate locally caught fish. But more frequently, the evidence of the existence of a problem is itself debatable, and the assignment of a cause becomes very difficult indeed. The techniques of statistical epidemiology must be applied, and in difficult cases these may be as controversial among practitioners as they are obscure to the public. The case of the "leukemia clusters" found in the vicinity of some nuclear power plants (but not only in such places) is a case in point.

Understanding Public Concern

The specification of possible future hazards from an untried technology is a still more uncertain task whose scientific basis is very thin indeed. The skills and presuppositions that make for success in the laboratory, the concentration on testable hypotheses and the dismissal of vague theories, can sometimes become counterproductive. A fine balance of attitudes becomes necessary; while it is impossible for scientists to chase down every scare story about disease or pollution, there have been important cases when evidence that was "merely anecdotal" eventually provided clues to a serious problem. One such was "Lyme Disease" in Connecticut, where an epidemic of rheumatic fever among children became a matter of public alarm while still unrecognized by the medical profession: it turned out to be due to a spirochaete parasite transmitted from deer by ticks. Also, the attitude of demanding scientific rigour in an argument about a suspected hazard in order that it be taken at all seriously may be misunderstood by a worried public as representing a lack of concern. For the public's worries may on occasion be ill-formed and perhaps also ill-informed; but they are not necessarily silly or irrational because of that. A concentration of focus on the hazards that can be clearly specified, and a dismissal of the others, can be interpreted as scientific hubris of the sort that has on occasion produced unacceptable environmental damage. Yet to expect scientists to go through the motions of assessing every conceivable hazards, however speculative or remote, is to misunderstand the workings of science. The need is for a balance, in which scientific uncertainties are managed by prudence and tact, and in which the differing perceptions of the different parties are recognized and respected.

It is likely that much of the acrimony that afflicted the first American debate over the hazards of recombinant-DNA research was caused by this difference of perceptions. The scientists had developed techniques whereby they had solved a host of new and challenging problems, and they saw no reason why it would be different for any of those to be encountered in the safe operation of these techniques. Their critics felt otherwise; they had been told of scientific problems of a novel environmental and epidemiological character, which the scientists themselves had deemed sufficiently serious to justify a research moratorium. With their heightened concern for ethical and theological issues, mixed with traditional town-gown politics, they felt justified in withholding their consent to the scientists' reassurances. If we are to get beyond that sort of mutual incomprehension, which could (in the present political climate) have disastrous effects on the development of biotechnology as a force for progress, we will need to think clearly about the management of uncertainty in this most difficult area.

The first step is for all sides to appreciate that, in this area, there can be none of the certainties that have traditionally been associated with scientific knowledge as it has been taught and popularized. For the management of uncertainties in this new area of science, so different from traditional laboratory work, we could do well to consider principles and techniques developed in other fields of practice. The first is a very general one, suggesting an enrichment of what is already enshrined in the practice of science. This is the maintenance of a critical presence, so that the quality of the work is ensured. In science this is accomplished through peer review, refereeing, and public discussions at seminars and public meetings. When scrutiny is focused on research projects and results, as in the case of science, it is inevitable and proper that only those with a trained technical competence should be part of the process. But when, as in the present case, the questions are of what might somehow happen, in a variety of areas of human experience (touching on issues of social affairs and ethics) and with greater or lesser degrees of probability or plausibility, the relevant competence is neither so tightly defined nor so exclusive in its scope.

In such circumstances, the presence of lay persons on regulatory agencies can be strongly defended. Professional scientists may be over-influenced by the intellectual excitement of successful technology and over-confident about the degree to which they are in control. And even where the scientists do have the necessary objectivity and detachment it would not be reasonable to expect the public to take this on trust. Lay representatives should not, of course, be selected for their lack of knowledge; it is clearly desirable that they should be able to understand what the scientists are talking about. It is only necessary that their first concern should be seen to be with human risks and benefits rather than with the advance of knowledge as such.

The problems of regulating hazards are in some ways similar to those encountered in the ordinary courts of law, where decisions must be made under conditions of uncertainty, and where scientific information, however certain in the abstract, is sifted like any other sort of evidence. The jury is not expected to deliver its verdict as if it were an indubitable truth; an irreducible element of doubt is explicitly allowed for. But then this doubt must be operated in relation to a burden of proof, so that (in our tradition) the accused is deemed innocent until (effectively) proved guilty. Such a principle is not foreign to scientific practice; whenever a statistical test is made, in relation to a particular level of a "confidence limit," a relative evaluation of the costs of the two possible errors (false-positives and false-negatives) is being implicitly invoked. When in a debate on regulatory policy the perspectives of laboratory-trained scientists and environmental activists are in direct conflict, this may be understood in terms of the implicit burden of proof which the two sides are applying to their uncertainties. Put most simply, is the process deemed safe until shown to be dangerous, or vice-versa? Where the issue of differing burdens of proof is to be made clear and explicit, then at least there would be some awareness of why conflict was occurring, rather than the mutual incomprehension that so frequently occurs in debates on risk and pollution issues.

Further, without attempting to copy in any way the management of uncertainty in the judicial process, we may adapt another of its principles in relation to the complexity of evidence. Working scientists know that the raw data from their equipment do not automatically constitute "facts"; these must be worked up (as by statistics) and then interpreted theoretically, before they can function as evidence in an argument leading to a conclusion, which is intended to be of sufficient reliability to count as a "fact." All this is accomplished largely informally, using techniques and inferences that are established by a working consensus of the relevant colleague community.

In the case of hazards, where the underlying data may be sparse and ambiguous, and no evidence can be conclusive, certain disciplined techniques can be of great assistance. These operate qualitatively in the elicitation of possibilities, complementary to the statistical techniques that lead to quantitative estimates of likelihood. Those concerned with the regulation of the new biotechnology have already begun to experiment with hazard-assessment techniques designed especially for this work. In the UK, development is in progress on a technique known as GENHAZ. This consists of a structured procedure for assessing risks which records the possible consequences of the widest possible range of unintended eventualities.[3] While this technique is extremely useful for calling attention to possibilities that might otherwise have been over-looked, it cannot by itself evaluate the quality of the information on which a particular scenario is based. Hence there are opportunities for much useful work on these techniques of making best use of the information that is available, and converting it into effective evidence in a hazard assessment.

Coping with the Scarcity of Information

One crucial problem is the paucity of relevant information. The fields of taxonomy and microbial ecology have not been in the mainstream of biological research, and the lion's share of funding has been going into work at the molecular or cell level. Indeed, there is now official recognition in the USA that systematics (now enjoying only about 0.3% of the funding of biomedical research) is in danger of going into rapid decline as its ageing experts retire with too few trained scientists to replace them.[4] Yet it is only through these relatively neglected fields that we will solve the problems that are created by the successful applications of the popular ones.

Until that situation is redressed, and it will require a long lead time for the expansion of teaching in those fields, those who assess hazards will do so in the absence of information that in some contexts might be considered crucial. Of course, it is in such situations that burdens of proof must operate, if not explicitly then implicitly, and where personal prejudgments can become decisive. A useful task would be to devise means for the disciplined management of this sort of uncertainty. Although we cannot embark on this here, we might offer a brief sketch of some of the steps in such a procedure. First, when some requisite information turns out to be not available, we may enquire why not. Although it is not always easy to find out about such things, it might be possible to discover whether someone had intended to do it but funding had been refused. The reasons for that could be a combination of uncertain feasibility, high cost or low priority. Thus if the necessary methodology of a project is inherently very difficult, funding agencies would require proof of very high priority for resources to be put into it.

Obtaining a profile of possible research that has not been done, a map of our ignorance as it were, could be an illuminating exercise in itself and could provide guidance for changed priorities. Second, perhaps the research had been done, but the results are confidential to some commercial firm or government agency. In some cases, as with defence-related biological research, deep secrecy is accepted as essential for national security; but otherwise, a regulatory programme is severely hampered if the very existence of confidential research results is itself a confidential item. Finally, given the absence of the required information, there could be an estimation of what difference would be made to a regulatory decision if it were available. This last procedure, aggregated over a number of cases, would help to focus attention on those questions on which further research is most urgently needed.

All this may appear to some to be an overly elaborate solution to what is, after all, a problem in scientific common sense. But the planned-release techniques raise rather special problems, so that (within the limits of feasibility) more conscious care is required in the analysis of their hazards. For example, a planned-release technique may involve an organism that is already dead or intended to expire after its work is completed. In that case, the task of assessment is simpler, for any possible harmful effects must be transferred across the barrier (not absolute, to be sure) of the short life and perhaps moribund condition of the organism. But sometimes the organism is intended to survive, at least for some time, in the wild. The question of regulatory principle is then whether it should be guaranteed not to displace any of the existing related biota, or (less restrictive) merely not to cause disturbance to the environment. In the former case, an exercise in biological fine tuning must be involved, so that the organism will be just strong enough but not too strong. There must also be a further exercise, in proving adequately (keeping in mind what we have already said about burden of proof) that the engineered organism will indeed have the desired property. Such studies could indeed strain our resources of microbial ecology in many cases; and yet if containment in the appropriate degree is to be guaranteed, it is hard to envisage any alternative.

We believe that with cases such as these, there will be many decisions where the scientific component is far from conclusive, and so where the skill and integrity of those making the decisions, as perceived by the public, will be crucial. It is for this reason that we advocate as much clarity and publicity as possible in the management of uncertainty in the control of the hazards of genetically engineered microorganisms.

Principles of Ecological Assessment

The hazards of genetically engineered organisms result from unplanned interactions between the organism and its environment, which may be good for it but bad for us. To control these interactions, we need to know about the organism and its genetic materials, and also about the relevant aspects of ecology. As we have already observed, in biology the study of the environment lags far behind the study of the organism. The field sciences have lacked scientific excitement for attracting researchers; they are in ways like a relic of the gentlemen-amateur science of the Victorian age and before. The crisis is now attracting notice, and it is becoming appreciated that unless present trends are reversed, this imbalance will increase, perhaps eventually approaching the point where there is a drastic shortage of personnel with the field-science skills necessary for assessing the hazards that the laboratory sciences are creating.

Even if we begin to redress this imbalance, there will always remain severe uncertainties in the predictive aspects of ecology. Some argue that the natural environment is so complex, involving interactions so numerous, so varied and so interrelated, that a "predictive ecology," on the analogy of physics, chemistry or molecular biology, is impossible. Others argue that it must and that (perhaps therefore) somehow it can be done. Whatever may be the eventual outcome of that debate, we need now to be able to make reasonable predictions to guide our regulatory decisions. By drawing on past experience, it is possible to draw up guidelines, focusing attention on those aspects of an organism in its environment that are crucial for assessment.

1. *Specificity.* As with conventionally produced organisms used in pest control, genetically engineered organisms should be screened against a wide range of possible targets and released only after their specificity has been established.
2. *Predictability.* Although unexpected effects have not been common following natural invasions, every effort ought to be made to reduce the possibility of unforeseen consequences following a deliberate release, to monitor for their occurrence and to contain them when they occur.
3. *Reproduction and spread.* As demonstrated by many weeds, high rates of both reproduction and dispersion are likely to mean difficulties in controlling a released organism.
4. *Scale of use.* Success in establishing a novel species will depend to some extent on the scale and frequency with which it is disseminated in the environment.
5. *Reversibility.* Some past invasions (such as the North American muskrat in Britain) have been reversible. Where, as with microbes, this is unlikely to be achievable, even greater reliance must be placed on a thorough assessment of an organism's range and specificity.

Even with such principles in force, there can still be no guarantee that any particular introduction will be totally benign. We should not forget, nor indeed will scientists be allowed to forget, that CFCs were brought into widespread use only after the best scientific advice was that these simple chemical compounds are truly inert and could therefore cause no environmental harm. Such cautions are even more appropriate when organisms are considered for large-scale routine commercial use; for then all the above criteria present very different problems. This difference will lead to a sixth ecological principle, of quality assurance in application; and we should argue for this in some detail.

Earlier we mentioned the difference between the conditions of traditional laboratory research and those of the control of hazards. There is an analogous difference between natural hazards and those that are crucially dependent on industrial practice. This difference erupted in a sharp debate on the regulation of biotechnology, and so it is particularly relevant to our concerns. The case, as described by Brian Wynne,[5] was the examination by the "Lamming Committee" on behalf of the European Commission of five hormones for use in the beef industry. Two, which were fully synthetic hormones, were deferred for further study, while three "nature-identicals" were judged acceptable under certain conditions of use. The Commission's own experts interpreted this as scientific approval, and on that basis were prepared to permit their use. In that way they would both support the beef industry and also encourage biotechnology. However, there ensued some intense political lobbying, with the result that the Council of Ministers rejected the experts' advice and banned the hormones.

The Commission's experts were indignant at this incursion of what one called a "wave of ill-informed emotive popular superstition." However, the nature of the problem is revealed in the conditions laid down by the Lamming Committee. These included: specified dose limits, non-edible site of injection, full veterinary supervision, and a minimum waiting period before sale of treated meat. In an ideal world, good farming practice would of course ensure that such conditions were met. But the real world is not ideal, and it is hardly irrational to query whether in the absence of close policing in all the agricultural districts of the Community, such standards of quality could be expected to hold. In the absence of such quality assurance on farming practice, the hazards of the hormones in question could not be assumed to be negligible.

Those who are familiar with technology and industry are keenly aware that quality assurance cannot be taken for granted; it is well known that the study and mastery of quality assurance by the Japanese some years ago was an essential part of their rise to dominance in so many fields. In the industrial context, quality relates mainly to reliability in performance; safety of products for consumers is generally assumed to be an unquestioned priority. But when hazardous substances are involved, in the context of industry or agriculture, quality will relate to risks as much as to reliability. And since it is employees, livestock or the natural environment that may be at risk, rather than consumers, the assurance of quality may become quite problematic. Hence for these circumstances, we suggest that the preceding ecological principles be supplement by another:

6. *Quality assurance in application.* The realities of practice, and of regulatory processes as well, must be included in any assessment of hazards in any context of agricultural or commercial application.

We recognize that it is not always easy to argue on the basis of imperfections in government practices, especially when such imperfections are not officially admitted. But to assume that quality assurance will automatically be maintained on all sides is to invite the sorts of damaging incidents that have brought the nuclear power industry down from its early bright promise to its present uncertain future.

Maintaining Public Confidence

The function of the above principles is two-fold: to ensure that safety in genetic engineering is maintained, and also to ensure that it is seen to be maintained. We have already commented on the tendency to optimism among the laboratory-based scientists, that manifested in the debates on recombinant DNA research in the 1970s. It is now more than 20 years since a public awareness of environmental problems became an issue for the conduct of industry and technological innovation. The confusions and extremism associated with the early struggles have abated, as both sides have matured in their understanding of the problems. However, public opposition can still be a potent force; now leading German manufacturers are building biotechnology research centres in the USA rather than at home.[6]

When the relevant public is really aroused, proposers and regulators face a most difficult situation. It is scientifically impossible to prove the impossibility of an unwanted event, and this may seem to be what is demanded by protestors. Equally, it is politically impossible to prove an "acceptability" of a risk, when a public sees it being imposed on them by an organization that lacks human concern and perhaps also technical competence. Hence the new politics of NIMBY ("Not In My Back Yard") is an important factor in any technological or industrial development.

It is clear that a sober and responsible mood now prevails among those responsible for the current work of managing the risks of genetic technologies. It is therefore possible for there to arise a new and mutually beneficial dialogue between the scientific and regulatory experts on the one hand, and representatives of special concerned groups and of the general public interest on the other. It is now understood that people generally do not attempt to make a quantitative assessment of the risks being imposed on them, but rather they assess those persons in authority who either create or regulate them. And in this they are not being irrational; just like a lay jury (and a lay judge) in our judicial system, they evaluate evidence to a great extent from the quality of the testimony of witnesses. Of course they can be mistaken, but the policy is prudent and effective both in practice and in principle. To continue the juridical analogy, all it needs is for safety to be respected and to be seen to be respected, and the problem of public confidence can be resolved.

A genuine policy of this sort brings its own benefits for the work of regulation and control. For if those in charge see it merely as a cosmetic exercise, placating or appeasing an ignorant public, then the work will eventually deteriorate. The prevention of accidents of any sort requires continuing commitment and morale at all levels, especially from the top down. Our culture and personal attitudes are not tuned to the sort of success that is measured by nothing happening. The best planned system of safety will atrophy and decay without regular injections of commitment. The natural consequence of a good safety record is complacency. Equally naturally this would be followed by corruption, as realized in the concealment of incidents that might spoil the good record. If those in charge of a safety programme believe that their scientific knowledge or personal integrity makes their system immune from danger, then the consequence of their attitude will manifest sooner or later, causing harm to all concerned.

What is needed, then, is a certain attitude of scientific humility, an awareness of partial ignorance and of the human fallibility of those operating the safety system. With this would go an appreciation of the role of independent observers and even critics, in keeping the work up to a good standard. Conversely, people who are experienced in public affairs would from their side recognize the genuinely good intentions, and the real special competences of those managing and regulating the system. There could then be a constructive dialogue (not necessarily always smooth) rather than a destructive polarized debate. There are precedents for this approach from the 1970s. One was in Cambridge, Massachusetts, after the shouting died down; a committee of citizens representing a cross-section of the population reviewed the problems at Harvard University, and eventually accepted regulations which were only a slight modification of those originally proposed. The other was the original Genetic Manipulation Advisory Group in Britain, where representatives of employees and of the public interest had equal members with the scientists, and (behind closed though leaky doors) engaged in a very effective common dialogue on the construction and operation of the regulatory system.

The effectiveness of a proper attitude is confirmed by recent experience of planned releases. Where public concerns have been respected, as in this country, there have been no problems; elsewhere, the record is mixed. But again, the respect for the public must be genuine, or else sooner or later the deception will become patent, with harmful consequences for all. As we have seen, the management of biohazards is a very different task from the scientific activity that created them. There the work has been successful in creating "public knowledge" through research on chosen laboratory problems; here the task is coping with the inherent uncertainties of environmental problems that are thrust upon us by the very success of our laboratory researches. As we have shown in this report, our knowledge of the hazards and how to control them is advancing along with our knowledge of the techniques of genetic engineering. This is a new sort of science, equally challenging; and on its success will depend the progress of biotechnology and all the advances in human welfare that depend on it.

NOTES

Reprinted by permission from J. R. Fincham and J. R. Ravetz, *Genetically Engineered Organisms: Benefits and Risks* (Toronto: University of Toronto Press, 1991), 121–42.

Editors note: The authors are Canadian as are the agencies referred to.

1. S. E. Stachel and P. C. Zambryski, "Genetic Trans Kingdom Sex," *Nature* 340 (1989): 190–91.

2. E. Sherr, "And Now Small Is Plentiful," *Nature* 340 (1989): 429.

3. C. W. Suckling, "G EN HAZ: An Attempt to Apply HAZOP to the Identification of Hazards in the Release of Genetically Engineered Organisms," *Royal Commission on Environmental Pollution,* May 31, 1989.

4. S. Nash, "The Plight of Systematists: Are They an Endangered Species?" *Scientist,* October 16, 1989.

5. B. Wynne, "Building Public Concern into Risk Management," in *Environmental Threats,* ed. J. Brown (London: Belhaven, 1989).

6. R. Zeil, "History Feeds German Fears on Gene Technology," *New Science*, August 26, 1989, 26–28.

7

Journey to the Origin: Biological Integrity and Agriculture

Miriam Therese MacGillis

At the outset, it is important for me to describe myself as a layperson in this gathering. My background is in art. I have no formal education in science or agriculture. But I am here with a point of view completely opposed to biotechnology in general and to its place within agriculture in particular. I say this out of my own experience. I left teaching art in the early 1970s, having been deeply moved by the specter of world hunger as it was looming on the global horizon. After the oil embargo of the same period, and in order to meet its balance of payments, the United States quadrupled the price of its export agricultural commodities. It was a response to the lawlessness of the global market system, whereby unilateral decisions on oil and food commodities wreaked havoc on the desperate economies of the non-industrial world. I remember the vivid realization of learning that the cattle of the feedlots of the United States were daily consuming grain adequate to feed the hungry peoples of the world. That fact determined the direction of my life, and I have been involved with the analysis of hunger and agriculture ever since.

In 1977, I heard a paper delivered by Thomas Berry. The context he outlined also set a direction in my life that has helped me to probe ever deeper the root causes of the kind of dysfunction that plays itself out in the global crises of our times.

Thomas Berry is a renowned historian of world cultures and in his two seminal books, *The Dream of the Earth* and *The Universe Story*, which he co-authored with Brian Swimme, he suggests that the root of our crises is contained within the cosmology that has shaped the total context of western thought. I suggest in this presentation that this cosmology also underpins the world of biotechnology and that it is both flawed and dangerous. I also suggest that a contemporary scientific understanding of the origin, nature, and functions of the cosmos would indicate that biotechnology is itself an extension of the same inadequate worldview and that it is taking us in a direction that is counter to the natural progression of the universe, the earth, and life.

The following chart is my simple attempt to model some of the assumptions that are inherent in our traditional western cosmology. Our origin story, rising several thousand years ago out of the Mediterranean world, provided a context of meaning that attempted to answer the ultimate question of the mystery of existence. In short, it provided a coherent set of meanings upon which the various structures of culture were formed. Some of them are:

That the divine is totally transcendent to the universe, hence perfect and unchanging.

The human has a transcendent destiny to be brought into union with the divine, but this union depends on the human transcending the cosmos. The cosmos itself does not have this spiritual transcendent destiny. It is a physical, material plane of existence and possesses no inherent spiritual substance.

The human is free to explore the physical world, analyze its physical energies, and redesign them to bring about some of its original perfection lost after the fall. Hence, this worldview might be described by this simple model:

Garden (Bliss)——History——Millennial Age (Bliss)

In this worldview, the ordinary conditions of life are perceived as temporary and abnormal. Thomas Berry suggests that this perception sets the stage for a growing pathological rage within the western psyche. This rage is directed towards the conditions in which life is actually granted to us. Historically, it has made it nearly impossible to develop the inner capacity to live creatively or graciously within the whole fabric of life. Instead, we have resisted all limitations imposed on us as abnormal, as a punishment from which we will one day be liberated. Our inner capacities have been stunted and our total intrusion into the fabric of life, as it has brilliantly evolved, is nearly total.

So it is the scientific story of evolution itself that suggests that our obsession with genetic engineering may well bring about a total undermining of the very life we commit to re-designing. I would like to suggest that as we review the process of evolution, seen now as a total evolution of the inner as well as outer dimensions of the universe, this context provides an essential correction to the direction that agri-biotechnology is pursuing.

The following overview of the evolution of DNA was prepared by Dr. Lawrence Edwards, Ph.D., of Genesis Farm.

EVOLUTION OF DNA

About 15 Billion Years Ago

The universe flared into existence. At first all was symmetric. Within much less than a millionth of a second the symmetry is broken as the primal four forces emerge. All subsequent relationships will be governed by these four. In particular, the nature of the electromagnetic force is now set. Even though there were no molecules in existence, the laws of chemistry are now in place. So even though no DNA was present, limits were in place on the strength of the hydrogen bonding and therefore on the diameter of the helix.

A Billion Years Later

The universe coalesces into galaxies and stars. The stars live by consuming primal hydrogen and helium and fusing them into new entities—lithium, beryllium, oxygen—all the chemical elements up to iron in weight. The larger stars exhaust their supplies of hydrogen and helium and can no longer sustain their existence as stars. They become supernova and, in that cataclysmic process, fuse to become the heavier chemical elements. Their bodies, rich in chemical elements, are strewn throughout the cosmos.

For Billions of Years After

Subsequent generations of stars form by gathering the chemically enriched gaseous clouds of hydrogen into themselves, fusing, becoming supernova, and again distributing more elements into the cosmos.

About 4.6 Billion Years Ago

A large star in our galactic neighborhood became a supernova.

About One Hundred Million Years Later

Our sun and solar system formed from the body of this supernova. For several hundred years the planets grew in size by accruing smaller asteroids in often violent collisions. The earth was often molten during this period. During this process the chemical elements born in the star and supernova combine to form simple molecules (e.g., water) and minerals (rocks, stones, etc.).

About 4.1 Billion Years Ago

The great bombardment was over. The solar system readied its present configuration of nine planets. The earth now cooled for the last lime, eventually enough so that steam could condense. It rained violently for eons creating the oceans.

About 4 Billion Years Ago

Probably during one or many of those thunderstorms, the first complex molecules were synthesized from the simple molecules and minerals. (No one knows; there are many theories.) Once created, those molecules self-organized themselves and others into creative possibilities. At least one of those possibilities worked. Over the eons those organizational capabilities resulted in the first living organisms. Probably the first genetic capability was through RNA (ribonucleic acid). Later, apparently, DNA (deoxy-ribonucleic acid) proved to be more effective and RNA was then used not for the storage of the genetic information, but only as a "messenger" between DNA and the enzyme production capability. (Again no one is certain of the process in those early years.)

DNA is a chain of four specific nucleic acids. A "word" in the language of DNA consists of a particular sequence of three of these acids. Thus, there are 4x4x4 or 64 possible words in the DNA language. Each word "speaks" of a particular amino acid. So a sentence of words specifies a sequence of amino acids, which is a protein (enzymes are proteins). In all life forms, the same correspondence exists between the DNA words and the particular amino acid. (In many cases there is more than one word possibility for a specific amino acid.)

This was all worked out 4 billion years ago!

Some Hundred Million Years Later

Simple bacteria emerge. A bacterium consists of a cell wall surrounding and containing protoplasm, a complex mixture of organic molecules including naked strands of DNA directing the maintenance of the bacterium. The protoplasm connects with the outside environment through the cell wall. Occasionally, the cell clones itself, a process directed by its DNA. Thus, DNA not only remembers how to create a new cell and how to maintain its existence, but also directs the processes.

About 2 Billion Years Ago

In response to the menace of oxygen, a new form emerges, the eukaryote cell, the cell with a nucleus. (Of course, the bacteria live on and prosper without a nucleus.) The cellular DNA is collected and stored in the nucleus as a double helix. This helix unzips during the reproduction process and then each strand duplicates itself in the daughter cells. During this unzipping the DNA is very susceptible to damage. Mutations occur primarily during this time. Apparently this susceptibility to damage is just right; more would result in higher death rates of the daughter cells, less would result in less ability to adapt. (Later, cells developed molecules to "walk" along (the DNA helix strand to find and correct errors.)

For the Last 2 Billion Years

DNA has learned, memorized, and directed the processes of life. Changes in the DNA of a particular species have been slow, in earth time. There have been periods of accelerated change, but these periods have still been long—hundreds or thousands of years when compared to human time. All changes were rigorously tested for compatibility with the organism's ecosystem.

15,000 Years Ago

Humans started consciously changing the DNA of other organisms through horticulture and domestication of animals. Those changes were made much more quickly than normal evolution, but still over many generations. There was not such rigorous testing of the changed organisms and this led to problems in some cases, e.g., exotics taking over an ecosystem. Often the changed organisms are not even independently viable and must be supported by human activities. But, overall, the changes were not large. For example, there was never the mixing of genes between species.

Today

Humans have learned many words and sentences in thee DNA language, the means to change sequences within a gene, and the ability to move sequences from one organism to another organism of a different species. Now the time scale of radical evolutionary changes is instantaneous. There is not the time or the incentive to thoroughly test the new organisms.

There is not even the knowledge of how to test such unknown creatures. While the goal of many of these manipulations is laudable, e.g., the curing of various inherited, debilitating diseases, most are driven by commercial goals. One can imagine taking a certain risk in order to improve the health of a certain segment of the population. We have done this before, e.g., fluoridation of drinking water and vaccinations. But often there have been unforeseen disadvantages to such activities. In any case, much of the genetic manipulation today is for profit. There are few, if any, redeeming qualities except a more efficient product, e.g., a longer lasting tomato.

We do not understand the consequences of genetic manipulations. We are launching yet another massive experiment on ourselves with little understanding of the long-term consequences.

In Conclusion

I would suggest that our refusal to live within the limitations of the unity of the whole, which has enabled the elegant miracles of life to unfold, is a dark extension of our mythology. Biotechnology is a commitment to myth. By refusing to acknowledge the superstition implied in our blind adherence to our vision of a world of bliss, we move deeper into a chaos from which life itself may be unable to recover.

Note

Reprinted by permission from J. F. Macdonald, ed., *Agricultural Biotechnology: Novel Products and New Partnerships* (Ithaca, N.Y.: National Agricultural Biotechnology Council, 1995), 101–10.

8

Three Concepts of Genetic Trespassing

Richard Sherlock

One of the most commonly used phrases in the debate over genetic modification of living organisms is the claim that in doing so human beings are "playing God." Though this phrase is obviously dependent for rhetorical popularity on the western religious traditions in which God is creator and lord of creation, this point alone is an inadequate response. Nor is it adequate to reply that we do not know what God desires, permits, or forbids in the matter of genetic modification.[1] These are easy responses to the theistic version of a claim that is of much broader importance than that found in specific western monotheistic versions. John Harris may very well be right that the claim "you are playing God," *simpliciter*, "is a non-starter" insofar as moral argument is concerned. It begs too many questions and is open to a number of easy objections.[2]

I submit, however, that a structurally similar argument is advanced in much of the radical critique of genetic engineering, especially transgenics. This argument is not so easily dismissed and deserves the respect of careful analysis. It can be developed within western theism, within variations of oriental religion, and importantly in the context of non-theistic naturalism. To capture this argument in crucially important versions I shall borrow a phrase from the radical critics and refer to the problem of "genetic trespassing." Thus Mae-Wan Ho writes that "transgenic technology transgresses both species integrity and species boundaries leading to unexpected systemic effects on the physiology of transgenic organisms produced as well as the balanced ecological relationships on which biodiversity depends."[3] This paper is an attempt to sort out crucial versions of this argument and see if they are sound. I will not try to solve all of the issues related to the concepts of biological or ecological integrity or trespassing. My hope is to shed some light on where the real issues lie around the concept of trespassing.

Broadly, the idea is that there are fixed genetic barriers that divide species or at most genuses. Crossing these lines is wrong in much the same way that crossing property lines is wrong. In doing so one has trespassed, i.e., crossed the line one should not cross. "Trespassing" is thus often used as a word like "murder" which already implies a moral wrong. What kind of wrong is a matter for dispute, though in the standard case of trespassing the idea of violating a pre-established, agreed upon, property boundary is the most obvious candidate. In the western theistic version God is said to have created specific "kinds" that are meant to remain apart. Other variants rely on ideas drawn from evolutionary biology, ecology, Hinduism or deism to ground a belief that fixed lines separate biological kinds, lines which should not be crossed. To see the flavor of this argument I quote at length from one of the most interesting of the writers in this vein:

> Crossbreeding uses natural reproductive mechanisms. Such mechanisms are only able to combine genetic material from the same or closely related species. For example, cauliflower can be crossbred with broccoli but not with zucchini. Furthermore, the natural reproductive mechanisms combine in a very precise and systemic manner. This process does not allow a random selection of genes from one organism to be inserted into the DNA of another. The DNA of a child, for example, combines a strand of DNA from the father and a strand from the mother. It is not made by randomly inserting a few genes from one parent into the DNA of another.
>
> The mixing of genes by crossbreeding is clearly subject to very definite rules— you can't mix unrelated species and you can't just drop in one gene on its own, you have to take the whole package of DNA. Where there are rules there are boundaries. For example, where a donkey breeds with a mare, the crossbreed—the mule—is sterile. Nature does not support further transformation or propagation of the mule DNA. Natural law has set a boundary.[4]

Though this is a useful starting point for examining the radical anti-biotechnology argument, the concept of "genetic trespassing" is ambiguous in two ways. First, as a conceptual matter, what fixed genetic boundaries exist in a manner similar to property lines? Second, exactly why genetic modification is wrong is not well discussed in the radical literature. The essence of the genetic trespassing argument is that there are fixed lines that demarcate a "natural" way of genetic existence as distinct from, and preferable to, deliberately manufactured existence. But why crossing these lines in a deliberate manner is wrong is not clear.

A careful review of the critical literature I believe will show that there are three versions of this claim about a "natural way" and the wrong of crossing such a line. Two of them can be understood with spatial analogies such as "trespassing." A third version is best understood differently while remaining rhetorically connected to the natural/unnatural dichotomy. Pace the Ticiatis, one cannot claim that crossing genetic boundaries is impossible. Rather one must claim that we *should* not cross such boundaries. Three versions of an argument to this conclusion follow.

HARD TRESPASSING

The first answer is that it is wrong per se and recognized as such by rational human beings. As most thinkers now admit, moral argument must begin with some widely shared convictions such as "torture of a small child is wrong," from which we can develop more general principles that are best able to account for our basic moral sensitivities such as embodied in the specific conviction about torturing children.[5] Is "genetic trespassing," however, such a widely shared moral conviction? The very fact that there is such a vigorous debate about it suggests otherwise. We do not, for example, debate the morality of child torture. But there is a vigorous discussion about the morality of genetic engineering of animals or plants for increased agricultural production or other human uses.

This argument is "essentialist" in the sense that it is premised on the idea that there are fixed "kinds" in nature, which are generally called "essences." Of course there is variation in a species (e.g., horses). But all horses share the essence of horse so that it is proper to class them together as we typically do. But essentialism per se does not have a moral conclusion. The debate about transgenics just referred to suggests that we are less certain about the moral implications to be drawn from transgenics than we are about some few core issues of ethics.[6]

However, the essentialist argument may be revised to hold that if we had some knowledge of science we would or should, on proper reflection, conclude that transgenics is wrong. To see the structure of what we shall call the "hard" trespassing argument we might represent it as follows:

1. Natural boundaries exist.
2. Natural boundaries are genetic.
3. Natural boundaries distinguish species.
4. Transgenics deliberately mixes genes from one species with another.
5. Thus, transgenics trespasses on species integrity.
6. Therefore, such transgenics is wrong.

Premises 4 and 5 are not the same, and 5 does not follow from any part of 1–4. Premise 4 merely states that lines between species are crossed in transgenics. This is partially true but trivial. Premise 5 holds that the core genotype of a species, that set of genes that makes it what it is in genetic terms, is undermined by transgenics.

Whether there are natural kinds in the living world is a much contested question. I shall not discuss that issue here. Even if one accepts the view that there are such natural kinds it is doubtful that the hard claim of genetic trespassing can be justified. If there are such kinds and further if these kinds are marked partially by genetic distinctions, genetic identity of all members of the species cannot be implied. Even with a robust view of genetic essentialism one must conclude that only an absolute core of genetic properties must be similar for two entities to be part of the same kind.

Consider a very substantial version of genetic essentialism, Mayr's view of a species as a "protected gene pool."[7] As a leading evolutionary theorist Mayr himself saw that species evolve and change. Variation is not the same as speciation and genetic variation does not per se threaten "species integrity," however that is understood. Thus, two features of even the most persuasive view of species as genetic "natural kinds" stand out: First, species are not fixed in an a temporal fashion. They may be "natural kinds" for the duration of their existence but change is inevitable.[8] Second, even supposing that genetic boundaries exist, genetic change as such does not cross a species boundary.

Many forms of genetic engineering may thus be only adding or deleting furniture in a room. The room, that is, the living entity, still remains a member of its kind. On the other hand, with enough genetic change a new kind results. This latter is what premise 5 claims: transgenics builds a new house by undermining the core integrity of a species and creating a new one. Most current transgenics comes nowhere near such a point. Furthermore, such changes as implied by premise 4 have been occurring since the beginning of life on Earth, and premise 5 is necessary for evolutionary change. Once this evolutionary natural history is admitted it seems difficult to maintain a hard view that fixed genetic boundaries exist that should never be crossed.

Thus, premises 4 and 5 are at once much too temporally narrow and spatially broad. Gene addition or deletion is not equivalent to crossing a species line or "trespassing on its integrity." Integrity can not be equated with sameness. Temporally, 4 and 5 overstate the fixity of genetic boundaries through time.

Finally, I note that the moral conclusion using the word "wrong" cannot be derived simply from a set of factual premises such as premises 1–5, even granting them in their strongest form. Natural history itself shows that premise 5 is essential to evolutionary development. Hence, the claim that deliberate transgenics is wrong because it violates nature involves a key conceptual problem. We know that premise 5 is an essential part of the natural history of the planet. Hence, to claim that it is unnatural is simply a contradiction.

GENETIC "WRONG WAY"

Many writers who deplore what I have called "genetic trespassing" frequently move from a purely essentialist claim about natural kinds and fixed lines, to a richer and somewhat more persuasive claim about the potential dangers of such trespassing. Thus, they attempt to solve a key problem for the hard trespassing argument: the lack of justification for the conclusion that "transgenics is wrong" or "transgenics must be stopped." A plausible claim of serious dangerousness would provide at least a relevant reason for such a conclusion. One fairly inflammatory version of this claim is provided by philosopher and student of eastern religion Ron Epstein: "What has gone on largely unnoticed is the unprecedented lethal threat of genetic engineering to life on the planet. It now seems likely, unless a major shift in international policy occurs quickly, that the major ecosystems that support the biosphere are going to be irreversibly disrupted and that genetically engineered viruses may very well lead to the demise of almost all human life."[9] If this is the situation we are in, then transgenics should be stopped.

Trespassing, however, may not initially appear to be the most appropriate analogy for the point the critics wish to make. If I walk across your open back yard while you are on vacation it is hard to see what significant harm has been done. I have technically trespassed but I have only created minimal harm. Without further qualifiers what I have done can not possibly be described as dangerous or as creating a danger for someone else. If I and others walked across your lawn so frequently as to cut a path a harm would result. If a great many lawns were crossed in this way a danger might have been created since some houses will be occupied and some objects in some yards will be harmful, e.g. garden chemicals or even pets. Suppose, for example, that there was a sign that read "BEWARE OF DOG" or "keep off: lawn has just been treated with chemicals." This might provide a reason to avoid that lawn. Cutting across would be the "wrong way" to someone's destination.

Two versions of the critical claim about the potential harms from transgenics are found in the literature and may be understood with the help of different metaphors than a standard account of "trespassing" would lead one to consider. The first version tries to capture the conclusion of the argument from "hard trespassing": stop deliberate gene transfer. We can explicate its essential nature with the metaphor of "the wrong way," taken from driving. Some critics, like Epstein, do not just want research moratoriums or more thoughtful analysis of risks. Rather they are opposed to transgenic technologies per se. But they oppose transgenics because they view it as so extraordinarily dangerous that it must be stopped altogether. The metaphor of "wrong way" captures this claim. There is no safe speed for driving on the wrong side of the road. The side of the road one should drive on is not fixed in nature like a part of the "great chain of being," as anyone who has been to England knows. But once a side has been chosen for the area you are driving in, there is no safe speed for going the other way.

This argument is not based on concepts of natural kinds or barriers. The driving rules are chosen, not given. Once chosen, however, some actions are so dangerous, given the rules, that they should never be undertaken. For the radical critics, deliberate transgenics is one such activity. We might summarize such an argument in the following form:

1. All species embody unique successful adaptations to changing environments and competitions since life began.
2. The persistence of successful adaptations requires genetic isolation.

3. The future success of life requires evolutionary adaptation.
4. Deliberate gene transfer breaks isolation.
5. Hence, gene transfer threatens life on Earth (from 2, 3, and 4).
6. Given 5, gene transfer should be stopped.

In other words, the form in which genetic development and isolation occur was not fixed or predetermined from the beginning. But once genetic integrity had developed in specific ways, the maintenance of integrity through isolation was essential for the stability of successful adaptations.

Once parsed out this way, problems with the argument of the radical critics can be clearly identified. The central difficulties are interconnected around premises 2 and 4. The successful adaptations referred to in premise 2 do require persistence, a phenomenon that Mayr calls "reproductive isolation": "the reproductive isolation of a species is a protective device that guards against the breaking up of its well integrated, co-adapted gene system."[10] But species emerge as well as persist. Their coming into existence will require evolutionary crossing of genetic lines (i.e., a breakdown of complete isolation). New species come into being as transformations of old species, which will partially involve genetic changes. In some instances, such as mutations, these may be abrupt. In most cases long generations of minute alterations are involved. But change and persistence go together. Thus, premise 2 is incorrect because it fails to capture both of the evolutionary principles of stability and transformation.

Premise 4 is incorrect for much the same reason. Gene transfer does break genetic isolation. But transgenics is not unique in this regard; other events can also do this such as horizontal gene transfer. Furthermore, genetic isolation is not the only threat to future life. So too is a failure to adapt to changing environments and species competition. However, over a long time frame, adaptation may involve most of the very same changes that transgenic technologies involve.

As a result of this analysis we can see that premise 5 is much too strong a conclusion as derived from premises 2 and 4. If premise 5 is unsound as it stands, then the conclusion in premise 6 does not follow, especially in this strong form.

Moreover, I would point out that many of the same writers that criticize animal transgenics as "unnatural" have also criticized deliberate agricultural use of genetically engineered microorganisms precisely because of concerns about horizontal gene transfer to other organisms at the micro-level in open environments. They were concerned that the isolation was not stable among microorganisms in soils. According to the critics, precisely because this natural transgenics occurs adding to it may be dangerous. Thus, many of the best critics admit that isolation and with it the "wrong way" argument can not be the whole story.[11]

By admitting that horizontal gene transfer happens and then refashioning an argument around this fact the critics have shifted ground, perhaps for good. Ironically they have done so in the direction of a sounder but less absolute position.

GENETIC SPEEDING

Less radical critics raise a slightly different and more substantive argument that can also be understood with the help of a driving metaphor: speeding. Transgenic technology, they argue, is going too fast in breaking genetic barriers. Most such transformations in evolutionary time have been slow to develop and complete. We are setting out to do in a generation what might have taken hundreds or thousands of years in the slow process of natural transformation. This is the essential nature of calls for moratoriums on genetically modified agriculture and germ line genetic engineering. It is also at the core of the discussion of the "precautionary principle" as it is applied to genetically modified agriculture.[12] The essential recommendation is that researchers and companies should slow down until we know much more about the consequences of what we are doing.

Furthermore, this long period of co-evolution allowed genetic modifications and new species to find an ecological niche in which to co-exist and thrive with the rest of the natural world that surrounded it. On standard accounts this growth was chaotic and many modifications failed to find a sustainable niche. But co-evolution over long time periods did ensure that those modifications which did survive were sustainable and useful.

The speeding argument too can be understood in a series of steps:

1. Life has developed in successive evolutionary transformations.
2. Persistence of life requires successful adaptations.
3. Successful and beneficial adaptations require co-evolution of species and their competitive environment or selective pressure.
4. Deliberate gene transfer bypasses co-evolution.
5. Therefore, deliberate transgenics is a dangerous speeding up of evolution.
6. Therefore we must slow down quickly.

As it stands, this sort of argument is flawed in three ways. First, evolution has not always proceeded as slowly as the conclusion implies. Mass extinction and what some have called "punctuations" of very rapid changes suggest otherwise. Second, speeding is not always wrong or dangerous. Sometimes speeding poses little increased risk, such as going 70 miles per hour in a stretch of a long straight interstate where the actual limit is 65. In a few instances (e.g., a medical emergency), one may actually have an obligation to speed. But if one must speed in a residential area around a hospital one bears a strong burden of justification when doing so. "My wife is in labor" might do very nicely.

This observation leads to a third criticism of the "speeding" argument as it stands. The "dangerousness" of speeding as referred to in the conclusion is highly context-dependent. As we just noted, one can speed by going 70 miles per hour. But the conclusion implies a different sense, namely, that speeding is dangerous. This seems to be because we are going too fast given our limited background knowledge of the context of evolutionary genetics. Thus, the most plausible version of the idea of genetic speeding must include an additional premise: that given our imperfect knowledge of the context of gene transfer technology we are going faster than the process of co-evolution typically involves. This is speeding given existing conditions. The revised argument would be:

1. Life has developed in successive evolutionary transformations.
2. Persistence of life requires successful adaptations.
3. Successful and beneficial adaptations require co-evolution and selective pressure.
4. Deliberate gene transfer bypasses co-evolution.
5. The substitute for co-evolution would be extensive knowledge of nature, which permits us to find stable and beneficial niches.
6. Without such knowledge, transgenics is a dangerous speeding up of evolution.

Steven Palumbi's important new book calls this "brute force genetic engineering" in which variation is drastically reduced by targeting for specific traits in the next generation of living organisms. He argues that: "brute force genetic engineering is like dropping guns on main street. Engineers drop the trait into the organism, loaded with potential impact, and everyone hopes for the best."[13]

The fundamental problem addressed in the revised speeding argument is the problem of species finding their niche in which to persist. Two basic approaches to a solution for the niche problem exist: 1. design or 2. randomness and selection. Design theory holds that a master designer cut the pieces of a puzzle of a complex system so that they will all fit together. On this solution to the niche problem, we are going too fast trying to change pieces of a puzzle when we do not know the plan from which it was cut or made.

The randomness and selection model is the standard biological paradigm. Literally millions of pieces are thrown up in the process of evolution. Most do not fit together to create a complex puzzle, a few do and they find a niche. But the living system is an extremely complex entity where niches develop and change slowly.

The speeding argument holds that transgenics is dangerous because it tries to create organisms that must find a niche faster than evolution, much more narrowly than randomness, and in an epistemic position much less complete than that of a master designer, e.g. God.

The revised version of the speeding argument still has two problems as applied to animal transgenics. It assumes that the context is one that will disturb co-evolution. The laboratory transgenics of the present and foreseeable future simply does not pose such a problem. Second, and more crucially, we have now moved from definite wrongs, either trespassing or going the wrong way, to possible harms. But as knowledge increases possible harms may very well diminish or vanish.

In other words, the argument is more relevant with respect to open system agricultural biotechnology, not research in properly run labs where the nature of the system can be carefully controlled and where accidents can be contained and limited. To use the analogy, in the lab the researcher can often act as God with complete or nearly complete control of what he or she is doing.

Though these are useful points to consider, the speeding metaphor faces a much more fundamental question. If you do not know the road at all a drive is unsafe at any speed. This conclusion, however, cannot be correct. Nature alters genotypes constantly. The speeding argument supposes that the natural rate of genetic change is the standard against which going too fast is to be measured. The natural rate of change is not zero. But this attempt to find a "scientific" or "technical" answer to a moral and policy question fails. Can rapid genetic change be criticized for being rapid any more than mass extinctions in the geologic past be thought to be uniquely terrible because they were massive?

Perhaps what we must conclude is not that transgenics is wrong because it is disruptive. It is problematic because it may be harmful to human beings either directly or indirectly. It is problematic only in the sense that we may not be giving ourselves and our institutions the time needed to adapt sensibly. Perhaps the key to the speeding version is that open system transgenics should proceed slowly enough that if a seriously disruptive event occurred or appeared likely we would be able to stop ourselves before worse harms came to be.

Finally, we might note how this discussion relates to the twin concepts of "playing God" and the non-theistic variant "genetic trespassing," with which we started this paper. First, on this account, the most plausible version of the complaint about playing God is best understood as an epistemological, not an ontological, claim. It is not about violating Divinely established rules, or assuming a power human beings should not have. It is about doing what God does, without being in God's epistemic position relative to the rest of creation. It is God's knowledge that grounds the use of His power without harm. I shall not follow up this point except to suggest that this version of the "playing God" claim may avoid some of the easy responses from those who are sanguine about biotechnology.

Second with regard to the problem of trespassing three views of its "wrongness" appear:

1. Trespassing may be wrong because a person crosses a property line and thus violates the right of a property holder. This would be my "hard trespassing" argument.
2. Trespassing may be very dangerous or hazardous, e.g. crossing a property line and ignoring the sign "vicious dog." This would be a reason to never cut across this property line because of the nature of the hazard involved. This would be the wrong way argument about transgenic technologies.
3. Trespassing may be risky. This is the problem of not knowing the possible harms that could result from going across a piece of property. Walking

slowly allows you time to avoid hazards that appear. This is the speeding argument I have advanced.

CONCLUSION

The most plausible version of the trespassing argument does not lead to the absolute conclusions desired by some. Severe hazards should always be avoided but not risks. Risks can be rationally accepted. But to do so we need to fully understand the likelihood and magnitude of the harm that might occur and the magnitude of the need that must be met by a certain technology. Whether agricultural biotechnology, in particular, is required to meet a severe enough need is a question for another paper.

NOTES

1. For one attempt to give positive meaning to the concept, i.e., promote biotechnology along these lines, see Ted Peters, *Playing God* (Routledge, 1997).

2. John Harris, *Clones, Genes, and Immortality* (Oxford, 1998).

3. The quotation is from Mae-Wan Ho and Beatrix Trapeser, "Transgenic Transgression of Species Boundaries and Species Integrity." The literature here is vast. For some titles, see John Fagan, *Genetic Engineering: The Hazards, Vedic Engineering: The Solutions* (Maharishi International University Press, 1995); Martha Crouch, "Biotechnology Is Not Compatible with Sustainable Agriculture," *Journal of Agriculture and Environmental Ethics* 8 (1992): 98–111; Miriam MacGillis, "Journey to the Origin: Biotechnology and Agriculture," *Agricultural Biotechnology: Novel Products and New Partnerships,* ed. J. F. Macdonald (National Agricultural Biotechnology Council, 1995): 101–5; Mira Foung, "Genetic Trespassing and Environmental Ethics" <SFSU.ED/~RONE/GE>.

4. Laura Ticciati and Robin Ticciati, *Genetically Engineered Foods: Are They Safe?* (NTC Publishing, 1998), 2–3.

5. One does not need to be a "hard intuitionist" to adopt the core of this position, though "hard intuitionism" will do just fine. See Baruch Brody, "Intuitions and Objective Moral Knowledge," *The Monist* 62 (1979): 446–56.

6. Perhaps the most vigorous statement of the general realist position about natural kinds is David Wiggins, *Sameness and Substance* (Harvard, 1980). Given genetic variation, one way to think of species is as a "cluster concept" such as described in R. Bambrough, "Universals and Family Resemblances," *Proceedings of the Aristotelian Society* 61 (1960): 207–21.

7. Ernst Mayr, *Animal Species and Evolution* (Harvard, 1963).

8. The question of the status of species is much contested. For some analysis, see Larry Arnhardt, *Darwinian Natural Right* (SUNY Press, 1998), 232–38; Michael Ruse, "Biological Species: Natural Kinds, Individuals, or What?" *British Journal for the Philosophy of Science* 38 (1987): 227–45; David Hull, "A Matter of Individuality," *Philosophy*

of Science 45 (1978): 335–60; A. L. Caplin, "Have Species Become Declasse?" *PSA* 1 (1980): 71–82.

9. Ron Epstein, "Redesigning the World: Genetic Engineering and Its Dangers," at the Web site <FU.EDU/~RONE/GE>.

10. Mayr, *Populations, Species, and Evolution*, 20.

11. The discussion of this topic is now more than 20 years old. Horizontal gene transfer as it is known is a well-established fact at the micro-level in soil. The technical literature is enormous. For one strong critic who has contributed to the technical literature but has written for the non-scientist as well, see Mae-Wan Ho, *Genetic Engineering, Dream or Nightmare: The Brave New World of Bad Science and Big Business* (Gateway Books, 1998).

12. The literature on the precautionary principle is now extensive and of uneven quality. See, for example, Carolyn Raffensperger and Joel Ticknor, eds., *Protecting Public Health and the Environment: Implementing the Precautionary Principle* (Island Press, 1999); Julian Morris, ed., *Rethinking Risk and the Precautionary Principle* (Butterworth, 2000); Indur Golkany, *Applying the Precautionary Principle to Genetically Modified Crops* (Washington University, 2000).

13. Steven Palumbi, *The Evolution Explosion* (New York: Norton, 2001), 79.

9

Ten Reasons Why Biotechnology Will Be Important in the Developing World

Martina McGloughlin

Biotechnology companies, national and international organizations, including the Consultative Group on International Agricultural Research (CGIAR), and numerous academics (e.g., Ruttan, 1999) have continued to argue for the need to increase agricultural productivity so that sufficient food supplies exist to meet the demand forthcoming from a swelling world population. Despite Altieri and Rosset's [chapter 10] assertion, population density is hardly the issue. In the absence of significant productivity gains, or expansion of agriculture into marginal lands (e.g., forests), there will not be sufficient food quantities to feed the projected levels of population. This simple reality is independent of income distribution or the location of the population. And hardly anyone, including Altieri and Rosset, will argue about the pragmatism of population projections. So in the absence of a good alternative—and in the face of a proven slowdown in the productivity gains from the Green Revolution—biotechnology is by default our best, and maybe, only way to increase production to meet future food needs.

My objective in this article is to challenge misconceptions often put forward about biotechnology. Within this context I challenge many of Altieri and Rosset's arguments which are not generally supported by existing scientific evidence. I follow their numbering of arguments to facilitate point-by-point comparisons.

1. The argument that hunger is a complex socioeconomic phenomenon, tied to lack of resources to grow or buy food, is correct. Equally correct is the argument that existing food supplies could adequately feed the world population. But how food and other resources (e.g., land, capital) are distributed among individuals, regions, or the various nations is determined by the complex interaction of market forces and institutions around the world. Unless our civic societies can come up

quickly with an economic system that allocates resources more equitably and more efficiently than the present one, 50 years from now we will be faced with an even greater challenge. Calorie for calorie there will not be enough food to feed the projected population of about 9 billion. With the purchasing power and wealth concentrated in the developed countries, and over 90 percent of the projected population growth likely to occur in developing and emerging economies, it is not difficult to predict where food shortages will occur. Unless we are ready to accept starvation, or place parks and the Amazon Basin under the plough, there really is only one good alternative: discover ways to increase food production from existing resources. Bottom line, Altieri and Rosset may want to argue against Western-style capitalism and market institutions if they so choose to—but their argument is hardly relevant to the issue of biotechnology.

2. The assertion that most innovations in biotechnology are not need driven is incorrect. Here are a few well-documented examples of biotechnology innovations targeting pressing needs:

- Development of a rice strain that has the potential to prevent blindness in millions of children whose diets are deficient in Vitamin A. Vitamin A is a highly essential micronutrient and wide-spread dietary deficiency of this vitamin in rice-eating Asian countries has tragic undertones: five million children in South East Asia develop an eye disease called xerophthalmia every year, and 250,000 of them eventually become blind. Improved vitamin A nutrition would alleviate this serious health problem and, according to United Nations Children's Fund (UNICEF), could also prevent up to two million infant deaths because vitamin A deficiency predisposes them to diarrhea diseases and measles. A research team led by Ingo Potrykus of the Swiss Federal Institute of Technology in Zurich, in collaboration with scientists from the University of Freiburg in Germany, has succeeded in producing the precursor to this vitamin, beta-carotene in rice (Potrykus, 1999).

- Development of rice strains with increased iron content and lowered antinutrients. Approximately 30% of the world's population suffers from iron deficiency, especially in less developed countries. Anemia characterized by low hemoglobin is the most widely recognized symptom of iron deficiency, but there are other serious problems such as impaired learning ability in children, increased susceptibility to infection and reduced work capacity. An adequate supply of iron is crucial during the first two years of life because of rapid body growth. Yet the body can use less than 20% of ingested iron. Most iron found in the soil is in the ferric state, an ionic form that can not be utilized until it is converted to the ferrous form. Plants can convert ferric to ferrous iron; however, humans lack the enzyme needed for such conversion. One approach to treating iron deficiency in people is to create plants that contain more iron. The gene for ferritin, an iron-rich soybean storage protein, has been introduced into rice under the control of an endosperm-specific promoter. Grains from transgenic rice plants contained three times more iron than normal rice. The bioavailabiiity of the mineral has been increased also through biotechnology. Seeds store the phosphorous needed for germination in the form of phytate, which is an anti-nutrient because it strongly chelates iron, calcium, zinc and other divalent mineral ions, making them unavailable for uptake. The same Swiss group that created beta-carotene rice has developed a series of transgenic rice lines designed to deal with this problem by introducing a gene that encodes phytase, an enzyme that breaks down phytate. In addition, sulfur-containing proteins enhance iron reabsorption, so to further promote the reabsorption of iron, a gene for a cystein-rich metallothionein-like protein has also been engineered into rice by Potrykus (Goto et al., 1999; Potrykus, 1999).

- Improvements to hybrid rice by introducing the gene of interest directly into maintainer or restorer lines. Early results at transforming rice with the nodulin gene indicate that this staple can be colonized by bacteria that fix nitrogen from the atmosphere. This would improve productivity in the absence of synthetic fertilizers, which are typically unavailable to resource-poor farmers in less developed countries (LDCs) (Dowling, 1998).
- Edible vaccines, delivered in locally grown crops, could do more to eliminate disease than the Red Cross, missionaries, and United Nations (UN) task forces combined, at a fraction of the cost (Arakawa et al., Tacket et al., Hag et al.).

All these and numerous other technologies are being advanced and directed towards resource-poor farmers and locations. Biotechnology is being advanced and directed towards resource-poor farmers and locations. Altieri and Rosset ignore the substantial technology pipeline and the efforts of thousands of scientists across the world to safeguard food safety and improve human nutrition and quality of life. They prefer to focus exclusively on the earliest biotechnology products that were broadly commercialized, Bt *(Bacillus thuringiensis)* and Roundup Ready technologies. Equally absent in Altieri and Rosset's arguments is an elementary understanding of market-economics and innovation dynamics.

In market-driven economies, need and profit are closely connected. Companies, large and small, profit only when they offer products and services that address needs and induce willingness to pay. Bt and Roundup Ready technologies have been adopted faster than any other agricultural innovation on record (Kalaitzandonakes, 1999). These adoption levels have taken place despite abundant supplies of conventional seed with which farmers can exercise their "age-old right to save and replant." The reason for the quick adoption, of course, is that farmers profit from the use of such technologies through reduced chemical sprays, improved yields, labor savings, shifts to reduced tillage systems and other benefits (Maagd et al. 1999; Abelson & Hines, 1999). Over half of all economic benefits generated by these technologies have gone to farmers, more than what has been appropriated by biotechnology and seed companies combined (Traxler & Falk-Zepeda, 1999; Falk-Zepeda, Traxler & Nelson, in press).

3. The argument that the integration of chemical pesticides and seed use has led to lower returns for farmers is incorrect. To support their argument Altieri and Rosset reference an obscure manuscript while they ignore several comprehensive studies that point to increased net returns and reduced chemical loads (Rice, 1999; Klotz-Ingram et al., 1999; Falk-Zepeda, Traxler & Nelson, in press, Gianessi, 1999; Abelson & Hines, 1999; USDA/ERS, 1999a; USDA/ERS, 1999b).

Because of their improved production economics, the introduction of Bt and herbicide resistant crops have forced tremendous competition in herbicide and insecticide markets. Prices of many herbicides and insecticides have been slashed by over 50% in these markets in order to compete with the improved economics of biotechnology seed/chemical solutions. Such price reductions have led to significant discounting of weed and insect control programs and have benefited even farmers who have not adopted biotechnology crops. Because of lower prices and reduced volumes of synthetic pesticides from the use of biotechnology crops, the agrichemicals sector has experienced significant financial losses over the last two to three years.

There is ample evidence to suggest that Altieri and Rosset's assertion that "the integration of seed and chemical industries appears destined to (deliver) lower returns" is incorrect. What is surprising, however, is the lack of rudimentary understanding of farm economics and decision making. Why would thousands of farmers adopt technologies that lead them to losses year after year while conventional seed and pesticide solutions are readily available and cheaper than before the introduction of biotechnology crops?

4. The assertion that "genetically engineered seeds do not increase the yield of crops" is misleading. Generally, Bt-type technologies are expected to increase yields while herbicide-resistant technologies are expected to reduce costs and input use. Conventional weed control programs applied on conventional seed may be as effective in controlling weeds as herbicide resistant plants and are expected to yield similarly. However, conventional weed treatment programs are expected, on average, to cost more and involve larger amounts of synthetic pesticides. In addition herbicide tolerant crops eliminate the need for pre-emergent spraying with far less benign herbicides. On the other hand, Bt-crops enjoy greater protection from hard to control insect pests relative to conventional plants that are applied to chemical insecticides. As a result, when insect pests exceed certain thresholds, Bt-crops are expected to yield better. Such effects will tend to vary from one region to another and from one year to another as insect pest pressures and weed infestations tend to be variable.

To effectively measure the yield and cost impacts of biotechnologies, one must control for all other variations (e.g., year-to-year weather and pest infestation variation, variability in seeding rates, differences in farming systems, and so on). Currently, a small number of studies have measured the yield impacts under proper statistical controls. In these few cases, adoption of herbicide resistance and insect resistance was generally associated with increases in yields and variable profits (Klotz et al., 1999; Falk-Zepeda, Traxler & Nelson, in press, Maagd et al., 1999; Abelson & Hines, 1999).

5. The assertion that "there are potential risks of eating (bioengineered) foods" is alarmist. Citing unspecified "recent evidence" Altieri and Rosset fail to acknowledge the extensive scientific evidence that consistently finds that the use of biotechnology methods and biotechnology products poses risks no different from those of other genetic methods and products. The Food and Drug Administration (FDA) has evaluated technical evidence on all proteins produced through biotechnology and which are currently in commercial food products. All of the proteins that have been placed into foods through the use of biotechnology and are currently on the market are non-toxic, sensitive to heat, acid and enzymatic digestion, and hence rapidly digestible, and have no structural similarities with proteins known to cause allergies (Thompson, 2000). Under their oversight structure, the FDA does not routinely subject foods from new plant varieties to premarket review or to extensive scientific safety tests, although there are exceptions. The agency has judged that the usual safety and quality control practices used by plant breeders, such as chemical and visual analyses and taste testing, are generally adequate for ensuring food safety. Additional tests are performed, however, when suggested by the product's history of use, composition, or characteristics.

Similarly, the argument that insertion of new DNA can alter the metabolism of plants or animals causing them to produce new allergens and toxins is deceptive. For one thing, these kinds of changes can happen through natural mutations or with any type of plant transformation (e.g., through traditional breeding or bioengineering). For another, newly developed plants (resulting from traditional breeding or bioengineering) are subjected to extensive testing that demonstrates that such plants look and grow normally, and have the expected levels of nutrients and toxins. Extensive scientific evidence suggests that there are no food safety issues with bioengineered plants (ibid). Presence of a substance that is completely new to the food supply or of an allergen presented in an unusual or unexpected way (for example, a peanut protein transferred to a potato) invokes greater scrutiny by the agency. This focus by the FDA on safety-related characteristics, rather than on the method by which the plant was genetically modified, reflects the scientific consensus that "the same physical and biological laws govern the response of organisms modified by modern molecular and cellular methods and those produced by classical methods," and, therefore, "no conceptual distinction exists" (National Research Council, 1989).

Finally, Altieri and Rosset assert that Roundup Ready soybeans are nutritionally inferior due to reduced quantities of isoflavons, known anti-cancer agents. Yet, to date there exist no studies that properly control for variations in the supply of water, light, minerals, pests and even germplasm, all of which are known to affect the amounts of isoflavons in soybeans, as they assess the effect of trans-genes on such amounts (e.g., Taylor & Hefle, 1999).

Over the years, scientists working on bioengineered crops have used strict scientific principles and thorough analysis to confirm for themselves and the public that the genes and techniques used are safe for the consumer and the environment. The most we can ask is that all foods produced by whatever method receive the same level of evaluation. Millions of people have already consumed the products of genetic engineering and no adverse effects have been reported or demonstrated. Scientists are confident in the validity of the system that regulates and oversees the food supply.

6. The argument that the new bioengineered varieties will fail, as pests develop resistance to the natural Bt-toxins produced by these varieties because they violate the basic principles of integrated pest management (IPM), is misleading. Pests tend to overcome any control mechanism, including those introduced through biotechnology, synthetic pesticides, or even the broader integrated approaches suggested by Altieri and Rosset. In biology no solutions are permanent. Once selection pressure is applied on a population, that population is effectively enriched for resistant organisms. That is why it is imperative to develop a multi-pronged approach. Integrating crop rotation and ecology with biotechnology is not only feasible but also the logical way to progress. Indeed biotechnology companies like Ecogen and AgraQuest use biotechnology to identify and enrich natural predators of damaging pests.

However, biotechnology supplies yet one more mode of defense. For instance, many variations and combinations of Bt genes are currently being produced to minimize pest selection pressure. Indeed, Altieri and Rosset are incorrect when they drive a parallel with the "one pest–one pesticide" paradigm. Biotechnology is striving for a "one pest–many genes" paradigm. Molecular biologists recognize the need to study and apply multiple and diverse mechanisms for controlling pests and pathogens to reduce selection pressure. Simultaneous or sequential deployment of different resistance genes has the same rationale as crop rotation. Pathogen evolution is less able to overcome a changing environment or an environment made inhospitable by an array of resistance genes.

There are many sources of resistance genes in addition to those found in nature. Combinations and re-combinations of genes may be used or completely synthetic genes can be developed. By having a range of gene products with subtle variations produced, for example through directed evolution (a technology that mimics the natural process of evolution and brings together advances in molecular biology and classical breeding), or, by creating suites of synthetic genes which the target pest would never encounter in nature, the selection for resistance is greatly reduced. Diverse mechanisms of action of gene products can also be employed to reduce selection pressure through a technique called gene pyramiding whereby genes with very different modes of action, such as chitinases, feeding inhibitors, maturation inhibitors, and so on, are used in combination. The probability of any single organism overcoming all of these diverse strategies is vanishingly small. Finally, use of refuges where conventional crops are planted alongside of bioengineered ones can further reduce pest selection. The recent refuge regulation introduced by the Environmental Protection Agency (EPA) targets long-term protection from selection and development of resistance among pests. In conclusion, not only can biotechnology be integrated with ecological and other pest management methods, it also supplies several new modes of action, thereby enriching IPM.

7. The argument that biotechnology crops have been commercialized without proper testing while posing risks to human health and the environment is incorrect. Biotechnology crops and foods have been massively tested over the years both in the laboratory and in controlled natural environments under the oversight of the EPA, the FDA and the Animal & Plant Health Inspection Service/United States Department of Agriculture (APHIS-USDA). Over 4,000 field tests have been performed in some 18,000 sites throughout the United States over the last 15 years for efficacy, performance and suitability for release in the environment. Thousands of similar field tests have been performed in other countries around the world. Volumes of data have been generated on the food safety of bioengineered foods as well, with no evidence of safety risks as indicated above.

Effective procedures of field testing and food safety assessment have been developed after careful consideration and subject to scientific standards (for example see National Research Council, 1989; Report of a Joint Food & Agriculture Organization/World Health Organization [FAO/WHO] Consultation, 1991; Organization for Economic and Cooperation on Development [OECD], 1993).

Altieri and Rosset fail to explain precisely how the FDA, EPA, APHIS-USDA and the vast majority of the scientific community, in undertaking more than 20 years of extensive assessment of biosafety claims, have been negligent. More importantly they should provide stronger scientific evidence in support of their arguments. Specifically:

- The argument that adoption of biotechnology crops is "creating genetic uniformity" inducing vulnerability to new matching strains of pathogens is incorrect. Transgenes are added to existing locally adopted germplasm and have no inherent influence on the genetic variation of the varieties planted. For example, there are over one thousand Roundup Ready varieties of soybeans cultivated in the United States alone. Hence, adoption of biotechnology has not increased the vulnerability of germplasm to homogeneous or other strains of pathogens and has not led to genetic erosion. Quite the opposite. Biotechnology tools are allowing traditional varieties to be revived and safeguarded (see for example Woodward et al., in this issue) or develop new genetic variation.
- The argument that herbicide resistant crops "reduce agrobiodiversity" is incorrect. While minimal restrictions are put on specific rotations (e.g., fol-

lowing Roundup Ready corn with Round-Ready soybeans), equally minimal planning can easily by-pass such restrictions. Indeed, herbicide resistant plants improve agrobiodiversity by encouraging minimum tillage and no-tillage cultivation systems. Unlike conventional tillage, which controls weed growth by plowing and cultivating, no-till agriculture depends on selective herbicides to kill weeds. The resulting vegetation detritus protects seedlings when they are most vulnerable. Soil erosion is reduced. Beneficial insects in the debris are protected. And the till-less technique reduces equipment, fuel, and fertilizer needs and, significantly, the time required for tending crops. It also improves soil-aggregate formation, microbial activity in the soil, and water infiltration and storage.

- Assertions that cultivation of herbicide resistant plants will result in "super-weeds" through gene flow are misleading and alarmist. Gene flow is the exchange of genetic information between crops and wild relatives. The movement of genes via pollen dispersal provides, in principle, a mechanism for foreign genes to "escape" from a genetically engineered crop and spread to weedy relatives growing nearby. Gene flow becomes an environmental issue when the associated trait confers some kind of ecological advantage. This is a particular concern in the case of herbicide resistance genes, for example, where transfer of the resistance trait to weedy relatives that are more difficult to control.

The risk of gene flow is not specific to biotechnology. It applies equally well to herbicide resistant plants that have been developed through traditional breeding techniques (e.g., STS soybeans). Moreover, gene flow is a constant concern of plant breeders who worry about unwanted genes flowing into their fields. It is widely recognized that the "superweed" concept is exaggerated. Resistance to a particular herbicide, if and when developed, implies that use of other herbicides may be necessary for effective control. Currently, there exist effectual alternative chemistries for most economically relevant weeds.

These arguments aside, gene flow is possible and could deem certain chemistries ineffective. The questions then become how possible is such gene flow and what are some alternative strategies that may be used to address the potential risk.

It is important to remember that for any transgene to spread (nuclear or plastomic), there must be successful hybrid formation between a sexually compatible crop plant and recipient species. The two species must flower at the same time, share the same insect pollinator (if insect-pollinated), and be close enough in space to allow for the transfer of viable pollen. Thus, the transfer of transgenes will depend on the sexual fertility of the hybrid progeny, their vigor and sexual fertility in subsequent generations, and the selection pressure on the host of the resident transgene.

There are also strategies to reduce the, however small, risk of gene flow from transgenic crops. One possibility is the use of male sterile plants, which works well but is limited to a few species. For the many crops in which chloroplasts are strictly maternally inherited, which is to say not transmitted through pollen, transformation of the chloroplast genome should provide an effective way to contain foreign genes. Henry Daniel and colleagues at Auburn University introduced a gene for herbicide resistance into tobacco, showed that it was stably integrated into the chloroplast genome, and demonstrated that transgenic plants contained only transformed chloroplasts. This result advances the potential for chloroplast transformation to be an effective strategy to manage the risk of gene flow (Daniel et al., 1998).

To test the theory of gene flow for herbicide tolerant genes introduced through chloroplast transformation, Scott and Wilkinson (1999) studied a 34-km region near the Thames River, United Kingdom where oilseed rape is cultivated in the vicinity of a native weed, wild rapeseed. Oilseed rape, the cultivated form of Brassica napus, and the wild rapeseed are capable of exchanging pollen to produce viable hybrids. The study was designed to determine whether oilseed chloroplasts could be transferred to wild rapeseed, and how long the hybrids and maternal oilseed plants would survive in the wild. To identify chloroplasts, the authors created primers specific to chloroplast DNA non-coding regions. In PCR experiments, oilseed chloroplasts produced a single amplification product of 600 bp, whereas wild rapeseed produced a 650 bp product. In all cases, the chloroplasts from hybrid plants contained the PCR product of the maternal line demonstrating that they are not transferred in pollen.

The authors studied the frequency of hybrid formation and viability of oilseed and hybrids in non-cultivated areas over a three-year period. Their studies show that oilseed has a very low survival rate outside cultivated fields. On average, only 12–19% of oilseed survived each growing season. At the same time, a very low level of natural hybridization was observed (0.4–1.5%). Taken together, the results indicate that there is a very low possibility of transgene movement into feral populations of maternal lineage. However, the persistence of the maternal line in the wild will be of limited duration.

Assertions about the impacts of Bt-crops on non-target insects are misleading. Reports of the potential for effects from these Bt-corn hybrids on monarch butterflies or other lepidoptera are not new. They have been reported in the scientific literature and regulatory review documents since at least 1986. The environmental protection agency has been provided data on the potential for impacts on non-target species from Bt pollen for years. Their analyses indicated that, when compared with the numerous other relevant factors, the impacts from such pollen were likely to be negligible. Despite popular belief, Losey et al. (1999) demonstrated nothing new other than that force feeding monarch caterpillars is still not as hazardous as using chemical insecticides.

Indeed, the use of Bt-crops may have a positive impact on biodiversity. Ongoing monitoring by companies of Bt-corn fields since their introduction shows that insect biodiversity and population densities in Bt-corn fields are significantly higher than in fields treated with chemical pesticide sprays. Bt-corn may help enhance beneficial insect populations that would otherwise be threatened by the use of pesticidal sprays. This could lead to benefits for, among others, insect eating birds and small mammals.

Strategies to minimize impact on non-target insects are also being developed. For example, the current generation of Bt-corn is aimed at reducing crop losses to an imported pest from Europe, the European corn borer. This pest eats corn stalks. Varieties of corn are already under development that could express Bt or other genes of similar effect only in corn stalks, and not in other parts of the crop (e.g., leafs, pollen). Likewise chloroplast transformation described above will eliminate expression in pollen. Such corn varieties would also eliminate entirely any risks to non-target organisms that might come from Bt containing pollen.

The issue of vector recombination and creation of new viruses has been considered by scientists independently and in specific forums. For example, the USDA-APHIS and the American Institute of Biological Sciences convened a workshop in 1995, to address risk issues associated with the possible generation of new plant viruses in transgenic plants expressing viral genes that confer virus resistance. Most workshop participants believed that current data obtained from laboratory and field research indicate the risk associated with the generation of new plant viruses through recombination is minimal and should not be a limiting factor to large-scale field tests or commercialization of transgenic plants expressing viral transgenes. Genomic viral RNA transcapsidated with coat protein produced by a transgenic plant should not have long-term effects, since the genome of the infecting virus is not modified. Similarly, synergistic interactions between an infecting virus and a viral transgene should not have long-term impacts on the agricultural production. The weight of opinion, though, was that, given time and opportunity, all viral recombinations are possible. With or without the use of transgenic plants, new plant viruses will develop that will require attention. Hence, this is an area where additional research is needed.

8. Many of Altieri and Rosset's "unanswered ecological questions regarding the impact of transgenic crops" are not unanswered. Indeed, there is a substantial body of knowledge and volumes of data on both the environmental and food safety of biotechnology crops and foods demonstrating their overall suitability. This is not to say that environmental and other impact assessment of biotechnology crops should not be expanded. Indeed, more impact assessment studies are needed to augment and expand the existing empirical evidence, answer any unanswered questions and put risks and benefits of biotechnology crops and foods in a proper perspective. This need is explicitly recognized in a recent report of a Task Force reporting to the Land Grant University and Extension Service administrators which placed high priority on assessment studies (ESCOP/ECOP Report, 2000).

9. Altieri and Rosset misrepresent the position of CGIAR and their research direction. Indeed Ismail Serageldin, chairman of the CGIAR, noted that, a priori,

biotechnology could contribute to food security by helping to promote sustainable agriculture centered on smallholder farmers in developing countries. Furthermore, they misrepresent the potential of "rotations, inter-cropping and biological control agents" as singular solutions for environmentally sound and productive agriculture. Despite Altieri and Rosset's indirect references to scientific evidence which they report has confirmed repeatedly the dramatic effects of such methods, the evidence in the published literature remains scant.

Crop rotation has been with us since the manor system of medieval times. And although there are no regulatory or technological barriers to its use, it has had only modest adoption by producers because of the limitation it places on resource management and because of its economics. In and of itself crop rotation has not proven to be the singular solution to our increasing food demand problem.

Use and commercialization of biological agents in crop production has also been limited despite decades of research both in the private and public sectors. Companies like Ciba, DuPont, American Cyanamid and various startups like Mycogen invested millions of dollars in research of biopesticides and biological agents and, ultimately, disposed them as uneconomical. Even companies that specialize in biological agents and biopesticides, like Ecogen and Agraquest, have focused primarily on high value markets with few chemical pesticide alternatives.

The most misleading aspect of Altieri and Rosset's argument, however, is the artificial dichotomy they draw between biotechnology and agroecology. As amply described above, biotechnology and agroecological approaches are synergistic and should be combined to improve the sustainability of our agriculture and food systems. Altieri and Rosset use an artificial dichotomy to mask an underlying issue: that there is "an urgent need to challenge the patent system and intellectual property rights intrinsic to the WTO." Ultimately, Altieri and Rosset are after market and political institutions that are unrelated to biotechnology.

10. Altieri and Rosset extend their artificial dichotomy further to pass judgement on what kind of agriculture we should have. "Small farmers using agroecological approaches and low input practices," who are presumably discovering better ways to yield more in environmentally benign and socially responsible ways, is the way to go. Again, there is nothing inherent in biotechnology that justifies the small versus large farm dichotomy. Biotechnologies are size neutral and can benefit smallholders and large commercial farmers alike. As Florence Wambugu, director of the International Service for the Acquisition of Agri-Biotech Applications (ISAAA) in Kenya, notes, the great potential of biotechnology to increase agriculture in Africa lies in its "packaged technology in the seed," which ensures technology benefits without changing local cultural practices. In the broader context, one must also question the wisdom of Altieri and Rosset's argument. In the presence of social, environmental and economic advantages they describe, why are smallholder agroecological production systems not quickly dominating?

While laudable in its intent to reduce environmental impact, much of Altieri and Rosset's philosophy is founded on a fallacy. They support a form of farming in which average crop yields on a variety of soils are about half those of intensive farming (Avery, 1999; Evans, 1998; Tillman, 1998). As populations rise, inefficient farming will destroy a much greater quantity of wilderness and its associated wildlife as farming infringes in those areas.

CONCLUDING COMMENTS

Altieri and Rosset's arguments are neither scientifically supported nor even really about biotechnology. Their arguments are primarily directed against Western-type capitalism and associated institutions (e.g., intellectual property rights, the WTO). Biotechnology is used as a Trojan Horse. They fail to acknowledge the scientifically proven potential of biotechnology and the ways it can contribute to environmental sustainability and food security. The developing and developed world will need and use biotechnology in many ways during this century. Those with political battles to fight may want to use other, more appropriate fora to fight them.

NOTE

This article was reprinted with permission from *AgBioforum*, Fall 1999, 9–21.

REFERENCES

Abelson, P. A., and P. J. Hines. 1999. The plant revolution. *Science* 285: 367–68.

Arakawa, T., et al. 1998. Efficacy of a food plant–based oral cholera toxin B subunit vaccine. *Nature Biotechnology* 16: 292–97.

Avery, D. 1999. In *Fearing Food,* edited by J. Morris and R. Bate, 3–18. Oxford: Butterworth-Heinemann.

Daniell, H., R. Datta, S. Varma, S. Gray, and S. Lee. 1998. Containment of herbicide resistance through genetic engineering of the chloroplast genome. *Nature Biotechnology* 16: 345–47.

Dowling, N., S. M. Greenfield, and K. S. Fischer, eds. 1998. *Sustainability of Rice in the Global Food System.* Philippines: International Rice Research Institute.

ESCOP/ECOP Task Force. 2000. *Agricultural Biotechnology: Critical Issues and Recommended Responses from Land Grant Universities.* Report to the Experiment Station Committee on Organization and Policy (ESCOP)/Extension Committee on Organization and Policy (ECOP).

Evans, L. T. 1998. *Feeding the Ten Billion.* Cambridge: Cambridge University Press.

Falk-Zepeda, J. B., G. Traxler, and R. Nelson. In press. Surplus distribution from the introduction of a biotechnology innovation. *American Journal of Agricultural Economics.*

FAO/WHO. 1991. Strategies for Assessing the Safety of Foods Produced by Biotechnology. Consultation. Geneva, Switzerland: World Health Organization.

Gianessi, L. 1999. *Agricultural Biotechnology: Insect Control Benefits.* Washington, D.C.: National Center for Food and Agricultural Policy.

Goto, F. T. Yoshihara, N. Shigemoto, S. Toki, and F. Takaiwa. 1999. Iron fortification of rice seed by the soybean ferritin gene. *Nature Biotechnology* 17: 282–86.

Haq, T., H. S. Mason, J. D. Clements, and C. J. Amtzen. 1995. Oral immunization with a recombinant bacterial antigen produced in transgenic plants. *Science* 268: 714–16.

Kalaitzandonakes, N. 1999. A farm-level perspective on agrobiotechnology: How much value and for whom? *AgBioForum* 2, no. 1: 61–64. Available on the World Wide Web: <www.agbioforum.missouri.edu>.

Klotz-Ingram, C., S. Jans, J. Femandez-Comejo, and W. McBride. 1999. Farm-level production effects related to the adoption of genetically modified cotton for pest management. *AgBioForum* 2, no. 2: 73–84. Available on the World Wide Web: <www.agbioforum.missouri.edu>.

Losey, J. E., L. S. Rayor, and M. E. Carter. 1999. Transgenic pollen harms monarch larvae. *Nature* 399: 214.

Maagd, R. A., et al. 1999. *Bacillus thuringiensis* toxin-mediated insect resistance in plants. *Trends in Plant Science* 4: 9–13.

National Research Council. 1989. *Field Testing Genetically Modified Organisms: Framework for Decisions.* Washington, D.C.: National Academy Press.

Organization for Economic Cooperation and Development (OECD). 1993. *Safety Evaluation of Foods Derived by Modern Biotechnology.* Paris: OECD.

Potrykus, I. 1999. Vitamin-A and iron-enriched rices may hold key to combating blindness and malnutrition: A biotechnology advance. *Nature Biotechnology* 17: 37.

Rice, M. 1999. *Farmers Reduce Insecticide Use with Bt Corn.* Iowa: Iowa State University.

Ruttan, V. W. 1999. Biotechnology and agriculture: A skeptical perspective. *AgBioForum* 2, no. 1: 54–60. Available on the World Wide Web: <www. agbioforum.missouri. edu>.

Scott, S., and M. J. Wilkinson. 1999. Low probability of chloroplast movement from oilseed rape *(Brassica napus)* into wild *Brassica rapa. Nature Biotechnology* 17: 390–92.

Serageldin, I. 1999. Biotechnology and food security in the 21st century. *Science* 285: 387–89.

Tacket, C. O., H. S. Mason, G. Losonsky, J. D. Clements, M. M. Levine, and C. J. Amtzen. 1998. Immunogenicity of a recombinant bacterial antigen delivered in a transgenic potato. *Nature Medicine* 4: 607–9.

Taylor, S. L., and S. L. Hefle. 1999. Seeking clarity in the debate over the safety of GM foods. *Nature* 402: 575.

Thompson, L. 2000. Are bioengineered foods safe? *FDA Consumer* 34, no. 1: 1–5. Available on the World Wide Web: <www.fda.gov.fdac/features,2000/100 bio.html>.

Tilman, D. 1998. The greening of the green revolution. *Nature* 396: 211–12.

Traxler, G., and J. Falck-Zepeda. 1999. The distribution of benefits from the introduction of transgenic cotton varieties. *AgBioForum* 2, no. 2: 94–98. Available from the World Wide Web: <www.agbioforum.missouri.edu>.

USDA/ERS. 1999a. *Impacts of Adopting Genetically Engineered Crops in the U.S.: Preliminary Results.* Washington, D.C.: United States Department of Agriculture (USDA), Economic Research Service.

USDA/ERS. 1999b. *Genetically Engineered Crops for Pest Management.* Agricultural Resource Management Study (ARMS). Washington, D.C.: USDA/ERS.

Wambugu, F. 1999. Why Africa needs agricultural biotech. *Nature* 400: 15–16.

Ye, X., S. Al-Babili, A. Kloti, J. Zhang, P. Lucca, P. Beyer, and I. Potrykus. 2000. Engineering the provitamin A (beta-carotene) biosynthetic pathway into (carotenoid-free) rice endosperm. *Science* 287: 303–5.

Ten Reasons Why Biotechnology Will Not Ensure Food Security, Protect the Environment, or Reduce Poverty in the Developing World

Miguel A. Altieri and Peter Rosset

Our objective is to challenge the notion of biotechnology as a magic bullet solution to all of agriculture's ills, by clarifying misconceptions concerning these underlying assumptions.

1. There is no relationship between the prevalence of hunger in a given country and its population. For every densely populated and hungry nation like Bangladesh or Haiti, there is a sparsely populated and hungry nation like Brazil or Indonesia. The world today produces more food per inhabitant than ever before. Enough food is available to provide 4.3 pounds for every person every day: 2.5 pounds of grain, beans and nuts, about a pound of meat, milk and eggs and another of fruits and vegetables. The real causes of hunger are poverty, inequality and lack of access to food and land. Too many people are too poor to buy the food that is available (but often poorly distributed) or lack the land and resources to grow it themselves (Lappe, Collins & Rosset, 1998).

2. Most innovations in agricultural biotechnology have been profit-driven rather than need-driven. The real thrust of the genetic engineering industry is not to make third world agriculture more productive, but rather to generate profits (Busch et al., 1990). This is illustrated by reviewing the principal technologies on the market today: (1) herbicide resistant crops, such as Monsanto's "Roundup Ready" soybeans, seeds that are tolerant to Monsanto's herbicide Roundup, and (2) "Bt" (*Bacillus thuringiensis*) crops which are engineered to produce their own insecticide. In the first instance, the goal is to win a greater herbicide market-share for a proprietary product and, in the second, to boost seed sales at the cost of damaging the usefulness of a key pest management product (the *Bacillus thurin-*

giensis based microbial insecticide) relied upon by many farmers, including most organic farmers, as a powerful alternative to insecticides.

These technologies respond to the need of biotechnology companies to intensify farmers' dependence upon seeds protected by so-called intellectual property rights which conflict directly with the age-old rights of farmers to reproduce, share or store seeds (Hobbelink, 1991). Whenever possible corporations will require farmers to buy a company's brand of inputs and will forbid farmers from keeping or selling seed. By controlling germplasm from seed to sale, and by forcing farmers to pay inflated prices for seed-chemical packages, companies are determined to extract the most profit from their investment (Krimsky & Wrubel, 1996).

3. The integration of the seed and chemical industries appears destined to accelerate increases in per acre expenditures for seeds plus chemicals, delivering significantly lower returns to growers. Companies developing herbicide tolerant crops are trying to shift as much per acre cost as possible from the herbicide onto the seed via seed costs and technology charges. Increasingly price reductions for herbicides will be limited to growers purchasing technology packages. In Illinois, the adoption of herbicide resistant crops makes for the most expensive soybean seed-plus-weed management system in modern history—between $40.00 and $60.00 per acre depending on fee rates, weed pressure, and so on. Three years ago, the average seed-plus-weed control costs on Illinois farms was $26 per acre, and represented 23% of variable costs; today they represent 35–40% (Benbrook, 1999). Many farmers are willing to pay for the simplicity and robustness of the new weed management system, but such advantages may be short-lived as ecological problems arise.

4. Recent experimental trials have shown that genetically engineered seeds do not increase the yield of crops. A recent study by the United States Department of Agriculture (USDA) Economic Research Service shows that in 1998 yields were not significantly different in engineered versus non-engineered crops in 12 of 18 crop/region combinations. In the six crop/region combinations where Bt crops or herbicide tolerant crops (HTCs) fared better, they exhibited increased yields between 5 and 30%. Glyphosphate tolerant cotton showed no significant yield increase in either region where it was surveyed. This was confirmed in another study examining more than 8,000 field trials, where it was found that Roundup Ready soybean seeds produced fewer bushels of soybeans than similar conventionally bred varieties (USDA, 1999).

5. Many scientists claim that the ingestion of genetically engineered food is harmless. Recent evidence, however, shows that there are potential risks in eating such foods as the new proteins produced in such foods could: (1) act themselves as allergens or toxins; (2) alter the metabolism of the food producing plant or animal, causing it to produce new allergens or toxins; or (3) reduce its nutritional quality or value. In the case of (3), herbicide resistant soybeans can contain less isoflavones, an important phytoestrogen present in soybeans, believed to protect

women from a number of cancers. At present, developing countries are importing soybean and corn from the United States, Argentina, and Brazil. Genetically engineered foods are beginning to flood the markets in the importing countries, yet no one can predict all their health effects on consumers, who are unaware that they are eating such food. Because genetically engineered food remains unlabeled, consumers cannot discriminate between genetically engineered (GE) and non-GE food, and should serious health problems arise, it will be extremely difficult to trace them to their source. Lack of labeling also helps to shield the corporations that could be potentially responsible from liability (Lappe & Bailey, 1998).

6. Transgenic plants which produce their own insecticides closely follow the pesticide paradigm, which is itself rapidly failing due to pest resistance to insecticides. Instead of the failed "one pest–one chemical" model, genetic engineering emphasizes a "one pest–one gene" approach, shown over and over again in laboratory trials to fail, as pest species rapidly adapt and develop resistance to the insecticide present in the plant (Alstad & Andow, 1995). Not only will the new varieties fail over the short to medium term, despite so-called voluntary resistance management schemes (Mallet & Porter, 1992), but in the process they may render useless the natural Bt-pesticide which is relied upon by organic farmers and others desiring to reduce chemical dependence. Bt crops violate the basic and widely accepted principle of integrated pest management (IPM), which is that reliance on any single pest management technology tends to trigger shifts in pest species or the evolution of resistance through one or more mechanisms (NRC, 1996). In general, the greater the selection pressure across time and space, the quicker and more profound the pest's evolutionary response. An obvious reason for adopting this principle is that it reduces pest exposure to pesticides, retarding the evolution of resistance. But when the product is engineered into the plant itself, pest exposure leaps from minimal and occasional to massive and continuous exposure, dramatically accelerating resistance (Gould, 1994). *Bacillus thuringiensis* will rapidly become useless, both as a feature of the new seeds and as an old standby sprayed when needed by farmers that want out of the pesticide treadmill (Pimentel et al., 1989).

7. The global fight for market share is leading companies to massively deploy transgenic crops around the world (more than 30 million hectares in 1998) without proper advance testing of short- or long-term impacts on human health and ecosystems. In the United States, private sector pressure led the White House to decree "no substantial difference" between altered and normal seeds, thus evading normal Food and Drug Administration (FDA) and Environmental Protection Agency (EPA) testing. Confidential documents made public in an on-going class action lawsuit have revealed that the FDA's own scientists do not agree with this determination. One reason is that many scientists are concerned that the large scale use of transgenic crops poses a series of environmental risks that threaten the sustainability of agriculture (Goldberg, 1992; Paoletti & Pimentel, 1996; Snow &

Moran, 1997; Rissler & Mellon, 1996; Kendall et al., 1997; Royal Society, 1998). These risk areas are as follows:

- The trend to create broad international markets for single products is simplifying cropping systems and creating genetic uniformity in rural landscapes. History has shown that a huge area planted to a single crop variety is very vulnerable to new matching strains of pathogens or insect pests. Furthermore, the widespread use of homogeneous transgenic varieties will unavoidably lead to "genetic erosion," as the local varieties used by thousands of farmers in the developing world are replaced by the new seeds (Robinson, 1996).

- The use of herbicide resistant crops undermines the possibilities of crop diversification, thus reducing agrobiodiversity in time and space (Altieri, 1994).

- The potential transfer through gene flow of genes from herbicide resistant crops to wild or semidomesticated relatives can lead to the creation of super-weeds (Lutman, 1999).

- There is potential for herbicide resistant varieties to become serious weeds in other crops (Duke, 1996; Holt & Le Baron, 1990).

- Massive use of Bt crops affects non-target organisms and ecological processes. Recent evidence shows that the Bt toxin can affect beneficial insect predators that feed on insect pests present on Bt crops (Hilbeck et al., 1998). In addition, windblown pollen from Bt crops, found on natural vegetation surrounding transgenic fields, can kill non-target insects such as the monarch butterfly (Losey et al., 1999). Moreover, Bt toxin present in crop foliage plowed under after harvest can adhere to soil colloids for up to 3 months, negatively affecting the soil invertebrate populations that break down organic matter and play other ecological roles (Donnegan et al., 1995; Palm et al., 1996).

- There is potential for vector recombination to generate new virulent strains of viruses, especially in transgenic plants engineered for viral resistance with viral genes. In plants containing coat protein genes, there is a possibility that such genes will be taken up by unrelated viruses infecting the plant. In such situations, the foreign gene changes the coat structure of the viruses and may confer properties, such as changed method of transmission between plants. The second potential risk is that recombination between RNA virus and a viral RNA inside the transgenic crop could produce a new pathogen leading to more severe disease problems. Some researchers have shown that recombination occurs in transgenic plants and that under certain conditions it produces a new viral strain with altered host range (Steinbrecher, 1996).

Ecological theory predicts that the large-scale landscape homogenization with transgenic crops will exacerbate the ecological problems already associated with monoculture agriculture. Unquestioned expansion of this technology into developing countries may not be wise or desirable. There is strength in the agricultural diversity of many of these countries, and it should not be inhibited or reduced by extensive monoculture, especially when consequences of doing so result in serious social and environmental problems (Altieri, 1996).

Although the ecological risks issue has received some discussion in government, international, and scientific circles, discussions have often been pursued from a narrow perspective that has downplayed the seriousness of the risks (Kendall et al., 1997; Royal Society, 1998). In fact, methods for risk assessment of transgenic crops are not well developed (Kjellsson & Simmsen, 1994) and there is justifiable concern that current field biosafety tests tell little about potential environmental risks associated with commercial-scale production of transgenic crops. A main concern is that international pressures to gain markets and profits are resulting in companies releasing transgenic crops too fast, without proper consideration for the long-term impacts on people or the ecosystem.

8. There are many unanswered ecological questions regarding the impact of transgenic crops. Many environmental groups have argued for the creation of suitable regulation to mediate the testing and release of transgenic crops to offset environmental risks and demand a much better assessment and understanding of ecological issues associated with genetic engineering. This is crucial, as many results emerging from the environmental performance of released transgenic crops suggest that in the development of resistant crops not only is there a need to test direct effects on the target insect or weed, but the indirect effects on the plant. Plant growth, nutrient content, metabolic changes, and effects on the soil and nontarget organisms should all be examined. Unfortunately, funds for research on environmental risk assessment are very limited. For example, the USDA spends only 1% of the funds allocated to biotechnology research on risk assessment, about $1–2 million per year. Given the current level of deployment of genetically engineered plants, such resources are not enough to even discover the "tip of the iceberg." It is a tragedy-in-the-making that so many millions of hectares have been planted without proper biosafety standards. Worldwide such acreage expanded considerably in 1998 with transgenic cotton reaching 6.3 million acres, transgenic corn reaching 20.8 million acres, and transgenic soybean 36.3 million acres. This expansion has been helped along by marketing and distribution agreements entered into by corporations and marketers (i.e., Ciba Seeds with Growmark and Mycogen Plant Sciences with Cargill), and in the absence of regulations in many developing countries. Genetic pollution, unlike oil spills, cannot be controlled by throwing a boom around it.

9. As the private sector has exerted more and more dominance in advancing new biotechnologies, the public sector has had to invest a growing share of its scarce resources in enhancing biotechnological capacities in public institutions, including the CGIAR, and in evaluating and responding to the challenges posed by incorporating private sector technologies into existing farming systems. Such funds would be much better used to expand support for ecologically based

agricultural research, as all the biological problems that biotechnology aims at can be solved using agroecological approaches. The dramatic effects of rotations and intercropping on crop health and productivity, as well as of the use of biological control agents on pest regulation, have been confirmed repeatedly by scientific research. The problem is that research at public institutions increasingly reflects the interests of private funders at the expense of public-good research, such as biological control, organic production systems and general agroecological techniques. Civil society must request more research on alternatives to biotechnology by universities and other public organizations (Krimsky & Wrubel, 1996). There is also an urgent need to challenge the patent system and intellectual property rights intrinsic to the World Trade Organization (WTO) which not only provide multinational corporations with the right to seize and patent genetic resources, but will also accelerate the rate at which market forces already encourage monocultural cropping with genetically uniform transgenic varieties. Based on history and ecological theory, it is not difficult to predict the negative impacts of such environmental simplification on the health of modern agriculture (Altieri, 1996).

10. Much of the needed food can be produced by small farmers located throughout the world using agroecological technologies (Uphoff & Altieri, 1999). In fact, new rural development approaches and low-input technologies spearheaded by farmers and non-governmental organizations (NGOs) around the world are already making a significant contribution to food security at the household, national, and regional levels in Africa, Asia and Latin America (Pretty, 1995). Yield increases are being achieved by using technological approaches, based on agroecological principles that emphasize diversity, synergy, recycling and integration, and social processes that emphasize community participation and empowerment (Rosset, 1999). When such features are optimized, yield enhancement and stability of production are achieved, as well as a series of ecological services such as conservation of biodiversity, soil and water restoration and conservation, improved natural pest regulation mechanisms, and so on (Altieri et al., 1998). These results are a breakthrough for achieving food security and environmental preservation in the developing world, but their potential and further spread depends on investments, policies, institutional support, and attitude changes on the part of policy makers and the scientific community, especially the CGIAR who should devote much of its efforts to the 320 million poor farmers living in marginal environments. Failure to promote such people-centered agricultural research and development due to the diversion of funds and expertise towards biotechnology will forego an historical opportunity to raise agricultural productivity in economically viable, environmentally benign, and socially uplifting ways.

NOTE

This article was reprinted with permission from *AgBioforum,* Fall 1999, 1–8.

REFERENCES

Alstad, D. N., and D. A. Andow. 1995. Managing the evolution of insect resistance to transgenic plants. *Science* 268: 1894–96.

Altieri, M. A. 1994. *Biodiversity and Pest Management in Agroecosystems.* New York: Haworth.

Altieri, M. A. 1996. *Agroecology: The Science of Sustainable Agriculture.* Boulder: Westview.

Altieri, M. A., P. Rosset, and L. A. Thrupp. 1998. *The Potential of Agroecology to Combat Hunger in the Developing World.* 2020 Brief no. 55. International Food Policy Research Institute, Washington, D.C.

Benbrook, C. 1999. "World Food System Challenges and Opportunities: GMOs. Biodiversity and Lessons from America's Heartland." Unpublished manuscript.

Busch, L., W. B. Lacey, J. Burkhardt, and L. Lacey. 1990. *Plants, Power, and Profit.* Oxford: Basil Blackwell.

Casper, R., and J. Landsmann. 1992. "The Biosafety Results of Field Tests of Genetically Modified Plants and Microorganisms." In *Proceedings of the Second International Symposium Goslar,* edited by P. K. Landers, 89–97. Germany.

Donnegan, K. K., C. J. Palm, V. J. Fieland, L. A. Porteous, L. M. Ganis, D. L. Scheller, and R. J. Seidler. 1995. "Changes in Levels, Species, and DNA Fingerprints of Soil Micro Organisms Associated with Cotton Expressing the *Bacillus thuringiensis* Var. Kurstaki Endotoxin." *Applied Soil Ecology* 2: 111–24.

Duke, S. O. 1996. *Herbicide Resistant Crops: Agricultural, Environmental, Economic, Regulatory, and Technical Aspects.* Boca Raton: Lewis.

Goldberg, R. J. 1992. "Environmental Concerns with the Development of Herbicide-Tolerant Plants." *Weed Technology* 6: 647–52.

Gould, F. 1994. "Potential and Problems with High-Dose Strategies for Pesticidal Engineered Crops." *Biocontrol Science and Technology* 4: 451–61.

Hilbeck, A., M. Baumgartner, P. M. Fried, and F. Bigler. 1998. "Effects of Transgenic *Bacillus thuringiensis* Corn Fed Prey on Mortality and Development Time of Immature Chrysoperia Camea Neuroptera: Chrysopidae." *Environmental Entomology* 27: 460–87.

Hobbelink, H. 1991. *Biotechnology and the Future of World Agriculture.* London: Zed.

Holt, J. S., and H. M. Le Baron. 1990. "Significance and Distribution of Herbicide Resistance." *Weed Technology* 4: 141–49.

James, C. 1997. Global status of transgenic crops in 1997 (ISAAA Briefs No. 5). Ithaca, N.Y.: International Service for the Acquisition of Agri-Biotech Application (ISAAA). Available on the World Wide Web: <www.isaaa.org/frbrief5.htiTi>.

Kendall, H. W., R. Beachy, T. Eismer, F. Gould, R. Herdt, P. H. Ravon, J. Schell, and M. S. Swaminathan. 1997. *Bioengineering of Crops.* Report of the World Bank Panel on Transgenic Crops, 1–30. Washington, D.C.: World Bank.

Kennedy, G. G., and M. E. Whalon. 1995. Managing Pest Resistance to *Bacillus thuringiensis* Endotoxins: Constraints and Incentives to Implementation. *Journal of Economic Entomology* 88: 454–60.

Kjellsson, G., and V. Simonsen. 1994. *Methods for Risk Assessment of Transgenic Plants.* Basil, Germany: Birkhauser Verlag.

Krimsky, S., and R. P. Wrubel. 1996. *Agricultural Biotechnology and the Environment: Science, Policy, and Social Issues.* Urbana: University of Illinois Press.

Lappe, M., and B. Bailey. 1998. *Against the Grain: Biotechnology and the Corporate Takeover of Food.* Monroe, Me.: Common Courage Press.

Lappe, F. M., J. Collins, and P. Rosset. 1998. *World Hunger: Twelve Myths.* New York: Grove.

Liu, Y. B., B. E. Tabashnik, T. J. Dennehy, A. L. Patin, and A. C. Bartlett. 1999. "Development Time and Resistance to Bt Crops." *Nature* 400: 519.

Losey, J. J. E., L. S. Rayor, and M. E. Carter. 1999. "Transgenic Pollen Harms Monarch Larvae." *Nature* 399: 214.

Lutman, P. J. W., ed. 1999. "Gene Flow and Agriculture: Relevance for Transgenic Crops." *British Crop Protection Council Symposium Proceedings* 72: 43–64.

Mallet, J., and P. Porter. 1992. "Preventing Insect Adaptations to Insect Resistant Crops: Are Seed Mixtures or Refugia the Best Strategy?" *Proceedings of the Royal Society of London Series B Biology Science* 250: 165–69.

National Research Council (NRC). 1996. *Ecologically Based Pest Management.* Washington, D.C.: National Academy of Sciences.

Palm, C. J., D. L. Schaller, K. K. Donegan, and R. J. Seidler. 1996. "Persistence in Soil of Transgenic Plant Produced *Bacillus thuringiensis* Var. Kustaki-Endotoxin." *Canadian Journal of Microbiology* 42: 1258–62.

Paoletti, M. G., and D. Pimentel. 1996. "Genetic Engineering in Agriculture and the Environment: Assessing Risks and Benefits." *BioScience* 46: 665–71.

Pimentel, D., M. S. Hunter, J. A. LaGro, R. A. Efroymson, J. C. Landers, F. T. Mervis, C. A. McCarthy, and A. E. Boyd. 1989. "Benefits and Risks of Genetic Engineering in Agriculture." *BioScience* 39: 606–14.

Pretty, J. 1995. *Regenerating Agriculture: Policies and Practices for Sustainability and Self-Reliance.* London: Earthscan.

Rissler, J., and M. Mellon. 1996. *The Ecological Risks of Engineered Crops.* Cambridge: MIT Press.

Robinson, R. A. 1996. *Return to Resistance: Breeding Crops to Reduce Pesticide Resistance.* Davis, Calif.: AgAccess.

Rosset, P. 1999. *The Multiple Functions and Benefits of Small Farm Agriculture in the Context of Global Trade Negotiations.* IFDP Food First Policy Brief no. 4. Washington, D.C.: Institute for Food and Development Policy.

Royal Society. 1998. *Genetically Modified Plants for Food Use.* Statement 2/98. London: Royal Society.

Snow, A. A., and P. Moran. 1997. "Commercialization of Transgenic Plants: Potential Ecological Risks." *BioScience* 47: 86–96.

Steinbrecher, R. A. 1996. "From Green to Gene Revolution: The Environmental Risks of Genetically Engineered Crops." *Ecologist* 26: 273–82.

United States Department of Agriculture (USDA). 1999. *Genetically Engineered Crops for Pest Management.* Washington, D.C.: USDA Economic Research Service.

Uphoff, N., and M. A. Altieri. 1999. *Alternatives to Conventional Modern Agriculture for Meeting World Food Needs in the Next Century.* Report of a Bellagio Conference. Ithaca, N.Y.: Cornell International Institute for Food, Agriculture, and Development.

III

FOOD BIOTECHNOLOGY

The food biotechnology discussed in part 3 refers to food products that have been derived by genetic engineering. These products can be divided into two categories: (1) genetically modified organisms and (2) products not containing the recombinant DNA which were derived from genetically modified organisms. Genetically modified organisms are referred to in the popular press as GMOs. The consumed organism contains the genetically engineered DNA and perhaps the recombinant protein. Examples of GMOs are tomatoes that remain fresh off the vine and corn that is insect resistant. Both organisms contain the engineered DNA. A good example of the second category is chymosin, which is a protein product, not a living organism. It is used to make cheese. It is isolated in the factory before distribution from a genetically engineered yeast, and hence it is referred to as a genetically modified product. But it is not a genetically modified organism and does not contain the genetically modified DNA.

EXAMPLES OF GENETICALLY
ENGINEERED PRODUCTS

This categorization is useful in evaluating the ethical, environmental, or social issues surrounding these products. A protein product is not a replicating organism. It does not possess the actual genetically engineered DNA from which it was derived. The engineered DNA cannot be ingested because it is not present with the protein. The protein product cannot accidentally and uncontrollably reproduce, since it is not an organism. Chymosin has been used for many years in the production of cheese without raising as much resistance from consumer groups as GMOs.

The Flavrsavr™ tomato is an interesting example of the first category, GMOs. A genetically engineered protein is not produced in the tomato; rather, a normal gene in the tomato is prevented from being expressed into a normal protein. The normal protein, polygalacturonase, is involved in the process of ripening. When this protein is absent from the tomato, the tomatoes do not rot as quickly in the store. It has a long shelf life. A person is eating a tomato that is missing a normal enzyme or protein. The normal RNA that codes for the polygalacturonase enzyme is destroyed by a genetic engineering technique. The tomato is genetically engineered to contain a gene that makes RNA that is complementary or homologous to the normal polygalacturonase RNA. This is referred to as "antisense" RNA. When the genetically engineered RNA hybridizes or forms a double-stranded RNA with the polygalacturonase RNA, cellular enzymes in the tomato see this double-stranded RNA as an abnormal structure and destroy it. In the process of destroying the double-stranded RNA it also destroys the normal polygalacturonase RNA molecules, and the polygalacturonase protein cannot be made. Consequently, the Flavrsavr™ tomato contains genetically engineered DNA and RNA, but no protein. It has the normal polygalacturonase enzyme removed. This explanation is important in evaluating the safety issues of this tomato as compared with other GMOs, in which recombinant proteins are present in the foods.

An example of a GMO that contains genetically engineered DNA, RNA, and protein is Bt corn. The corn plants are genetically engineered to contain a gene coding for a foreign protein called Bt protein, which is a selective insecticide. This protein is normally produced by a bacteria called *Bacillus thuringiensis*. These bacteria produce the Bt protein to protect them against insects. When the gene for this protein is put into the corn plant, the protein is produced in the corn and insects are killed when exposed to the corn. The corn is resistant to many types of insects, and insecticide chemicals do not need to be applied to the corn crop as often. This has an advantage for the environment because the use of chemical insecticides is greatly reduced. Moreover, the groundwater is less contaminated.

Bt corn has different issues than the tomato because the corn produces a foreign recombinant protein and the tomato does not. It is conceivable that the Bt protein in the corn could be unhealthy to people eating the corn. Some people could conceivably develop an allergy to these foreign proteins. Recombinant proteins might exert some unforeseen harmful biochemical reaction. The Flavrsavr™ tomato does not have this problem because no foreign protein is present. Federal regulations indicate that these issues must be tested and that the overall safety should be verified. Extensive safety studies have been done with Bt protein to show that it is safe for human consumption, but new proteins will also need to be evaluated for these safety issues.

An unexpected benefit of Bt corn is less fungal growth on the corn as a result of less damage from insects. Typically, a certain amount of corn has fungal growth that produces harmful substances called mycotoxins. Consequently, Bt in corn enhances the safety of corn for animal and human consumption by reducing mycotoxin concentration.

Cotton has also been genetically engineered to resist some insect pests. A 1998 study revealed that the number of pesticide treatments for bollworm/budworm across six major cotton states was reduced from about five- to twofold. Some farmers felt that the two treatments were not even needed but were a matter of habit and caution.

Some have asked if GMO DNA or protein could be transported or produced inside the human body if people eat the GMO. DNA and proteins are in all plants, and we eat these chemicals at every meal as a natural part of our diet. Yet foreign DNA from plants is not observed in people or animals. To date, genetically modified plant DNA has not been found in milk, meat, or eggs. Animal digestive systems effectively digest DNA and use it as broken-down nutrients. Protein might conceivably cross into humans through the digestive tract, so extra care is taken to measure this. To date, no genetically modified plant proteins have been transferred into animals. Nevertheless, new genetically modified proteins should be thoroughly tested under government regulations.

Another safety issue in food biotechnology that is addressed by government regulations is the possible alteration of nutritional value of genetically engineered organisms or plants, that is, the nutrients are measured to determine if they have been changed. The effects on food processing, wildlife safety, disease susceptibility, and allergenic potential are also measured. The U.S. Department of Agriculture and the Food and Drug Administration regulate these various safety issues.

The containment of the genetically engineered DNA in the intended original plant is a real scientific risk. The transfer of transgenic or foreign DNA to an unintended plant is undesirable and might present a risk to the environment. Many plants reproduce by pollination. The characteristics of pollen that facilitate widespread distribution by insects or wind are the same characteristics that could allow the spread of the transgene to other plants. Hypothetically, pollination occurs, seeds are formed, and the resulting plant then carries the transgene. Pollination obviously occurs between plants of the same species, so attempts to prevent the spread of corn transgenes to other corn crops, for example, may be difficult. This is not necessarily bad, except for the control of the transgene beyond the intended foods. Outcross-pollination can also occur between related plants of different species, but the resulting progeny plants are usually sterile. In wheat, outcrosses between wheat and jointed goatgrass weed appear relatively often, but the progeny are usually sterile. Some hybrid progeny, however, have limited female fertility, especially when mated back to the original goatgrass. A recent study at Ohio State University showed 60 percent fertility in outcross progeny of half-wild, half-domesticated radishes, and this outcrossing appeared to continue through subsequent generations. Thus there are some natural barriers to outcross-pollination for spread of the transgene, but the biological systems can be complex and may not be thoroughly studied. Researchers are trying to devise methods to prevent this accidental spread of the transgene.

A large number of crops have been genetically modified with the intent of improving production and limiting the use of insecticides and herbicides (see the list below). Only a few of these crops have been field tested for regulatory approval and production. Genetically modified products such as soybean, cotton, corn, tomato, chymosin, and yeast products are currently in commodities in the United States. Genetically engineered animals, however, are not found in production for a number of reasons. Once the technology was developed for making genetically engineered plants, the uses for plants far outnumbered the uses for animals because plants can be grown faster, consumers have less of a problem with genetically engineered organisms lower on the phylogenic scale, and initial costs are lower.

Over the last decade a strong debate about genetically modified food has developed. As we saw in the case of agriculture, concerns have mounted in the wake of increasing globalization and the possibility of new or expanded food-borne illnesses such as "mad cow disease" (technically Creutzfeldt Jacob syndrome), though as in this case they have nothing to do with biotechnology. Nevertheless they play into a generalized public concern: what is happening to our food? Is it still safe? Are we being told everything?

The first question that concerns us about food, especially genetically modified food, is the question of safety. Can human beings consume it without undue harm? The Food and Drug Administration's seminal answer reprinted below and reaffirmed in 2001 is, generally, yes. The FDA views genetically modified food (GMF) as not differing in principle from food developed by "traditional methods of plant breeding." We may be faster in producing changes with biotechnology, but we are not doing anything fundamentally different. There are exceptions noted by the FDA, but they are basically committed to the notion of "substantial equivalence." In this view GMFs are "substantially equivalent" to naturally occurring foods, and thus no labeling or premarket safety testing is required. An example helps to explain this point. Monsanto Corporation has developed a brand of soybeans known as "Roundup Ready," meaning that Monsanto's popular agricultural herbicide Roundup will not kill these soybean plants. The farmer can spray all of his soybean plants knowing that he will not harm his crop. Whether this promotes greater use of pesticides is debated, but it is not the issue under study here. The FDA argues that since the actual food, the soybean, is no different from regular soybeans (proper pesticide removal being the same in both cases), the GM variety does not need to be safety tested or labeled. The safety of soybeans is already known and the new variety is equivalent. No new data are required.

This notion of biological equivalence underlies the FDA approach to GMF, including the exceptions. When the food is not substantially equivalent because of new nutritional or compositional qualities that would affect the use or digestability of the food, testing might be required and at the very least labeling would be mandated. Furthermore, if the food product is so modified that its common name no longer reflects its qualities, then at least labeling will be required.

In general, the FDA believes that GMFs are safe for human consumption. Only when changes are made that affect the core characteristics of the food are any safety or labeling concerns raised. For example, consider a case that was very prominent in the 1990s, synthetically produced bovine growth hormone (BGH). BGH was given to dairy cattle to increase milk production per animal. Relying on an analysis of test data provided by the maker of BGH, Monsanto, the FDA ruled that milk produced from cows treated with BGH is safe because it is essentially equivalent to milk produced without synthetic BGH. BGH naturally occurs in cows and in the past it had to be extracted from the pituitary gland of cows. Recombinant technology allows sufficient amounts of the BGH to be produced at low enough cost to be useful for dairy producers, especially large ones.

The FDA judgment was that the BGH milk had no increased levels of any chemicals associated with BGH. Hence the FDA mandate was complete. The substances occurred naturally and were thus presumed safe in general. No change was found in the milk, and hence no labeling was required.

Consider what was left out of the FDA analysis. It did not consider the possible effect on the cows themselves that animal rights persons would be concerned about. It did not consider changing social structure of agriculture or the way in which BGH might exacerbate a divide between large and small producers or between rich and poor nations. These were beyond the FDA mandate and did not need to be considered.

Critics like Cummins (chapter 13) are hostile to genetically modified food. They believe that not enough safety testing has been done on such foods. They point to examples of things going very badly where genetically modified foods and food supplements harmed human beings. Some of these are dietary supplements over which regulators have no control. Others are theoretical but unproven risks. Some of the damage is to the environment and other damage is said to be done to organic farmers through pollen spreading from genetically modified food plants in nearby fields.

Cummins calls for a moratorium based on examples of problems but he is very critical of all biotechnology, believing in particular that it is unnecessary and unnatural and has proven itself harmful. The FDA, by contrast, believes that with proper regulatory oversight genetically modified food can be useful and important for human beings.

Crops that have been genetically modified include alfalfa, apple, apricot, arabidopsis, asparagus, cabbage, canola, carnation, carrot, cauliflower, celery, chrysanthemum, corn, cotton, cranberry, cucumber, eggplant, flax, foxglove, grape, horseradish, kiwi, lettuce, licorice, lotus, morning glory, muskmelon, oil seed rape, orchard grass, papaya, pear, peas, petunia, poplar, potato, rice, rye, soybean, strawberry, sugar beet, sunflower, sweet potato, tobacco, tomato, tulip, walnut, wheat, and white spruce.

LABELING

It is useful to categorize genetically engineered crops by pest or herbicide resistance, or by nutritional enhancement for purposes of labeling the products. An example of nutritional enhancement is rice that has had an enzyme modified in its synthesis pathway to produce vitamin A. The strategy is that vitamin A supplementation in rice for people whose primary diet is rice would provide much needed nutritional value. Although this could add nutritional value to the rice, it also poses marketing problems. The Food and Drug Administration (FDA) currently requires labeling on products if (1) nutrients or composition are changed, (2) there is a safety issue such as an allergenic effect, and (3) identity is changed so that traditional names of the plant do not apply. When a crop requires unique labeling, producers cannot easily keep the crop separated from the other crops processed in large volumes. A separate processing plant would conceivably be required to process the specially labeled crops; therefore, from a commercialization point of view it is desirable to not engineer crops that require such labeling. These types of business decisions limit the utility of genetically modified foods until regulations are changed, the demand is increased, or the food is marketed in countries that have no such restrictions.

Another business issue of labeling is what information to put on the label. Completely accurate labeling would require all the scientific identification to be placed on the label. For example, the Flavrsavr™ tomato would be labeled with "DNA coding for antisense polygalacturonase RNA." This description would likely not be very helpful for consumers because most people would have no idea of its meaning. Another confusing label at the other extreme would be a simplistic acronym like "GMO." It would mean little to people except that the crop has in some way been genetically modified, and it would not indicate anything about the safety of the plant. The GMO identification would only benefit those who categorically oppose genetically modified foods regardless of any benefit or potential harm. It might allow them to use examples of potentially harmful genetic modifications to propagandize and generalize all GMOs as harmful, which would damage farmers and producers. Clearly, the labeling of genetically modified foods presents interesting issues, which perhaps can be addressed by the public having confidence in regulatory agencies that determine safety to people and to the environment.

One of the issues that have caused the greatest concern about genetically modified food is the supposed "right to know" that food is genetically modified. Like food additives, whether they are hazardous or not, the government requires the labeling of anything that is added to food. This requirement is based on a fundamental right of autonomy, that is, the right of people to be able to lead life as they choose. This is part of what it means to have liberty, to lead our lives as we wish. It follows that we have the right to full knowledge of what we are purchasing, including food that is genetically modified. In chapter 14 this position is argued by Halloran and Hansen, scientists at Consumers Union. They argue that genetically modified food is different (a point disputed by the FDA) and may be harmful in some cases (a point admitted by the FDA). Consumers must know about the food they purchase or choose not to purchase. They also add that individuals may have personal moral or religious objections to consuming genetically modified food.

These personal preferences can only be respected by a system of mandatory labeling.

Critics of such labeling are not so sure that it is either practical or morally required. Labeling may not be possible for practical reasons. It is estimated that as much as 70 percent of the food on store shelves has some connection to genetic modification. Flour, for example, may be made from genetically modified corn or wheat and it may turn up in baked goods or cake mixes. Dairy products may partially come from cows treated with growth hormone.

How would a label work and what would it reflect? Experts believe that growth hormone does not affect the composition of the milk at all. Would milk need to be labeled because some people are concerned about animal suffering? Would those who do not like corporate agriculture require a label so that they only consume food that reflected their social and political concerns? Clearly, requiring labels to reflect any connection with biotechnology would be impractical.

Thompson (chapter 15) does not believe that positive labeling should be required and suggests that it may be morally harmful. Labeling will be expensive and the cost will be borne by all consumers, even the majority who do not want it. Since studies show that only a small number want labeling on any ground, the majority will be taxed for the few.

Thompson agrees with the FDA that genetically modified food is biologically safe. Thus we cannot hold that since everyone is implicitly concerned about physical safety, labeling benefits everyone. But physical safety is not the only concern that individuals might reasonably have. Individuals may be concerned about the political and social effects of transgenic agriculture. They may be concerned about religious or cultural purity of food or about whether the food meets religious dietary laws such as kosher or halal rules. They may have a personal distaste for genetically modified food, what some have called a "yuk" factor, as some persons have regarding raw fish in a sushi bar.

These are not unreasonable preferences and people are not wrong to hold them. They are, however, minority preferences. Food specifically labeled as meeting kosher standards costs slightly more. Those wanting that certification of religious purity are willing to pay the additional costs. Shouldn't the same principle apply to genetic modification?

Thompson's preferred solution is voluntary negative labeling, much as organic food is now labeled as organic. With uniform standards food could be labeled as "GMO free." It may cost slightly more but those with unique political, moral, or religious concerns would be willing to bear the costs of their nonstandard preferences. Issues surrounding genetically modified food are likely to grow in importance as more crops and animals are modified for more purposes. The current debate is likely to increase, not decrease. If we can resist the temptation to grasp at easy solutions, we may be able to fashion a practical solution that promotes (1) useful new foods, (2) the autonomy of consumers, and (3) the justice that only imposes absolutely necessary or willingly borne costs on others.

11

The Scientific and Health Aspects of Genetically Modified Foods: Rapporteurs' Summary

Ian Gillespie and Peter Tindemans

This summary reports the views of the two rapporteurs on common ground emerging during the Edinburgh Conference, both on matters of substance and on how to move debate forward.

The Edinburgh Conference drew together 400 participants from a variety of backgrounds. The aim was to identify common ground on whether and how applications of GM technologies in the food and crops sector serve the needs of society. This report concentrates on developing this common ground.

The focus is the safety of the tens of GM crops now in use for food. Environmental impacts, trade and developmental effects, and ethical and societal concerns were not considered at length, but could not altogether be separated or hierarchically ordered. However, the various issues need to be addressed separately if they are to be analytically tractable.

The conference critically reviewed—from the various perspectives represented—different approaches to assessment of the risks and benefits of GM food. A strong sense emerged that there was a need to take steps to rebuild trust among the various actors, particularly governments, industry, scientists, regulatory agencies and the public.

There were a significant number of points on which there was general agreement amongst the majority if not all of the participants. Unsurprisingly, there were also many issues where opposing views were put. In some cases, these were a result of different interpretations of the available evidence; in others, disagreement was more fundamental. Finally, there were points where there was neither clear agreement nor disagreement, since there is currently a lack of knowledge.

POINTS OF AGREEMENT

In considering how society deals with GM food, the circle of debate needs to be widened, including bringing in workers in laboratories, factories, farms, etc. The debate needs to become more open, transparent and inclusive. Openness and transparency are also required in the policy process. The general public—consumers and citizens—not only have the right to know, but they also have valid points of view, which need to be effectively voiced, understood and given their due weight in the decision-making and policy process.

Many consumers already eat GM foods, though they do not necessarily know they are doing so. No peer-reviewed scientific article has yet appeared which reports adverse effects on human health as a consequence of eating GM food.

The concept and practice of assessment of risk—including a consistent international approach to the use of the concept of substantial equivalence and to a form of the precautionary approach—have been valuable tools, and should remain so—as long as they continue to be kept under regular open review.

Benefits as well as risks posed by GM foods should be evaluated. GM foods have the clear potential to bring real benefits to developing countries. However, that potential has yet to be fully realised and will only be so if the technology is put to use under appropriate conditions. Population growth and poverty are real challenges for global food production, both in terms of quality and quantity of food. But GM biotechnology will not be the whole answer to increasing demands for food; it can only be a part of it.

Consumers in industrialised countries have yet to appreciate benefits from the first generation of GM food products. Benefits might include reduced food prices and, in some cases, health benefits—for example, reduced rates of use of pesticides or carcinogenic compounds by farmers. There has, however, been no quantification of tangible benefits to consumers as yet. So-called second-generation GM food products will offer more tangible potential health benefits. Past experience teaches important lessons on how to conduct the debate in the future.

The continued use of antibiotic resistance marker genes in GM food crops is unnecessary given the existence of adequate alternatives, and should be phased out. There was broad consensus not to use technologies that deliberately render seeds sterile if farmers—particularly in least developed countries—do not have a realistic choice about which seed varieties they can use. However, the use of "terminator" genes for the purpose of genetic containment was widely acknowledged as a useful safety measure.

Partnerships between the public and the private sector must develop further if the potential benefits of GM food are to be delivered to those who most need them. These should seek to combine public and private investment, technical know-how, technology transfer and local knowledge.

Consumers need to be able to exercise choice; for this they need information on the way products have been manufactured. Almost all participants recognised the value of labelling in enabling consumer choice.

POINTS OF DISAGREEMENT

Some participants regard human health aspects of GM foods as inseparable from wider issues, like impact on the environment, trade and socio-economic factors and people's belief systems. Others favour distinct and specific ways of assessing the various potential impacts.

Some see genetic modification as part of a continuum in the development of tools for plant breeding. For them, GM is just another step in the process, albeit a powerful one. Others see genetic modification as a fundamental change in the way new crops are produced. For them, this fundamental difference necessitates new ways of assessing safety.

There is disagreement whether GM foods in animal feed present a problem either for animal or human health.

There is no consensus as yet on the level of agency at which risks and benefits of GM food should be decided. While some would favour a global framework for developing, marketing and using GM technologies and the products based thereon, others adhere strongly to national sovereignty and to making their own judgements on risks and benefits. There is equally disagreement about whether GM crops have a role in global agriculture in the long term.

There is as yet no agreement on the detailed process of assessing consumer concerns about GM foods.

The need for traceability of GM material is controversial. Some consider it a necessary complement to a priori risk assessment. Others claim that risk assessment can be sufficiently improved so as to avoid post-market monitoring (the practicalities of which are uncertain).

CURRENT LACK OF KNOWLEDGE

While there seems to be agreement that the social process of risk handling needs to be "open, transparent, and inclusive" and should clearly acknowledge scientific uncertainties and take into account the validity of social concerns, there is no consensus on how this should be done in practice.

There remains uncertainty about the potential long-term effects of GM food on human health and on worker safety (as a result of exposure during production).

Current methods for testing toxicity and allergenicity (for example, the potential for unknown allergens to be transferred between species during genetic modification) leave some uncertainties and need to be improved.

There is still uncertainty over long-term environmental effects, potential complex ecological interactions and impacts on biodiversity. The impact in tropical zones is particularly uncertain, as most field trials have been carried out in temperate zones.

Though feeding trials in animals may in some cases offer supplementary safety guarantees, it is unclear whether for GM foods they will be applicable or useful.

A POSSIBLE WAY FORWARD

To tease out the issues and present them in a coherent form, the key points raised are discussed under the following headings:

- *Benefits versus risks*: this describes the consensus emerging from the conference that the benefits of GM food needed to be considered as well as the risks.
- *Management of GM technologies*: this presents a synthesis of the participants' views on managing risks from GM foods.
- *The role of stakeholders*: this describes how stakeholders can move forward together.
- *An international programme* as a potential way forward.

BENEFITS VERSUS RISKS: A FIRST BALANCE

The trade-off between benefits and risks for any one GM food may vary across different regions of the world and between different economies—from least developed countries, to emerging economies, to countries in transition, and to fully industrialised countries. Within regions, the balance may be different for different segments of the population. For the least developed countries in particular more food of improved nutritional quality will be needed as population growth continues. Distribution problems are real, as are post-harvest losses. But it is not realistic to assume that a strategy that does not incorporate increased local production of cheap food for consumers will work.

In judging new technologies one should also bear in mind that the potential of present-day agriculture to meet future demands is under strain. The area of available arable land is decreasing with urban development, and productivity growth is levelling off (partly due to the diminishing potential of "traditional" breeding technologies). GM technologies, if developed under appropriate conditions, offer the potential of providing solutions, though it should be clear that GM technologies, and biotechnology as such, can only be part of a range of answers to the problem of providing sufficient food of the necessary nutritional quality to the developing world. Policy measures are needed in a whole range of domains. GM technology should not take the place of efforts to eradicate the sources of poverty.

GM food–based oral vaccines and nutritional supplements offer potentially great benefits. Though other technologies may be available, biotechnology potentially offers more practical and affordable options for least developed countries in the short and medium term. To deliver on this may require a redirection of research and development efforts, so that the needs of developing countries are given a higher priority. This does not imply that on the individual or the societal level the judgements about what might constitute "acceptable" risks need to be different. If developing countries are to choose to use GM technologies, strict safety procedures are in every way as necessary for them as for the industrialised world, and special efforts will be required by all countries to ensure the capacity exists for risk analysis to be carried out in the context of the countries using the GM product. But, in the end, the balance of benefits and risks may result in different decisions in various societies.

For highly industrialised countries the core issues are different. Producers are concerned with efficiency gains and corporate profits—but also with reduced environmental stress. However, tangible benefits to consumers, which should come through lower prices, reduced health risks (owing to less exposure to pesticides or carcinogenic chemicals) or improved health, are generally perceived to be small. The exception may be when GM foods address special dietary problems of specific segments of the population. The potential for developing nutrient-fortified foods and edible vaccines to increase the resistance of elderly people to infectious disease is a case in point. The question remains unanswered of how safety of such products (e.g., "nutraceuticals") should be assessed—in particular, whether this should be more akin to assessment of pharmaceuticals than of novel foods.

There was almost complete consensus that consumers in all parts of the world should (where possible) have the opportunity to exercise choice on whether or not to consume GM foods. Mandatory labelling was widely, but not comprehensively, supported as a means to help achieve this.

Though labelling might allow choice, it would not in itself help answer the question of whether there were long-term impacts of GM food—beneficial or detrimental—on human health. Appropriate testing and monitoring measures would be necessary for this purpose. Further investigation is needed on the need for and the practicalities of tracing GM food products throughout the food chain.

A differentiated approach is also required to tease out the possible impacts of GM food and crops on the sustainability of ecosystems as well as more generally on societies and economies (including concerns about, for example, the structure of agriculture and the future of rural areas, or about market concentration in the biotechnology industry). Assessing impact on, in particular, the sustainability of ecosystems requires long-term data collection, which is not per se yet available. Meaningful comparisons need to be made with traditional varieties of crops. Impacts will, by definition, be different for different countries or regions. The issue does need to be studied further and would benefit from long-term study programmes.

MANAGEMENT OF GM TECHNOLOGIES

We cannot rule out all risk for all eternity for any area of human activity. The challenge is to take a sufficiently precautionary approach to investigating scientifically risks that may occur. We need to communicate effectively in our societies about these assessments. We need to be clear about how we decide about acceptability. And we need systems in place, trusted by citizens, for managing risks that encompass those measures that become effective after decisions on acceptability have been taken.

There is, of course, a need to constantly remain vigilant. As every new technology is based on harnessing different aspects of nature in varying combinations, testing techniques and protocols need to be addressed regularly and developed further as appropriate. But, in making an assessment of risks and benefits, novel products and their production technologies should be compared with products of existing technologies and not just looked at in absolute terms.

The present situation can be summarised as follows. With regard to health issues, tests on toxicity and allergenicity have been and are being conducted. So far, none has shown significant toxic or allergenic harm. No peer-reviewed article on clinical trials or epidemiological study reporting adverse effects on human health has yet appeared. Where there have been indications of potential unacceptable effects, the present mechanisms have enabled us to identify them and prevent such products from coming to the market. Yet the example of the current uncertainty about whether genetic modification could lead to the transfer of unknown allergens demonstrates the need to be alert and continually to refine protocols.

It is recognised that a large number of field trials of GM crops have taken place without apparent adverse effects and this suggests that we may be able to manage the risks at least under the specific conditions encountered. However, though the majority of trials have been evaluated at least to some extent in terms of safety, evaluation has not followed identical protocols and most trials have been carried out in temperate zones. There is a need to improve the monitoring of such trials and to do more work on pre-release assessment of new as well as existing crops. This needs to be done in environments in which GM crops are to be grown.

Views differ as to whether modern GM technologies are simply yet another stage in the progress of approaches available for genetic transfer. Some see GM technology as the most recent point on the continuum of development from the earliest breeding technologies through breeding assisted by cellular techniques, to modern interventions at the molecular level. Others see GM technology as something fundamentally different from what has gone before. All agree, however, that open, transparent, inclusive systems of risk assessment, management and communication are essential.

The systems of risk assessment, management and communication that countries have developed to deal with the safety aspects of food in general and how they are applied have recently come under increased scrutiny. This has not just focused on GM issues and it has been further complicated by the difficulties in disentangling public concerns focused on health and safety issues from environmental, socio-economic and ethical considerations.

Both specific assessment methods applied to GM foods and crops and the design of the overall system of assessing, managing and communicating risks have been criticised. Yet, even though there is consensus that risk handling systems need to be improved, evidence to date by way of the responses taken to deal with unexpected consequences showing up in trials and other experiments indicates that actual risk levels have not been increased. In general, experimenters have learnt how to better cope with them. Mechanisms have been developed which appear—so far as we can tell—to have addressed such emerging unexpected events and to have assigned responsibility for dealing with them. Responses have included referring a problem back to science, requiring further testing, and action for regulatory authorities. A case in point is the now general consensus to phase out the use of antibiotic resistance marker genes.

The challenge now is to monitor and adapt risk handling systems so that they are able not just to continue to respond to unexpected events and deal with scientific uncertainty, but to handle broader questions that emerge in society about GM foods and crops, in an effort to help to restore confidence in regulatory regimes.

If this is to be done, the methods used to appraise possible adverse effects on health and the environment will have to be regularly reviewed and judged to be adequate for their intended purposes. A number of proposals for concrete action came out of the conference:

- Much of the allergenicity and toxicity testing of GM products now done is based on gene products expressed in recombinant micro-organisms rather than in the target crop plant. There is a case for reviewing this practice, and for considering the applicability and usefulness of animal testing as a predictive tool.
- Immediate improvement can and should be made to the decision tree in use for testing against allergenicity (which is not specific for GM foods), so as to incorporate in vivo and in vitro tests.
- There is a need to review the principle of substantial equivalence. The OECD has carried out an ongoing review of the concept over its five years of use, but the conference was of the view that a more fundamental reassessment is necessary. The means for carrying out a transparent review—which should acknowledge the need to include the various interest groups—should be worked out between the various international bodies active in the field;

namely, the new Codex Alimentarius Task Force, the OECD, the FAO and the WHO. Ideally, the review process should be completed and the results widely disseminated within two years.

- The concept of a precautionary approach to risk assessment, recognised in the Cartagena Protocol on Biosafety, has the potential to be worked up into a practical way to accommodate the new approaches consumers, the public at large, special interest groups and scientists request from a risk assessment and management system.

In developing the system of risk analysis, it would be useful to categorise risks according to likelihood. For example, risks might be assigned a probability (where empirical results allow such an estimate), or labelled as hypothetical (where they may exist in theory, but there are no data to assess probability), and designated speculative (where there is no convincing theoretical base, but where further R&D may be warranted).

Reviewing methods and principles used in safety assessment should not imply automatically more or tighter regulation. There was wide support for the conclusion that an assessment of the benefits is equally necessary to arrive at final judgements. Regulation should balance the potential risks and benefits of the technologies in the circumstances prevailing in countries or regions of use. It is a policy process in which the costs of regulation also must go into the equation.

Variations in health risk assessments may be substantially less than for the environment. Environmental risk/benefit assessment will almost always involve a strong regional or local component. Results may not be completely transferable between one environment and another. Health risk assessment is more universal— and is recognised as such in international approaches to medicine regulation, allowing more sharing and consistency of results obtained in different countries. Though consumption patterns may differ significantly, and basic health and nutritional levels vary, in general the same food will have similar impacts on human beings regardless of location.

Finally on this issue, the conference concluded that risk analysis systems are only likely to generate public trust if based on transparency, provision of information (on monitoring, research results, etc.), and greater inclusiveness of the various stakeholders. There then needs to be clarity about how stakeholders' perspectives are taken into account in the policy process.

THE ROLE OF STAKEHOLDERS

Tackling the issues mentioned above demands a major commitment from governments. Focusing research and development priorities on longer-term domestic and global agricultural needs will be particularly challenging.

So, too, will organising a more inclusive public debate on the risks and benefits of new GM technologies. The conference was clear that independent scientific advice—even if it is contrary to the generally accepted view—has a role to play in a fully open process.

Both governments and scientists should do more to provide the public with clear, understandable and relevant information. That does not necessarily mean that what scientists say must be taken at face value, or that scientific arguments are the only ones that count when the final decisions are made, but decisions—which are the politicians' business—must be informed by the best available scientific advice. It is also important that scientists work both on internal mechanisms in the scientific community and mechanisms that reach out to the wider public, in order to review the state of scientific knowledge at regular intervals with the aim of reducing or specifying areas of uncertainty.

More open access to information will be essential to convince concerned consumers that there is nothing to hide in making safety assessments of GM foods. Data on safety assessments, on field monitoring and on post-marketing assessment ought to be made much more widely and easily available than is presently perceived to be the case. This needs to include as much as possible of the extensive data sets held within the private sector. There is a real challenge for industry, academia and government to deliver on this.

Many good examples from a variety of countries and regions were put forward at the conference on how to involve the public—whether through interest groups or independently. This practical experience provides a basis for social and natural scientists working together to design better ways of building open, transparent and inclusive processes of analysis and decision making that might restore public trust. Many groups have legitimate views and experience that should be drawn upon—farmers, for instance, have diverse approaches and concerns in the different global economies.

In promoting the public good, governments work routinely with industry. GM technologies in the agricultural domain are recognised as an area where there is much scope for enhanced public-private co-operation. The private sector needs to take part in the debates as it plays a major role in developing and marketing GM technologies. But, as not all development and dissemination of technology, including biotechnology, can or should be carried out under conditions of commercial profit, relying on the private sector alone will not be sufficient to harness the full potential of GM food. There is considerable scope for developing further public-private partnerships.

This is also true for risk assessment. Improved sharing of comparable data and risk assessments and co-operative international publicly funded research may contribute to reducing the overall cost burden of effective risk analysis. Rising costs could be expected to act as a driver for industry concentration.

For partnerships between the public and private sectors to serve effectively the multiple interests at stake, it will be important to develop a more subtle approach to ownership issues. For example, the most appropriate balance between patent rights and plant variety protection rights continues to require careful assessment and consideration. Special exemptions on intellectual property rights protection for crops widely used by farmers and government agencies of developing countries, or more general policies of sharing GM technologies, have been proposed as part of a number of private/public ventures. Ongoing efforts by the appropriate bodies to review current arrangements in order to develop equitable solutions for these concerns should be strongly encouraged. They are essential for reducing the controversies around GM foods and crops. Similarly, issues of liability between companies, farmers and breeders must be settled.

AN INTERNATIONAL PROGRAMME AS A WAY AHEAD

This is an area in which there is already substantial activity by a number of international bodies, including the FAO, the WHO, the OECD, the CGIAR, the ILSI and others. Earlier this year, the Cartagena Protocol on Biosafety was agreed in Montreal. Any new international initiatives must offer added value and complementarity to what is and can be carried out effectively by existing parties and frameworks.

At the detailed level, a three-pronged international programme might show collective commitment to action:

- In the light of the urgency of the problem of producing food for a larger population in increasingly smaller areas, many participants referred to the alarming decline of conventional agricultural research. Any effort to harness the potential of GM crops should therefore include a fundamental reconsideration by governments of the level of funding of agricultural research through the existing international frameworks.
- A collaborative and comparative testing programme on health and environmental issues of GM technologies, involving all parties—including farmers—under appropriate conditions, could through its international visibility contribute to viable approaches that could then trickle down to national practices without involving the transfer of autonomy. For such a programme it would be essential, and feasible given the prevailing consensus on the general framework to be used in the assessment of health effects, to start working on internationally agreed scientific protocols. The development of criteria for environmental risk assessment in different geographical zones is ongoing.
- Participation in such a programme could be linked to training possibilities for professionals and scientists in developing countries, the need for which was repeatedly stressed during the conference.

There remains, however, the persistent need for some sort of over-arching international initiative to knit all of these strands—ongoing and proposed—together if the global benefits of this technology are to be maximised and risks minimised.

At the very least, the international debate begun at this conference should be continued and broadened in an attempt to inform international policy making. The inherent scientific uncertainties, and the necessary involvement of actors from a broad range of disciplines seeking a way forward together, bears a strong resemblance to the current debate regarding climate change. In that context, the International Panel on Climate Change (IPCC) might provide a possible model.

In terms of immediate next steps, if there is political will, a feasibility study into creating an ongoing expert forum for debate of the issues around GM food might quickly be drawn together by the various international organisations and leading countries in this field.

NOTE

Reprinted by permission of the Organization for Economic Cooperation and Development.

12

The Safety of Foods Developed by Biotechnology

David Kessler
Michael R. Taylor
James H. Maryanski
Eric Flamm
Linda S. Kahl

By virtue of its broad regulatory jurisdiction over foods, drugs, biologicals, and medical devices, the Food and Drug Administration (FDA) directs its efforts to ensure that safety and other public health issues are properly addressed as the exciting fruits of new biotechnology come to the market. The FDA has already approved human drugs and vaccines, diagnostic devices, and food processing enzymes produced through recombinant DNA techniques and other tools developed by the recent and continuing revolution in the biological sciences.

Recombinant DNA techniques are now being used to develop new plant varieties that will be sources of foods, such as fruits, vegetables, grains, and their by-products. These techniques enable developers to make specific genetic modifications in plants, including modifications that introduce substances into plants that could not be introduced by traditional methods. To ensure the safety of the resulting foods and to foster innovation, the FDA is taking the initiative, before foods from such plants are ready to enter the market, to see that there is an agreed upon scientific basis to evaluate the safety of whole foods and animal feeds derived from new plant varieties. Here, we summarize our regulatory framework and our approach to safety assessment and discuss its scientific basis (1).

Our safety assessment approach addresses new varieties of food crops developed by both traditional and newer methods of genetic modification and provides guidance on how safety issues should be addressed. The approach identifies scientific or regulatory issues where developers may need to consult the FDA. In developing this approach, we have examined many techniques now available to plant breeders and the types of food safety issues that might arise as a result of those techniques (2–7). However, we focus here primarily on issues associated with foods derived from new plant varieties developed by recombinant DNA techniques. Such foods are now approaching the point of commercial introduction.

We consume in our diet a great diversity of chemical substances. In some cases, these substances are macroconstituents of the daily diet, such as potato starch or wheat gluten. Other substances are microconstituents, such as most flavors, enzymes, vitamins, and minerals. The major classes of food constituents include carbohydrates (mostly monosaccharides, disaccharides, oligosaccharides, and polysaccharides, including gums, starches, and celluloses), fats (mostly triglycerides containing fatty acids of varying chain lengths and degrees of saturation), enzymes and other proteins and peptides, minerals, DNA and RNA, essential oils, waxes, vitamins, pigments, and alkaloids (2). Developers introduce hundreds of new varieties of food plants into commerce every year. Most have improved agronomic characteristics, such as higher yield. Varieties are also being developed with enhanced quality characteristics, such as improved nutritional or processing attributes. To develop new varieties, breeders use all the techniques at their disposal to generate ever more advantageous combinations of genetic traits. Breeding techniques include hybridizations between plants of the same species, between plants of different species, and between plants of different genera; chemical and physical mutagenesis; interspecies and intergeneric protoplast fusions; somaclonal variation resulting from regeneration of plants from tissue culture; and in vitro gene transfer techniques. Traditional methods of plant breeding of some major crops have yielded dramatic changes in food composition, including increases in major plant constituents. For example, traditional breeding resulted in the transformation of the kiwi fruit from a small berry native to Asia to the recognizable variety in our grocery stores. Hundreds of similar or more subtle improvements in the agronomic, food processing, or other attributes of food crops have been achieved without any significant adverse impact on the safety of foods (2–4).

Recombinant DNA techniques which are being used to achieve the same types of goals as traditional techniques offer plant breeders a number of useful properties. First any single gene trait (and potentially multigene traits) whose chromosomal location or molecular identity is known can be transferred to another organism irrespective of mating barriers. Second this transfer can be accomplished without simultaneously introducing undesirable traits that are chromosomally linked to the desirable trait in the donor organism. Thus the techniques have great power and precision. Currently more than 30 different agricultural crops developed with recombinant techniques are being tested in the field trials. With these techniques food crops are being developed to resist pests and disease, to resist adverse weather conditions, to tolerate chemical herbicides and to have improved characteristics for food processing and nutritional content. The genes conferring these traits usually encode proteins that are responsible for the new trait or that directly or indirectly modify carbohydrates or fats in the plant to bring about the desired characteristics. In addition genes encoding antisense messenger RNA have been introduced to decrease gene expression and thereby bring about the desired new phenotype.

FDA APPROACH TO REGULATION

The United States today has a food supply that is as safe as any in the world. Most foods predate the establishment of national food laws. And the safety of these foods has been accepted on the basis of extensive use and experience over many years or even centuries. Foods derived from new plant varieties are not routinely subjected to scientific tests for safety, although there are exceptions. For example potatoes are generally tested for the glycoalkaloid solanine. The established practices that plant breeders use in selecting and developing new varieties of plants, such as chemical analysis, taste testing and visual analysis, rely primarily on observations of quality, wholesomeness and agronomic characteristics. Historically these practices have been reliable for ensuring food safety (2–4).

The Federal Food, Drug and Cosmetic Act (8) gives the FDA authority to ensure the safety of whole foods. This act places a legal duty on those who develop and sell food to assure the safety of the products they offer to consumers and provides the FDA with a range of legal tools to enforce this duty. The FDA can take action to remove a food from commerce if there is even a reasonable possibility that a substance added by human intervention might be unsafe (8 section 402a1). The FDA also has authority to require formal premarket review and approval of substances intentionally added to food if there is a question about safety (that is, if the substance is not generally recognized as safe, GRAS).

Because of the limited nature of most modifications likely to be introduced, the FDA would waste its resources and would not advance public health if it were routinely to conduct formal premarket reviews of all new plant varieties. We will require such reviews before marketing, however, when the nature of the intended change in the food raises a safety question that the FDA must resolve to protect public health. The FDA also has a responsibility to provide scientific guidance on how the safety of foods from new plant varieties should be evaluated, regardless of whether formal FDA review and approval are required.

SAFETY ASSESSMENT: SCIENTIFIC BASIS

Our safety assessment approach, like that of others (2–6), addresses important food safety issues that pertain to the host plant, donor organisms, and new substances that have been introduced into the food. The host plant is a benchmark for considering modifications that may affect the safety of food derived from new varieties. Potential new substances considered in this safety assessment are proteins, carbohydrates, and fats and oils because these are the substances that will be introduced or modified in the first plant varieties developed by recombinant DNA techniques.

The Host Plant

The host is the plant that is genetically modified and is the recipient of any newly introduced traits. In general, it is a species commonly used as a source of food. We expect that developers will consider information consistent with currently accepted scientific practices, such as the potential adverse effects of an altered metabolic pathway in the plant, the inheritance of the introduced genetic material as a single Mendelian trait, and the genetic stability of the new plant variety. In principle, factors that favor stability and facilitate subsequent genetic manipulation include a minimum number of copies of the introduced genetic material and a single site of insertion. How critical these factors are for maintaining stability is unclear because virtually all plants have multi-gene families.

Developers should consider changes in the concentrations or bioavailability of important nutrients for which a food is widely consumed. For example, if a new tomato variety contained no vitamin C, consumers would need to be informed of that fact through appropriate labeling (for example, a change in the common name). Most plants produce a number of toxicants and antinutritional factors, such as protease inhibitors, hemolytic agents, and neurotoxins, presumably as a means of resisting natural predators. The concentrations of toxicants in most species of domesticated food plants (for example, corn and wheat) are so low as to present no health concern. In others (for example, potato and rapeseed), breeders routinely screen new varieties to ensure that toxicant concentrations are within an acceptable range. In some cases (for example, cassava and kidney bean), proper preparation, such as soaking and cooking, is required to produce food that is safe to eat (2–4).

Additionally, plants, like other organisms, have metabolic pathways that no longer function because of mutations that occurred during evolution. Products or intermediates of some of these pathways may include toxicants. In rare cases, such silent pathways may be activated by the introduction or rearrangement of regulatory elements, or by the inactivation of repressor genes by point mutations, insertional mutations, or chromosomal rearrangements. Similarly, toxicants ordinarily produced at low concentrations in a plant may be produced at higher levels in a new variety as a result of such occurrences. However, the likelihood of such events occurring in food plants with a long history of safe use is low. The potential of plant breeding to activate or upregulate pathways synthesizing toxicants has been effectively managed by sound agricultural practices, as evidenced by the fact that varieties with unacceptably high levels of toxicants have rarely been marketed (2–4).

Therefore, the toxicants that are of concern in any particular species are those that have been found at unsafe concentrations in some lines or varieties of that species or related species. In many cases, characteristic properties (such as a bitter taste associated with alkaloids) are known to accompany elevated concentrations of specific natural toxicants, and the absence of bitter taste may provide an assurance that these toxicants have not been elevated to unsafe levels; in other cases, analytical or toxicological tests may be necessary.

The Donor

The donor (plant, microorganism, or animal) is the source of the new trait. We expect that producers will consider information consistent with currently accepted scientific practices that might relate to the presence of unintended toxicants, such as history and derivation of molecular constructs (for example, passage through microbial hosts), known activities of any introduced regulatory sequences (for example, environmental, developmental, and tissue-specific effects on promoter activity), and the potential for inadvertently introducing undesirable substances (for example, due to the expression of extraneous open reading frames).

Toxicants known to exist in the donor, related species, or progenitor lines may be transferred to the new plant variety, for example, during hybridization of a cultivated variety with a wild, poisonous relative. The possibility that donor-derived toxicants could occur in food derived from a genetically modified plant should be considered. One of the questions raised most frequently about the use of recombinant DNA techniques to develop improved food crops concerns the safety for consumption of substances (now primarily proteins, carbohydrates, and fats and oils) that will be introduced into foods such as fruits, vegetables, grains, and their by-products. Here we discuss the scientific issues pertaining to these substances in food derived from the new plant variety.

Proteins

Proteins (including antisense modifications that modulate expression of native proteins) make up the largest group of substances being introduced into food through recombinant DNA techniques. Our approach in evaluating uncertainty associated with proteins is first to ask, "Does the protein have a safe history of use in food, or is it substantially similar to such a food component?" The scientific issues pertaining to proteins that are derived from other food sources, or that are substantially similar to proteins that are derived from food sources, are known toxicity, allergenicity, and dietary exposure. Thousands of proteins have been safely consumed in the human diet. In fact, the eukaryotic cell contains 5,000 or more different polypeptides. Genetic polymorphism, the occurrence of more than one allele of a gene, also contributes to the diversity of proteins in the diet. For example, six alleles of beta-galactosidase have been identified in 39 widely used inbred lines of corn (9). Variation may also occur as a result of posttranslational processing (for example, glycosylarion or methylation pattern of the host plant).

This variation is also seen in enzymes derived from microorganisms used in food processing. Enzymes that have the same fundamental catalytic activity may differ in DNA sequence, protein structure, and functional properties (10). For example, alpha-amylases from different organisms may differ in optimal conditions of use, such as temperature and pH, and substrate affinity.

Generally, enzymes that are substantially similar to enzymes known to be safely consumed (including minor variations in structure or function) would not raise safety concerns (5, 6, 11). For example, in food crops a gene coding for an enzyme whose catalytic activity confers herbicide resistance may be isolated from a plant or bacterium, subjected to site-directed mutagenesis to enhance its herbicide resistance, and introduced into the desired host plant to substitute for the biochemical activity of the plant enzyme that is sensitive to the herbicide. However, some enzymes produce toxic substances (for example, the enzymes that convert cyanogenic glycosides to cyanide) that would raise safety questions.

As discussed above, a variety of proteins are present in the diet and have a history of safe consumption. In general, proteins that are currently consumed or substantially similar to proteins already in the diet do not pose specific safety concerns. A seed storage protein, for example, may be transferred from one plant species to another to improve nutritional quality. However, a number of groups of proteins present in common food sources are known to be toxic or antinutritional (for example, lectins and protease inhibitors). Because processing (such as soaking or cooking) may reduce or eliminate the toxic effects of these proteins, many foods that contain these toxic substances are poisonous if eaten raw but safe when properly prepared. Sound scientific practices dictate that such toxic proteins not be introduced into food or animal feed components of new plant varieties.

Many foods produce an allergenic response in some individuals. Foods that commonly cause allergenic responses include milk, eggs, fish, crustacea, mollusks, tree nuts, wheat, and legumes. Although only a small fraction of the thousands of proteins in the diet have been found to be allergenic, all known food allergens are proteins. The transfer of proteins from one food source to another might therefore confer on food from the host plant the allergenic properties of food from the donor plant. For example, the introduction of a peanut allergen into corn might make that variety of corn newly allergenic to people ordinarily allergic to peanuts.

In some of these foods, the protein responsible for the allergenicity is known (for example, gluten protein in wheat). In such cases, the precision of methods such as recombinant DNA techniques allows the developer to determine whether the allergenic determinant has been transferred from the donor to the new variety. In many foods, however, the protein responsible for the allergenicity is not known. In these cases, well-designed in vitro tests, such as serological tests, may provide evidence that the suspected allergen was not transferred or is not allergenic in the new variety; a separate issue is whether any new protein in food has the potential to be allergenic to a segment of the population. At this time, we are unaware of any practical method to predict or assess the potential for new proteins to induce allergenicity. Uncertainty may exist about the safety for consumption of a protein that has not been a constituent of food previously (or has no counterpart in food that would serve as a basis for comparison of safety). The degree of testing these new proteins should be commensurate with any safety concern raised by the objective characteristics of the protein. Generally, the function of proteins that have been introduced into food by recombinant DNA techniques is well known, and these proteins are not known to exert toxic effects in vertebrates. If such well-characterized proteins do not exhibit unusual functions, safety testing will generally not be necessary.

However, certain groups of proteins are known to be toxic to vertebrates. These include bacterial and animal toxins, hemagglutinins, enzyme inhibitors, vitamin-binding proteins (avidin), vitamin-destroying proteins, enzymes that release toxic compounds, and selenium-containing proteins (12). For such substances, testing may be the only means available to ensure safety.

Carbohydrates

Developments that affect carbohydrates will often be modifications of food starches, presumably affecting the content of amylose and amylopectin, as well as the branching of amylopectin. Such modified starches are likely to be functionally and physiologically equivalent to starches commonly found in food and thus would not suggest any specific safety concerns. However, if a vegetable or fruit is modified to produce high concentrations of an indigestible carbohydrate that normally occurs at low concentrations or to convert a normally digestible carbohydrate to an indigestible form, nutritional questions may arise.

Fats and Oils

Some alterations in the composition or structure of fats and oils, such as an alteration in the ratio of saturated to unsaturated fatty acids, may have significant nutritional consequences or result in marked changes in digestibility. Such changes may warrant labeling that describes the new composition of the substance. Additionally, safety questions may arise as a result of the presence of fatty acids with chain lengths greater than fatty acids with cyclic substituents, fatty acids with functional groups not normally present in dietary fats and oils, and fatty acids of known toxicity, such as erucic acid.

NONCLINICAL SAFETY TESTING

Animal feeding trials of foods derived from new plant varieties are not conducted routinely. However, in some cases testing may be needed to ensure safety. For example, substances with unusual functions or that will be new macroconstituents of the diet may raise sufficient concern to warrant testing. Tests could include metabolic, toxicological, or digestibility studies, depending on the circumstances.

Developers may also need to conduct tests on the "wholesomeness" of foods derived from new plant varieties as a means of ensuring that the food does not contain higher levels of unexpected, acutely toxic substances. Such tests may provide additional assurance to consumers that food developed by new technology is as safe as food derived from varieties already in their grocery stores. However, animal tests on whole foods, which are complex mixtures, present problems that are not associated with traditional animal toxicology tests designed to assess the safety of single chemicals. Potential toxicants are likely to occur at very low concentrations in the whole food, and the tests may therefore he inadequately sensitive to detect toxicants. Efforts to increase the amount of whole food ingested by the test animals in order to increase the sensitivity and attempt to establish a traditional margin of safety (for example, a 100-fold safety factor) may not always be possible. When tests are contemplated, careful attention should be paid to test protocol, taking into account issues such as nutritional balance and sensitivity.

FDA's science-based approach for ensuring the safety of foods from new plant varieties focuses safety evaluations on the objective characteristics of the food: the safety of any newly introduced substances and any unintended increased concentrations of toxicants beyond the range known to be safe in food or alterations of important nutrients that may occur as a result of genetic modification. Substances that have a safe history of use in food and substances that are substantially similar to such substances generally would not require extensive premarket safety testing. Substances that raise safety concerns would be subjected to closer inquiry. This approach is both scientifically and legally sound and should be adequate to fully protect public health while not inhibiting innovation.

NOTES

Reprinted by permission of the American Association for the Advancement of Science. *Science,* 26 June 1992, 1747–50.

1. For a full discussion, see FDA's "Statement of Policy: Foods Derived from New Plant Varieties," *Fed. Regist.* 57, May 29, 1992.

2. International Food Biotechnology Council, *Regul. Toxicol. Pharmacol.* 12, no. 3 (1990): pt. 2.

3. Nordic Council of Ministers and the Nordic Council, *Food and New Biotechnology* (Scantryk, Copenhagen, 1991).

4. *Approaches to Assessing the Safety of Crops Developed through Wide-Cross Hybridization Techniques* (proceedings of a Food Directorate Symposium, November 22, 1989, Ottawa, Canada) (Ottawa: Food Directorate, Health Protection Branch, Health and Welfare Canada, 1990).

5. "Strategies for Assessing the Safety of Foods Produced by Biotechnology" (Geneva: World Health Organization, 1991).

6. *Report on Food Safety and Biotechnology* (Paris: Group of National Experts on Safety in Biotechnology, Organization for Economic Cooperation and Development, Paris, in press).

7. D. D. Hopkins, R. J. Goldburg, S. A. Hirsch, *A Mutable Feast: Assuring Food Safety in the Era of Genetic Engineering* (New York: Environmental Defense Fund, 1991).

8. 21 U.S.C. §321 et seq.

9. D. B. Berkowitz, *Biotechnology* 8 (1990): 819.

10. M. Vihinen and P. Mantsala, *Crit. Rev. Biochem. Mot. Biol.* 24 (1989): 329.

11. M. W. Pariza and E. M. Foster, *J. Food Prot.* 46 (1983): 453.

12. W. G. Jaffe, *Toxicants Occurring Naturally in Foods* (Washington, D.C.: National Academy of Sciences, 1973), 10, 128.

13

Hazards of Genetically Engineered Foods and Crops: Why We Need a Global Moratorium

Ronnie Cummins

The technology of genetic engineering (GE), wielded by transnational "life science" corporations such as Monsanto and Novartis, is the practice of altering or disrupting the genetic blueprints of living organisms—plants, animals, humans, microorganisms—patenting them, and then selling the resulting gene-foods, seeds, or other products for profit. Life science corporations proclaim, with great fanfare, that their new products will make agriculture sustainable, eliminate world hunger, cure disease, and vastly improve public health. In reality, through their business practices and political lobbying, the gene engineers have made it clear that they intend to use GE to dominate and monopolize the global market for seeds, foods, fiber, and medical products.

GE is a revolutionary new technology still in its early experimental stages of development. This technology has the power to break down fundamental genetic barriers—not only between species but between humans, animals, and plants. By randomly inserting together the genes of non-related species—utilizing viruses, antibiotic-resistant genes, and bacteria as vectors, markers, and promoters—and permanently altering their genetic codes, gene-altered organisms are created that pass these genetic changes on to their offspring through heredity. Gene engineers all over the world are now snipping, inserting, recombining, rearranging, editing, and programming genetic material. Animal genes and even human genes are randomly inserted into the chromosomes of plants, fish, and animals, creating heretofore unimaginable transgenic life forms. For the first time in history, transnational biotechnology corporations are becoming the architects and "owners" of life.

With little or no regulatory restraints, labeling requirements, or scientific protocol, bio-engineers have begun creating hundreds of new GE "Frankenfoods" and crops, oblivious to human and environmental hazards, or negative socioeconomic impacts on the world's several billion farmers and rural villagers. Despite an increasing number of scientists warning that current gene-splicing techniques are crude, inexact, and unpredictable—and therefore inherently dangerous—biotech governments and regulatory agencies, led by the US, maintain that GE foods and crops are "substantially equivalent" to conventional foods, and therefore require neither mandatory labeling nor pro-market safety-testing. This Brave New World of Frankenfoods is frightening.

There are currently more than four dozen genetically engineered foods and crops being grown or sold in the US. These foods and crops are widely dispersed into the food chain and the environment. Over 70 million acres of GE crops are presently under cultivation in the US, while up to 500,000 dairy cows are being injected regularly with Monsanto's recombinant bovine growth hormone (rBGH). Most supermarket processed food items now "test positive" for the presence of GE ingredients. In addition several dozen more GE crops are in the final stages of development and will soon be released into the environment and sold in the marketplace. According to the biotechnology industry almost 100% of US food and fiber will be genetically engineered within 5–10 years. The "hidden menu" of these unlabeled genetically engineered foods and food ingredients in the US now includes soybeans, soy oil, corn, potatoes, squash, canola oil, cotton seed oil, papaya, tomatoes, and dairy products.

Genetic engineering of food and fiber products is inherently unpredictable and dangerous—for humans, for animals, the environment, and for the future of sustainable and organic agriculture. As Dr. Michael Antoniou, a British molecular scientist, points out, gene-splicing has already resulted in the "unexpected production of toxic substances . . . in genetically engineered bacteria, yeast, plants, and animals with the problem remaining undetected until a major health hazard has arisen." The hazards of GE foods and crops fall basically into three categories: human health hazards, environmental hazards, and socioeconomic hazards. A brief look at the already-proven and likely hazards of GE products provides a convincing argument for why we need a global moratorium on all GE foods and crops.

TOXINS AND POISONS

Genetically engineered products clearly have the potential to be toxic and a threat to human health. In 1989 a genetically engineered brand of L-tryptophan, a common dietary supplement, killed 37 Americans and permanently disabled or afflicted more than 5,000 others with a potentially fatal and painful blood disorder, eosinophilia myalgia syndrome (EMS), before it was recalled by the Food and Drug Administration. The manufacturer, Showa Denko, Japan's third largest chemical company, had for the first time in 1988–89 used GE bacteria to produce the over-the-counter supplement. It is believed that the bacteria somehow became contaminated during the recombinant DNA process. Showa Denko has already paid out over $2 billion in damages to EMS victims.

In 1999, front-page headline stories in the British press revealed Rowett Institute scientist Dr. Arpad Pusztai's explosive research findings that GE potatoes, spliced with DNA from the snowdrop plant and a commonly used viral promoter, the Cauliflower Mosaic Virus (CaMv), are poisonous to mammals. GE-snowdrop potatoes, found to be significantly different in chemical composition from regular potatoes, damaged the vital organs and immune systems of lab rats fed the GE potatoes. Most alarming of all, damage to the rats' stomach linings—apparently a severe viral infection—most likely was caused by the CaMv viral promoter, a promoter spliced into nearly all GE foods and crops.

Dr. Pusztai's pathbreaking research work unfortunately remains incomplete (government funding was cut off and he was fired after he spoke to the media). But more and more scientists around the world are warning that genetic manipulation can increase the levels of natural plant toxins in foods (or create entirely new toxins) in unexpected ways by switching on genes that produce poisons. And since regulatory agencies do not currently require the kind of thorough chemical and feeding tests that Dr. Pusztai was conducting, consumers have now become involuntary guinea pigs in a vast genetic experiment. As Dr. Pusztai warns, "Think of William Tell shooting an arrow at a target. Now put a blindfold on the man doing the shooting and that's the reality of the genetic engineer doing a gene insertion."

INCREASED CANCER RISKS

In 1994, the FDA approved the sale of Monsanto's controversial GE recombinant Bovine Growth Hormone (rBGH)—injected into dairy cows to force them to produce more milk—even though scientists warned that significantly higher levels (400–500% or more) of a potent chemical hormone, Insulin-Like Growth Factor (IGF-1), in the milk and dairy products of injected cows, could pose serious hazards for human breast, prostate, and colon cancer. A number of studies have shown that humans with elevated levels of IGF-1 in their bodies are much more likely to get cancer. In addition the US congressional watchdog agency, the GAO, told the FDA not to approve rBGH, arguing that increased antibiotic residues in the milk of rBGH-injected cows (resulting from higher rates of udder infections requiring antibiotic treatment) posed an unacceptable risk for public health. In 1998, heretofore undisclosed Monsanto/FDA documents were released by government scientists in Canada, showing damage to laboratory rats fed dosages of rBGH. Significant infiltration of rBGH into the prostate of the rats as well as thyroid cysts indicated potential cancer hazards from the drug. Subsequently the government of Canada banned rBGH in early 1999. The European Union has had a ban in place since 1994. Although rBGH continues to be injected into 4–5% of all US dairy cows, no other industrialized country has legalized its use. Even the GATT Codex Alimentarius, a United Nations food standards body, has refused to certify that rBGH is safe.

FOOD ALLERGIES

In 1995 a major GE food disaster was narrowly averted when Nebraska researchers learned that a Brazil nut gene spliced into soybeans could induce potentially fatal allergies in people sensitive to Brazil nuts. Animal tests of these Brazil nut-spliced soybeans had turned up negative. People with food allergies (which currently afflict 8% of all American children), whose symptoms can range from mild unpleasantness to sudden death, may likely be harmed by exposure to foreign proteins spliced into common food products. Since humans have never before eaten most of the foreign proteins now being gene-spliced into foods, stringent pre-market safety-testing (including long-term animal feeding and volunteer human feeding studies) is necessary in order to prevent a future public health disaster. Mandatory labeling is also necessary so that those suffering from food allergies can avoid hazardous GE foods and so that public health officials can trace allergens back to their source when GE-induced food allergies break out.

Unfortunately the FDA and other global regulatory agencies do not routinely require pre-market animal and human studies to ascertain whether new allergens or toxins, or increased levels of human allergens or toxins we already know about, are present in genetically engineered foods. As British scientist Dr. Mae-Wan Ho points out, "There is no known way to predict the allergenic potential of GE foods. Allergic reactions typically occur only some time after the subject is sensitized by initial exposure to the allergen."

DAMAGE TO FOOD QUALITY AND NUTRITION

A 1999 study by Dr. Marc Lappe published in the *Journal of Medicinal Food* found that concentrations of beneficial phytoestrogen compounds thought to protect against heart disease and cancer were lower in genetically modified soybeans than in traditional strains. These and other studies, including Dr. Pusztai's, indicate that genetically engineering food will likely result in foods lower in quality and nutrition. For example the milk from cows injected with rBGH contains higher levels of pus, bacteria, and fat.

ANTIBIOTIC RESISTANCE

When gene engineers splice a foreign gene into a plant or microbe, they often link it to another gene, called an antibiotic resistance marker gene (ARM), that helps determine if the first gene was successfully spliced into the host organism. Some researchers warn that these ARM genes might unexpectedly recombine with disease-causing bacteria or microbes in the environment or in the guts of animals or people who eat GE food, contributing to the growing public health danger of antibiotic resistance—of infections that cannot be cured with traditional antibiotics, for example new strains of salmonella, e-coli, campylobacter, and enterococci. EU authorities are currently considering a ban on all GE foods containing antibiotic resistant marker genes.

INCREASED PESTICIDE RESIDUES IN THE SOIL AND ON CROPS

Contrary to biotech industry propaganda, recent studies have found that US farmers growing GE crops are using just as many toxic pesticides and herbicides as conventional farmers, and in some cases are using more. Crops genetically engineered to be herbicide-resistant account for 70% of all GE crops planted in 1998. The so-called benefits of these herbicide-resistant crops are that farmers can spray as much of a particular herbicide on their crops as they want—killing the weeds without damaging their crop. Scientists estimate that herbicide-resistant crops planted around the globe will triple the amount of toxic broad-spectrum herbicides used in agriculture. These broad-spectrum herbicides are designed to literally kill everything green. The leaders in biotechnology are the same giant chemical companies—Monsanto, DuPont, AgrEvo, Novartis, and Rhone-Poulenc—that sell toxic pesticides. These companies are genetically engineering plants to be resistant to herbicides that they manufacture so they can sell more herbicides to farmers who, in turn, can apply more poisonous herbicides to crops to kill weeds.

GENETIC POLLUTION

"Genetic pollution" and collateral damage from GE field crops already have begun to wreak environmental havoc. Wind, rain, birds, bees, and insect pollinators have begun carrying genetically altered pollen into adjoining fields, polluting the DNA of crops of organic and non-GE farmers. An organic farm in Texas has been contaminated with genetic drift from GE crops on a nearby farm and EU regulators are considering setting an "allowable limit" for genetic contamination of non-GE foods, because they don't believe genetic pollution can be controlled. Because they are alive, gene-altered crops are inherently more unpredictable than chemical pollutants—they can reproduce, migrate, and mutate. Once released, it is virtually impossible to recall genetically engineered organisms back to the laboratory or the field.

DAMAGE TO BENEFICIAL INSECTS
AND SOIL FERTILITY

Earlier this year, Cornell University researchers made a startling discovery. They found that pollen from genetically engineered Bt corn was poisonous to monarch butterflies. The study adds to a growing body of evidence that GE crops are adversely affecting a number of beneficial insects, including ladybugs and lacewings, as well as beneficial soil microorganisms, bees, and possibly birds.

CREATION OF GE "SUPERWEEDS"
AND "SUPERPESTS"

Genetically engineering crops to be herbicide-resistant or to produce their own pesticide presents dangerous problems. Pests and weeds will inevitably emerge that are pesticide- or herbicide-resistant, which means that stronger, more toxic chemicals will be needed to get rid of the pests. We are already seeing the emergence of the first "superweeds" as GE herbicide-resistant crops such as rapeseed (canola) spread their herbicide-resistance traits to related weeds such as wild mustard plants. Lab and field tests also indicate that common plant pests such as cotton boll worms, living under constant pressure from GE crops, will soon evolve into "superpests" completely immune to Bt sprays and other environmentally sustainable biopesticides. This will present a serious danger for organic and sustainable farmers whose biological pest management practices will be unable to cope with increasing numbers of superpests and superweeds.

CREATION OF NEW VIRUSES AND BACTERIA

Gene-splicing will inevitably result in unanticipated outcomes and dangerous surprises that damage plants and the environment. Researchers conducting experiments at Michigan State University several years ago found that genetically altering plants to resist viruses can cause the viruses to mutate into new, more virulent forms. Scientists in Oregon found that a genetically engineered soil microorganism, *Klebsiella planticola*, completely killed essential soil nutrients. Environmental Protection Agency whistle blowers issued similar warnings in 1997 protesting government approval of a GE soil bacterium called *Rhizobium melitoli*.

GENETIC "BIO-INVASION"

By virtue of their "superior" genes, some genetically engineered plants and animals will inevitably run amok, overpowering wild species in the same way as introduced exotic species, such as kudzu vine and Dutch elm disease, which have created problems in North America. What will happen to wild fish and marine species, for example, when scientists release into the environment carp, salmon, and trout that are twice as large, and eat twice as much food, as their wild counterparts?

SOCIOECONOMIC HAZARDS

The patenting of genetically engineered foods and widespread biotech food production threaten to eliminate farming as it has been practiced for 12,000 years. GE patents such as the Terminator Technology will render seeds infertile and force hundreds of millions of farmers who now save and share their seeds to purchase ever more expensive GE seeds and chemical inputs from a handful of global biotech/seed monopolies. If the trend is not stopped, the patenting of transgenic plants and food-producing animals will soon lead to universal "bioserfdom" in which farmers will lease their plants and animals from biotech conglomerates such as Monsanto and pay royalties on seeds and offspring. Family and indigenous farmers will be driven off the land and consumers' food choices will be dictated by a cartel of transnational corporations. Rural communities will be devastated. Hundreds of millions of farmers and agricultural workers worldwide will lose their livelihoods.

ETHICAL HAZARDS

The genetic engineering and patenting of animals reduce living beings to the status of manufactured products and will result in much suffering. In January 1994, the USDA announced that scientists had completed genetic "road maps" for cattle and pigs, a precursor to ever more experimentation on live animals. In addition to the cruelty inherent in such experimentation (the "mistakes" are born with painful deformities, crippled, blind, and so on), these "manufactured" creatures have no greater value to their "creators" than mechanical inventions. Animals genetically engineered for use in laboratories, such as the infamous "Harvard mouse" which contains a human cancer-causing gene that will be passed down to all succeeding generations, were created to suffer. A purely reductionist science, biotechnology reduces all life to bits of information (genetic code) that can be arranged and rearranged at whim. Stripped of their integrity and sacred qualities, animals who are merely objects to their "inventors" will be treated as such. Currently, hundreds of genetically engineered "freak" animals are awaiting patent approval from the federal government. One can only wonder, after the wholesale gene-altering and patenting of animals, will GE "designer babies" be next?

NOTE

This article is published by permission of the author.

14

Why We Need Labeling of Genetically Engineered Food

Jean Halloran and Michael Hansen

OVERVIEW

Food is different from other consumer products. It's something we literally take inside ourselves, it's necessary on a daily basis for growth and life, and bound up in our cultures and traditions, so we care about it intensely. Consumers, therefore, have a fundamental right to know what they are eating, and that it is safe. Most developed countries have adopted laws that reflect this view, requiring labeling to show ingredients (e.g., broccoli, beef), processing (e.g., frozen, homogenized, irradiated), conformance to standards of identity (e.g., peanut butter must be made from peanuts), and additives (e.g., sulfites, preservatives). Some countries require fat, protein, carbohydrate and vitamin content of food to be labeled as well.

All of this labeling serves the consumer's right to know, and is above and beyond underlying national programs to assure the safety of food from such things as hazardous pesticide residues and additives, and disease-causing bacteria.

Consumers want to know what they are eating both as a matter of taste and preference, and for many health-related reasons. They may want to eat fish to improve their chances of avoiding heart disease, or avoid fish because they are concerned about depletion of certain species in the oceans or about mercury contamination. They may seek out carbohydrates because they are training for a marathon, or avoid them because they want to lose weight. They may eat bananas because they want a good source of potassium, or may avoid bananas because even one bite causes them to go into anaphylactic shock (as is the case with some people with severe food allergies). Body builders may want red meat, vegetarians will avoid it, and Muslims will avoid pork but not lamb. A mother may look for

221

apple juice for her child because it is a natural drink, or avoid it because it gives the child a stomach ache. Every day, millions of consumers world-wide read millions of food labels and make millions of decisions like this for themselves and their families.

Consumers also have a right to know if food is genetically engineered. In this case too, it may be for taste and preference, or for important health related reasons. Some food producers claim that genetically engineered food is basically the same ("substantially equivalent" is the description used) as conventionally produced food. But this is not the case; some individuals can have unpredictable mild to severe allergic reactions; it can have unanticipated toxic effects; and it can change the nutrition in food. In addition consumers express a wide variety of religious, ethical and environmental preferences in their food choices, and they cannot do this without comprehensive labeling.

The countries of the European Union have recognized this, and have introduced regulations requiring labeling of all genetically engineered food. In the United States, where genetically engineered corn, soybeans and potatoes are being grown commercially, repeated public opinion surveys show consumers overwhelmingly want labeling, but thus far the government has failed to require it. In 1997, a survey sponsored by Novartis found that more than 90 per cent of Americans want labeling (Feder, 1997). Most countries have not considered the issue yet. Of the large chemical/biotechnology companies that are developing these foods, some, like Novartis, support labeling, but most, like Monsanto and other major developers, oppose it.

The Codex Alimentarius Commission, an agency of the United Nations World Health Organization and Food and Agriculture Organization, has been considering whether to adopt a guideline recommending that all countries require labeling of genetically engineered food. Codex guidelines are not binding, but are often adopted by developing countries and can be used to settle trade disputes (if a country adopts a Codex standard, that standard cannot be challenged as protectionist). Consumers International is urging the Codex Alimentarius to recommend full mandatory labeling of all genetically engineered foods. This paper discusses eight important reasons why.

1. GENETICALLY ENGINEERED FOOD IS DIFFERENT

A strawberry can be given a flounder gene that makes it frost resistant, a bacterial gene that confers antibiotic resistance, and a virus gene that "turns on" the other added genes. Under normal circumstances, a strawberry can only acquire genetic material from other strawberries—that is, plants of the same or closely related species. With genetic engineering, however, scientists can give strawberries genetic material from trees, bacteria, fish, pigs, even humans if they choose to. Consumers International believes that any plant or animal food to which genes have been added from a source other than the species to which the food belongs should be required to be labeled, to tell the consumer that this has been done.

Some people, mostly scientists and corporations involved in the development of genetically engineered food, argue that the strawberry with the foreign genes is not really different but "substantially equivalent" in the language of Codex and international regulation and therefore needs no label.

Consumers, however, through their organizations, through comments to regulators, and through opinion surveys, have repeatedly expressed the view that this strawberry, and all other genetically engineered foods, are not "substantially equivalent," but sufficiently different that, like irradiated foods and foods containing additives, they should be labeled. Since labelling laws are created to meet consumer needs, consumer opinion should be respected.

2. GENETICALLY ENGINEERED FOOD CAN CAUSE TOXIC EFFECTS

The fact that genetic engineering can go seriously wrong was shown by one of the very first products introduced into the market. An amino acid (a protein building block) called tryptophan is sold in a number of countries including the United States as a dietary supplement. In the late 1980s, the Showa Denko company of Japan began making tryptophan by a new process, using genetically engineered bacteria, and selling it in the United States.

Within months thousands of people who had taken the supplement began to suffer from eosinophilia myalgia syndrome, which included neurological problems. Eventually at least 1,500 were permanently disabled and 37 died (Mayeno and Gleich, 1994).

As doctors encountered this syndrome, they gradually noticed that it seemed linked to patients taking tryptophan produced by Showa Denko. However, it took months before this was taken off the market. Had it been labeled as genetically engineered, it might have accelerated the identification of the source of the problem.

Showa Denko refused to cooperate in any U.S. government efforts to investigate the cause of the problem. However, the tryptophan that caused the problem was determined to contain a toxic contaminant which appears to have been a by-product of the increased tryptophan production of the genetically engineered bacteria (Mayeno and Gleich, 1994).

There are many ways besides this in which genetic engineering could go awry and result in hazardous toxins in food. Many common plant foods such as tomatoes and potatoes produce highly toxic chemicals in their leaves, for example. Any responsible company working with such plants would check for changes in toxin levels. But not all companies are equally responsible, and as the Showa Denko example shows, a serious hazard can be missed.

Government agencies cannot be counted on to prevent unexpected problems. World-wide, government premarket safety reviews of genetically engineered products range from relatively thorough in the European Union to no review at all in much of the world. In the United States, premarket safety reviews are voluntary.

We can expect that in the future genetically engineered food will be developed and grown in many countries with no premarket safety reviews. Unless all such products are labeled, it will be difficult to determine the source of any toxin problems originating in such food.

3. GENETICALLY ENGINEERED FOOD CAN CAUSE ALLERGIC REACTIONS

In the United States, about a quarter of all people report that they have an adverse reaction to some food (Sloan and Powers, 1986). Studies have shown that 2 percent of adults and 8 percent of children have true food allergies, mediated by immunoglobin E (IgE) (Bock, 1987; Sampson et al., 1992).

People with IgE mediated allergies have an immediate reaction to certain proteins, ranging from itching to potentially fatal anaphylactic shock. The most common allergies are to peanuts, other nuts and shellfish.

Genetic engineering can transfer allergies from foods to which people know they are allergic, to foods that they think are safe. In March 1996, researchers at the University of Nebraska in the United States confirmed that an allergen from Brazil nuts had been transferred into soybeans. The Pioneer Hi-Bred International seed company had put a Brazil nut gene into soybeans to improve their protein content for animal feed. In an in-vitro and a skin prick test, the engineered soybeans reacted with the IgE of individuals with a Brazil nut allergy in a way that indicated that the individuals would have had an adverse, potentially fatal reaction to the soybeans (Nordlee et al., 1996).

This case has a happy ending. As Marion Nestle, the head of the Nutrition Department at New York University, summarized in an editorial in the respected *New England Journal of Medicine*, "In the special case of transgenic soybeans, the donor species was known to be allergenic, serum samples from persons allergic to the donor species were available for testing and the product was withdrawn" (Nestle, 1996: 726). However, for virtually every food, there is someone allergic to it. Proteins are what cause allergic reactions, and virtually every gene transfer in crops results in some protein production. Proteins will be coming into food crops not just from known sources of common allergens, like peanuts, shellfish and dairy, but from plants of all kinds, bacteria and viruses, whose potential allergenicity is uncommon or unknown. Furthermore, there are no fool-proof ways to determine whether a given protein will be an allergen, except tests involving serum from individuals allergic to the given protein. Nestle continues, "The next case could be less ideal, and the public less fortunate. It is in everyone's best interest to develop regulatory policies for transgenic foods that include premarketing notification and labelling" (Nestle, 1996: 727).

To protect consumer health from the effects of unrecognized or uncommon allergens, all genetically engineered food must be labeled. Otherwise there will be no way for sensitive individuals to distinguish foods that cause them problems from ones that do not. This need is particularly urgent, since one of the potential consequences is sudden death, and children are the part of the population most at risk.

4. GENETIC ENGINEERING CAN INCREASE ANTIBIOTIC RESISTANCE

Despite the precise sound of its name, genetic engineering is actually a messy process, and most attempts end in failure. While the gene to be transferred can be identified fairly precisely, the process of inserting it in the new host can be very imprecise. Genes are often moved with something that is the molecular equivalent of a shotgun. Scientists coat tiny particles with genetic material and then "shoot" these into thousands of cells in a petri dish before they get one where the desired trait "takes" and is expressed.

Because the transferred trait, such as ability to produce an insecticide in the leaves of the plant, is often not immediately apparent, scientists generally also insert a "marker gene" along with the desired gene into the new plant. The most commonly used marker is a bacterial gene for antibiotic resistance. Most genetically engineered plant food contains such a gene.

Widespread use of antibiotic resistance marker genes could contribute to the problem of antibiotic resistance. The genes may move from a crop into bacteria in the environment, and since bacteria readily exchange antibiotic resistance genes, move into disease-causing bacteria and make them resistant too.

Antibiotic resistance genes could even be transferred in the digestive tract to bacteria. An example of this is the genetically engineered Bt maize plant from Novartis which includes an ampicillin resistance gene. Ampicillin is a valuable antibiotic used to treat a variety of infections in people and animals. A number of European countries, including Britain, have refused to permit the Novartis Bt corn to be grown, because of concern that the ampicillin resistance gene could move from the corn into bacteria in the food chain, making ampicillin a less effective weapon against bacterial infections.

But there are already foods in the market made using plants with antibiotic resistance marker genes. Without labeling, consumers cannot choose not to buy them.

5. GENETIC ENGINEERING CAN ALTER NUTRITIONAL VALUE

Genetic engineering can alter nutritional value of foods in positive ways. For example, canola oil has been engineered to have a different profile of fatty acids, so that they contain less of the fat molecules that tend to build up in people's arteries. Scientists are also working on increasing the vitamin C content in some foods. However, it is also possible that nutritional content could be reduced as an unexpected side effect of some other genetic engineering. Labeling is needed to make sure consumers are properly informed.

6. GENETICALLY ENGINEERED FOOD CAN CREATE ENVIRONMENTAL RISK

The most widely grown genetically engineered crops, accounting for 99 percent of the land under transgenic cultivation world-wide, are engineered for herbicide tolerance, insect resistance, and virus resistance (James, 1997). Each of these poses environmental risks.

Herbicide-tolerant crops are varieties on which herbicides can be used to kill weeds, without killing the crop itself. These varieties encourage farmers to use more herbicides, which frequently pollute groundwater and can cause various other forms of ecological damage.

Insect-resistant crops almost all contain a gene from the bacterium *Bacillus thuringiensis* (Bt) which causes the plant to produce an endotoxin throughout the plant, including leaves and fruit. Bt corn, cotton, potatoes, tomatoes and rice are all being grown in various parts of the world.

While Bt crops at first glance appear to be ecologically sound, because they need less chemical pesticides, they have serious drawbacks. Crops that continuously produce Bt endotoxin quickly speed up the process of the spread of resistance to the Bt endotoxin among the pests feeding on the crops.

A recent computer model developed by a scientist at the University of Illinois in the U.S. predicted that if all U.S. farmers grew Bt corn, resistance would develop in a single year! Scientists at the University of North Carolina in the U.S. have already found Bt resistance genes in wild populations of a moth pest that feeds on corn (Gloud et al., 1997).

The Bt endotoxin, produced by the Bt bacteria, is a staple of organic farming since it is a relatively harmless natural pesticide. It is also widely used by conventional farmers who use integrated pest management to minimize the use of more toxic chemicals. Scientists predict that Bt will become less and less useful, however, within a few years of widespread planting of Bt crops.

The Bt crops may also be toxic to beneficial insects. Researchers from Swiss Federal Research Station for Agroecology and Agriculture found, for example, over 60% mortality of green lacewings that ate moth larvae that had fed on Bt corn.

Virus-resistant crops almost all contain genes that can mix with genes from other viruses that naturally infect the plant to create new gene combinations, some of which can give rise to new or deadlier viruses. U.S. and Canadian work has shown that wild viruses can hijack genes from engineered crops at rates far higher than previously suspected. The concern was great enough that the U.S. Department of Agriculture held a meeting in October 1997 to discuss possible restrictions aimed at reducing the risk of creating harmful new plant viruses due to the use of virus-resistant crops (Kleiner, 1997).

Another serious concern is "gene pollution." If the gene for herbicide tolerance escapes into wild relatives of crop plants that are weeds, it could result in a new generation of herbicide-tolerant superweeds. In fact, researchers in both Norway (Jorgensen and Andersen, 1995) and the United States (Hileman, 1995) have already demonstrated that the gene for herbicide tolerance moved from cultivated canola to close relatives in nearby fields, such as wild mustard.

If the gene for the production of the Bt endotoxin moves into wild plants, they could become resistant to butterfly, moth and beetle pests, just like the Bt crops. This could upset established ecological balances by either causing the wild plant to flourish excessively and become a plant pest, or by reducing the butterfly or moth population that previously fed on the newly toxic plant.

Gene pollution would be especially problematic in many developing countries where the center of origin for many crops is. In these areas, traditional crop varieties could become "polluted" with genes from the genetically engineered crops and biological diversity will suffer. The rate of gene flow between genetically engineered plants and their wild relatives may be higher than previously thought. Researchers in the southern United States demonstrated that more than 50% of the wild strawberries growing within 50 meters of a strawberry field contained marker genes from the cultivated strawberries. Researchers in the central U.S. found that after ten years more than a quarter of the wild sunflowers growing near fields of cultivated sunflowers had a marker gene from the cultivated sunflowers (Kling, 1996).

These problems illustrate the need for great caution in introducing and using genetically modified plants. But even with this, consumers have a right to know about the environmental impact of the foods they buy so that, if they wish they can exercise their own preferences and avoid—or choose to buy—food that has been produced in a particular way.

7. GENETIC ENGINEERING CAN AFFECT DIETARY PREFERENCES

Consumers make decisions about what they eat for a wide variety of religious, ethical, philosophical and emotional reasons. Most major world religions have some rules or traditions as to food. Jews and Muslims do not eat pork; Christians often avoid meat on Fridays or during Lent; many Buddhists are vegetarians.

Many other individuals have food preferences that are not related to an organized religion but which reflect deeply held personal beliefs, such as wanting to protect the environment.

Consumers International supports labelling of genetically engineered food in order to allow consumers the opportunity to exercise their religious and ethical preferences. For example, some people will want to avoid lamb which contains pig genes (a product which is not yet on the market, but is well within the current capabilities of science). For this, labelling would be essential.

8. SCIENCE IS FALLIBLE

When a new technology of food production emerges, all the problems it may cause may not be foreseen. When pesticides were first synthesized and used widely in the 1950s, they were heralded as a miracle cure for pest problems. Only later did we discover that some of them could also cause birds to lay eggs with shells that collapsed, humans to get cancer, and insects to become resistant to them.

Genetic engineering is shuffling the deck of genes in ways that are entirely new, and creating living things that have never before existed. Consumers International believes consumers have a right to be cautious about using these, if they wish. The right to choose can be exercised only if proper information is provided—on labels or the food itself.

REFERENCES

Reprinted by permission of Consumers Union.

Bock, S. A. 1987. Prospective appraisal of complaints of adverse reactions to foods in children during the first 3 years of life. *Pediatrics* 79: 683–88.
Feder, B. J. 1997. Biotech firm to advocate labels on genetically altered products. *New York Times*, February 24.
Gould, F., A. Anderson, A. Jones, D. Sumerford, D. G. Heckel, J. Lopez, S. Micinski, R. Leonard, and M. Laster. 1997. Initial frequency of alleles for resistance to *Bacillus thuringiensis* toxins in field populations of *Heliothis virescens*. *Proceedings of the National Academy of Sciences* 94: 3519–23.
Green, A. E., and R. F. Alison. 1994. Recombination between viral RNA and transgenic plant transcripts. *Science* 263: 1423–25.
Hileman, B. 1995. Views differ sharply over benefits, risks of agricultural biotechnology. *Chemical and Engineering News*, August 21, 1995.

James, C. 1997. Global status of transgenic crops in 1997. *ISAAA Briefs no. 5.* Ithaca, N.Y.: International Service for the Acquisition of Agri-biotech Applications (ISAAA).

Jorgensen, R., and B. Andersen. 1995. Spontaneous hybridization between oilseed rape *(Brassica napus)* and weed *Brassica campestris:* A risk of growing genetically engineered modified oilseed rape. *American Journal of Botany* 81: 1620–26.

Kleiner, K. 1997. Fields of genes. *New Scientist,* August 16.

Kling, J. 1996. Could transgenic supercrops one day breed superweeds? *Science* 274: 180–81.

Mayeno, A. N., and G. J. Gleich. 1994. Eosinophilia myalgia syndrome and tryptophan production: A cautionary tale. *TIBTECH* 12: 346–52.

Nestle, M. 1996. Allergies to transgenic foods: Questions of policy. *New England Journal of Medicine* 334, no. 11: 726–27.

Nordlee, J. A., S. L. Taylor, J. A. Townsend, L. A. Thomas, and R. K. Bush. 1996. Identification of a brazil-nut allergen in transgenic soybeans. *New England Journal of Medicine* 334, no. 11: 688–92.

Sampson, H. A., L. Mendelson, and J. P. Rosen. 1992. Fatal and near-fatal anaphylactic reactions to food in children and adolescents. *New England Journal of Medicine* 327: 380–84.

Sloan, A. E., and M. E. Powers. 1986. A perspective on popular perspectives of adverse reactions to foods. *Journal of Allergy and Clinical Immunology* 78: 127–33.

15

Ethical Issues in Food Biotechnology

Paul Thompson

THE PHILOSOPHY OF FOOD SAFETY

Once these risk-based tests are passed, why would the safety of food biotechnology raise any serious ethical questions at all? The establishment answer to this question is that the critics of biotechnology are hysterical or liars (or both) and that the public is ill informed, easily misled and otherwise fairly apathetic about the safety of food biotechnology. But the establishment answer is wrong. Risk, defined as probability of food-borne illness or injury, is too narrow to encompass the full range of issues that are traditionally associated with food safety. Socially based management of food risks has undergone three broad phases. For most of human history the problem has been one of classification. Is it food or not?

When medical scientists began to appreciate the importance of germs and other contaminants as a cause of disease, a more subtle approach arose which stressed the elimination of impurities. Only recently have scientists begun to appreciate the complexity of whole foods, recognizing that many components of foods may have toxic properties under certain circumstances and that toxicants themselves may be responsible for benefits that offset the disease risk associated with their presence in food. This has introduced a tendency to think of food safety as an optimization problem that gives rise to the risk-based approach described above and supports it with impressive scientific credentials.

My own view is that the optimization approach is correct when safety is conceived narrowly, but that alternative views are entirely reasonable. For those food safety decisions that must be made collectively (and given the complexity of our food system, that is most of them), norms of democracy demand an accommodation for all reasonable points of view. This is not to say that everyone gets whatever they want, but it does entail making collective or political decisions that must be made in a manner that does not foreclose the possibility of reasonable disagreement and that when divergent viewpoints can be accommodated without imposing unreasonable costs or inconveniences on others, they should be. The case for applying these general norms of democracy to the issue of food safety and biotechnology begins by considering each of the three historical approaches to food safety in more detail and by gaining an appreciation of the concomitant values that complicate food safety decisions. Given the diversity and complexity of these values, norms of democracy weigh heavily in favor of a policy that preserves key elements of consumer sovereignty and consent. Labeling is one means of supplying consumers with information and protecting the autonomy of their decision making with respect to diet.

Classification

Human beings recognized the existence of poisonous foods in prehistory. By the time of the Greeks, such knowledge served as the basis for Socrates' famous meditations on death and duty prior to drinking hemlock. The lack of an ability to distinguish foods with toxic constituents from microbial toxins confounded early efforts to manage the risk of ingesting poisons through a dietary regimen, but trial and error eventually ruled out acutely poisonous plants and animals. It is less sure but seems likely that humans came early to the knowledge that eating the wrong thing could have delayed effects and increase chronic health risks. Clearly early beliefs about food safety were a mix of superstition, speculation and hard-won experience. The Pythagorean cult of which Plato was a member proscribed the eating of beans, though it is unclear whether the deadly effect of castor beans or the problem of flatulence was the occasion for this food rule. Generally such rules classify plants and animals into food and non-food groups and in some cases specify finer distinctions for preparation or serving foods. Whatever else might be said about these culturally transmitted food rules, in every known case, following them results in food consumption that is safer than a diet of randomly sampled plants and animals. This is an unexceptional fact; a food culture that included acutely toxic elements would disappear rather quickly.

Dietary rules that distinguish food from non-food, as well as stipulating which foods may be consumed together, are a part of every culture. In some cases these rules are relatively weak prescriptions of taste: one does not mix ice cream and onions, and croutons with red wine is not a breakfast dish. In some cultures dietary rules take on religious significance and become a serious matter indeed. Semitic rules against the eating of pork are among the best known food taboos. Food safety is thus but one function of a cultural dietary regimen (Harris, 1974). While adherents to a given system of food classification may associate violation of the rules with sickness, injury or death, they may also associate it with religious, spiritual and social forms of risk. Religious, ethnic and regional food rules persist today because they are constitutive of social, cultural and personal identity, because they reinforce feelings of well-being and order and because people love them, find them pleasing, satisfying and take gratification in following them.

Purification

The anthropologist Mary Douglas (1975) has used the term "purity" to describe a traditional society's adherence to a system of food rules such as are described above. What I have in mind here is a way of looking at food safety that begins with the advent of the germ theory of disease in the nineteenth century. The germ theory held that disease was the result of invisible infectious agents or germs and that strategies of purification could control disease before it starts. The germ theory gave rise to a strategy for food safety that deployed technical means for preventing the entry of infectious agents into human food, for destroying them once they are there and eventually for treating those afflicted by germs. Refrigeration, sterilization, irradiation and temperature monitored cooking are all weapons in the arsenal of purification (Douglas and Wildavsky, 1982).

Belief in the efficacy of purification spawned legislation that regulated the food industry and created governmental agencies for enforcing the first food safety laws. It was the beginning of the modern conception of food safety. Upton Sinclair's *The Jungle* (1906) created an uproar with its description of unsavory practices in the Chicago packing plants, among them an episode in which a hapless immigrant worker is literally ground into sausage as the line rolls on without missing a beat. In a ghoulish way, this episode from *The Jungle* is indicative of the way that food safety regulation based on a strategy of purification carries the water for cultural foodways, just as Mary Douglas would predict. Prion diseases that were unknown to the readers of *The Jungle*, aside from grinding up a human being along with the sausage introduce little additional health risk for consumers. *The Jungle* was effective because however little regard middle class Americans may have had for immigrant workers, they were queasy about consuming parts of them in their hot dogs. The food safety regulations that came on line in most industrialized countries prior to World War II used the phrase "safe and wholesome." They prohibited the use of dogs, cats and rodents in meat products despite the fact that all these animals are used for food in some non-European cultures. Purification introduced science, technology and government into the pursuit of food safety, but retained many cultural norms in defining when a food is pure.

Purification was still the model when discussion began about a new class of food safety problems in the 1950s. The U.S. Delaney Clause, for example, appears to be a model of the purification approach. The law requires the U.S. Food and Drug Administration to ban any additive found to cause cancer in humans. The law is accompanied by a list of traditional food ingredients deemed "generally recognized as safe" (GRAS). It is imminently plausible to think that many of the legislators who voted for this law, if not Congressman Delaney himself, were thinking that foods (including the GRAS list) are non-risky and that hazards are associated with contaminants. They may well have thought of themselves as instructing the regulators to seek out the contaminants and ban them. As eminently plausible as this thinking may seem, it is utterly incompatible with the new thinking on food safety that is represented by the optimization paradigm.

Optimization

Scientists and regulators have adopted the risk-based view of food safety. The conceptual distinction between purity and risk-based approaches to food choice hangs on what interpretation is put on "no risk," the criterion applied to food additives under the Delaney Clause. A thorough purificationist (if such ever existed) believes that whole foods consumed since time immemorial bear no risk. This cannot mean that no harm will ever come from eating them, since food pathogens and unexplained poisonings and reactions have been around since time immemorial. "No risk" must mean no human-caused risk, no risk introduced into the food system as a result of intentionally introducing additives into whole foods. What counts as "intentional introduction"? If something not on the GRAS list is put into food, it is an intentional introduction and it must be proved harmless.

Few (if any) scientists or regulators think of risk in these terms. For them "no risk" means zero risk, a quantity that can be reached only when the probability of harm associated with an action or choice is zero. This is an assumption that makes the Delaney Clause intellectually incoherent. At best, epidemiological studies and animal trials will reveal no statistical evidence of carcinogenicity, but "no statistical evidence" is far short of proving zero probability. More detailed studies and better tests can always overturn this result and that has indeed been the case. In addition, the purificationist interpretation of the Delaney Clause clearly admits circumstances in which GRAS substances with risks known or strongly suspected to be higher than those of banned additives are allowed, clearly a suboptimal situation.

What is worse, the risk-based approach has begun to uncover evidence that many common foods fail the zero-risk test. The work of Bruce Ames (1983) is typical of this new view, though it was clearly on the horizon well before Ames became an advocate of it. Ames's bio-chemistry work has shown that virtually all foods contain mutagens—substances that increase the statistical rate of "errors" or minor changes in the order of DNA base pairs as a cell duplicates itself. Mutagens are thought to precipitate cancerous cell growth, though their very ubiquitousness shows that any link between mutagenicity and carcinogenesis is complex. Cancer would be everywhere if mutagens caused cancer, tout court. Ames has promulgated the view that since naturally occurring mutagens far outnumber those associated with chemical additives and pesticide residue, it is unlikely that the rate of chronic diseases in the human population will be significantly reduced by strategies that target food additives or chemical residues.

Identification and Purification versus Risk-Based Optimization

Since mutagens are everywhere in human foods, the new view eliminates the importance of the food/non-food distinction. Foods (tomatoes, beets, beef) are at least as likely to introduce the fatal mutagen into the body as are additive or pesticidal non-foods. People do not get cancer all the time because somehow these mutagens are kept in check, either by the body's defense mechanisms or perhaps even by each other. The risks of chronic diseases such as cancer cannot be controlled by purifying foods, for the foods themselves are not benign. Instead we must seek a balance point, not yet well understood, in which the mutagenicity of what we eat is held in check (as much as possible) by all the other factors (good nutrition, exercise, cognitive stimulation) that create health.

This way of conceptualizing food health risks fits hand in glove with the risk/benefit trade-off thinking that had already begun to emerge in the wake of attempts to regulate agricultural insecticides, herbicides and other chemically based pest control technologies. Some of these technologies, notably rodenticides and insecticides which used to control pest infestations of harvested grain, had been introduced in order to prevent contamination—a clear application of the purification philosophy. It was evident that when adequate substitutes were unavailable a ban on these technologies would be followed by a resurgence of pest infestations and the attendant health problems. In such cases trade-offs are inevitable and the trade-off involves food safety, not merely economic losses. Sometimes the interests of human health will be better served by accepting the risks associated with the technology, sometimes it will be better to accept the pests.

Both the work of Ames and the trade-off logic of chemical technology support a reconceptualization of food safety along the lines of risk optimization. Risk comes to be seen as pervasive and the presumptions of purification (that risk is due to impurities only) come to be seen as naive. An advocate of optimization argues that some risk must be accepted and the norm that should be applied to the problem of food safety is to define an optimum, a balance point and to regulate food production and processing so as to approximate the optimum as closely as possible. Optimization is clearly a tricky business; it is not simply a matter of minimizing risk. It requires answers to questions like "How should we compare a statistically high level of risk to a few agricultural field laborers, to a very low probability of harm that may fall on food consumers?" Optimization strategies for food safety demand a high level of philosophical and ethical sophistication in their treatment of risk (Sagoff, 1985), but the point here is simply to note how dramatically the optimization approach differs from that of either classification or purification.

When seen as strategies that are bound by the limited knowledge of their respective moments in history, classification and purification can be interpreted as worthy attempts at risk optimization that have become obsolete. This amounts to the claim that ethnic or religious food regimes and purification technologies were conceived and adopted with the object of striking the proper balance between risk and benefit for food safety. I do not think that this claim is plausible, though I know of no historical or anthropological research that could disprove it. What seems more likely is that the normative philosophy of optimization evolved at the same time as the scientific theories and political problems to which it is so admirably suited. If that is right, then the values and experience of previous generations and of many in the present day do not provide adequate support for a strategy of optimization and if this is true, then reasonable, intelligent people are likely to conceptualize food safety more along the lines of classification and purification than as an optimization problem.

FOOD SAFETY AND ETHICS

The most obvious way to move from risk-based optimization to an ethical philosophy for food safety is to interpret it as an extremely sophisticated extension of utilitarianism. The main problem with risk-based optimization is a traditional problem of utilitarianism: too little attention to consent. The consent based approach harks back to a system of *caveat emptor*. Individuals accept risks to which they have given informed consent, but coercion or concealment of relevant information is morally unacceptable. Critics of the consumer's right to know argue that this ethic is unrealistically demanding for a modern food system, but the justifiable demands of such rights are less onerous than is sometimes thought. The existing food system in most parts of the world is surprisingly close to what the rights view would hold as ideal.

Utilitarianism and Risk-Based Optimization

Utilitarianism is the system of ethics that evaluates an action, rule or public policy in light of its consequences for all affected parties. Traditional English utilitarians such as Jeremy Bentham or J. S. Mill assumed that the total value of all consequences, beneficial or harmful, could be summed and proposed the decision rule of "promote the greatest good for the greatest number." The emphasis on optimization, rather than maximization, signifies an appreciation of (if not a solution to) the complexity of quantifying and consistently ranking the myriad of goods and evils that bear on food choice (Finkel, 1996). One strength of the utilitarian approach is that in emphasizing the prediction of consequences from action or policy, it represents the most obvious way to bring predictive science to bear on ethical decision making. Indeed, many scientists gravitate so easily to the view that public policy should apply the best science to predict the consequences of several policy options, and then choose the option with the best outcomes, that they fail to notice that they are applying a philosophical framework at all.

Frequently we are happy to allow people in decision making positions to decide on our behalf, for when they can be trusted to look after our interests, it saves us the time and expense of staying fully informed about what the latest science has to say about this or that. The modern food system has evolved under circumstances in which the public was generally happy to have many decisions about the ingredients or composition of food left to the experts (Knorr and Clancy, 1984). A utilitarian would interpret this state of affairs as ethically justified because the cost of staying informed far outweighs any benefits the information would bring to the mass of people. But although the cost of staying informed clearly plays a key role in the willingness of the public to entrust the safety of food to experts, it is not necessary to adopt a utilitarian analysis of this fact. It is equally plausible to assert that in view of these costs, consumers consented to a set of social arrangements in which trustees were given responsibility to decide on the basis of the public interest. Furthermore they did so under circumstances in which they were frequently capable of choosing otherwise, as alternatives to processed foods and foods produced using agricultural chemicals were generally available. The fact that many people did not choose these alternatives makes no difference; what guarantees consent is that they could have opted otherwise, had they wanted to. This is a horse of a different moral color, as the logical force of the moral judgment now rests on giving consent, whereas for the utilitarian it rests in selecting the most attractive ratio of benefit and cost, irrespective of consent.

Science-Based Policy and Its Discontents

The argument for optimization does not lie in its alliance with utilitarian ethics, but with the fact that this approach best synthesizes current scientific thinking with respect to the probability of food-borne illness and injury. Nevertheless, the problems with a too narrow emphasis on risk management and trade-offs are also generally the problems of the utilitarian approach. Utilitarianism has always been criticized because it appears to make short work of rights in too many instances, depriving people of the opportunity to make their own choices in the name of doing what is good for them. In the case of food safety policy, the claim that policy should be risk-based or simply science-based neglects the fact that science provides little insight into many of the dimensions that influence individual food choices. None of these dimensions provide reasons to ban or regulate food biotechnology, but any of them might well provide an individual with defensible reasons for preferring not to eat these new foods.

Is "food safety" a purely scientific concept, to be defined and controlled exclusively by food scientists, or is "food safety" a term of ordinary language? If it is the latter, then it becomes relevant to see what non-scientists mean when they use the concept of safe food. Food journalist Robin Mather puts it this way in writing for a popular audience:

> Because food is so important to us on so many levels, we must generally trust blindly in government's promises of a safe food supply and in the safe practices of those who produce the foods we buy. We wade doggedly through complicated, confusing and often contradictory information about the foods we eat. The reason is that most of us realize on a fundamental level that food choice is one of the last arenas in which we have some measure of control.

Here Mather characterizes "control" as a primary dimension of food safety. Most of her book discusses the social organization of agricultural production, but she links the social impact of biotechnology to this paragraph on food safety by noting how difficult it is for food consumers to "track where our food comes from" (Mather, 1995).

Building on Mather's comment, the problem with the risk-based approach is not that it is wrong as far as it goes, but that it does not go far enough, ignoring the "many levels" on which food is important to us. Mather's complaint against government regulation of food safety is that it is being used to undermine "one of the last arenas in which we have some measure of control." Of course, the Department of Transportation may be "undermining our control" when they mandate speed limits or airline security checks. In one sense, every public policy affects the control of an individual's control over their lives. This is not in itself a powerful argument. What must be shown is that food consumers have rationally defensible ends in view in wanting alternatives to biotechnology-derived foods. There are at least four such ends.

Religious and Ethnic Beliefs

As already noted, the cultural history of food beliefs has produced a rich array of religious and ethnically based beliefs about what is and is not food. These beliefs are imperfectly correlated with scientific probabilities concerning illness or injury, at best. Yet no one challenges the right of religious and ethnic minorities to stipulate food rules that are far more restrictive than those stipulated in legal codes. Kosher and halal practices, among the most widespread, demand both dietary restrictions and special procedures for slaughter and preparation. These rules remain under the constant supervision of the ecclesiastical authorities of each group. Clearly it is up to these authorities whether the use of genetic engineering or other forms of biotechnology are consistent with their traditional dietary rules. Furthermore, Judaic and Islamic religious authority is widely distributed across many rabbis and mullahs, each of whom might interpret the code differently. There is no hierarchy that can answer the question once and for all. Any intentional or de facto attempt to decide this question on their behalf would violate minority rights that are well established throughout the industrialized world and covered by the International Declaration on Human Rights.

Latent Purificationism

Few laypersons and even many scientifically trained individuals can make ready sense of the risk-based optimization approach that is *de rigueur* among toxicologists, epidemiologists, biochemists and food regulators. The purification model that has been around for a century is more consistent with educated common sense. The idea that crossing genes is a violation of purity rules comes readily to many people and they are at least psychologically justified in feeling somewhat queasy about genetically engineered food. Even many scientists can identify with this queasy feeling. A 1993 dictionary of technical terms for biotechnology includes an entry for "yuk factor." The yuk factor is "a flippant term for the very real observation that the public, indeed many scientists, judge the ethical acceptability of experimental procedures and biological manipulations in accordance with a scale of personal distaste"(Bains, 1993).

Do the queasy need more justification than that? Historically, simply *preferring* to pass up a food innovation has been sufficient justification for being allowed to do so. Though we find it hard to believe today, resistance to pasteurized milk once ran deep. However, those who resisted and who resented the fact that their product of choice was not available on the local grocery shelf had (admittedly expensive and inconvenient) alternatives and they were never placed in the position of having to guess whether the milk on the grocery shelf was pasteurized or not. Fluoridation of public water supplies also made tempers flair, but bottled water was generally available and again no one had to wonder. Queasiness and latent purificationism regarding food biotechnology are clearly present among members of the lay public and they are all the reason some people need to avoid genetically engineered food.

Distrust of Science

A subset of the queasy are angry too. Many of them purchase organic foods, often at great inconvenience to themselves. They write angry letters to the *Sierra* magazine (a popular publication of the club, a powerful American environmental organization), complaining about "Franken-foods," signaling their distrust of science by linking food biotechnology to Mary Shelley's Frankenstein. Some, like Christian Scientists and Mennonites, may combine resistance to science with religious faith. Some may press their anger beyond rationality. The so called Unabomber terrorized scientists and executives of technology companies in the USA for over a decade. His mail bomb targets included molecular biologists and his anonymous manifesto, published in the *Washington Post*, leveled vitriolic criticism against biotechnology. But in a world that has given us disasters like Bhopol and Love Canal, where the safety of chemical and then nuclear technologies was badly oversold and where scientists performed experiments on humans without informed consent, not only in Nazi death camps but also in Tuskegee, Alabama, is it really so surprising that some people are a little reluctant to accept scientific assurances about the safety of biotechnology?

For some, skepticism about science is learned from science itself. A key thesis in Ulrich Beck's influential book *Risk Society* is that public suspicion of science is a learned response owing to the emphasis our educational system put upon scientific skepticism in the face of uncertainty (Beck, 1992). Clearly it is impossible to isolate oneself completely from science and its impact on our world. Claiming a right to do that would be preposterous. But as already argued, people have never been forced to choose between total reliance on science and subsistence agriculture. Food choices and alternatives have been the norm, if not a right, and people are justifiably resentful (and suspicious!) of the forces that threaten this valued *status quo*.

Solidarity

As will be discussed, some of the most potent arguments against agricultural biotechnology are about its social consequences. Critics argue that family farms will be lost and that farmers will lose control of their operations, becoming little more than serfs as one farmer recently put it. (This comment was entered into the public record by an Ontario farmer attending the annual meeting of the National Agricultural Biotechnology Council on campus at Rutgers University, June 1996.) Whether this is true or not, those who believe it may wish to avoid foods from agricultural biotechnology as a form of protest. Just as consumers boycotted grapes to show solidarity with field workers or meat products to show solidarity with packers, some consumers may wish to boycott biotechnology as a way to show solidarity with traditional or small farmers. This appears to be Mather's (1995) primary reason for opposing biotechnology, for example. The last chapter of her book is entitled "Voting With Your Buck" and it gives details of how consumers can make food choices that support small-scale, sustainable agriculture, rather than "bioengineered foods."

If consumers have a right to "vote" with their food purchases, it is certainly less secure than Mather's writing implies. It is far from clear that producers of a targeted good are obligated to help their opponents identify them on supermarket shelves, for example. Yet food consumers who were willing to put themselves to a bit of inconvenience have generally been able to express their politics through their purses in the past. They have often paid a significant price premium in order to do so and it is not unreasonable to think that those who accept animal welfare. environmental or social criticisms of biotechnology will want to do so again in the future. Like religious or ethnic beliefs or simple queasiness and distrust of science, solidarity motives do not constitute irrational objections to biotechnology.

FOOD LABELS AND CONSENT

Many of the things that people want to know about food have nothing to do with science and are only marginally related to safety (as conceived as the probability of illness or injury). But people want to feel good about their food choices and if this means knowing that their Champagne comes from France rather than California or that their hot sauce is made in Texas, not New York City, then having the ability to discriminate on the basis of such information contributes to their feelings of well-being and satisfaction. To the extent that "safety" connotes a feeling of security and well-being, such information contributes to food safety. Perhaps it's better not to stretch the word "safety" this way, but what is clear is that people will feel suspicious of and at-risk from individuals or groups who try to deprive them of information that they deem valuable.

The point of food safety policy is not merely to make foods safe, but to provide the public with reasonable assurances of food safety. Ironically, it may not be possible to accomplish this latter objective without providing information that has little bearing on the probability of harm from consuming the food in question. Thus there are calls for mandatory labeling that would identify foods derived from biotechnology (Group of Advisors, 1996). As will be shown presently, this is a policy approach that has problems of its own. First it is crucial to be absolutely clear about the ethical argument.

Bad Arguments for Mandatory Food Labels

Most of the arguments that have actually been proffered for labeling genetically engineered foods do not stand up to close examination. It will be useful to review several of them.

Biotechnology is unsafe. The most straightforward argument for requiring labels proclaiming the use of biotechnology would be if there are demonstrated health risks associated with consumption of products derived from biotechnology, as there are for alcohol and tobacco products. Such labels would warn consumers of such risks. Clearly from what has been discussed before in this chapter, no evidence exists for health risks, so this argument relies on a false premise.

Biotechnology is unnatural. The argument here is that since biotechnology may have been used to produce a whole food, meats or grains, products that are not readily recognizable as processed may be confused with natural foods, meats or grains. Consumers who want a certain kind of product (e.g., one in which biotechnology has not been used in any way) and have no reliable way of recognizing such products are indeed deprived of any right they have to satisfy that preference with unlabeled produce, meats and grains. Yet it is difficult to see how "naturalness" comes into play. On this point, Roger Straughn is right. The abuse of natural/artificial distinctions is extensive enough that we ought not encourage more of it.

People want genetically engineered foods to be labeled. Data on public opinion demonstrate an abiding interest in labeling of genetically engineered foods in virtually every population surveyed. While such surveys might constitute a sufficient political argument for mandatory labels if there were no countervailing concerns, they do not establish ethical reasons for requiring that the producers of biotechnological foods label their products. There is a fallacy in applying any kind of survey data to reach such a conclusion. Given the question, Would you like more information (or labels) on X? many people are likely to respond affirmatively in regard to what X is, or whether they have any legitimate interest in having the information. At a minimum, this argument needs supplementation with an account of why consumers might have reasonable preferences which they would exercise if the information were available.

Biotechnology is irreligious, impure or harms animals, the environment and small farmers. The previous section provided an account of the religious, the queasy, the skeptical and the politically active which demonstrates the legitimacy of their concerns. Perhaps if that account were added to the survey data documenting a desire for labels, it would demonstrate a legitimate reason for requiring labels. It is reasonable for people to believe any or all of these things about biotechnology (though I do not) and people do have a right that protects the exercise of these beliefs from interference by others. But arguments intended to convince others that biotechnology is irreligious, impure or bad for animals, the environment or small farmers do not in themselves provide a sufficient ground for mandatory labels. At most, they entail that the food system must not be manipulated through conspiracy or public policy in a manner that effectively forces them to buy and eat genetically engineered foods. Mandatory genetic engineering labels would be necessary only if they were the sole means to protect the food system from such manipulation, but they are not.

Alternatives to mandatory labels. The ability to avoid genetically engineered foods is what matters most. Those who wish to avoid processed foods do so by choosing and preparing whole foods. Those who wish to avoid fluoridated water do so by buying bottle water that is labeled as free of fluoride. Those who wish to avoid pesticide residue do so by purchasing food that is labeled as "organic," "green" or otherwise free of pesticides. In each of these cases, the principle of informed consent is protected, but in none of them are the offensive products the object of mandatory labeling laws. Indeed they are not labeled at all.

The principle of consent is protected in each of these cases by the availability of alternatives. These alternative foods give food consumers the right of exit from a system of food transactions that they find objectionable (Hirschmann, 1970). If there are identifiable alternatives to the products of biotechnology, then consumer sovereignty and the principles of consent are protected. There are several ways in which the principle of exit can be protected and the most obvious of them all involve labels that identify a product as "biotech free."

Voluntary Negative Labels

The most straightforward approach is a voluntary label that may be placed on products where no biotechnology has been used. The label would be negative in that it would proclaim the absence of biotechnology, rather than its presence. If a sufficient number of products begin to use negative or "no biotech" labels (for the time being, most products outside the dairy case could wear them), then those who wish to avoid biotechnology can do so. Biotechnology differs from food processing, fluoridation or chemical additives in that it is virtually impossible to trace on a *post hoc* basis. Negative labels would only be effective if government or trade associations undertook the costs necessary to monitor every stage of production where growers or processors might be tempted to cheat.

Organic or "Green" Labels

Another approach is to specify that "green" or "organic" products may not utilize biotechnology, so that the concerned may opt out of the soon-to-be-biotechnologically dominated mainline food system by shifting over to a segment that already exists. This is the solution that appears most likely at present. It would not involve the start-up costs of establishing a new market niche and it appears that many of those who initially wish to avoid biotechnology are quite willing to segment themselves into the green section of the supermarket. Yet this solution is far from ideal. Some of the most attractive products from agricultural biotechnology are those that promise virus or disease resistance, or that permit the elimination of chemical pesticides. The green and organic buyers should be the most enthusiastic buyers of these environmentally friendly products and the move to a "green equals biotech free" policy both stigmatizes these products and undercuts their potential market.

Mandatory Labels: For and Against

Alternatives to mandatory labels exist and they are capable of resolving the most potent ethical problems. Yet an advocate of mandatory labels might protest that if negative labels provide exit, so do mandatory "positive" labels (e.g., labels that identify a food as being produced through biotechnology). They do so directly and there is less chance that people will be confused or misled into thinking that they are avoiding genetically engineered foods when they are not. What is more, the cost falls on the biotechnology industry, rather than those who want to avoid it.

The most serious ethical objection to mandatory labels is that they would stigmatize products of biotechnology unjustly. Although there are reasonable concerns that may lead some to avoid genetically engineered foods, it is at least as reasonable to accept them as beneficial additions to the human diet. Given this fact, a policy that would groundlessly sway people who are neither religious, queasy, untrusting nor politically active in the manner described is questionable. It would also groundlessly reduce the commercial viability of genetically engineered foods and this could plausibly be interpreted as interference in the rights of the food industry, its investors and non-profit biotechnology researchers. If this were purely a philosophical problem, however, it must be admitted that the case for mandatory labels is very nearly as strong as the case against them. Other circumstances, such as strong demonstrated political demand for mandatory labels, might plausibly be advanced to tip the balance in favor of the mandatory alternative. This might especially be the case for societies where a majority or large plurality of religious believers decide against biotechnology. However, technical obstacles make mandatory labels problematic, no matter how attractive they are from an ethical standpoint.

Mandatory labels for genetically engineered foods would be very difficult to enforce. Currently, laws that regulate the use of drugs or chemical food additives are enforced by testing product samples for the presence of the regulated substance. No test currently exists that would reliably detect whether genetic engineering (much less other forms of biotechnology such as cloning or embryo transfer) has been used on processed foods. In general, the technical literature on the question is sparse and the informal testimony that can be reported from personal communications is inconsistent. Some scientists believe that the polymerase chain reaction (PCR) would make identification of whole foods modified using transgenes "easy," but others are not so sure. How such testing might be done has never been described in detail, much less applied to foods taken from grocery shelves. Biotechnology companies go out of their way to introduce marker genes so that they can determine whether someone is using their technology illegally. While such markers are indeed easy to find, even without PCR, some believe that a company wishing to avoid detection would have little trouble doing so.

The alternative to testing is inspection of the production process. Farmers and food producers are currently monitored for sanitary and environmental compliance, but inspections at the farm level are infrequent and violations are common. The enforcement of sanitary and environmental regulations aims to keep infractions within limits of tolerance and it is possible that similar results could be achieved for genetically engineered foods. If the boosters of food biotechnology are right, however, within a few years biotechnology will be pervasive in the food system. A monitoring approach would require that bulk commodities such as rice, corn and soybeans (all currently on world markets in genetically engineered varieties) would have to be segmented into biotech and non-biotech shipping channels. As it stands now, one farmer's biotech variety will be mixed with another traditional variety during the handling process. Segmentation might work if there is a price premium to be gained from it (the negative labels approach), but if not it will be far easier to just label everything as biotech, just to be on the safe side.

The result is a system that defeats the purpose of labeling, namely to safeguard individual consent.

CONCLUSION

Regulators and scientists generally approach food safety as a problem of optimizing the trade-offs between the nutritional, aesthetic and economic benefits that are derived from food and the probability that consuming any given food or diet of foods will result in illness, injury or other outcome detrimental to health. Given the seriousness of food-borne health hazards and the power of food science to anticipate and manage these hazards, this is an eminently defensible approach from both an ethical and a scientific standpoint. However, in one of the supreme ironies of science and public policy, the more aggressively regulators and scientists promote the wisdom of the risk-based approach, the less effective it becomes (Thompson, 1995).

One reason it is ineffective is that regulators and scientists become entrapped in an indefensible political position when they follow the logic of risk-based or science-based food safety policy too literally. Some industry scientists and sympathetic regulators promote the view that scientifically assessed probabilities of injury are the sole criterion on which food choice should be made (or what is the same thing, that such risk information is the only information that consumers have the right to demand). A moment's reflection reveals the absurdity of this view, but what is worse than its absurdity is the way that it inculcates suspicions in the public mind: what are they trying to hide?

When viewed from the risk-based perspective on food safety, biotechnology scores well. Scientists and regulators must not abandon the view that their primary responsibility is to ensure that biotechnology does not endanger public health, but there are ways to do this without coercive manipulation of the food system. Consumer concern about food biotechnology is not irrational. As long as it is possible to accommodate the desire for alternative choices without unduly stigmatizing the products of biotechnology, we should do so. The ethical basis for this prescription resides in the importance of minority rights, consumer sovereignty and the principles of informed consent.

NOTE

Reprinted by permission of the author from *Ethical Issues in Food Biotechnology* (London: Chapman & Hall, 1996), 65–80.

BIBLIOGRAPHY

Ames, Bruce. 1983. Dietary Carcinogens and Anti-Carcinogens. *Science,* 156–63.
Bains, W. 1993. *Biotechnology from A to Z.* New York: Oxford University Press.

Beck, Ulrich. 1992. *Risk Society.* Translated by M. Ritter. London: Sage.

Douglas, Mary. 1975. *Implicit Meanings.* London: Routledge.

Douglas, Mary, and Aaron Wildowsky. 1982. *Risk and Culture.* Berkeley: University of California Press.

Finkel, A. M. 1996. Comparing Risk Thoughtfully. *Risk, Health, and the Environment* 7: 323–59.

Group of Advisors on the Ethical Implications of Biotechnology, European Commission. 1996. Ethical Aspects of the Labeling of Food Derived from Modern Biotechnology. *Politics and the Life Sciences* 14: 117–19.

Harris, M. 1974. *Cows, Pigs, Wars, and Witches.* New York: Random House.

Knorr, D., and K. Clancy. 1984. Safety Aspects of Processed Foods. In *Food Security in the United States.* Edited by L. Busch and W. Lacy. Boulder: Westview.

Mather, R. 1995. *A Garden of Unearthly Delights: Bioengineering and the Future of Food.* New York: Dutton.

Sagoff, Mark. 1985. *Risk Benefit Analysis in Decisions Concerning Public Health.* Dubuque: Kendall Hunt.

Sinclair, Upton. 1906. *The Jungle.* New York: Doubleday.

Straughan, Roger. 1995. Ethical Aspects of Crop Biotechnology. In *Issues in Agricultural Bioethics,* edited by T. B. Mepham, 163–76. Nottingham: Nottingham University Press.

IV

ANIMAL BIOTECHNOLOGY

As already mentioned, foreign DNA is put into animal cells to generate a transgenic animal—an animal that is genetically engineered to contain foreign genes that are anticipated to be beneficial if the DNA becomes incorporated into the genome or numbers of generations. However, the DNA must be incorporated into the earliest progenitor cells such as the sperm, egg, or single-cell fertilized embryo for this to happen. If the DNA were incorporated into a lung cell, obviously the foreign DNA would last as long as the progenitor cells survive. Eventually the original lung cell, or even the progenitor cells from this lung cell, would die out and the presence of the foreign DNA would not be permanent.

MICROINJECTION FOR GENERATING TRANSGENIC ANIMALS

How are transgenic animals generated so that the foreign DNA is permanent? Three basic methods are currently in use. The most conventional method for putting DNA into early reproductive cells is to inject the DNA using microsurgical instruments and needles into the nucleus of a newly fertilized single-cell embryo. The sperm and the egg have already fused, but the resulting embryo has not yet divided. One nuclei from the sperm and one nuclei from the egg can be seen through a microscope and then either of the nuclei is injected with a solution of DNA. Cellular repair enzymes recognize the foreign DNA in the nucleus and insert it into the genome of the nucleus. When integration occurs, it becomes a permanent part of the cellular DNA. When a cell divides, all of the cellular DNA is replicated. Because this is the first cell of the organism and every cell of the eventual animal will originate from this single cell, the foreign DNA will be present in every cell of the animal. Compared to other methods of generating transgenic animals, this method is relatively easy to do. The disadvantage is that the location where the foreign DNA integrates is random and the production of the protein is subjected to the microenvironment of the chromosome.

EMBRYONIC STEM CELLS TO GENERATE TRANSGENICS

Another method of generating transgenic animals uses embryonic stem cells—early cells in the developing embryo that have the potential to develop into any cell of the body. When a single-cell embryo divides, the progeny cells for several generations remain undecided as to what type of cell they will become. Eventually the embryo cells become committed to certain pathways that lead to blood cells, skin cells, liver cells, or any one of the many types of cells of the body. Cells that are uncommitted to differentiate have the potential to develop into any cell; hence they are called embryonic stem cells. Once a cell starts down some path for differentiation, it is no longer an embryonic stem cell. However, it has not terminally differentiated. Picture this as a tree with a trunk and a primary branch. The primary branches divide into secondary branches. Eventually they branch out to the very tips. The tips can be compared to the terminally differentiated cell. The trunk can be compared to the embryonic stem cells, which have the potential to develop out to any tip of the tree. The primary and secondary branches are committed to some extent, but they still have the potential to divide into different tips of the tree. These somewhat differentiated cells are called pluripotent, meaning they have many potential pathways for differentiation, but they may not have totipotency, meaning total freedom to differentiate into any cell of the whole organism. Embryonic stem cells are totipotent or maybe pluripotent, but other stem cells are pluripotent.

APPLICATION OF EMBRYONIC STEM CELLS

In certain strains of mice, totipotent embryonic stem cells can be cultivated in culture to divide and grow in numbers without differentiating; they remain totipotent during cultivation. DNA can be put into large populations of these totipotent cells, cultivated, and then screened to identify cells that have taken up the DNA. DNA can be put into many of these cells at one time by various methods, such as by electroporation, not a single cell at a time, as with the microinjection method. In electroporation, many cells are mixed with many molecules of foreign DNA and a pulsating electrical field causes the DNA to enter into many cells. Each individual cell does not need to be manipulated as with the microinjection procedure. Because the cells can be grown in culture outside the body, scientists can analyze a large population of cells for the transgene or foreign DNA while in cell culture rather than wait for animals to be born. With the microinjection method, the injected embryos are implanted in the foster mother, and then the newborns are screened for the presence of the foreign DNA. If a 10 percent success rate is realized with the microinjection procedure, ten animals must be born on average before one animal is identified with the transgene. If the totipotent embryonic stem cells are used, cells in culture are screened for the transgene so that 100 percent of the animals born from these cells are transgenic.

GENETICALLY ENGINEERING SPECIFIC GENES USING EMBRYONIC STEM CELLS (HOMOLOGOUS RECOMBINATION)

Perhaps the greatest benefit of using embryonic stem cells to make transgenic animals is that the integration of the foreign DNA can be targeted to specific locations in the genome of the animal. This works through the principle of homologous recombination. When engineering a foreign DNA, the scientist makes the flanking or terminal ends of the foreign DNA identical or homologous to specific sequences in the genome of the host cell. This allows the scientist the ability to target a specific site in the genome for insertion. Any DNA of interest can be put between these homologous sequences of the foreign/engineered DNA. The host cell recognizes the homologous ends of the foreign DNA and "repairs" the foreign DNA by inserting it into the site of homology within the cell's genome. This can be viewed as "like attracting like" because the flanks of the foreign DNA can combine with identical sequences of the host genome. This is called homologous recombination; however, the event is very rare. When the homologous foreign DNA is put into the totipotent stem cells and they are grown in culture, a few cells out of the millions of cells will have DNA that has undergone homologous recombination. By certain techniques these millions of cells can be screened while in culture to identify cells having undergone homologous recombination.

The reason for performing the homologous recombination procedure is to get a consistent expression of the transgene as well as to modify genes within the animal. If the integration site is located within a DNA sequence coding for a certain protein, the production of that protein can be prevented or destroyed. By using homologous recombination, foreign DNA is inserted in a targeted gene, which can be used to modify or remove the function of the gene. This is a powerful biological research tool to determine the function of genes within the organism. These are called "knockout" experiments because the functions of genes are knocked out specifically to see what functions of the organisms are removed. Another reason to perform a homologous recombination experiment is to slightly modify a gene within the animal but not to destroy it, such as in repairing a mutation. To do this, the transgene would be identical to the native gene except for a small change in some critical nucleotides. Many genetic diseases are due to small nucleotide changes, maybe even a single nucleotide, which if corrected would eliminate the abnormality.

GENERATE WHOLE ANIMAL FROM EMBRYONIC STEM CELL

How are these totipotent embryonic stem cells used to generate the whole transgenic animal? We are describing certain details of this procedure to help you understand some of the ethical issues associated with human stem cell research, as well as animal cloning. Unfertilized eggs that are at the stage of being ready for fertilization are obtained from a female animal. (Figure 17 shows the steps.) Using microsurgical equipment (as used in the microinjection procedure), scientists remove the chromosomes of the eggs by sucking out the DNA with a needle while viewing through a microscope. This process is called "enucleation," or removing nuclear material containing the DNA. The eggs no longer have chromosomal DNA. Next, a totipotent stem cell is transferred to the egg. The egg cell is much larger than the embryonic stem cell, so this is relatively easy to do using the microsurgical equipment. As can be seen in figure 17, the next step is to get the chromosomes of the embryonic stem cell into the egg. Electrical pulses are used to fuse the outer shell or membrane of the embryonic stem cell with the membrane of the egg. In this fusion process the embryonic stem cell chromosomal DNA is released into the egg. The "activation process" is next. Some chemicals plus the electrical pulses are used to make the egg behave as though it is now fertilized. Without this activation step, the egg behaves as though it is only an unfertilized egg. With the fertilization step, the egg will begin to grow and divide as an embryo. The final step is to put the embryo in a pseudopregnant female animal. This is an animal that has been treated with the appropriate hormones and bred with a vasectomized male animal so that she will physiologically behave as though she is pregnant, even though she has not been impregnated with sperm as a prelude to fertilization. The pseudopregnant female can then be implanted with the genetically engineered embryos. At this stage in the procedure, if everything was done properly and the physiology of the cells is working as anticipated, the embryo will develop into a fetus and will be delivered as a newborn animal. Each animal born from this procedure will be transgenic for the foreign DNA. If the homologous recombination procedure was performed with the embryonic cells in culture, all resulting animals will have the DNA integrated in an exact location of the genome of the animal.

Dolly the sheep was generated using nonembryonic stem cells by the method just described. As in the embryonic stem cell procedure used to generate animals, the nonembryonic stem cells were multiplied in cell culture before being used to

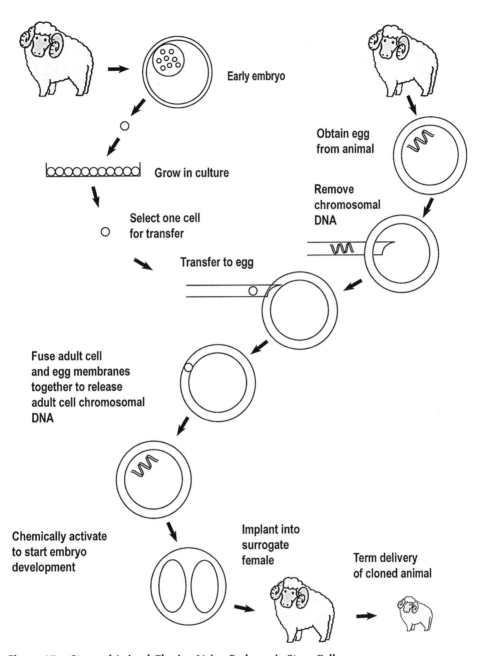

Figure 17. **Steps of Animal Cloning Using Embryonic Stem Cells**

generate Dolly. Because homologous recombination procedures require the ability to multiply large numbers of the cells outside the body, the use of culture-multiplied cells to generate Dolly will probably result in homologous recombination procedures in other mammals besides mice. The ability to modify or knock out the functions of specific genes in agriculturally important animals will be of great practical value for agriculture. If the ethical concerns of human embryo research were overcome, homologous recombination procedures could probably be used to accurately correct genetic defects in human embryos for development into normal children. In summary, homologous recombination procedures have been useful in studying genetics in mice and could provide accurate genetic manipulation valuable to agriculture and medicine. Theoretically, targeted integration by homologous recombination is a much better option than random integration obtained with direct microinjection of DNA.

DESCRIPTION OF ANIMAL ORGAN TRANSPLANTS

Transgenic animals may be used as sources of organs for transplant into human beings, a process known as xenotransplantation or xenografting. The root of the word means different or strange, so xenotransplantation is transplantation of tissues from one species to another. Xenotransplantation has been tried on and off for over thirty years with very limited success. The most compelling difficulty has always been the immediate rejection of the donor organ by the recipient animal. With the need for donor organs for human beings growing yearly, the interest in xenotransplantation has increased in recent years. If immediate rejection could be avoided, xenotransplantation might see substantial growth. Recently leading researcher Robin Weiss identified five possible strategies for solving the immediate rejection problem. One has proven to be ineffective in animal experiments, and another cannot be done at present. Of the remaining three, two approaches involve transgenic animals.

Many experts regard the use of transgenic animals as the most promising avenue for overcoming immediate rejection of xenograft organs. In recent years a number of researchers have suggested that genetically altering certain animals, especially swine, with morphologically compatible organs may overcome crucial immunological hurdles. If so, xenotransplantation may succeed either as a bridge for patients waiting for a human transplant or possibly as a longer-term replacement itself. As the Food and Drug Administration readies approval of the first experiments using genetically altered pig organs, crucial scientific and moral issues are also being addressed.

SCIENCE PROSPECTS AND RISKS

Here, we shall only note the most salient features of this topic. Initially, primates might be considered the donor animal of choice. But for reasons of ethical distaste and scientific merit, pigs and not primates are at the center of current research. Many people see primates as closer to human beings and would recoil at their being used in large numbers as donor animals. In addition, solid scientific reasons exist for excluding primates. First, chimpanzee organs are too small for use in adult human beings, and larger primates do not exist in sufficient numbers to solve the shortage problem. Second, primates are known to carry viral infections that are or might be fatal to humans with little prospect for cure. Primate-to-human transmission is thought to have been the source of HIV, and data exist showing that simian immunodeficiency virus has been transmitted to laboratory personnel who work with primates. Several other infections may come from primate organs and other as yet unknown organisms may pose risks. Finally, at present transgenic manipulation is problematic in primates, so the sorts of alterations that are or may be helpful in avoiding organ rejection are not being pursued in primates.

Pigs are thought to avoid some of these problems. The organs are of the right size and physiologically compatible. They can be transgenically altered and they are thought to harbor few infectious agents that are a danger to human beings. Hence, they have become the focus of transgenic work leading to xenotransplantation.

Organ transplants, especially from widely divergent species, are subject to three different sorts of rejections that occur in three stages.

Hyperacute Rejection

Hyperacute rejection (HAR) is the first stage of rejection and exhibits rapid, violent rejection within minutes or hours. HAR is mediated by two sets of proteins—complement proteins. Unlike many mechanisms of the immune system, complement proteins are always present; they do not need to be synthesized in response to a foreign invader, such as a bacteria or virus. As a result, complement proteins are often the first line of defense, particularly with transplantation of a foreign organ. If the transplanted tissues have certain proteins, then the complement is not activated to destroy the foreign tissue. Human tissues have these certain proteins, so transplant from one human to the next does not involve rejection by complement, but transplant from one species to another does involve a rapid, violent attachment by complement. The strategy to avoid this rapid attachment by complement is to genetically engineer donor animals to have these certain human proteins in their organs. When the animal organ is transplanted, the transgenic proteins are recognized by the body and complement does not act to destroy the organ. It is important to realize that the animal is not completely altered genetically. The genetic addition of a couple of types of proteins does not physiologically or anatomically alter the animal; it is basically normal.

Acute Vascular Rejection

The next level of tissue rejection is acute vascular rejection. This second stage of rejection takes place usually within days or a few weeks of the transplant. This rejection involves changes in the cells lining the blood vessels. Their normal anti-coagulant state of preventing blood clots is changed to a procoagulant state that favors blood clots. This results in complex pathways of tissue damage and blood clotting in which the viability of the transplanted organ or tissues declines rapidly. Overcoming this complex acute vascular rejection process will require much more development and probably multiple strategies.

Risk of Infection

Zoonosis is infection passed from animals to humans. When infection from animal to human occurs through transplanted tissue, it is referred to as xenosis. Examples of zoonosis are infections from mad cow disease, influenza from chickens or pigs, hantavirus from rodents, or HIV originating from monkeys. Mad cow disease originated from a sheep disease called scrapie. The sheep disease was passed to cows when the cows were fed sheep by-products as a source of protein. The cows contracted the neurologic brain disease from the sheep. Some people were then exposed to the cow tissues, probably the neurological tissue, in some unknown way and contracted the disease, which is called Creutzfeldt-Jacob syndrome in its human form. The great influenza epidemic of 1918 started when a pig influenza virus was passed to people. The virus was much more dangerous to people than pigs, and an estimated 20–25 million people died. Some believe that the high mortality among soldiers brought World War I to an end. Evidence indicates that HIV infection was passed from primates to people. These examples illustrate the rare but profound effect zoonotic infections have on the human race.

Because primates are likely to have viruses that may cross-infect human beings, primates are not at this time being considered as a source of organs for transplantation. As mentioned above, pig organs are being developed for organ donation. But does the use of pigs present risk of xenosis? The answer is yes, but the extent of the risk is not known. Pigs, as well as most animals, have DNA sequences that are characteristic of viruses related to HIV, which are called endogenous retroviruses. It has long been known that these internal DNA sequences of endogenous retroviruses can recombine or associate with viruses injected into mice to result in a new virus containing portions of the endogenous virus and the injected virus. New hybrid viruses might emerge through this close association or recombination in the animal. Since pigs are known to have these endogenous viral-like sequences, the risk is that new viruses with different characteristics might emerge.

Transplantation of animal organs into human beings might also increase the risk of xenosis as compared with human exposure to the virus but not to infected organs. Grafted tissues could serve as a "culture factory" inside the body from which the virus might pass to the host or recipient by cell-to-cell contact. This is one of the most efficient ways for some viruses to infect a cell, since the membranes of the cells are touching and the virus does not have to reinfect each new cell. Also, transplant patients are typically treated with drugs that suppress the immune system, which could allow new viruses to replicate without an immuno-logical challenge.

THE MORAL STANDING OF ANIMALS

Over the last thirty years a robust debate has developed in several fields about the moral standing of animals. This debate is the background to any serious discussion of the moral issues involved in animal transgenics. Three basic approaches to the moral standing of animals are developed in this literature. The first approach is demonstrated by D'Silva in chapter 16. What is wrong in our treatment of sentient animals is the pain and suffering to which they are subjected. The higher animals, the ones that are eaten, trapped, and farmed, for example, are sentient creatures, as are human beings. It is nearly universally acknowledged as wrong to inflict pain and suffering on a handicapped child in a medical experiment that cannot benefit the child. How then can it be acceptable to subject a dog to the same experiment?

The argument is that what makes it wrong to use the child is the pointless pain and suffering inflicted on the child. The child cannot give consent, and so individual autonomy is not at issue. If a child is irreversibly handicapped, we cannot be thwarting the growth of the child. The only reasonable ground of our rejection is the pain and suffering of the child, and that ground seems to encompass all sentient creatures. D'Silva presents this point of view with specific reference to issues of animal transgenics.

D'Silva clearly rejects animal transgenics on utility grounds. By contrast, R. G. Frey offers a more nuanced discussion of many of the same issues that emerge from a utilitarian view of ethics. Frey presents two widely admitted theses. The first is D'Silva's utility thesis: it is wrong to inflict pain and suffering for no purpose on a sentient creature. Second, it is widely agreed that where utility grounds exist for conducting an experiment (to develop an AIDS vaccine, for example), it is preferable to do the testing on animals first.

Frey explores the tension between these two beliefs. What would give animal life less value than human life per se? Frey believes that the second principle, use animals first, is accepted even in the case of handicapped children or newborns. Thus it cannot be the case that conscious experience, and hence suffering, is always thought to be greater in the case of human beings. Nor can autonomy or the quality of life be the ground of a nearly absolute principle that we should use animals first. The rule is clearly biased in favor of using animals before humans. In this way it is rooted in a preference for our own species. To use the language of our current discussion, it is "speciesist," a term that refers to moral bias against animals in the way that racism denotes bias against racial minorities. Frey ultimately concludes that we cannot give a good ground for this bias apart from a religious view of human beings as special creations from God. This would provide an adequate grounding for the second principle. The problem is that while the second principle is very widely shared, this particular grounding is not. Frey does not solve this problem but alerts us to it in a careful manner.

Suppose, in regard to what is wrong in our use of humans in experiments, we considered not only the pain and suffering to which they are subjected but the manner in which such use stunts or deforms their growth and development. They cannot live up to their potential because of the way in which we manipulate them. What is wrong with Frankenstein is the way in which he is deformed, not just his pain and suffering.

This is the point developed by Bernard Rollin, who has greatly contributed to the current discussion of the moral status of animals and our uses of them. Rollin is sensitive to the harms that genetic technologies may cause for animals, especially the harm of pain as in the case of BGH. In addition, Rollin believes that animals have natures that we can allow to develop or can deform. What is wrong with factory farming in his view is the manner in which we force animals to live in ways that are unsuited to their nature, like forcing a child to live in a closet. What is wrong is the stunted growth of the child. By analogy what is wrong with the typical dairy is forcing the dairy cows to live in small stalls.

What if we could fit the "nature" of the animals to the environment in which we have them grow? This is Rollin's intriguing suggestion. Could we use transgenics to change animal natures so that the way we use them would fit their nature? Transgenics could be used in a way that harms animals or human beings. In such a case it would be wrong. But it does not need to be. Animal transgenics could be used to reduce harm to animals. In such instances it would not be wrong.

One way to resolve questions about the moral status of animals is to deny that they have any moral standing whatsoever. This third view of the problem of the moral standing of animals attempts to find some capacity that human beings have and animals do not, a capacity that is relevant to moral standing. Among philosophers the most important proponent of this view for many years has been Carl Cohen. The article reprinted in this volume is a classic statement of his position. Cohen believes that the key moral question about animals is whether they have rights. In order to answer this question, we first need to know what it is about human beings that justifies our belief that human beings have this special moral standing. Cohen gives an argument developed from the philosopher Immanuel Kant, one of the most important philosophers who ever lived. Humans confront moral choices, make moral decisions, and develop moral rules that regulate their behavior in a structure of rights and responsibilities. Since animals, in Cohen's view, do not develop or follow moral codes, they cannot be said to have moral rights. In order to have rights you must have responsibilities. Since animals do not have the latter they cannot have the former.

Many thinkers, such as Franz deWaal, reject the idea that animals do not follow some moral codes, believing instead that some higher animals do have moral natures and follow some patterns of general moral behavior, though perhaps not as robust as those of human adults. Whether or not this idea is accepted, the most important question for Cohen is, What about the many human beings who lack a capacity for moral choice (e.g., the senile)? Cohen responds that this supposed problem rests on a mistake. It treats an "essential feature of humanity" as if it could also be used to sort or divide human beings. But might not this answer be too short? If responding to moral rules is an "essential" quality of human beings, does this not imply that to be a human being one must have this capacity or at least be able to have it in the future? But this would obviously not include the senile. Cohen may not here give a full and complete response to the challenge of handicapped human beings.

Animal Rights

The question of the moral standing of animals may be answered in a way that builds to a very different conclusion from the initial positions of Rollin and Cohen. Immanuel Kant is justly famous in the field of ethics for the "categorical imperative," which he believed was the foundational principle of ethics. Kant argued that there were two equivalent versions of this principle. The first version is employed by Cohen: always act so that the maxim or principle of your action could be willed by yourself or others as a general law. This is the root idea behind Cohen's idea that to act morally you must follow rules you believe to be universal. Cohen argued that since animals do not follow such rules, they cannot have the standing conferred by a moral order.

Kant, however, offered a second version of his principle: act always to treat humanity, whether in yourself or others, as an end in itself, never as a means only. Roughly this means that human beings are "above all price." People use things to pursue their projects. It should never be the other way around. Second, human beings are rational agents capable of making their own decisions. Therefore, they should be thought of as having a unique dignity that deserves our respect even when we would not do what they are doing. We must never manipulate or use people for ends that they have not chosen.

Holland's chapter is a fine review of some key positions in the discussion of the moral standing of animals. He concludes with this principle of respect or dignity as a way of critiquing the proposal made by Rollin. He argues that even such a seemingly benign use of transgenics makes the animal serve as a means to a general end of the reduction of suffering. In this way he follows Kant's position on suicide. Kant thought that suicide was wrong because in part it violates the "end in itself" principle: the individual sacrifices himself to a general prospect of the reduction of suffering. Thus he uses a person as a means to en end. Holland argues that this is just what is being proposed by Rollin—using an animal so that a reduction in suffering becomes the goal—and the animal becomes a means to serve that goal.

Notice that Holland's position seems to be based partly on the view that Cohen denies: higher animals are rational creatures. If they are, they are to be treated with respect for the lives they lead now, not as clay to be shaped to a life we think is better.

Another way to ground respect for animals is theological. They should be treated with generosity and respect because they are divinely created and endowed with worth. According to the biblical tradition all creatures are divine creations for which human beings have a stewardship responsibility. Furthermore, not a sparrow falls that God is not concerned about. His care even extended to the "lilies of the field," which are creations more perfect than Solomon's temple.

This is the general position from which Linzey defends a complete rejection of animal transgenics. Animal biotechnology is equated with animal slavery as a complete denial of God's goodness as creator of human beings and animals. No human being can own an animal species because God is the true owner. Hence any patenting of animals is "idolatrous." As Linzey argues, "We have no right to appropriate God's own." Linzey is the leading Christian thinker working on animal rights, and in the selection included here he provides a strong defense of the view that animal biotechnology represents a moral problem of "playing God."

XENOTRANSPLANTATION: ETHICAL ISSUES

The scientific uncertainties and moral quandaries of xenotransplantation are so serious that some leading researchers such as Bach and his coauthors (chapter 22) have called for a moratorium on actual transplant studies while a national commission sorts out the issues and develops a public consensus about them. These issues are well reflected in our readings and are only summarized here.

The first issues concern the moral standing of and significant risks to the donor animals themselves. Since we have already surveyed readings about the moral standing and welfare of transgenic animals per se, nothing new is added. The issue uniquely raised by Bach and colleagues is the possibility that viruses that do not cause disease in the donor animal could be altered in human beings and return to cause serious disease in the donor animal population. Even apart from animal welfare issues, a "transgenic" virus in an animal with an established commercial use could be devastating to the industry. Notice that this concern is not rooted in a notion of animals' rights or moral standing, as in Holland or Rollin. Rather, Bach and his coauthors assume that the animals in question are there for human use. Their concern is for the potential effects of the virus on the animals while being used and on the industry that uses them.

The second major moral issue in xenotransplantation concerns the tension between individual benefits from a transplant and the risk of societal harms that may follow if an animal retrovirus is transmitted into the human population via transplanted organs. This is a variant of the classical problem of benefiting one and harming many, which is especially prominent in the utilitarian forms of ethics. In the transplant case the benefits and harms are partially unknown, yet there is an asymmetry. The individual needing the transplant will presumably die without it and thus may be willing to undertake a highly risky procedure which, if it fails, will only shorten his life by weeks or at most a few months. The other members of society who are placed at risk are not in extremis. For them, the risk may not be worth the theoretical future benefit that may result if they ever need a transplant.

The danger of new retrovirus infections raises a serious set of issues about what ought to be required of transplant recipients in such a situation. Both they and their significant others will need to be fully informed of the known and even theoretically potential risks. They will also need to be extensively followed and monitored, as noted by Bach and colleagues and Vanderpool. But how intrusive should the monitoring be? Because the period of incubation of retroviruses can be prolonged, as in the case of HIV, will patients be required as a part of the consent form to submit to a lengthy period of possibly intrusive monitoring? Given the unknown risks to be evaluated in the first studies, should patients be forbidden to drop out of the studies until the monitoring period is over?

These might be thought of as the minimum "moral dues" to be paid by patients in the first several years of xenotransplant studies. But what if the possible harms materialize and a new retrovirus infection develops in transplant recipients? What restrictions on their freedom are justified to prevent harm to third parties? Can sexual partners or others close to the patient be required to submit to monitoring? Would job restrictions or other forms of quarantine be justified to prevent what will only at first be potential harms, even supposing that a retrovirus develops? Should these sorts of policies be framed in terms of the very small likelihood of a retrovirus problem emerging and being transmitted beyond one or a few patients or as a means of avoiding the worst possible outcome, even if the risks of occurrence are very minimal?

The final set of issues raised by both Bach et al. and Vanderpool in their respective ways is how as a society we should resolve these sorts of social issues. As they point out, it cannot be enough to simply gain the informed consent of patients and their significant others. The risks are also borne by third parties, and whatever policies we adopt will affect future patients as well. Simply showing that as of now the risk-benefit ratio favors transplantation for any specific patient is also not enough. We need to have a thorough discussion of what policies we should adopt and what requirements may be made of patients if the retrovirus risks materialize. The readings in part 4 are a good place to begin thinking about these issues.

16

A Critical View of the Genetic
Engineering of Farm Animals

Joyce D'Silva

Advocates of transgenic farm animals often claim that genetic engineering is really not any different from the selective breeding that has been going on for centuries. What modern genetic engineers do is simply faster and more accurate and therefore of enormous potential benefit to the farmer. Some even claim that by using transgenic techniques in combination with cloning they may soon be able to produce super-herds of identical, high-performance animals—a boon to farmer and supermarket alike. However, as I explain below, there are flaws at every stage of the argument that genetically engineered animals will be a boon to the farmer and the food processing industry. But before undertaking this explanation, I shall examine the adverse animal welfare effects of traditional selective breeding.

TRADITIONAL SELECTIVE BREEDING
AND ANIMAL WELFARE

In traditional breeding methods genes cannot be exchanged between unrelated species. However, if we look at the results of selective breeding, it can be seen that it is far from being a harmless technique. Selective breeding has a long history indeed, but refinements in the last 30 years have probably resulted in greater changes to farm animal physiology than in the previous 200 years. Moreover, the results have been far from innocuous for the animals concerned. Let us examine some of these animals in order that by seeing what selective breeding has achieved we can perhaps deduce what genetic engineering may achieve in the decades to come.

Take the modern broiler chicken which is bred for meat. These birds now grow to slaughter-weight in just 6 weeks, about half the time it took 30 years ago, and that 6 weeks is being reduced by about one day a year. Presumably there must be a stopping point somewhere around the corner—but we do not know where! As for the chickens themselves, the rapid growth rate has led to severe physiological problems.

The Agriculture and Food Research Council (AFRC) estimates that up to 80 percent of broilers suffer from leg problems, ranging from mild deformity to incapacitation. It seems muscle (that is, meat) development has been at the expense of skeletal strength. Some deformed and incapacitated chickens, unable to walk even to the feed and drink provided, sink down onto the floor to endure a slow death unless put out of their misery by an alert and merciful stockperson. Lame chicks spend more time resting on blisters and hock burns caused by the ammonia which builds up in the soiled litter underfoot.

Broilers are also increasingly succumbing to congestive heart failure because their cardiovascular system simply cannot keep up with their rapid growth rate. So, much of the chicken's short life may be spent in discomfort, pain or even agony. So much for the achievements of the selective breeding of broilers!

Let us now look at the modern turkey—a creature domesticated comparatively recently. It is nowadays virtually impossible for male turkeys to mount females as they have been bred to develop huge, meaty and profitable breasts, which make mounting impossible. Instead the breeder males are "milked" every few days and the females are artificially inseminated. It has been said that this is probably the only accepted form of bestiality!

But is it not gross to have developed such creatures at all? Pigs also suffer an increasing degree of leg problems, lameness and heart failure due, at least in part, to their faster than ever growth rates and quicker, heavier muscle development.

The modern high-yielding dairy cow produces ten times as much milk as her calf would have needed, had it been left to suckle from her. As a result, there has been an increase in udder injuries and diseases parallel with the increase in milk production. Currently, over a third of dairy cows in the EU suffer from painful mastitis every year.

So to claim that traditional breeding techniques are accepted and acceptable is just not true. They may have become accepted in farming circles—sadly even in many veterinary circles—but they are not considered acceptable by those whose prime concern is the welfare of the animals.

TECHNICAL MERIT OF
TRANSGENIC FARM ANIMALS

Let us look now at the claim that genetic engineering is far more accurate than traditional breeding. Theoretically, the possibility may exist for accuracy to be achieved; occasionally it is achieved; but for every "bull's eye" there are a myriad of failures. One of the chief areas of research activity has been with growth hormone genes to increase growth rates and to act as muscle promoters, because lean meat is more desirable and profitable than fat. But whilst adding an extra growth hormone gene to an animal may make it grow faster and leaner, it may have other effects as well.

Dr. Rexroad of the United States Department of Agriculture (USDA) has written how his transgenic lambs with added growth hormone developed a lethal form of diabetes. The lambs described by Dr. Ward showed degeneration of the liver and kidneys at necropsy which may have reflected diabetes-associated degenerative changes, and Rexroad et al. conclude, not surprisingly, that early death of the transgenic lambs prevented development of lines of transgenic sheep. Similar results were obtained by Nancarrow and others, including Dr. Ward: of 12 transgenic animals expressing the growth hormone gene, all died before one year of age, with clear evidence of damage to liver, kidney and cardiac functions along with abnormal plasma levels of glucose and insulin indicating a diabetic condition.

Let us continue to let the researchers speak for themselves. Groups led by both Bolt and Pursel, for example, have been experimenting with added growth hormone genes in pigs at the USDA's facility at Beltsville. Accuracy is hardly the correct adjective to describe the results here. After the eggs have been micro-injected with several hundred copies of the human or bovine growth hormone gene, they are transferred into the oviducts of recipient females. On average, fewer than one pig per litter will be transgenic. Of these transgenic animals only 60 per cent will actually express the foreign gene.

The researchers are frank about the "bad news." They report the existence of several health problems in transgenic pigs due to the excessively high levels of growth hormone. For example, pigs expressing high levels of growth hormone tend to be lethargic, they exhibit indications of muscle weakness, and some are susceptible to stress. Others tend to lack coordination in their gait, probably because their feet are rather tender. So far, all of Bolt's group's transgenic gilts that express the growth hormone transgene have been anoestrus and their reproductive tracts are infantile. Boars also tend to lack libido, but can be used for breeding with the help of electro-ejaculation and artificial insemination. A number of transgenic pigs have died from gastric ulcers before they reached one year of age. Others have had lesions on the stomach lining when slaughtered for carcass evaluation. Some of the pigs show evidence of arthritis and the authors remark frankly that to date the technology has not produced a single animal with a beneficial transgene.

Vernon Pursel's group, also at the USDA's facility at Beltsville, records the problems encountered with their transgenic pigs. Having examined, killed and carried out necropsies on both transgenic pigs and control pigs, Pursel reports that the most common clinical signs of disease associated with transgenic expression include lethargy, lameness, uncoordinated gait, exopthalmos (bulging eyeballs) and thickened skin. Some of the transgenic pigs also showed other severe health problems ranging from gastric ulceration, severe joint inflammation and degenerative joint disease to heart problems and pneumonia. Interestingly, Pursel notes that centuries of selection for growth and body consumption may limit the ability of the pig to respond to growth hormone.

In 1991, at the Conference of the British Society for Animal Production, Dr John Clark of the AFRC station at Edinburgh is reported as declaring that in experiments conducted in the UK in 1990, 11,399 pig eggs were injected with a foreign gene, yet just 67 transgenic animals were produced, and in sheep, of 4,500 injected eggs only 34 transgenic animals resulted. Pursel reports that in 3 years of gene transfer studies in pigs, only 8 per cent of the 7,000 injected eggs developed to birth and about 7 per cent of those born were transgenic—an efficiency of 0.6 per cent.

Thus, to date, genetic engineering is a "hit and miss" technology and, sadly, many of the misses have disastrous effects on the animals concerned. Such experiments look likely to continue, and indeed, increase in number in the years to come.

CLONING FARM ANIMALS

Scientists working on transgenic animals are looking for a certain kind of genetic perfection—the right promoter, the right gene for this or that trait, the perfect combination—the grade A result. Once achieved, the next step would be to clone—to produce identical copies of—"genetically perfect" transgenic animals by transferring gene-containing cell nuclei from such an animal into egg cells. An example of the results of such manipulations could be a herd of genetically identical cows. Let us examine the results of research in this area, as reported by the scientists themselves.

Dr. N. L. First of the University of Wisconsin, for example, speaks of an era of exciting possibilities for rapidly propagating and tailoring animals to meet product and environmental demands. Dr. First (and he is not alone in this) also speculates on the use of cloning combined with genetic engineering and embryo transfer to achieve what he describes as well-organised systems for widescale production, sale and transfer of genetically superior animals, all tailor-made to supermarket specifications regarding carcass quality, fat coverage and joint size. Everything tailor-made to consumer preference! Genetic engineering optimists view the production of identical superior animals as the ideal answer to the requirements of the supermarket. Professor Peter Street of Reading University proclaims: "In effect, with this implanted, designed embryo, if we then are able to manipulate the feeding system, we can design the whole carcass, if you like, from embryo to plate to meet a particular market niche." One senior researcher in this area has told me that he believes every farmer's ideal is to have his own brand of animal and a contract with Marks & Spencers!

But are there not inherent dangers in cloning? Already researchers have found that calf fetuses produced by nuclear transfer have abnormal growth rates. The cows giving birth to these extra-large calves are therefore more likely to experience difficulties, require more caesarean sections and as a result more of the cloned calves die. Those scientists who have made further clones from the initial clones, that is multiple-generation cloning, have found lower pregnancy rates and higher abortion rates. So far, only third-generation calves have been achieved. Furthermore, the other inherent danger in cloning is that the production of genetically identical animals means that such animals are not only identically super-fast growing, super-lean, et cetera, but also identically vulnerable to the same pathogens. Thus, one strain of disease to which all cloned animals were highly vulnerable could "wipe out" the entire herd. (In an analogous way, this has already happened with crops.)

Were agriculture to become dependent on cloned animals, then there would, of course, also be a significant loss of genetic diversity, which could spell long-term disaster for the livestock industry.

DISEASE-RESISTANT FARM ANIMALS

I have so far explored genetic engineering for enhanced productivity, but research is also being undertaken to develop farm animals with inherent resistance to diseases which result in, for example, massive vaccination programs for poultry and the huge prophylactic and therapeutic use of antibiotics. To reduce dependence on vaccination and antibiotics would probably be preferable both to the animal and for consumer perceptions of animal products and probably to the welfare of the consumer of animal foods.

An overall criticism of this work is that most of the diseases being worked on are those that are the endemic diseases of the factory farm. There is a real danger inherent in this work: by genetically engineering these animals to be resistant to the endemic diseases of the factory farm, are we not condemning them to a continued existence in those same factory farm conditions—conditions often of filth and overcrowding and the gustation of natural instincts and physiological needs?

Donald Salter, of Michigan State University, has written of his ambition regarding the chicken which is to produce a transgenic "super chicken" using the new genetic engineering techniques. This "super chicken" would have germline inserted genes that would convey resistance to a variety of different pathogens that affect the "livability" and productivity of chickens.

For Duane Kramer and Joe Templeton, of Texas A&M University, the *raison d'être* for their work on disease resistance in mammals is that animal health problems account for around USSH.5 billion in losses per year. They admit, however, that there are no publications documenting the generation of a disease-resistant cow, pig or sheep through recombinant DNA technology. These researchers introduced interferon genes into cattle, but the two successful pregnancies ended in terminations. They warn that some genes will be detrimental to early development and will have to be used with regulators that will delay expression of the gene until the desired time. They point out that we still know very little about why animals are susceptible or resistant to various diseases. In other words, this work is still at the "shot in the dark" stage and it may be some time before disease-resistant farm animals are found on our farms. In the interim, the agripharmaceutical industry can heave a sigh of relief.

MOLECULAR "PHARMING"

Another important development in transgenic animal engineering is "molecular" pharming—the production of useful proteins in the milk, blood or eggs of animals. This has been forecast to be the most economically important aspect of biotechnology implementation for animal agriculture. Dr. Floyd Schanbacher of the Ohio Agricultural Research and Development Center points out that the market saturation for milk and eggs would make alternative markets for these products especially attractive. He speculates that were the technique to prove successful in cows so that a cow could produce 3 grams of a highly valuable recombinant protein a day, her daily production would have a value of over US$200,000. He concludes that to him it is obvious that a strong commercial incentive is driving the development of the technology. He foresees that, if successful, this technology will allow dairy cows to be used not just to produce high-value pharmaceutical proteins, but also bulk recombinant proteins of lower value to be used as nutritional supplements or to produce milk specifically high in certain food processing qualities. Otto Postma, of Gene Pharming Europe, declares the cow to be an extremely suitable production medium for making these proteins, with no safety or technical problems.

What are the drawbacks of these developments? There is still technical uncertainty about the integration of the gene and its level of expression. Trials with dairy cows are inevitably costly and slow as it is 2 years before another generation can be observed. This means it will take time for a viable herd of such transgenic animals to be developed. Dr. Schanbacher forecasts that the development of the technology will require the sacrifice or biopsy of significant numbers of dairy animals. In addition, it is recognised that changing just one component of milk, particularly the proteins, may have dramatic effects on the nature of the milk which could render it unsuitable for human consumption, or indeed consumption by future calves.

Producing human proteins in the blood of animals has been shown to be possible, but has inherent disadvantages. For example, not too much blood must be extracted otherwise the animal will die. Also, the presence of the products of foreign genes in the animals' bloodstream could have severe biological effects on the animals themselves.

Producing human blood-clotting factors in the milk of sheep or cows could guarantee a product free from human immunodeficiency virus (HIV) or hepatitis agents. However, it would also be necessary to exclude other infectious agents which could be present in the milk.

In theory, if foreign genes can successfully be targeted to the mammary gland they should carry less risk to the welfare of the host animal as the mammary glands are relatively isolated from the rest of the body's system. In fact, however, some welfare problems have arisen. For example, researchers at the USDA's facility in Beltsville have found that high expression of the murine whey acidity protein in the mammary glands of transgenic pigs had adverse effects on the physiology of the udder.

This work is still in its early stages both as regards the successful expression of economically viable levels of the recombinant protein in the milk, and as regards the extraction and purification of the proteins from the milk. Some experiments in molecular pharming have shown that the end-product has significant differences in structure to a similar protein produced in cells in the laboratory. Pursel has forecast problems if the human protein has to be separated from its animal equivalent.

Proponents of molecular pharming triumphantly proclaim that it can only be good for the animals concerned. Of course, creatures as valuable as these will be looked after well. But if the techniques are successful and commercial production takes off, I find it hard to envisage that transgenic sheep, goats or cows will be allowed to roam the fields. Surely they are far more likely to be kept in the most sterile conditions possible, surrounded by stainless steel, and well away from possible sources of infection such as bedding material, or indeed other sheep, goats or cows. So we will have a type of hygienically superior "factory farm" developed in the laboratory.

Of course, if developments like the "DNX project" to produce human hemoglobin in pigs' blood get off the ground, then we know where we are as regards the welfare of the transgenic animals. Projections state that if 100,000 pigs were to be killed and desanguinated yearly, they would yield US$300,000 million worth of human haemoglobin.

I think it is also important to note that producing high-value pharmaceutical proteins via the mammary glands of transgenic animals is not the only way to obtain these products. Tissue culture of insect or mammalian cells can be used and is already being used. In the future it may be possible to produce such proteins in plants—surely a far better idea?

Although potentially the most benign of the genetic engineering technologies now being applied to farm animals, molecular pharming is obviously not without its problems. At the moment it appears to be greatly overrated by its proponents who seem to see it as a cure-all solution for many human deficiency conditions. Whether or not this proves to be the case is not known at the present time.

Philosophically speaking, I find it hard to be at ease with the description of such animals as "bioreactors." Truthfully, we are all bioreactors of a sort, but we know we are much more than that; so too, I believe, are farm animals.

EMBRYO TRANSFER IN FARM ANIMALS

I would like to make a few comments about embryo transfer as this technique is integral to most transgenic work because once the gene manipulation has taken place, the resulting embryos have to be implanted into recipient animals.

To produce large numbers of eggs for this transgenic work, female animals may be injected repeatedly with hormones to induce super-ovulation. Of itself this is stressful to the animal. In some instances, of course, eggs are recovered from the ovaries of slaughtered heifers at the abattoir. This at least relieves the pressure on living animals. The new technique of locating ovarian follicles by the use of ultra-sound and removing fertilized eggs from the ovaries of living animals has now been developed. This may sound innocuous, but is not so if it means animals have to be interfered with on a weekly or bi-weekly basis to procure a sufficient number of eggs. The Institute for Animal Production in the Netherlands estimates that each cow should yield 900–1,000 ova (eggs) of which 15–200 will be fertilized. Egg collection can take bizarre turns. Doctor Bob Stubbings of Guelph University has extracted eggs from 7 month-old calf foetuses and fertilized them in his laboratory. He is planning to remove one ovary from each new-born calf to increase his egg collection. In his opinion, "Cows get along fine with only one ovary."

The resulting implantation of the embryo may cause distress and discomfort to the constrained cow, hopefully anaesthetized by an epidural. The partially or fully anaesthetized sheep, goat or pig will endure surgery of varying degrees of severity.

In this case, animals are being viewed routinely, not as bioreactors, but as reproductive machines. It seems strange to me to see veterinary surgeons involved in this sort of work at all. How do routine hormonal injections, physical interferences and surgical operations, all for non-therapeutic purposes, fit in with the veterinary oath, an oath which is taken by all new veterinary surgeons in the UK?

GENETICS UNBOUND

Several people appear to oppose animal genetic engineering because they fear the techniques could also be used on humans for nefarious ends, for example, positive eugenics. This is a possibility, but it is not an idea I wish to pursue. Far more sinister in the short term is the possibility that farm animals may be developed that are so different physiologically or psychologically from animals as we now know them, that they will be unrecognizable. The wingless chicken is not a figment of the imagination. It would after all fit so neatly into an even smaller battery cage! And we all know battery hens cannot fly in their cages. Even worse is the prospect voiced by many, including Professor John Owen of Bangor University College, who speaks of the tremendous scope for breeding docility in species such as pigs and poultry which we want to keep in conditions that go against their natural instincts. Even philosophers like Bernard Rollin of Colorado State University seem to find this acceptable. He asserts that while it is wrong to cage a burrowing animal so that it cannot burrow, there is nothing wrong in principle in changing its nature so that burrowing no longer matters to it. This seems to me to be, at the very least, warped thinking. If related to the possibility of similar work with humans we may see why we feel an intrinsic revulsion to changing a creature's nature.

So what are we facing? An enormous effort to overcome animals' inbuilt genetic constraints on growth and productivity. As if the world's health was not already being seriously compromised by over-indulgence in animal flesh and fats. A huge effort to produce tailor-made, high-yielding and probably identical creatures. A major effort to use animals as bio-reactors to produce pharmaceuticals and other proteins.

Surely what we, as human beings, have not faced up to in all of these endeavours is our relationship with the rest of the animal kingdom. It seems to me to be a totally anthropocentric view rooted in the medievalism of Saint Thomas Aquinas who said: "By divine providence animals are intended for Man's use." When divine providence is not being used to justify the genetic engineering of animals, then some evolutionary ethic of "the strongest calls the tune" or "might is right" seems to obtain. But is such an uncivilised view-point not merely fascism operating under the new name of speciesism? Modern genetic engineering is heralded as the technology of the twenty-first century, a technology with enormous potential, yet so far it is being used to engineer transgenic farm animals in ways which reflect the medieval mind-set rather than that of modern liberalism.

In conclusion, it seems to me that we have forgotten that these animals are each and every one individual sentient creatures capable of experiencing a state of well-being but also capable of suffering. As long as we view them solely as commodities or bioreactors we are indeed being blind to their wholeness and to their individual being.

NOTE

Reprinted by permission from Joyce D'Silva, "A Critical View of the Genetic Engineering of Food Animals," in *Animal Genetic Engineering,* ed. P. Wheale (London: Pluto, 1996), 97–109.

17

The "Frankenstein Thing": The Moral Impact of Genetic Engineering of Agricultural Animals on Society and Future Science

B. E. Rollin

Shortly after I had accepted the invitation to address the Iowa State University Agricultural Bioethics Symposium, I remarked to a friend of mine (a nonscientist) that I was going to address a conference on genetic engineering of animals. "Ah," he said, "the Frankenstein thing!" I didn't pay much mind to his remark until perhaps a week later, when, while perusing the new acquisitions in our library, I encountered an extraordinary newly published, five-hundred-page volume entitled *The Frankenstein Catalog: Being a Comprehensive History of Novels, Translations, Adaptations, Stories, Critical Works, Popular Articles, Series, Fumetti, Verse, Stage Plays, Films, Cartoons, Puppetry, Radio and Television Programs, Comics, Satire and Humor, Spoken and Musical Recordings, Tapes and Sheet Music Featuring Frankenstein's Monster and/or Descended from Mary Shelley's Novel.*[1] The entire book is precisely a descriptive catalogue, a list and very brief descriptions of the works mentioned in the title. Amazing though it is that anyone would publish such a book, its content is even more incredible, for it in fact lists 2,666 such works (including 145 editions of Shelley's novel), the vast majority of which date from the mid-twentieth century. All of this obviously indicates that in the Frankenstein story is an archetypal myth or category which somehow speaks to or for twentieth-century concerns and which could perhaps be used to shed light on the social and moral issues raised by genetic engineering of animals. My intuition was confirmed while visiting Australia and discussing with an Australian agricultural researcher the, to him, surprising public hostility and protest that his research into teratology in animals had provoked. "I can't understand it," he told me. "There was absolutely no pain or suffering endured by any of the animals.

271

All I can think of," he said, "is that it must have been the Frankenstein thing." And in its cover story on the fortieth anniversary of the Hiroshima bombing, *Time* magazine again invoked the Frankenstein theme as a major voice in post–World War II popular culture, indicating that it was society's way of expressing its fear and horror of a science and technology that had unleashed the atomic bomb.[2]

Given this pervasive reaction, it seems valuable to explore the social and moral concerns about research into genetic engineering of agricultural animals using the Frankenstein myth as a framework for our discussion. As I shall try to show, the social concerns and the genuine moral concerns are not always identical and are, in fact, sometimes confounded and not clearly separated in the public mind and, indeed, in the minds of many scientists. Furthermore, some of the deepest and most genuine moral concerns encapsulated in the Frankenstein story are undoubtedly least discussed and explored either by the scientific community or the public.

Before pursuing this inquiry, it is worth pausing to stress that, in general and not just in the case of genetic engineering, both the scientific community and the general public often miss the mark in their attention to the ethical issues growing out of scientific activity. My good friend, the late Dr. Bernard Schoenberg, associate dean of the Columbia University College of Physicians and Surgeons, used to remark that while the public and the medical community alike were spilling a great deal of ink on issues such as disconnecting the respirator from Karen Ann Quinlan, almost no one was discussing the far more fundamental moral issue of fee for service in medicine! By the same token, when the Baby Fae affair occurred, generating much debate, neither scientists nor the public seemed to realize that there was little moral difference between this case and any case of killing an animal for possible human benefit or for research. For that matter, it was hard to see the moral difference between harvesting hearts from baboons and harvesting heart valves from pigs—something which has been standard practice for some time and yet something which no one had raised as an ethical question. The practical rather than moral difference, of course, was in the sensational nature of the story—transplanting hearts from animals plucks at primordial emotions. As I told the press, this issue was not from a conceptual point of view worth discussing in isolation from the general question of whether science has the right, in far less dramatic cases, to expend animal lives, sometimes with far more suffering, anxiety, and fear on the part of the animal than those experienced by the anesthetized baboon.

Again, and in the same vein, I recently gave the keynote speech at an Australian conference on the moral issues in animal experimentation. In the course of my talk, I pointed out that merely citing a list of human benefits engendered by research on animals does not in itself logically serve to morally justify that invasive use of animals, any more than a listing of benefits which emerged from medical research on political prisoners, concentration camp inmates, slaves, criminals, and the like would justify doing such research without obtaining noncoerced, informed consent. Despite this obvious point, many researchers, in their talks, continued to base their defense of animal use solely on benefits to man, as indeed the U.S. medical research community has tended to do.

Unfortunately, the general public is usually too ignorant about science to be able to sort out the genuine moral issues emerging from scientific activity and in practice tends to rely on the media to do the job for it. The media, in turn, is of course less interested in conceptual or factual accuracy than in selling papers, as one reporter candidly told me during the Baby Fae case. So that, as we shall see in the case of genetic engineering, what gets presented to the public as major moral issues are often not moral issues at all. At the same time, scientists are themselves often unable to discriminate the ethical issues implicit in or arising out of their own activity, and essentially wait to have the issues defined for them by the public, or by the same people who define them for the public, so that the issues do not get adequately dealt with from the scientific side either. The failure of scientists to discriminate moral issues in science in turn raises doubt about what I have called "the ideology of science"—in essence, the set of philosophical principles, positions, assumptions, presuppositions, and values that scientists tend to acquire unconsciously along with their scientific knowledge in the course of their training. This pervasive ideology is rooted in the logical positivism and behaviorism of the 1920s, and suggests that science deals only with what is observable and verifiable—with "facts." Since statements about values, including moral values, are not verifiable, they are alleged not to fall within the scientist's purview, at least in his or her capacity as scientist. This is often codified as the slogan that science is "value-free," and is accompanied by the claim that, although values perhaps enter into the use to which science is put by society, values never enter into science itself. Value judgements, including ethical ones, are often viewed by scientists as emotive responses and matters of individual preference or taste, and hence not as rationally adjudicable; after all, *de gustibus non disputandum est*. Thus, philosophically, many scientists see nothing wrong with ignoring moral issues or even with being emotional about moral issues, since their unspoken philosophical training leads them to believe that moral issues are nothing but emotional issues.

In actual fact, as I have taken pains to demonstrate elsewhere, science is not value-free and includes ethical values.[3] Indeed, all science is permeated with valuational presuppositions. Surprisingly, perhaps, the very notion of what will count as a fact, as a legitimate object of investigation, or as data relevant to a given question rests squarely upon valuational presuppositions. Consider, for example, the Scientific Revolution, during which the commonsense, sense-experience-based physics and cosmology of Aristotle were replaced by the rationalistic, mathematical, geometrical physics of Galileo and Newton. The discovery of new data or new facts is not what forced the rejection of Aristotelianism—on the contrary, empirical observations all buttressed Aristotle's idea of a world of qualitative differences! What led to the rejection of Aristotelianism was essentially a change in value—a discrediting of information provided by the senses, as Descartes does so

well in his *Meditations,* and a correlative valuing of the rational and mathemati-
cally expressible over the empirical, of Plato's philosophy over Aristotle's. This
was nicely expressed in Galileo's claim that, in essence, an omniscient deity
would have to be a mathematician and create a mathematical unity underlying
apparent diversity.

Few of you would go along with one of my acquaintances, an accomplished
medical researcher and Rhodes scholar, who heatedly informed me that the question of the use of animals in science is simply a scientific, not a moral, question,
and that, indeed, science has nothing to do with ethics. In an attempt to show him
that he had not thought out the logic of his position, I pointed out that if science
is indeed constrained only by scientific concerns, why don't we use children for
research, since they are better models for humans than are animals? His reply,
amazingly enough, was, "Because they won't let us." And none of you who have
watched the obviously morally based changes in scientific opinion on whether
race differences and intelligence are legitimate objects of study, on whether
homosexuality is a disease or an alternative lifestyle, and on whether alcoholism
or wife-beating is sickness or badness can truly deny that science is rooted in
moral valuational assumptions.[4]

In any event, my main concern thus far has been with showing that our understanding of moral issues does not usually keep pace with the scientific progress
that generates these issues. And if I should stress to you any urgent message at
all, let it be that scientists ignore or shunt off these issues until they assume crisis
proportions at the scientists' own peril. In the final analysis, public money pays
for science and ever increasingly demands accountability. A failure on the part of
any area of science to clearly define the moral issues growing out of its activity
and to deal with them puts that area's very existence in peril, as the case of animal
research around the world dramatically illustrates. Furthermore, in a moral variation on Gresham's Law, bad moral thinking can drive good moral thinking out of
circulation. Thus a failure on the part of scientists to articulate the genuine moral
issues in genetic engineering or any area leaves open the very real possibility of
false and irrelevant, but sensationalistic, issues occupying the public mind and
being used as a basis for social policy. And, as we shall now see as we return to
the "Frankenstein thing" as a basis for discussing genetic engineering of animals,
the same sort of thing can happen here.

A nice illustration of my moral Gresham's Law may be found in the fact that
probably the most socially pervasive component of the Frankenstein metaphor as
it applies to genetic engineering of animals is also the least interesting morally.
This component may be characterized in terms of the classic line from old Frankenstein genre movies that "there are certain things man was not meant to know"
(or to do or to explore). In other words, there is certain scientific knowledge or
activity, or application of scientific knowledge, that in and of itself is taboo, irrespective of its consequences. In the case of genetic engineering of animals, this
would most likely be attributed by those who hold such a view to the creation of

chimeras or crossing of species lines, to major modifications within a species which are phenotypically apparent (such as genetically manipulating for leglessness in farm animals), or even, as press and public reaction to the Fox-Rifkin lawsuit against the U.S. Department of Agriculture (USDA) indicates, to introduction of genetic material derived from humans into animals, or, presumably, the introduction of animal-derived genetic material into humans. (As I suggested earlier, a similar strain of thought arose during the Baby Fae case; numerous people seemed to have perceived unspecified ethical difficulties in a human having an animal part.)

The pattern of thinking represented in this sort of version of the "Frankenstein thing," though widespread, does not represent a genuine moral issue and does not raise moral questions requiring social adjudication. It appears to me to have a variety of nonmoral sources which are typically confused with moral concerns.

One such source is most certainly theological: the Judeo-Christian notion that God created living things "each according to its own kind," with the clear implication, expressed both in nineteenth-century and contemporary opposition to Darwin, that species are fixed, clearly separated from one another, and immutable—and furthermore, ought to be. A nontheological, historically influential, philosophical vector buttressing this view in Western thought is Platonized Aristotelianism, which again postulates fixed natural kinds, again immutable and clearly demarcated from one another. Indeed, Aristotle defends this view on the grounds that its contrary would make knowledge impossible. (An opposite tendency also found in Aristotle, which suggests an infinite continuum and graduation in species, has been all but ignored.) But, of course, such theological and philosophical prejudices are not in themselves legitimate bases for moral questioning of genetic engineering, though they help explain certain people's kneejerk bias against it. And, of course, to a religious person, anything that violates any of his or her religious tenets must be seen as morally problematic.

But reservations against "meddling with species" stem from sources beyond theology and Aristotle. They stem also from a common but scientifically unsophisticated and rather muddled understanding by a virtually scientifically illiterate public of species as being, as it were, the building blocks or atoms of the biological world, out of which the biological world is built and upon which it rests. To tinker with species is, in this view, to tinker with the stability of nature, to (in some unarticulated way) shake the entire Great Chain of Being, as Coleridge's Ancient Mariner did when he killed the albatross. The fact that species are, in current biological theory, dynamic rather than static, stop-action views of a continuing evolutionary process is ignored by such critics. These critics also ignore the fact that the notion of (genetic) species is highly complex and problematic, and that it has been rejected by some biologists such as Kensch in favor of notions like subspecies, races, Rassenkreis, or Formenkreis as not being the fundamental taxonomic unit.[5] (On the other hand, the fact that most biologists do treat species as the fundamental taxonomic unit and as being "more real" than other such units, as Michael Ruse puts it, lends support to such critics.[6]) If subspecies is the fundamental unit, incidentally, then we have been genetically engineering biological reality with no fuss for thousands of years. For that matter, if one takes seriously the currently standard definition of a species as a naturally interbreeding population, then one could argue that certain subspecies we have genetically engineered by breeding, such as the Great Dane and Chihuahua, in fact constitute separate species.

Incidentally, as I have argued elsewhere, much of the debate about the reality or nonreality of species rests upon a deep and ancient philosophical mistake, the attempt to classify all phenomena as being either *nomos* or *physis,* nature or convention.[7] In actual fact, it appears that species represent something of both: what species we find in the world depend on the scientific-theoretical lenses with which we examine the world. Given current theories of evolution and molecular genetics, such procedures as DNA matching and serological evidence from protein matching give us an objective method of species classification. But, at the same time, we must recognize that these objective tests are based on accepted biological theories and that, given an alternative biological theory—one, for instance, oriented far more to whole-organism function or ecological place than to the molecular basis of life—we would probably generate a completely different taxonomy, complete with a totally different set of objective tests.[8]

Another factor which appears to me to foster the belief in inviolability and sacredness of species is the environmental movement. It is a psychologically small, albeit conceptually untenable and logically vast, step to go from concern that species not be allowed to become extinct to the idea that we ought not change them. Or perhaps, a bit more reasonably, the movement of thought is rather from the idea that species ought not be allowed to vanish as a result of what humans do to the idea that they ought also not change at our hands. Built into the environmental movement is, in short, a "nature knows best, hands off nature" mindset, but that is more an attitude than a reasoned position.

In my view, as I have argued elsewhere, species are not the sorts of things which are legitimate objects of moral concern.[9] It makes little sense to me to assert that it would be permissible to shoot ten Siberian tigers as long as there were plenty of Siberian tigers or to suggest, as one of my environmental ethicist colleagues has written, that if a species of endangered moss is in the migratory path of a species of plentiful elk, it is not only permissible but obligatory to save the moss by shooting the elk.[10] In my view, as I shall discuss later in detail, only sentient individuals are legitimate objects of moral concern; species count morally only insofar as they represent a group of individuals, and the last ten Siberian tigers are no different as a moral issue than are any other ten Siberian tigers.[11] There is certainly a great loss in species becoming extinct, but it is fundamentally, perhaps, an aesthetic one, analogous perhaps to our repulsion at trampling a flower. Ethics is relevant only insofar as one is morally obligated not to destroy aesthetic objects, or to deprive future generations of having them in their *Umwelt.*

In any case, I think we can conclude from all of the above that the first aspect of the "Frankenstein thing," namely, that "there are certain things we simply ought not do, and species modification by genetic engineering is one of them," does not represent a defensible moral claim even if it may be so perceived by large numbers of people. To respond to this pervasive idea, however, the research community needs to do a great deal of public education, necessarily preceded by self-education in ethical issues.

Any rational attempt to extract a genuine moral issue from the first aspect of the "Frankenstein thing" we have discussed must be based in a second aspect of the "Frankenstein thing," to which we now turn. Crucial to most versions of the Frankenstein myth is the danger to humans that grows out of unbridled scientific curiosity. Thus, the dictum that "there are certain things that are just wrong to do" becomes replaced in this aspect of the myth by the dictum that "there are certain things that are wrong to do because they must or will inevitably lead to great harm to human beings." The archetypal image of this is Dr. Frankenstein's monster on a rampage—terrorizing, hurting, killing, and harming the innocent. Despite the scientist's noble intentions (Dr. Frankenstein's purpose, in the novel, was to help humanity), his activity was morally wrong not (or not merely) because of hubris but because of his unjustifiable failure to foresee the dangerous consequences of his actions or even to consider the possibility of such consequences and take steps and precautions to limit them. And to this objection, of course, twentieth-century science and technology is quite vulnerable. We have tended to believe that if we can do something, we should, and we forge ahead as quickly as possible, damn the torpedoes.

And we have also tended to believe, as part of the ideology of science discussed earlier, that scientists are not morally responsible for the pernicious uses to which their explorations are put; the responsibility for these consequences allegedly belongs to politicians, governments, military agencies, or corporations. There are, of course, notable exceptions to this claim. as Asilomar nobly illustrates[12] but in the main, scientists are vulnerable to this criticism, as any of us who have served on university biosafety or surveillance committees knows all too well. The recent discovery of killer bees on the loose in California represents another example of unjustifiable negligence on the part of scientists, who of course imported and bred these insects apparently without proper regard for the dangers involved.

What, if any, are the potential dangers inherent in genetic manipulation of animals in agriculture? This is certainly a legitimate issue which should be addressed by all of those working in the area. Even a cursory examination of the area suggests a number of possibilities that should be raised, explored, and assessed in terms of likely risk, and for which mechanisms of minimizing the risk should be devised before embarking upon genetic engineering of animals utilizing new principles of biotechnology.

I would suggest that any country contemplating such work establish formal mechanisms to ensure that the social questions associated with potential risks growing out of genetic engineering of animals be fully evaluated and made known to the public, much in the way recombinant DNA work has been dealt with in the United States. I have recommended to the USDA the establishment of something analogous to the National Institutes of Health's (NIH's) Recombinant DNA Advisory Committee (RAC) to assess potential risks and other ethical and social issues associated with genetic engineering of agricultural animals. This ought to proceed in a number of stages. First, a fairly large committee consisting of scientists, attorneys, public policy people, ethics people, and members of the public should delineate the issues and suggest broad guidelines for assessing and minimizing risk. If possible, levels of risk should be identified and broad characterizations of types of research and applications thereof delineated. Subsequently, local committees analogous to human research committees, animal research committees, and biosafety committees, with significant public membership, should be appointed at institutions engaged in research or application of genetic engineering of animals. As much accurate publicity as possible should accompany all aspects of this process, both to dispel irrational components of the "Frankenstein thing" and to show responsiveness to legitimate concerns. Such committees should also engage an entirely different set of ethical questions which we will outline shortly in discussing the last component of the Frankenstein myth.

The sorts of hazards, risks, and potential dangers associated with genetic engineering of agricultural animals appear to be the following (doubtless, most of you can supplement my list significantly). At this stage, I believe that it is vital to err on the side of caution, to look at and consider every possible danger, however apparently unlikely. It is usually far easier to prevent than to amend, especially in an area like agriculture in which vast amounts of money or food are at stake when a technological tool or procedure becomes integral to an operation and is later found questionable or unsuccessful. The use of antibiotics in feeds provides a clear example, as do overly intensified and overly capitalized systems in pork production and crop decimation growing out of unanticipated disease and genetic uniformity.

The first set of potential dangers emerging from the new forms of genetic engineering of agricultural animals obviously stems from the rapidity with which such activity can introduce wholesale change in organisms. Traditional genetic engineering, of course, was done by selective breeding over long periods of time, during which time one had ample opportunity to observe the untoward effects of one's narrow selection for isolated characteristics. But with the techniques we are discussing here, we are doing our selection "in the fast lane." This leads to two sorts of potential danger.

First of all, there may be untoward consequences affecting the organism which one is rapidly changing. The characteristic consequence is that genetic engineering may have implications that are unsuspected. Thus, for example, when wheat was genetically engineered for resistance to blast, that characteristic was looked at in isolation, and the genetic basis for this resistance was encoded into the organism. The back-up gene for general resistance was, however, ignored. As a result, the new organism was very susceptible to all sorts of viruses which, in one generation, mutated sufficiently to devastate the crop.

Second, the isolated characteristic being engineered into the organism may have unsuspected harmful consequences to humans who consume the resultant animal. Thus one can imagine genetically engineering, for example, faster growth in beef cattle in such a way as to increase certain levels of hormones, which, when increased in concentration turn out to be either carcinogens for human beings when ingested over a thirty-year period or teratogens, in a manner similar to diethylstilbestrol. The deep issue here is that one can of course genetically engineer traits in animals without a full understanding of the mechanisms involved in phenotypic expression of the traits, with resulting disaster. This in turn suggests that it would be prudent to be cautious in one's engineering until one has at least a reasonable sense of the physiological mechanisms affected.

A second set of risks growing out of genetic engineering of the sort we have been discussing replicates and amplifies problems already inherent in selection by breeding—namely the narrowing of a gene pool; the tendency towards genetic uniformity; the emergence of harmful recessives; the loss of hybrid vigor; and, of course, the greater susceptibility of organisms to devastation by pathogens, as has been shown in genetic engineering in crops. (On the other hand, genetic engineering can have the opposite effect in making available to the gene pool greater variety than ever before as in the case of artificial insemination making new genetic material available to beef breeders.)

A third set of risks arises out of the fact that in certain cases when one changes animals, one can thereby change the pathogens to which they are host. This can occur in two conceivable ways. First, if one were genetically engineering for resistance to a given pathogen in an animal, one could thereby unwittingly be selecting for new variants among the natural mutations of that microbe to which the modified animal would not be resistant. These new organisms could then be infectious to these animals, other animals, or humans. Second, even if one were changing the animal in nonimmunological ways, one could be changing the pathogens to which it is host by changing the microenvironment where they live. This in turn could result in these pathogens becoming dangerous to humans or to other animals. Thus, in changing agricultural animals by accelerated genetic means one runs the risk of affecting the pathogenicity of the microorganisms that inhabit the organism in unknown and unpredictable ways. And the more precipitous the change, the more inestimable the effects on the pathogens are likely to be.

A fourth set of risks is environmental and ecological and is associated with the possibility of radically altering an animal and then having it get loose in an environment which was not anticipated. While this certainly seems like a minimal danger when dealing with intensively maintained cattle or chicken, it could surely pose a real problem with extensively managed swine, or with rabbits, or even with extensively managed cattle. Bitter experience teaches us that such dangers cannot be estimated, even with species whose characteristics are well known (witness what happened with rabbits and cats in Australia and with the mongoose in Hawaii), a fortiori, an ignorance of what would happen with newly engineered creatures is even more certain.

We have talked briefly of the potential risks of genetically engineering agricultural animals involving the animals themselves, the general human population, other animals, and the environment. A fifth set of risks is relevant to a special subgroup of the human population—namely, those who will actually be doing the experimentation on and genetic manipulations of the animals. Common sense tells us (and there is ample evidence to support this claim) that people working directly with dangerous materials are at greater risk than is the general population. The last smallpox death in England resulted from a virus contracted in a laboratory, and standard precautions are taken universally to protect people working with dangerous substances. But the need for extra vigilance in dealing with new situations is well illustrated by the deaths caused twenty years ago by Marburg virus, which did not affect laboratory workers who had been handling live monkeys, but which killed those people who had been collecting cells from dead animals for cell culture. In the case of genetic engineering, people handling the vectors used to introduce the genes could conceivably be at risk.

For any significant risks which we have discussed, or for others I may have omitted and which might pose real danger, the imperative for risk management can be generated without recourse to ethical considerations; rational self-interest and prudence would dictate that one not be cavalier about them. Thus even if a person has absolutely no concern for anyone but himself and his loved ones, he would wish to see anything that might do massive harm controlled, since he and his might just as easily as anyone else fall victim to its effects. Therefore, in my view, following up an insight of Kant's, it is difficult to separate moral from prudential reckoning in such areas. Only when we consider the third and final aspect of the "Frankenstein thing" do we in fact encounter something that requires purely moral deliberation and decision, because morality and self-interest are very unlikely to coincide in these cases. In other words, one is unlikely to do the right thing for prudential reasons and, in actual fact, moral behavior in this area is likely to exact costs in self-interest. It is to these questions we now turn.

The final aspect of the Frankenstein myth is more difficult than the others to find in many of the popular renditions of the myth but was in fact a central theme in Mary Shelley's novel. This dimension concerns the plight of the creature engendered by abuse of science. In the novel, the creature is innocent yet isolated: shunned, mocked, abused, and persecuted in a plight not of its own making. Seeking love and companionship, it finds only hatred and rejection. One can find traces of this concern for the monster in the classic Frankenstein movie, and it is in fact a central theme in the recent remake of King Kong. Translated into the arena of genetic engineering of agricultural animals, this aspect of the myth, in essence, raises the question of the moral status of animals, of the rights of these animals—

certainly the most difficult of the moral questions we have examined in our discussion. And it is so difficult for a complex of reasons worth briefly detailing.

In the first place, when it comes to trying to get a purchase on our obligations to other creatures, we get little help from common sense, our intuitions, ordinary practice, the law, or even traditional moral philosophy. Common sense and ordinary practice say little about our obligations to animals other than enjoining us to avoid cruelty, hardly a great help since most animal suffering and death is not the result of cruelty. (The great emphasis on cruelty to animals and love for animals is the major failing in the traditional animal welfare movement. Most scientists and agriculturists are not cruel, yet they invasively use countless numbers of animals. On the other hand, loving something is neither necessary nor sufficient for treating it morally. I certainly don't love most of the human surgeons I know; I don't even like them; yet, I am bound to treat them morally. By the same token, many people who love their pets mistreat them in countless ways, from providing improper diets to denying them exercise.) Our intuitions on animals are incoherent; the same people who condemn branding of cattle may dock the ears and tails of their dogs. The law is of no help—in the eyes of the law animals are property, either private property or community property. The Animal Welfare Act, reflecting irrational social prejudice, does not consider rats, mice, or domestic farm animals to be animals; for purposes of the act, a dead dog used in research is an animal, a live mouse is not. And traditional moral philosophy is of no help either, since for most of its history it was virtually mute on the subject of our obligations to other creatures. More has been written on this subject, in fact, in the past ten years than in the previous three thousand.

All of this is further complicated by a major component of the same ideology of science that we discussed earlier and that I have explored in detail elsewhere.[13] From about 1920 until the mid-1970s, behaviorism was a major component of scientific ideology, and it was dogma to assert that we could not scientifically know that animals were conscious or even that they felt pain. Indeed, this is still dogma in many quarters—a USDA inspector recently told me that a medical researcher had informed him that dogs lack a sufficiently highly developed cerebral cortex to experience pain, and I have heard variations on this theme over and over. A leading veterinary pain expert told me that the majority of veterinarians still view anesthesia as a way of restraining the animal. (This was confirmed for me when I was lecturing at a leading veterinary school early in my involvement with this issue and naively remarked that at least veterinarians could not doubt that animals felt pain, or else why would they study anesthesia and analgesia? Up jumped the associate dean, livid with rage. "Anesthesia and analgesia have nothing to do with pain," he shouted. "They are methods of chemical restraint." Analgesia is virtually never used on laboratory animals and very rarely used in clinical veterinary practice. Ironically, rodents are the most infrequent recipients of analgesia, yet most pain and analgesic research is done on rodents, so the dose response curves are well known.

Obviously, much scientific research, agricultural practice, and, indeed, ordinary activity rest on exploiting animals, so that it is far easier and more comfortable not to think about animals in moral categories. Nonetheless, common sense has never denied that what we do to animals matters to them; that they have needs and interests, physical and psychological, and that they can suffer, physically and psychologically, when those needs and interests are thwarted and infringed upon. During the past decade, society has just begun to realize the implications of its own assumptions about animals and has begun to be aware that we do have moral obligations to them. Hence the rise of the animal rights movement, a massive international stirring which cannot be ignored, which questions much of our traditional treatment of animals, and which has been called "the Vietnam of the 80s."

At all events, a growing number of people in the scientific community are beginning to think seriously about the moral status of animals. In the past eight years, I have lectured to over thirty veterinary schools all over the world on these issues, as well as to biomedical scientists of all sorts, attorneys, agriculturalists, psychologists, government officials, farmers, ranchers, and scores of other groups. I have testified before Congress and state legislatures and have served as a consultant to various agencies of three national governments. In the course of this decade, I have tried to develop an ethic to guide us on the uses of animals, one that I believe follows logically from moral assumptions we all share by virtue of living in democratic societies. In other words, rather than generate my own ethic and attempt to force it on others, I have attempted, following Socrates, to extract from others (though they may not and often do not realize it) what their own moral assumptions entail about animals. Such an ethic is necessary not as a blueprint for instant social change in all areas but, as Aristotle put it, as a target to aim at and as a yardstick to measure our current conduct. Without such an ethic, as my colleague Dr. Harry German, surgeon and researcher, has beautifully put it to me, we tend to confuse what we are doing with what we ought to be doing.

Given the constraints of time, I can only present the briefest sketch of this ideal for animals. For those of you who wish to pursue the topic more deeply, I would refer you to my book *Animal Rights and Human Morality*.[14] Stated boldly, I ask people to consider whether they can present any rationally defensible grounds for excluding animals from the moral arena or from the scope of moral concern and deliberation. Surely animals are more like children than like wheelbarrows in that they can be hurt and that what we do to them matters to them. None of the standard, historically pervasive differences that have been cited to exclude animals from the moral arena will bear rational scrutiny. The claims that man has a soul and animals do not, that man is evolutionarily superior to animals, that man is superior in force to animals, that man is rational and animals are not, do not suffice to exclude animals from the moral arena and from falling under the purview of our socially pervasive moral concepts. In other words, given the logic of our moral ideas, there is no way to preclude extending them to animals. And this is not difficult to do.

In democratic societies, we accept the notion that individual humans—not the state, the Reich, the *Volk*, or some other abstract entity—are the basic objects of moral concern. We attempt to cash out this insight, in part, by generally making many of our social decisions in terms that would benefit the majority, the preponderance of individuals (i.e., in utilitarian terms of greatest benefit to the greatest number). In such calculations, each individual is counted as one, and thus no one's interests are ignored. But such decision making presents the risk of riding roughshod over the minority in any given case. So democratic societies have developed the notion of individual rights, protective fences built around the individual which guard him or her in certain ways from encroachment by the interests of the majority. These rights are based upon plausible hypotheses about human nature—about the interests or needs of human beings that are central to people, and whose infringement or thwarting matters most to people (or, we feel, ought to matter). So, for example, we protect freedom of speech, even when virtually no one wishes to hear the speaker's ideas. Similarly, we protect the right of assembly; the right to choose one's own companions and one's own beliefs; and also the individual's right not to be tortured even if it is in the general interest to torture, as in the case of a criminal who has stolen and hidden vast amounts of public money. And all of these rights are not simply abstract moral notions but are built into the legal system.

The extension of this logic to animals is clear. Animals too have natures (i.e., fundamental interests central to their existences, whose thwarting or infringement matters to them). This set of needs and interests—physical and psychological, genetically encoded and environmentally expressed—which make up the animal's nature I call the animal's telos, following Aristotle. It is the pigness of the pig, the dogness of the dog. Such a notion is not mystical; it follows, in fact, from modern biology. Thus, it ill serves the issue at hand when scientists sneer at this notion, as one person at the NIH did recently, by suggesting that an animal's only nature is to serve us and die. According to the logic of our position, animals' basic interests as determined by their telos ought also to be morally and legally protected; this is the cash value of talking about rights. This. then, is what I take to be the logical extension of our socially sanctioned, moral notions when applied to animals, and when one cannot cite a morally relevant difference between people and animals which would forestall such application. Obviously animals do not have the same rights as humans, even ideally, since they do not have the same natures. So it will not do to ridicule the position by saying that I am urging that turtles have the right to vote or dogs have freedom of bark.

I have devoted much of my recent activity to attempting to actualize this ethic as far as is practically possible into veterinary medicine and research uses of animals, where its relevance is evident. But what does it tell us about genetic engineering of animals? Let me first of all clear up a misconception which has arisen about my notion of telos. It has been asserted by some opponents of genetic engineering that in my view telos is inviolable, and it is immoral to change it. I have never said that. What I argue is that, given an animal's telos, certain interests which are part of that telos ought to be inviolable. Thus, given a burrowing animal, it is wrong to cage it so that it can't burrow. But 1 have never asserted that there is anything wrong with changing the telos of a burrowing animal so that burrowing no longer matters to it.

The proper application of these ideas to genetic engineering of farm animals is made quite interesting by the fact that so much of our current intensive agricultural use of animals involves forcing animals into environmental contexts for which their natures are not suited. As a result, we must perpetually depend on highly artificial devices like debeaking in chickens and extensive uses of drugs and chemicals as well as contend with "production diseases." While extensive agriculture has its own problems, at least the problems are, as it were, natural to the animals rather than being created by the humanly devised management system. Ideally, from the point of view of the animals' welfare, I would like to see society back off from ever-increasing intensification. (This would, I think, have certain social and economic benefits as well.) But in all likelihood, increasing intensification is here to stay. So the main moral challenge to those involved in genetic engineering of agricultural animals is to avoid modifying the animal for the sake of efficiency and productivity at the expense of the animal's happiness or satisfaction of its nature. Economic pressures, of course, in the main, militate against my recommendation. This is why I asserted earlier that this is truly a moral challenge. Also militating against this is the fact that hitherto the animal's welfare (except insofar as it affects economic productivity of an entire operation) has entered neither into intensive agricultural decision making nor into research serving that decision making. (This was freely admitted to me by a group of high-ranking USDA officials.)

Nonetheless, given the increasing public concern about the welfare of all animals including agricultural animals, as well as the strong moral arguments in favor of concern for animals, it is imperative that this moral vector enter into agriculture. And certainly genetic engineering is an excellent place for this vector to be felt. The basic principles that should guide thinking in this area are not hard to see. Obviously, as a minimal principle, the animals should suffer no more as a result of genetic intervention than they would have without it. Ideally, they should suffer less and be happier. Thus, in my view, it would be grossly immoral (as has actually been suggested) to use genetic engineering to change chickens into wingless legless and featherless creatures who could be hooked to food pumps and not waste energy. Similarly (as has also been suggested), it would be wrong to manipulate the genome of pigs to produce leglessness, with the animals after all still having all the psychological urges to move. On the other hand if genetic engineering is used to genuinely suit the animal to its stipulated environment and therefore eliminate the friction between telos and environment which clearly results in suffering, boredom, pain, stress, and disease, and this conduces to the animal's happiness, it does not appear morally problematic. Thus, if one were to genetically alter chickens' physical and psychological needs so that all evidence (such as results of preference testing, physiological signs of stress, behavioral signs of stress, individual animal productivity, and health) indicated that the animals were happy this would be morally acceptable according to the theory I have been expounding, though many people, myself included, would certainly not be quite comfortable with it, probably on aesthetic grounds.

Obviously, therefore, these considerations of the animal's welfare, independently of the effect on humans, should be weighed and considered before a piece of genetic engineering is undertaken. And such consideration should be part of the formal charge of the committee we discussed earlier. Thus, if someone were to suggest using genetic engineering[15] to create larger beef cattle, the researcher should be required to show that there is good reason to believe that the animal's joints could withstand the extra stress and that no new suffering would be engendered by such genetic manipulation. In this way, we can at least begin to assure that the animals' interests are weighed along with ours. In the case of totally virgin territory, as in the creation of chimeras, an even stronger burden of proof should be put on the proposer to demonstrate that his manipulation would not lead to suffering.

In sum, in my view, the genetic engineering of animals in and of itself is morally neutral, very much like the traditional breeding of animals or, indeed, like any tool. If it is used judiciously to benefit humans and animals, with foreseeable risks controlled and the welfare of the animals kept clearly in mind as a goal and a governor, it is certainly morally nonproblematic and can provide great benefits. On the other hand, if it is used simply because it is there, in a manner guided at most only by considerations of economic expediency and "efficiency" or by quest for knowledge for its own sake with no moral thinking tempering its development, it could well instantiate the worst rational fears encapsulated in the "Frankenstein thing." To those of you upon whom the primary responsibility for this choice rests, let me conclude by reminding you that, though Frankenstein was in fact the name of the scientist, virtually everyone thinks it is the name of the monster.

NOTES

Reprinted by permission from "The Frankenstein Syndrome: The Moral Impact of Genetic Engineering in Farm Animals," in *The Genetic Engineering of Animals,* ed. J. W. Evans and A. Hollander (New York: Plenum, 1985), 292–308.

1. D. F. Glut, *The Frankenstein Catalog* (Jefferson, N.C.: McFarland, 1984).

2. *Time,* July 29, 1985, 54–59.

3. B. E. Rollin, *The Teaching of Responsibility* (Hertfordshire, U.K.: Universities Federation for Animal Welfare, Potters Bar, 1983); "The Moral Status of Research Animals in Psychology," *American Psychologist* 40, no. 8 (1985): 920–26.

4. B. E. Rollin, "On the Nature of Illness," *Man and Medicine* 4, no. 3 (1979): 157ff.

5. J. R. Baker, *Race* (London: Oxford University Press, 1984).

6. M. Ruse, *The Philosophy of Biology* (London: Hutchinson, 1973).

7. B. E. Rollin, "Nature, Convention, and Genre Theory," *Poetics* 10 (1981): 127–43.

8. Ibid.

9. B. E. Rollin, *Animal Rights and Human Morality* (Buffalo: Prometheus, 1981).

10. H. Rolston, "Duties to Endangered Species," *BioScience,* 1984.

11. Rollin, *Animal Rights.*

12. See the foreword of this volume for more on Asilomar.

13. B. E. Rollin, "Animal Consciousness and Scientific Change," in *New Ideas in Psychology* (in press); R. E. Rollin, "Animal Pain," in *Advances in Animal Welfare Science* (in press).

14. Rollin, *Animal Rights.*

15. I wish to thank Linda Rollin, M. Lynn Kesel, David Neil, Murray Nabors, Robert Ellis, George Seidel, and Dan Lyons for dialogue and criticisms.

18

On the Ethics of Using Animals for Human Benefit

R. G. Frey

$\mathbf{A}$ good deal of interest is today focused upon what we do to animals both as experimental subjects, whether in medical research or other scientific disciplines, and as sources of food, clothing, and money. Much of this interest is motivated by a concern over medical and farming practices that inflict pain and suffering and eventually death upon animals, and a number of people, quite apart from those who are members of the animal liberation movement, think that such practices require a moral defense. Whilst the author agrees with that view, it is also believed that such a moral defense can be given of at least a good many of our practices with regard to animals, and an attempt has been made, in both books and articles (see Bibliography), to present the philosophical basis of such a defense. Here, one aspect of this defense is examined which is fundamental to all of our interactions with and treatment of animals but which is on the whole little discussed. In what follows, the terms of discussion are kept as broad as possible, so as to include much more than merely agricultural practices; for the issue to be discussed is itself a broad one. It is this: how do we factor into our ethical thinking about our treatment of animals our views about the comparative value of human and animal life?

Much of what we do to animals raises obvious ethical issues. For example, how miserable can we make animals' lives, in order to eke out more life or an enhanced quality of life for ourselves? Can human benefit justify the infliction of any degree of pain and suffering, no matter how substantial? What kinds of invasive techniques that result in permanent damage to, or genetic alteration in, animals are warranted by our concern for our own health and well-being or, for example, by our concern for enhanced milk yield and the profits this can bring? How many healthy animal lives may be used and sacrificed in order to understand and cure human ailments and forestall human deaths, or to satisfy human demand for meat or leather products, or to satisfy some other human need for animals, such as occurs with pets and zoos? And does the particular species of animal used matter?

The realization that, in order to make ourselves better off, we often make other creatures vastly worse off cannot escape anyone who interacts with animals today, including those who make use of other people's interactions with animals in order to supply us with food, clothing, health care and companionship. All are aware that we in part purchase our quality of life (indeed, our very lives and health) at the expense of animals. Morally, then, since we are accountable for what we do, we must surely justify our use of animals in any regard for human benefit. Under certain conditions, it would seem that our use of animals for human benefit can be justified; but it is no easy task to set out and defend any such justification. In what follows, an attempt is made to bring out one of the difficulties involved in this defense of animal research and of the general use of animals for human benefit. The discussion is not focused exclusively on agriculture, but ranges over a number of uses including, in particular, our use of animals in medical and scientific research, an area which most people consider vital to their quality of life and to its continuation at an acceptable level.

Many issues are involved in this defense of animal use, several of which have been explored at length in the author's previous publications (see Bibliography). The objectives of this chapter are to indicate why we cannot escape this task of moral justification and to show how a common type of justification has serious implications for humans.

Most importantly, this issue about justification is not, at its most fundamental level, about the infliction of pain on animals. Pain is pain, as much an evil for an animal as for a human, and it is a form of speciesism or discrimination to pretend otherwise. Where pain is concerned, there would seem to be no difference between burning a cat alive and burning a child alive, and the deliberate infliction of pain on a feeling creature requires justification. Yet, fundamentally, what concerns us is not the painful use of animals for human benefit but their use at all for human benefit. For it seems obvious that, only if we can justify using animals for human benefit in the first place, can we go on to question the justification of their painful use for human benefit. To be sure, it may be that pain is what draws the eye, as well as the heart, in contemporary discussions of animal welfare, and that is an important issue; but these matters do not remove the fact that the issue of use at all is the more fundamental. If we cannot satisfactorily deal with this issue, then it would seem to follow that we must, if we are morally serious, give up using animals. So how are we to deal with this question?

Reference has been made to the matter of accountability: sometimes, today, medical and other scientific researchers try to lift accountability from our backs by some well-known ploys, all of which appear to come to nothing. Four of these ploys deserve mention, before turning to the fifth and most important one.

1. Since animals show no reluctance to use each other, why should we show

any moral reluctance in using them? Humans are reflective creatures, at best, reflective moral creatures, and the fact that we cannot ask creatures incapable of moral thought to consider their own actions morally in no way shows that we cannot ask precisely this of intelligent human beings and so of scientists. Everything that is done to animals in laboratories, everything that is done to them to enhance, either directly or indirectly, our quality of life is done deliberately, intentionally, and we are accountable for what we deliberately, intentionally do.

2. Why cannot scientists shift accountability onto the back of the law by claiming that, given approved protocols, project licences, and inspectorate certificates, the law permits them to do what they are doing? Why is that not the end of the matter? The short answer is that, morally, we often impose stricter standards of conduct on ourselves than merely what the law permits. For example, there is no general legal duty of rescue in Anglo-American law: one may legally stand on shore and watch a man drown. But what would we say of such a person? What would we say of them, morally, when they insisted that they followed the letter of the law? The fact is that there are cases where we expect more of ourselves and others, even if we are legally permitted to do what we are doing. Activities involving animals include: depriving them of their lives, genetically altering them, caging them, intentionally burdening their lives with maladies of all kinds, caring for them assiduously in order to restore them to the uses of their owners, whether farmers, pharmaceutical concerns, or research laboratories. If someone said the law permitted these things and made no attempt to explain why they thought that what they were doing was morally appropriate, then morally serious persons would become concerned. For the fact is that what really matters is not what the law decrees but what each of us, as morally serious persons, can live with.

3. If our child were dying of hunger, would we not favour animal farming, whether painful or not, so long as we had access to the result? If our child were dying of an AIDS related disease, would we not favour unlimited animal experimentation, and in whatever species of animal, in the search for a cure? Of course, we would. But this is why we do not do our reflective moral thinking in the heat of the moment. It is highly problematic as to whether we can extrapolate from what we favour in such circumstances to what we should come to decide in those calm periods when we exercise our mature, detached, reflective judgment. A terminally ill patient might well favour spending the entire National Health Service budget on the search for a cure; but none of us in our calm moments, including the terminally ill patients themselves when able to achieve a certain intellectual and emotional detachment from their condition, would think this a wise allocation of scarce resources. Detached moral thinking is hard at the best of times, let alone at the worst of times.

4. Do not people in agriculture, do not scientists and veterinarians actually benefit animals? Do they not look after their animals, care for them, see to their health and comfort and, where possible, improve their lot? This claim, however, misses the point. Suppose these things are true: the point is that they are undertaken with,

ultimately, our own benefit in mind. Our pet comes to mean something to us, and though we want it made well to some extent for its own sake, it is actually our pet that we want made well. Farmers' animals are the ones by which they make money. Laboratory research animals will be cared for because the experiments conducted require healthy animals to eliminate contaminating factors. This in no way denies, for example, that a particular veterinarian cannot be in the business of making animals well for the animals' sake or that all veterinarians cannot on some occasion or other have this objective; but there must be very few people who take other people's pets to be cured or farmers who call in help for the sick cattle of other farmers or research laboratories that pay the expenses for the upkeep of another laboratory's research animals. Veterinarians treat all these animals in a way which, if successful, restores the animals to the roles they are playing in human society, and they are clearly aware of what they are doing. So we can hardly pretend that our systematic use of animals in all the myriad ways we use them is selfless and solely for their benefit.

Thus, it seems that the above ploys do not enable us to evade providing a moral justification for our general use of animals. Scientists, for example, who want to continue to use animals, might feel compelled to resort to a stronger argument, an argument that enables them to avoid the thrust of any charge of moral accountability. Such an argument is, it might be thought, readily available: it consists in denying that animals count morally, denying that they are part of the moral community and so warrant our moral concern. In fact, however, if any such argument as this were ever widely accepted among scientists, it would not seem to be so at present. For even the very practices of scientists themselves today give the lie to it.

What is it about animals that does not count morally? Two candidates come to mind, their pain and suffering, and their lives. As for pain and suffering, ethics review committees, professional societies, government regulatory bodies, all kinds of research guidelines, peer review policies of journals and granting agencies, all seek to ensure a number of things. They urge the control of animal pain and suffering; they urge that it be limited insofar as possible; they urge that it be mitigated by drugs where feasible; they often urge that the animal be killed before it recovers from the effect of the drugs; and they seek the justification of the infliction of pain and suffering in the course, and by the nature, of the experiment or project proposed. Professional oversight committees, including governmental ones, can now call into question research where these matters, though approved in the research protocol, do not receive appropriate attention. All this is plain fact.

If, however, one thinks that animal suffering counts, it seems bizarre to think that animal lives do not. For all the worry about pain and suffering is in part simply the worry about the way in which these things impose very negative drawbacks to a life, whether in humans or animals. Unless we thought those lives had some value, it is hard to see why we should care about ruining them or drastically lowering their quality or why we should go to the great lengths we do in order to cite the actual or potential benefits that are supposed to justify their sacrifice.

If one thinks that animal life has even *some* value, then its deliberate destruction or the intentional lowering of its quality is something in which morally serious people can, and doubtless should, take an interest. What those who deny animals moral status must show is that there is a genuine moral difference between the human and animal case, where pain and suffering and the destruction of lives are concerned. But exactly what is the difference between burning a man and burning a cat, infecting the man with a degenerative disease and infecting the cat with one, and bleeding both the man and cat to death? One can say that it is worse to do these things to the man, but what the present position must hold is that doing these things to the cat is of no moral concern whatsoever. So far as the cat is at issue, even though suffering ensues, the quality of life is drastically lowered and a killing takes place, nothing here is of moral concern. Put differently, these things done to a human would be wrong; done to a member of a different species, according to the present view, they are not wrong. We are owed an explanation of how species membership alone can make this difference. Where pain and suffering are concerned, there would seem to be no difference between the human and animal cases (and it does not matter if one substitutes for the cat another feeling subject, such as a rodent); where the destruction of lives of at least some value is concerned, the burden is upon those who destroy these valuable things to explain how species membership can distinguish morally between two relevantly similar acts of killing.

When we enquire as to what justifies all that is done to animals, the answer usually given, directly or indirectly, overtly or covertly, is human benefit. This in turn has spawned a vast literature on what counts as a benefit, how we tell which benefits justify animal suffering and sacrifice, how large a benefit must be to justify this level of sacrifice, how exactly the repeated duplication of research is a benefit and whether there are some benefits, including important medical ones, that are simply not worth the terrible price in sacrifice of valuable lives that they require. Issues such as these confront us with a whole series of challenges. They raise the question of whether a benefit, however small, can justify the repeated destruction of valuable lives. They raise the question of whether a piece of research should be permitted or continued, if it is very costly in pain and suffering and loss of life, when deciding upon what counts as "very costly" cannot always be assumed to be beyond our powers.

In this way, the temptation to deny that animal lives have value can seem overwhelming to researchers: it lets them off the moral hook, enables them to carry on work free of moral doubts, and releases them from the difficult task of showing how benefit to one species can justify the deliberate, intentional harm or destruction of another. As indicated, this temptation must be resisted.

Of course, one can always insist that our morality simply must permit us to sustain our lives and to enhance their quality. But no one doubts this. The real question is the limits to which we may go in order to sustain and enhance our lives. Is everything permissible in the search for our preservation? (This question has an analogue in ordinary life: may one sacrifice another person in order to save oneself?)

In short, important ethical issues arise in using and disposing of animals, and the morality of these things must be argued for and not just assumed or held to consist in paying attention to the infliction of pain and suffering and following laboratory policies for the proper care and custody of animals. Killing, the destruction of valuable lives, still confronts us, and this remains true whatever the size of the benefit that is supposed to lie before us as a result of the killing. There would appear to be no way to avoid facing up to this killing.

It is this issue of killing and its justification to which the discussion now turns. It is appreciated that this may be thought to take us somewhat away from a concern with animals in agriculture, but it would appear that the issue that underlies a discussion of killing, namely, how we separate the human from the animal case, is actually vital to discussion of the interactions of humans and animals in any domain.

There would seem to be widely held views that: (1) animal lives have some value, (2) not all animal lives have the same value, and (3) human life is more valuable than animal life. The author considers that (4) is an important datum, and animal liberationists would seem to be correct in insisting that we cannot give a speciesist justification of this claim of greater value. There is insufficient space here to go into all the discussion that surrounds the claim that human, though not animal, life is sacred. But two things need to be said. First, it is very convenient indeed that many of our ethical codes based on the Christian religion give us dominion over the remainder of creation or nature. One could be forgiven at marveling how wondrous it is that Christianity gives not merely the power but also the right to slaughter the rest of creation or nature for our own ends, especially when other religions forbid any such thing. However, it is increasingly argued by many Christians that Christianity confers on us no such right; instead of dominion, we are urged to a notion of stewardship of the rest of creation and so to a notion that, at least in intention, is resistant to rapacious human will. Second, many people in science, scientific research and agriculture are no longer religious and do not think their moral views are grounded in, or flow out, of a religion and they may well see no need for another round of debates about God's existence in order to address moral and social problems that presently confront us. Yet, they do need and seek a justification for what they do.

What is suggested, then, is that the justificatory boundaries of using and destroying animal lives must include, in addition to an account of what makes lives valuable, a non-speciesist account of why it is, as virtually all of us believe, that human life is more valuable than animal life. To say that an infant without a brain or an adult in an irreversible coma has a more valuable life than the lives of any of the great apes is implicitly to raise the question of why the infant and adult have more valuable lives. Ultimately, it can seem that we fall short of our task simply to revert to religious assertions of greater value. The view might be proposed that a non-religious, non-speciesist account of the value of a life can be given which explains why a human life is more valuable than an animal life. This claim is now discussed.

In the example discussed above, of killing the man and killing the cat, one might respond that killing the man is worse. However, we have seen that what makes it worse cannot consist in species membership. So what does make it worse? Whatever the full answer, part of it seems bound up with our belief that the man's life is more valuable than the cat's life. This view allows that the cat's life has value, so it does not collapse into the extreme position of denying animal life value. What it maintains is that the cat's life does not have the same value as the man's life. Importantly, however, this belief will itself be guilty of species discrimination unless we can show that it is something other than species membership that makes the man's life more valuable. If we could show this, then we could claim a genuine moral difference between killing a man and killing a cat. We could then use this difference in the argument from benefit to justify some of our uses of animals. What, then, makes a life valuable, and why is the man's life more valuable than the cat's life?

The answer to these questions argued for elsewhere is this (see Bibliography): the value of a life is a function of its quality, its quality of richness, and its richness of capacities or scope for enrichment. In fact, in many areas of medical ethics today, quality-of-life views of the value of a life are common and they make the value of a life turn upon its content or experiences. Since the cat and rodent have experiences, and an unfolding series of experiences at that, they have lives of some quality or other and so of some value. Yet, everything that observation and science teach us of their lives indicates that their richness is not comparable to the richness of ours precisely because the capacities for enrichment are not nearly as diverse as ours. Obviously, we share many activities with cats and rodents: we eat, sleep and reproduce. But such activities come nowhere near exhausting the richness of lives with the diverse elements of ours, from music, art, literature, culture generally, and friendship to love, science and all the many products and joys of reflection. And even this does not begin to take account of how we fashion our lives into and so live out lives of striving to exemplify excellences of various sorts, whether as veterinarians, philosophers, athletes or policemen. These are ways of living that can be and typically are the source of value to us as well. In short, no cat or rodent has ever lived at the level we live, and nothing we know about them leads us to believe that one will. Accordingly, we judge the value of the respective lives to be different, not on the basis of species, but on the basis of quality and richness.

Autonomy helps us in living out the life we want. To direct one's own life to secure what one wants; to make one's own choices in the significant affairs of life to assume responsibility over a domain of one's life and so acquire a certain sense of freedom to act; to decide how one will live and to mould and shape one's life accordingly; opportunities such as these open up areas of enrichment in a life with

consequent effect upon that life's quality and value. (Equally, of course, it is possible that nothing of this kind will issue from the exercise of one's autonomy: just
because a life's value can be augmented through the exercise of autonomy it does
not follow that it is inevitably or always so augmented.) The point is that, arguably, animals are deprived of these ways of augmenting the value of their lives.

Suppose we grant that a cat has keener senses of smell and hearing than we do:
is it not possible that these raise its quality of life to such an extent that the value
of the cat's life, even though it lacks a great number of our capacities, approaches
our own? That is, is it not possible that, say, the exercise of a single capacity by
the cat can confer on it a quality of life that the exercise of all of our capacities
confers upon us? It seems we must allow that this is possible, but there appears to
be no reason whatever for believing it to be true. And there would seem to be even
less reason for believing it as we descend the ladder of animals that are common
research subjects or farm animals. (If we ascend the ladder to primates whether
confined or in the wild, the case for a greater richness to their lives may become
more compelling.)

It is not affirmed that exactly what is relevant to discussing the quality of our
lives is relevant to discussing the quality of animal lives. Let there be differences—we must be presented with some reason for thinking that these differences
enable the animal's life to approach the richness of ours, given all our additional
capacities. This reason must cohere with what observation and science enable us
to know of the capacities present in animals of the kind in question. One is not
simply licensed to invent capacities of animals that observation and science cannot confirm.

The outcome of quality-of-life views of the value of lives for our present concerns should be obvious: normal adult human life is of greater value than animal
life and that is why it is worse to kill the human. The greater value is not dependent upon claims about mere species membership but upon quality and richness,
and our view of richness does not require us to use criteria for assessing the richness of human lives as if they applied without revision, or at all, in the case of
animals. Let these criteria be whatever one chooses: what evidence is there that,
on those very criteria, animal lives approach human lives in richness? But should
we not give animals the benefit of the doubt? Yes, if the case is close; but what is
the evidence that the case is close?

The complication in the argument referred to at the outset should now be apparent. We have, on the basis of quality and richness, a case for believing that human
life is more valuable than animal life, but every one of us knows that not all human
lives are of the same degree of quality and richness. All of us know or know of
tragic cases where the quality of a life, adult or child, has fallen drastically. In
fact, among some, a question can arise about whether that life is worth preserving.
A seriously defective newborn, with all kinds of maladies, is an example: do we
operate on it incessantly, in order to eke out a few more days of life? And what
of adults in the final stages of a terminal, degenerative disease: do we continually

seek to put off the moment of death? All of us know that there are some human lives of such a tragic quality that we would not wish them upon anyone, and it is mere pretense to claim how valuable those lives are. (Euthanasia and physician-assisted suicide have become important subjects of public debate just because of cases of this sort.) In the abstract, each human life is sometimes said to be precious; but medical people and the rest of us have long recognized that in practice some human lives are of a quality so tragically low as to deviate massively from the quality of normal adult life. To assert how valuable such lives are can at times lack credibility, as even the person who is living that life seeks earnestly, urgently, and clear-headedly to die.

However, if not all human lives are of the same value, because not only their richness but their scope for enrichment is severely truncated (indeed, in the case of infants without a brain, absent), then we have to face the possibility that some human lives may reach so tragic a level as to fall below the quality of life of a normal, healthy animal. In this eventuality, our argument would force us to conclude that the animal's life is more valuable than the human's life. And this has implications for anyone who operates, as most of us do, an argument from benefit as the justification of our use of animals. Take medical experimentation on retinas, undertaken for human benefit: what justifies our using an animal of a higher quality of life to a human of a lower quality of life in the experiment? If one operates with the argument from benefit, then one realizes that the benefit can be obtained by experimenting upon humans or upon animals. So what is the argument for always using animals, when the quality of life of some humans falls below that of the animals to be used? Consistency in the operation of the argument from benefit forces us to face this dreadful prospect.

What we need then, in every case, is something that ensures that a human life of any quality, however low, exceeds in value an animal life of any quality, however high, in circumstances in which we have not even begun to explore the tragic depths to which the quality of human lives can plummet. There would not appear to be any such thing.

It is stressed that medical experiments upon these unfortunate humans are not being advocated. What is suggested is that envisaging such experiments is inescapable if we use the argument from benefit and rely upon the comparative value of human and animal life to indicate why we must always use the animal. For the comparative value of human and animal life will not always favour the human. And what is true here of medical experimentation is true of all of our encounters with, and uses of, animals. If the justification we offer for what we are doing runs through the argument from benefit and the comparative value of human and animal life, in order to show why what we do to animals may not be done to humans, then the logic of our position forces us to envisage using some humans, admittedly, the weakest, most unfortunate ones among us, in the same or very similar ways. As discussed, the only way to avoid this outcome is to cite something that guarantees that any and all human lives, no matter how tragically blighted, exceed in value any and all animal lives, no matter how wonderfully high in quality, and the author knows of no such thing.

It seems, then, our usual justification for what we do to animals in the name of human benefit has an outcome that needs to be faced up to. If we continue to justify what we do to animals in this way, then we cannot avoid envisaging the use of humans in the same or similar ways; if we simply cannot bring ourselves to envisage using unfortunate humans in these same or similar ways, then we seem forced to forgo our use of such a justification in the case of using animals, unless we can cite what makes the human case, each and every human case, special. For otherwise we seem to be caught out by a kind of inconsistency: we continue to use a quality-of-life argument so far as it enables us to make use of animals, but the moment we see its implications for humans we then retract that argument.

It should be noticed that, in response to the query of why we simply cannot favour our own kind, the answer is not that we cannot; the answer is rather to enquire after the basis of that favouritism. If we find that it turns on certain properties or characteristics, as it almost always does, then what do we say of those humans who lack those properties or characteristics? And except for species membership, what property or characteristic are we going to cite that every human has but no animal has?

So why not just endorse speciesism and be done with it? That is, why not just favour "our own kind"? The problem now is to decide how we limit our determinations of our own kind. Suppose that "own kind" is defined as "white male": it can be assumed that today there would be an outcry at his favouring this class in moral judgements. So how are we to determine that "human species" is any the less discriminatory, that it is a perfectly acceptable form of favouritism? It is important to remember here that, no matter what other property or characteristic we cite to advantage ourselves, we have always to contend with two problems, namely, the fact that all humans may not have the property or characteristic in question and the fact that some animals may.

Here, it may be said, is the reason why we need to believe in God and to accept a religion. We could then tell a story that provides the magical ingredient we require, that is, an ingredient that ensures that a human life of any quality, however low, exceeds in value an animal's life of any quality, however high. Each of us must decide for ourselves whether we are satisfied by this move, and this seems in part to turn upon how well we can argue the case for the religion in which this move occurs.

Finally, it is clear that this is not the only account of the comparative value of human and animal life, though in the author's opinion it is consonant with much of the thinking today in the various areas of medical ethics. (Equally, this is not the only account of the richness of life.) Some may urge that we have to rethink entirely our relation to the animal kingdom. Certainly, if we adopt some Eastern religion or some form of quasi-religious metaphysic, then we might come to have a different view of animals and of how we stand to them. Indeed, we might come to take a different view of our relations to the animal kingdom (as well as the inanimate environment) without any specifically religious impulses at all. This much is clear to all of us, through poetry, through cultural differences we encounter among the individuals that make up our society, and through exposure to the art of different ages and cultures. From these different, possible views of our relations to animals, different, possible accounts of the comparative value of human and animal lives may flow. But from the mere fact that there are different, possible accounts of this comparative value nothing follows *per se* about the adequacy of any single one. Argument must establish the soundness of such an account, and if, for example, one's claims about comparative value turn upon one's adoption of a religion or religious metaphysic, then it is that religion or metaphysic that must be subjected to scrutiny. Merely to be able, within the fabric of that religion or metaphysic, to relate things the purport of which is to have us believe all manner of things of animals is not the kind of scrutiny of a view or metaphysic that is required. Such scrutiny, of course, must be applied to the view advanced here of the comparative value of lives: no exemption can be tolerated. For the truth of the matter is that there is no established or correct view of the matter.

NOTE

Reprinted by permission from R. G. Frey, "On the Ethics of Using Animals for Human Benefit," in *Issues in Agricultural Biotechnology,* ed. T. B. Mepham (Nottingham: Nottingham University Press, 1995), 335–44.

BIBLIOGRAPHY

Animal parts, human wholes: On the use of animals as a source of organs for human transplants. In *Biomedical Ethics Reviews,* ed. J. M. Humber and R. F. Almeder. Clifton, N.J.: Humana Press, 1987, 87–107.

Animals, science, and morality. *Behavioral and Brain Sciences* 13 (1990): 13, 22.

Autonomy and the value of animal life. *The Monist* 70 (1987): 50–63.

The ethics of the search for benefits: Animal experimentation in medicine. In *Principles of Health Care Ethics*, ed. R. Gillon (New York: Wiley, 1993), 1067–75.

Frey, R. G. *Interests and Rights.* Oxford: Clarendon, 1980.

———. *Rights, Killing, and Suffering.* Oxford: Blackwell, 1983. These two volumes provide background reading on the ethical theory underpinning the author's views.

Moral standing, the value of lives, and speciesism. *Between the Species* 4 (1988): 191–201.

The significance of agency and marginal cases. *Philosophica* 39 (1987): 39–46.

Vivisection, medicine, and morals. *Journal of Medical Ethics* 9 (1983): 94–104.

19

The Case for the Use of Animals in Biomedical Research

Carl Cohen

Using animals as research subjects in medical investigations is widely condemned on two grounds: first, because it wrongly violates the *rights* of animals,[1] and second, because it wrongly imposes on sentient creatures much avoidable *suffering*.[2] Neither of these arguments is sound. The first relies on a mistaken understanding of rights; the second relies on a mistaken calculation of consequences. Both deserve definitive dismissal.

WHY ANIMALS HAVE NO RIGHTS

A right, properly understood, is a claim, or potential claim, that one party may exercise against another. The target against whom such a claim may be registered can be a single person, a group, a community, or (perhaps) all humankind. The content of rights claims also varies greatly: repayment of loans, nondiscrimination by employers, noninterference by the state, and so on. To comprehend any genuine right fully, therefore, we must know *who* holds the right, *against whom* it is held, and *to what* it is a right.

Alternative sources of rights add complexity. Some rights are grounded in constitution and law (e.g., the right of an accused to trial by jury); some rights are moral but give no legal claims (e.g., my right to your keeping the promise you gave me); and some rights (e.g., against theft or assault) are rooted both in morals and in law.

The differing targets, contents, and sources of rights, and their inevitable conflict, together weave a tangled web. Notwithstanding all such complications this

299

much is clear about rights in general: they are in every case claims, or potential claims, within a community of moral agents. Rights arise, and can be intelligibly defended, only among beings who actually do or can make moral claims against one another. Whatever else rights may be, therefore, they are necessarily human; their possessors are persons, human beings.

The attributes of human beings from which this moral capability arises have been described variously by philosophers, both ancient and modern: the inner consciousness of a free will (Saint Augustine[3]); the grasp, by human reason, of the binding character of moral law (Saint Thomas[4]); the self-conscious participation of human beings in an objective ethical order (Hegel[5]); human membership in an organic moral community (Bradley[6]); the development of the human self through the consciousness of other moral selves (Mead[7]); and the underivative, intuitive cognition of the rightness of an action (Prichard[8]). Most influential has been Immanuel Kant's emphasis on the universal human possession of a uniquely moral will and the autonomy its use entails.[9] Humans confront choices that are purely moral; humans—but certainly not dogs or mice—lay down moral laws, for others and for themselves. Human beings are self-legislative, morally *autonomous*.

Animals (that is, nonhuman animals, the ordinary sense of that word) lack this capacity for free moral judgment. They are not beings of a kind capable of exercising or responding to moral claims. Animals therefore have no rights, and they can have none. This is the core of the argument about the alleged rights of animals. The holders of rights must have the capacity to comprehend rules of duty, governing all including themselves. In applying such rules, the holders of rights must recognize possible conflicts between what is in their own interest and what is just. Only in a community of beings capable of self-restricting moral judgments can the concept of a right be correctly invoked.

Humans have such moral capacities. They are in this sense self-legislative, are members of communities governed by moral rules, and do possess rights. Animals do not have such moral capacities. They are not morally self-legislative, cannot possibly be members of a truly moral community, and therefore cannot possess rights. In conducting research on animal subjects, therefore, we do not violate their rights, because they have none to violate.

To animate life, even in its simplest forms, we give a certain natural reverence. But the possession of rights presupposes a moral status not attained by the vast majority of living things. We must not infer, therefore, that a live being has, simply in being alive, a "right" to its life. The assertion that all animals, only because they are alive and have interests, also possess the "right to life"[10] is an abuse of that phrase, and wholly without warrant.

It does not follow from this, however, that we are morally free to do anything we please to animals. Certainly not. In our dealings with animals, as in our dealings with other human beings, we have obligations that do not arise from claims against us based on rights. Rights entail obligations, but many of the things one ought to do are in no way tied to another's entitlement. Rights and obligations are not reciprocals of one another, and it is a serious mistake to suppose that they are.

Illustrations are helpful. Obligations may arise from internal commitments made: physicians have obligations to their patients not grounded merely in their patients' rights. Teachers have such obligations to their students, shepherds to their dogs, and cowboys to their horses. Obligations may arise from differences of status: adults owe special care when playing with young children, and children owe special care when playing with young pets. Obligations may arise from special relationships: the payment of my son's college tuition is something to which he may have no right, although it may be my obligation to bear the burden if I reasonably can; my dog has no right to daily exercise and veterinary care, but I do have the obligation to provide these things for her. Obligations may arise from particular acts or circumstances: one may be obliged to another for a special kindness done, or obliged to put an animal out of its misery in view of its condition—although neither the human benefactor nor the dying animal may have had a claim of right.

Plainly, the grounds of our obligations to humans and to animals are manifold and cannot be formulated simply. Some hold that there is a general obligation to do no gratuitous harm to sentient creatures (the principle of nonmaleficence); some hold that there is a general obligation to do good to sentient creatures when that is reasonably within one's power (the principle of beneficence). In our dealings with animals, few will deny that we are at least obliged to act humanely—that is, to treat them with the decency and concern that we owe, as sensitive human beings, to other sentient creatures. To treat animals humanely, however, is not to treat them as humans or as the holders of rights.

A common objection, which deserves a response, may be paraphrased as follows:

> If having rights requires being able to make moral claims, to grasp and apply moral laws, then many humans—the brain-damaged, the comatose, the senile—who plainly lack those capacities must be without rights. But that is absurd. This proves [the critic concludes] that rights do not depend on the presence of moral capacities.[11]

This objection fails; it mistakenly treats an essential feature of humanity as though it were a screen for sorting humans. The capacity for moral judgment that distinguishes humans from animals is not a test to be administered to human beings one by one. Persons who are unable, because of some disability, to perform the full moral functions natural to human beings are certainly not for that reason ejected from the moral community. The issue is one of kind. Humans are of such a kind that they may be the subject of experiments only with their voluntary consent. The choices they make freely must be respected. Animals are of such a kind that it is impossible for them, in principle, to give or withhold voluntary consent or to make a moral choice. What humans retain when disabled, animals have never had.

A second objection, also often made, may be paraphrased as follows:

Capacities will not succeed in distinguishing humans from the other animals. Animals also reason; animals also communicate with one another; animals also care passionately for their young; animals also exhibit desires and preferences. Features of moral relevance—rationality, interdependence, and love—are not exhibited uniquely by human beings. Therefore [this critic concludes], there can be no solid moral distinction between humans and other animals.[12]

This criticism misses the central point. It is not the ability to communicate or to reason, or dependence on one another, or care for the young, or the exhibition of preference, or any such behavior that marks the critical divide. Analogies between human families and those of monkeys, or between human communities and those of wolves, and the like, are entirely beside the point. Patterns of conduct are not at issue. Animals do indeed exhibit remarkable behavior at times. Conditioning, fear, instinct, and intelligence all contribute to species survival. Membership in a community of moral agents nevertheless remains impossible for them. Actors subject to moral judgment must be capable of grasping the generality of an ethical premise in a practical syllogism. Humans act immorally often enough, but only they—never wolves or monkeys—can discern, by applying some moral rule to the facts of a case, that a given act ought or ought not to be performed. The moral restraints imposed by humans on themselves are thus highly abstract and are often in conflict with the self-interest of the agent. Communal behavior among animals, even when most intelligent and most endearing, does not approach autonomous morality in this fundamental sense.

Genuinely moral acts have an internal as well as an external dimension. Thus, in law, an act can be criminal only when the guilty deed, the *actus reus*, is done with a guilty mind, *mens rea*. No animal can ever commit a crime; bringing animals to criminal trial is the mark of primitive ignorance. The claims of moral right are similarly inapplicable to them. Does a lion have a right to eat a baby zebra? Does a baby zebra have a right not to be eaten? Such questions, mistakenly invoking the concept of right where it does not belong, do not make good sense. Those who condemn biomedical research because it violates "animal rights" commit the same blunder.

IN DEFENSE OF "SPECIESISM"

Abandoning reliance on animal rights, some critics resort instead to animal sentience or their feelings of pain and distress. We ought to desist from the imposition of pain insofar as we can. Since all or nearly all experimentation on animals does impose pain and could be readily forgone, say these critics, it should be stopped. The ends sought may be worthy, but those ends do not justify imposing agonies on humans, and by animals the agonies are felt no less. The laboratory use of animals (these critics conclude) must therefore be ended—or at least very sharply curtailed.

Argument of this variety is essentially utilitarian, often expressly so;[13] it is based on the calculation of the net product, in pains and pleasures, resulting from experiments on animals. Jeremy Bentham, comparing horses and dogs with other sentient creatures, is thus commonly quoted: "The question is not, Can they reason? nor Can they talk? but, Can they suffer?"

Animals certainly can suffer and surely ought not to be made to suffer needlessly. But in inferring, from these uncontroversial premises, that biomedical research causing animal distress is largely (or wholly) wrong, the critic commits two serious errors.

The first error is the assumption, often explicitly defended, that all sentient animals have equal moral standing. Between a dog and a human being, according to this view, there is no moral difference; hence the pains suffered by dogs must be weighed no differently from the pains suffered by humans. To deny such equality, according to this critic, is to give unjust preference to one species over another; it is "speciesism." The most influential statement of this moral equality of species was made by Peter Singer:

> The racist violates the principle of equality by giving greater weight to the interests of members of his own race when there is a clash between their interests and the interests of those of another race. The sexist violates the principle of equality by favoring the interest of his own sex. Similarly the speciesist allows the interests of his own species to override the greater interests of members of other species. The pattern is identical in each case.[13]

This argument is worse than unsound; it is atrocious. It draws an offensive moral conclusion from a deliberately devised verbal parallelism that is utterly specious. Racism has no rational ground whatever. Differing degrees of respect or concern for humans for no other reason than that they are members of different races is an injustice totally without foundation in the nature of the races themselves. Racists, even if acting on the basis of mistaken factual beliefs, do grave moral wrong precisely because there is no morally relevant distinction among the races. The supposition of such differences has led to outright horror. The same is true of the sexes, neither sex being entitled by right to greater respect or concern than the other. No dispute here.

Between species of animate life, however—between (for example) humans on the one hand and cats or rats on the other—the morally relevant differences are enormous, and almost universally appreciated. Humans engage in moral reflection; humans are morally autonomous; humans are members of moral communities, recognizing just claims against their own interest. Human beings do have rights; theirs is a moral status very different from that of cats or rats.

I am a speciesist. Speciesism is not merely plausible; it is essential for right conduct, because those who will not make the morally relevant distinctions among species are almost certain, in consequence, to misapprehend their true obligations. The analogy between speciesism and racism is insidious. Every sensitive moral judgment requires that the differing natures of the beings to whom obligations are owed be considered. If all forms of animate life—or vertebrate animal life?—must be treated equally, and if therefore in evaluating a research program the pains of a rodent count equally with the pains of a human, we are forced to conclude (1) that neither humans nor rodents possess rights, or (2) that rodents possess all the rights that humans possess. Both alternatives are absurd. Yet one or the other must be swallowed if the moral equality of all species is to be defended.

Humans owe to other humans a degree of moral regard that cannot be owed to animals. Some humans take on the obligation to support and heal others, both humans and animals, as a principal duty in their lives; the fulfillment of that duty may require the sacrifice of many animals. If biomedical investigators abandon the effective pursuit of their professional objectives because they are convinced that they may not do to animals what the service of humans requires, they will fail, objectively, to do their duty. Refusing to recognize the moral differences among species is a sure path to calamity. (The largest animal rights group in the country is People for the Ethical Treatment of Animals; its codirector, Ingrid Newkirk, calls research using animal subjects "fascism" and "supremacism." "Animal liberationists do not separate out the *human* animal," she says, "so there is no rational basis for saying that a human being has special rights. A rat is a pig is a dog is a boy. They're all mammals."[14])

Those who claim to base their objection to the use of animals in biomedical research on their reckoning of the net pleasures and pains produced make a second error, equally grave. Even if it were true—as it is surely not—that the pains of all animate beings must be counted equally, a cogent utilitarian calculation requires that we weigh all the consequences of the use, and of the nonuse, of animals in laboratory research. Critics relying (however mistakenly) on animal rights may claim to ignore the beneficial results of such research, rights being trump cards to which interest and advantage must give way. But an argument that is explicitly framed in terms of interest and benefit for all over the long run must attend also to the disadvantageous consequences of not using animals in research, and to all the achievements attained and attainable only through their use. The sum of the benefits of their use is utterly beyond quantification. The elimination of horrible disease, the increase of longevity, the avoidance of great pain, the saving of lives, and the improvement of the quality of lives (for humans and for animals) achieved through research using animals are so incalculably great that the argument of these critics, systematically pursued, establishes not their conclusion but its reverse: to refrain from using animals in biomedical research is, on utilitarian grounds, morally wrong.

When balancing the pleasures and pains resulting from the use of animals in research, we must not fail to place on the scales the terrible pains that would have resulted, would be suffered now, and would long continue had animals not been used. Every disease eliminated, every vaccine developed, every method of pain relief devised, every surgical procedure invented, every prosthetic device implanted—indeed, virtually every modern medical therapy is due, in part or in whole, to experimentation using animals. Nor may we ignore, in the balancing process, the predictable gains in human (and animal) well-being that are probably achievable in the future but that will not be achieved if the decision is made now to desist from such research or to curtail it.

Medical investigators are seldom insensitive to the distress their work may cause animal subjects. Opponents of research using animals are frequently insensitive to the cruelty of the results of the restrictions they would impose.[15] Untold numbers of human beings—real persons, although not now identifiable—would suffer grievously as the consequence of this well-meaning but shortsighted tenderness. If the morally relevant differences between humans and animals are borne in mind, and if all relevant considerations are weighed, the calculation of long-term consequences must give overwhelming support for biomedical research using animals.

CONCLUDING REMARKS

Substitution

The humane treatment of animals requires that we desist from experimenting on them if we can accomplish the same result using alternative methods—in vitro experimentation, computer simulation, or others. Critics of some experiments using animals rightly make this point.

It would be a serious error to suppose, however, that alternative techniques could soon be used in most research now using live animal subjects. No other methods now on the horizon—or perhaps ever to be available—can fully replace the testing of a drug, a procedure, or a vaccine in live organisms. The flood of new medical possibilities being opened by the successes of recombinant DNA technology will turn to a trickle if testing on live animals is forbidden. When initial trials entail great risks, there may be no forward movement whatever without the use of live animal subjects. In seeking knowledge that may prove critical in later clinical applications, the unavailability of animals for inquiry may spell complete stymie. In the United States, federal regulations require the testing of new drugs and other products on animals, for efficacy and safety, before human beings are exposed to them.[16,17] We would not want it otherwise.

Every advance in medicine—every new drug, new operation, new therapy of any kind—must sooner or later be tried on a living being for the first time. That trial, controlled or uncontrolled, will be an experiment. The subject of that experiment, if it is not an animal, will be a human being. Prohibiting the use of live animals in biomedical research, therefore, or sharply restricting it, must result either in the blockage of much valuable research or in the replacement of animal subjects with human subjects. These are the consequences—unacceptable to most reasonable persons—of not using animals in research.

Reduction

Should we not at least reduce the use of animals in biomedical research? No, we should increase it, to avoid when feasible the use of humans as experimental subjects. Medical investigations putting human subjects at some risk are numerous and greatly varied. The risks run in such experiments are usually unavoidable, and (thanks to earlier experiments on animals) most such risks are minimal or moderate. But some experimental risks are substantial.

When an experimental protocol that entails substantial risk to humans comes before an institutional review board, what response is appropriate? The investigation, we may suppose, is promising and deserves support, so long as its human subjects are protected against unnecessary dangers. May not the investigators be fairly asked, Have you done all that you can to eliminate risk to humans by the extensive testing of that drug, that procedure, or that device on animals? To achieve maximal safety for humans we are right to require thorough experimentation on animal subjects before humans are involved.

Opportunities to increase human safety in this way are commonly missed; trials in which risks may be shifted from humans to animals are often not devised, sometimes not even considered. Why? For the investigator, the use of animals as subjects is often more expensive, in money and time, than the use of human subjects. Access to suitable human subjects is often quick and convenient, whereas access to appropriate animal subjects may be awkward, costly, and burdened with red tape. Physician-investigators have often had more experience working with human beings and know precisely where the needed pool of subjects is to be found and how they may be enlisted. Animals, and the procedures for their use, are often less familiar to these investigators. Moreover, the use of animals in place of humans is now more likely to be the target of zealous protests from without. The upshot is that humans are sometimes subjected to risks that animals could have borne, and should have borne, in their place. To maximize the protection of human subjects, I conclude, the wide and imaginative use of live animal subjects should be encouraged rather than discouraged. This enlargement in the use of animals is our obligation.

Consistency

Finally, inconsistency between the profession and the practice of many who oppose research using animals deserves comment. This frankly ad hominem observation aims chiefly to show that a coherent position rejecting the use of animals in medical research imposes costs so high as to be intolerable even to the critics themselves.

One cannot coherently object to the killing of animals in biomedical investigations while continuing to eat them. Anesthetics and thoughtful animal husbandry render the level of actual animal distress in the laboratory generally lower than that in the abattoir. So long as death and discomfort do not substantially differ in the two contexts, the consistent objector must not only refrain from all eating of animals but also protest as vehemently against others eating them as against others experimenting on them. No less vigorously must the critic object to the wearing of animal hides in coats and shoes, to employment in any industrial enterprise that uses animal parts, and to any commercial development that will cause death or distress to animals.

Killing animals to meet human needs for food, clothing, and shelter is judged entirely reasonable by most persons. The ubiquity of these uses and the virtual universality of moral support for them confront the opponent of research using animals with an inescapable difficulty. How can the many common uses of animals be judged morally worthy, while their use in scientific investigation is judged unworthy?

The number of animals used in research is but the tiniest fraction of the total used to satisfy assorted human appetites. That these appetites, often base and satisfiable in other ways, morally justify the far larger consumption of animals, whereas the quest for improved human health and understanding cannot justify the far smaller, is wholly implausible. Aside from the numbers of animals involved, the distinction in terms of worthiness of use, drawn with regard to any single animal, is not defensible. A given sheep is surely not more justifiably used to put lamb chops on the supermarket counter than to serve in testing a new contraceptive or a new prosthetic device. The needless killing of animals is wrong; if the common killing of them for our food or convenience is right, the less common but more humane uses of animals in the service of medical science are certainly not less right.

Scrupulous vegetarianism, in matters of food, clothing, shelter, commerce, and recreation, and in all other spheres, is the only fully coherent position the critic may adopt. At great human cost, the lives of fish and crustaceans must also be protected, with equal vigor, if speciesism has been forsworn. A very few consistent critics adopt this position. It is the reductio ad absurdum of the rejection of moral distinctions between animals and human beings.

Opposition to the use of animals in research is based on arguments of two different kinds—those relying on the alleged rights of animals and those relying on the consequences for animals. I have argued that arguments of both kinds must fail. We surely do have obligations to animals, but they have, and can have, no rights against us on which research can infringe. In calculating the consequences of animal research, we must weigh all the long-term benefits of the results achieved—to animals and to humans—and in that calculation we must not assume the moral equality of all animate species.

NOTES

Reprinted by permission from Carl Cohen, "The Case for the Use of Animals in Biomedical Research," *New England Journal of Medicine* 315 (1986): 865–69.

1. T. Regan, *The Case for Animal Rights* (Berkeley: University of California Press, 1983).

2. P. Singer, *Animal Liberation* (New York: Avon, 1977).

3. Augustine, *Confessions*, bk. 7 (A.D. 397) (New York: Pocket Books, 1957), 104–26.

4. Thomas Aquinas, *Summa Theologica* (A.D. 1273), Philosophic texts (New York: Oxford University Press, 1960), 353–66.

5. G. W. F. Hegel, *Philosophy of Right* (1821; London: Oxford University Press, 1952), 105–10.

6. F. H. Bradley, Why should I be moral? in *Ethical Theories,* ed. A. I. Melden (1876; New York: Prentice-Hall, 1950), 345–59.

7. G. H. Mead, The genesis of the self and social control, in *Selected Writings,* ed. A. J. Reck (1925; Indianapolis: Bobbs-Merrill, 1964), 264–93.

8. H. A. Prichard, Does moral philosophy rest on a mistake? in *Readings in Ethical Theory*, ed. Sellars W. Hospers (1912; New York: Appleton-Century-Crofts, 1952), 149–63.

9. I. Kant, *Fundamental Principles of the Metaphysic of Morals* (1785; New York: Liberal Arts Press, 1949).

10. B. E. Rollin, *Animal Rights and Human Morality* (New York: Prometheus), 1981.

11. C. Hoff, Immoral and moral uses of animals, *N Engl J Med* 302 (1980): 115–18.

12. D. Jamieson, Killing persons and other beings, in *Ethics and Animals*, ed. H. B. Miller and W. H. Williams (Clifton, N.J.: Humana Press, 1983), 135–46.

13. P. Singer, Ten years of animal liberation, *New York Review of Books* 31 (1985): 46–52.

14. J. Bentham, *Introduction to the Principles of Morals and Legislation* (London: Athlone, 1970).

15. K. McCabe, Who will live, who will die? *Washingtonian Magazine*, August 1986, 115.

16. U.S. Code of Federal Regulations, title 21, sec. 505(i), Food, drug and cosmetic regulations.

17. U.S. Code of Federal Regulations, title 16, sec. 1500.40–2, Consumer product regulations.

20

Artificial Lives: Philosophical Dimensions of Farm Animal Biotechnology

Allan Holland

The category of domesticated animals, animals whose natures have been genetically modified by humans, raises a quite special set of philosophical questions, so far insufficiently attended to by philosophers, which are compounded when the question of their further modification by the new techniques of genetic engineering are raised. (In this context, "genetic engineering" is defined as the genetic modification of an animal by the direct manipulation of its DNA, sometimes also called "transgenic" manipulation.) The resulting animals, whether modified by "conventional" or "transgenic" means, have been subject to what Darwin called "artificial selection," which is what justifies the reference to "artificial lives" in the title of this chapter. The contrast here is with wild animals, whose natures are "as nature made them." Admittedly, not all farmed animals are domesticated, deer and salmon being obvious exceptions, but the categories overlap sufficiently to justify the focus on domesticated animals as such. There are distinct questions to be raised concerning both their philosophical status and ethical treatment, but before proceeding to engage with some of the harder questions, two preliminary perspectives will be outlined.

THE CALL OF THE WILD?

One approach to the treatment of domesticated animals is to compare their status with that of animals in the wild. In the light of such a comparison, concern for animals held in some form of "captivity," whether in zoos, on farms, in laboratories or as pets, can strike people as excessive. After all, life in the wild is tough.

As J. S. Mill famously observed: "In sober truth, nearly all the things which men are hanged or imprisoned for doing to one another are nature's every-day performances" (Mill, 1968). And Stephen Bostock, in his recent book on zoos, is drawn to remark of zoo animals in particular that in view of the protection which they enjoy from violent death and disease they have "by no means a bad bargain" (Bostock, 1993). If that is true of zoo animals, most of whom are wild, how much more true is it likely to be of farm animals, who are bred for captivity? (Or at least bred *in*, and therefore, even if unconsciously, selected for, captivity.) There do not appear to have been any serious attempts at an actuarial estimate, but a general impression is that, in comparison with domesticated animals, wildlife breeds more but lives less—even if we include in the calculations domestic animals bred for early slaughter.

Contrary, however, to this initial train of thought, the notion that the wretched lives lived by many wild animals may serve to justify some of the more appalling conditions experienced under domestication is an implausible one. Having just witnessed a young rabbit being mauled to death by the fangs of a weasel, does this really justify our going home and visiting some serious injury or even death on a domesticated animal, on the grounds that this would be no worse than it might experience in the wild? Above all, this notion neglects to take into account that in the latter case we have a choice, and that the domesticated animal exists as a result of our choice. Of course, every animal comes into being as the result of some other animal's "choice," humans most particularly, but domesticated animals are brought into existence specifically to serve human purposes, and there is as yet no argument as to why their lives should be judged by the yardstick of those who are not. Furthermore they live their lives on human terms, not on theirs, which is why their circumstances differ so markedly, for example, even from those of "tame" robins or urban foxes.

SYMBIOSIS?

The alleged human "dominion" over nature, however, can seem a mite exaggerated. How shamelessly do groundsel and shepherd's purse, cats and robins exploit the newly tilled soil which the poor gardener has laboured to render into the proverbial "fine tilth," for living space, waste disposal and food, respectively! And how voraciously do the slugs and cabbage whites, mice and blackbirds devour the fruits of the gardener's labours! It is certain that the bee devouring the nectar gives no thought to the good of the plant but seeks only its own good; but Darwin himself observes how it is a "great gain to the plant to be thus robbed" (Darwin, 1901).

Who is in control here, the flower or the bee? One might raise the same question over the relation between humans and their domesticated animals. Indeed, it might be argued to be parochial to single out artificial selection, selection by "man," from natural selection, which is effectively selection by everything else, so as to construe the former as singularly dominating in its character. In the biotic community at large, there are innumerable forms and agents of selection ranging from humans and hawks through to insects and viruses. As Darwin again observed: "it is certain that insects regulate in many cases the range and even the existence of the higher animals, whilst living under their natural conditions" (Darwin, 1899). In evolutionary terms also, various kinds of domestic animal may be thought of as comparatively successful, certainly so far as numbers are concerned; one might think, for example, of domestic poultry in this connection. How far then should we be concerned about their "plight"?

But here again, this symbiotic perspective on the relation between humans and domesticated animals, while it wears considerable promise as an account of how that relation might be developed, and what we might ideally strive to make it, is considerably lacking as an account of what it actually is. Although we know that there are various pitfalls awaiting the attempt to draw up criteria for evolutionary success, enough is probably known to be able to say that numbers, in themselves, are a poor indicator. Moreover, the argument singularly neglects to register that the domestic animals' lives depend crucially on ephemeral human intentions and purposes rather than on the relative constancy of natural processes. In many cases the very form of life itself, and not just the life of any individual animal, depends upon a culturally contingent prop which humans provide, and in the absence of which it would collapse. It is true of course that natural processes are not entirely reliable and that natural catastrophes occur. But domesticated animals are subject to these things too. The relationship, therefore, between humans and their domesticated animals is an unequal one in a quite special way.

It is worth remarking that both perspectives so far examined involve, in different ways, an appeal to what is natural to establish legitimacy. The first proposes that we should look to the wild and judge the lives of domesticated animals by what we find there; and if we do, we shall conclude that farm animals by and large have no bad life. The second invites us to construe the institution of domestication as itself a natural phenomenon, with the suggestion, again, that this somehow insulates it from criticism. But although each perspective is useful, neither seems to offer a fully satisfactory account of the nature of our responsibilities regarding domesticated animals in general and farm animals in particular. It is important to delve deeper and consider, among others, approaches which cite the natural, in order to criticize, rather than defend, the status of domesticated animals.

THE UTILITY PRINCIPLE

If the case of farm animals in particular is examined, it is undoubtedly true that domestication for agricultural purposes puts considerable strains on the bodies and lives of many of the animals concerned, stemming both from the environments in which they are kept and the genetic endowment which they are given. There are, still, farm animals deprived of one or more of the five freedoms identified thirty years ago by the Brambell Committee (1965) as minimum requirements for an adequate life, namely the freedom to turn around, groom, get up, lie down and stretch. There are many more animals who are restricted in one way or another as regards their physical constitution, behavioural repertoire or psychosocial needs. Focusing, for brevity, on physical constitution only, Webster has suggested that a Friesian dairy cow yielding 35 litres of milk per day has a work load equivalent to that of a cyclist doing the Tour de France (Webster, 1990). Another example is the extent of leg problems in broiler chickens, highlighted recently by the UK Government's Farm Animal Welfare Council, which "ranged from slight abnormalities in gait, to the worst cases in which birds were only able to move with great difficulty, using their wings to balance" (RSPCA, 1992). Among the causes of this condition is mentioned "the extremely fast growth rate of the modern bird."

There is no doubt, too, that transgenic manipulation is capable of adding considerably to the strains which domesticated animals undergo. In this connection, forms of genetic engineering designed to increase rates of growth—which is usually what is sought, rather than increase in size (Seidel, 1986)—are commonly cited. Another recent example, should it prove successful, is work at the Rowett Institute in Aberdeen, which is on the way to identifying the gene which is said to control the timing of ovulation in sheep (*The Times,* July 12, 1993). If the gene could be deleted, then ewes might ovulate and be capable therefore of lambing all the year round. How should such developments be judged?

There are those who are content at this point to reach for their utilitarian "account book." A utilitarian is one who believes that matters of right and wrong are decided by weighing the balance of satisfactions, interests or benefits of all concerned—and in this context those concerned might include non-human animals. It is a position espoused by many, including some philosophers, and is commonly appealed to in matters involving human-animal relations. On this approach, even if the impact of genetic engineering on animal welfare overall is found to be a negative one, the technology might still be justified by the benefits which accrue to humans. This is essentially because the satisfactions and benefits, dissatisfactions and costs, are thought all to be commensurable and capable, therefore, of being balanced off and traded against each other. The most desirable course of action is the one which secures the greatest overall benefit. Since the issues raised by this approach are far too large to address fully here, suffice it to say that the position seems profoundly unsatisfactory at the theoretical level and counter-intuitive in its results; two arguments are offered in support of this claim.

One is that benefits and costs should never be allowed to count uncritically in the weighing and balancing. Both the source and the nature of a benefit can make a difference to whether the benefit is an acceptable one. To take the simplest form of utilitarian coinage, pleasure, it is clear, for example, that pleasures may be wrongfully or shamefully gained. The receipt of a large sum of money would be regarded as a benefit by most people; but many would find it unacceptable, if it were the proceeds of a robbery. In short, the balance of advantages cannot decide acceptability, since it presupposes it.

The second argument is that calculation of cost and benefit cannot ignore the question of the relation between the two. A benefit which is obtained by paying a cost cannot be regarded in the same light as one which is not. Certainly, to take a case from the sphere of human relations, we appear to think the advantages of motorized transport, more especially the freedoms which this represents, outweigh the loss of life which is thereby caused. We would view the matter very differently, however, if the advantages of motorized transport had to be secured in advance by the sacrifice of a certain number of lives, rather in the way that, according to Greek legend, Agamemnon was obliged to sacrifice his daughter Iphigeneia to secure a fair wind for his fleet in the assault on Troy.

THE KANTIAN PRINCIPLE

It is this kind of case which points people away from the simple calculation, balancing and trading off of pleasures and pains, and more in the direction of a "Kantian" system of morality which bids us never to treat one another as means only, but as "ends" (Kant, 1948). The central proscription here is against using the beneficial circumstances of one subject to compensate for, or justify, the adverse circumstances of another. And if this principle applies in the case of human relations with one another, it is hard to see why an analogous principle should not apply in the case of human-animal relations also. Although the language of "rights" will be avoided, which tends to draw down difficulties of a distracting kind, this principle is what chiefly lies behind the appeal to animal rights.

It is true that Kant himself would not have approved of this move because he did not think that animals possessed "rational natures"; but it is not necessary to follow him in this belief. Indeed it will be assumed for the purposes of this discussion that farm animals live not just biological lives but also biographical lives—in other words that they are not just alive, like trees, but that they are sentient, experiencing subjects of a life (Rachels, 1990). Although this will not necessarily entail their having rationality in the sense that Kant has in mind—which involves essentially the capacity to follow a moral rule—it would seem to entail their having degrees of rationality "after their kind." Moreover it is not obvious that rationality of *any* kind is the *only* reason there may be for treating another with respect, for treating that other as an "end." Another, older, view sees a common significance, although not the same significance, in every creature's organized form of life. This is the view which may be traced back to Aristotle, that every creature is possessed of a *telos* (which literally translated in fact means "end" or "goal") which constitutes the fulfilment of its particular form of life. An extension of Kant's principle would bid us respect a creature's *telos*. According to this view, other things being equal, one ought not willingly and avoidably to subject any creature to conditions which deny it the ability to express its *telos*. This is the view of Michael Fox among others who holds that "transgenic manipulation is wrong because it violates the genetic integrity or *telos* of organisms or species" (Fox, 1990). It should be noticed that there are two elements in this claim; one is the enunciation of a principle of respect for *telos*; the other is the identification of *telos* with *genetic* integrity. Both need justification.

So far as the first is concerned, it has to be recognized that in extending Kant's principle beyond the human domain we lose the distinctive *justification* which he provides for the original injunction to treat rational natures as ends. However, there is no shortage of attempts to supply alternative bases for the extended principle, and it is assumed that some alternative justification may be found. Regan suggests a duty of respect for any creature which has a good of its own (Regan, 1979). Others cite the complexity and improbability of the nature of (non-human) living beings and the fact that they are "centers of relations independent of human will" (Colwell, 1989). It is also worth mentioning that any support for the maintenance of biodiversity is likely to involve, in some form, endorsement of a principle of respect for forms of life.

So far as the second element is concerned, given that *telos* was identified long before the formulation of any theory of genetics, it would seem that the relation between genetic integrity and *telos* can only be a contingent one. The implication of accepting this point would be that genetic manipulation does not *in itself* constitute a violation of *telos*; whether such a violation has occurred would depend upon the *result* of such manipulation.

TELOS

Many practicing scientists, however, do not take kindly to the concept of *telos*, regarding such talk as "mystical," or at best the conceptual residue of a scientific outlook which has been superseded. Rollin, perhaps, does not help matters when he lays claim to having been responsible for reintroducing the original Aristotelian concept (Rollin 1986a—more colourfully, he speaks of having roused it from "dogmatic slumber"). In fact, he would seem to be mistaken in equating the concept which he introduces with the Aristotelian one. *Telos*, in Aristotelian science, is a substantial explanatory factor: it is with reference to the *telos* of a creature, the full flourishing stage of its existence ("that for the sake of which it came to be"), that one explains the nature and existence of all of its major characteristics (in effect, all except those which are intrinsic to the matter from which it is made). Furthermore, these so-called final causes operate by virtue of their being the best way for things to be: "and there are the things which stand to the rest as their end and good; for what the other things are for tends to be best and their end" (Aristotle, 1979). Thus the reason why carnivores have sharp front teeth lies in their need to bite into their prey in order to make a meal, not absurdly, in Aristotle's view—in some mechanical cause such as the fact that their gums at the front are narrow and can only support a narrow protuberance. Rollin's concept of *telos*, explained as comprising the "interests and nature of an animal which it possesses as a member of a species" (Rollin 1986b; cf. Rollin 1989), retains only a part of the older concept. He has jettisoned the causal role which makes it an embedded part of an older scientific world view; which is why indeed he is able, with some justice, to protest its innocence. (Of course, it is allowed that teleological explanations still work in a sense, but this is thought to be because Darwinian theory explains how it is as if a substantial teleological factor were operating when in fact it is not.)

Rollin claims, in fact, that the concept of *telos* is as well founded as the concept of species, and is not something at which practising scientists need flinch. In its reduced sense, this may be true; but there is still some ground to be fought over, since there are many differing accounts of species which may be given, and which are glossed over by Rollin when he claims that the current divisions of biological science all presuppose "the existence of kinds of animals" (Rollin, 1986a). We are touching here on the question with which philosophers are apt to perplex themselves (and others): are species real? The question is whether even the members of *natural* species have a species-specific nature, and thus, in general, whether one can even speak of the *natures* of animals, which is what the reduced concept of a *telos* seems to amount to. A number of plausible modern accounts of the species concept suggest not. On the so-called cladistic concept of species, for example, according to which a species is simply the lineage between two speciation events, what binds together the members of a given species is not some common "nature" but genealogical relations (Ridley, 1989). (Similar implications hold for Mayr's (1987) concept of a species as an interbreeding population, and Ghiselin's (1987) characterization of a species as an historical individual.) Because of the tenacity of the gene and the apparent tendency of evolutionary selection pressures to focus populations of organisms around norms, individual members of these populations which are conveniently called "species" will tend to exhibit very similar "natures"; but this remains a contingent fact. They are not members of the same species because they have similar natures. (All this is consistent with the fact that, because historical lineages may be difficult, or sometimes impossible to reconstruct, similarity of nature, physiological or genetic, may be used as an operational criterion of species identity.)

The idea that every animal has a species-specific nature is therefore somewhat problematic. But what can be said is that each animal has an *individual* nature, namely the physical constitution, behavioural repertoire and psychosocial capacities which it has inherited from its parents. Moreover, and importantly, this is a notion which applies equally in the case of wild and domesticated animals. Thus, if it is important that an animal should not needlessly be denied the means to express its nature, this is as true for the domesticated animal as it is for the wild. Physical, behavioural and psychological traits which have been selected for by humans are as real, and the implications of denying their expression as serious, as they are in the case of those which an animal has by nature. Darwin's researches on the question of the similarities and differences between wild and domesticated animals (not, so far as the author is aware, superseded) reveal no more than this: that there is in general a greater uniformity in natures, i.e. consistency from one individual to the next, among wild animals and a greater variability among domesticated animals (Darwin, 1899), which does nothing to show that the domesticated nature is in any sense less "real." Whether one wants to be as assiduous in one's respect for the individual nature of a domesticated animal in *all* its ramifications is another matter. Some existing breeds of farm animal, for example, may be thought less than ideal *from the point of view of the animals themselves,* and this could be grounds for allowing the breed to die out. Such a decision need not indicate a lack of respect for the animal itself but, rather, respect for its potential offspring about whom one may not unreasonably judge that it would be better if they did not exist.

THE NATURAL TELOS PRINCIPLE

When it comes to working out the impact of a respect for *telos* principle on the practice of genetic engineering, it has to be admitted that Fox's treatment of the concept of a *telos* is a little confusing. For he illustrates it with examples of *both* wild *and* domesticated animals, referring to the "'wolfness' of a wolf and the 'pigness' of a pig," but appears to object to the genetic manipulation of an animal's *telos* chiefly on the grounds that it involves a change in "that which is natural" (Fox, 1990). But given that the "pigness" of a pig, unlike the "wolfness" of a wolf, is not entirely natural but the outcome of modifications which humans have introduced, the appeal to what is natural appears to offer no defense whatever against the genetic manipulation of domestic animals, which is clearly contrary to Fox's intention.

There are indeed writers who find this result acceptable, including Colwell, who sees intrinsic value in *natural* organisms, species and systems, but sees "no ethical justification for any bar on genetic alteration of domesticates, by whatever technical means." Although recognizing that humans have a normal duty of care towards any individual animal, wild or domesticated, Colwell reasons that domesticated species have already been "devalued" and that transgenic manipulation is logically no different from conventional genetic modification (Colwell, 1989). Some reason would need to be given, however, for going along with Colwell, and in general for letting one's whole approach to this issue turn upon the cult of the natural. The simple affirmation of the value of the natural does not give it, since it would involve an elementary logical error to infer from the claim that what is natural has intrinsic value and that what is not natural does not. Perhaps there is the feeling that a creator is entitled to do what he or she likes with his or her creation. But we do not, however, take this view with regard to our children, and this despite the fact that the genetic engineers play nowhere near as large a role in connection with their creations as do parents with their children. Moreover, we would normally think that at least similar standards of conduct apply with regard to children, whether or not we have created them. Indeed, we would normally think that the fact we have created them engenders even greater obligations of care and responsibility.

Another response to Colwell, as well as some form of rescue for the position advanced by Fox, is afforded by the Dutch biologist, Verhoog, who calls into question whether it is actually possible to subvert the natural *telos* of an animal: "we misuse the word *telos*," says Verhoog, "when we say that human beings can 'change' the *telos* of an animal or create a new *telos*" (Verhoog, 1992). On this account, a domesticated animal still has a natural *telos*—that part of its nature which may be attributed to its natural endowment—even though it has been over-laid to a greater or lesser extent by the history of domestication undergone by its more recent ancestors. But whatever else has been done by way of supplanting an animal's original natural endowment, it has not been supplied with a new *telos* of its own, but has had a human *telos* imposed upon it from the outside. What this means, in turn, is that the process of domestication itself is open to critical scrutiny, for the invasion of an animal's natural *telos* which it has already perpetrated, and Verhoog's point is valuable for reminding us of this possibility. It is too readily assumed by some of the more ardent defenders of the new technology that if transgenic manipulation can be shown to introduce no significantly new issue when compared with conventional forms of genetic modification, then everything has been done which needs to be done by way of a defence. Such a position is far too complacent. For what is possible is the realization that precisely *because* the "trajectory," as we may call it, of conventional engineering does lead naturally on to transgenic manipulation, practices formerly thought to be acceptable may need to be re-examined. This would not necessarily lead to a rejection of all forms of domestication, or indeed of transgenic manipulation, but it would entail the operation of a much more stringent criterion of acceptability than has ever hitherto been envisaged: only those forms of domestication, or of transgenic manipulation, would be admitted which permitted the animal in question to express its natural *telos*.

It is hard to be persuaded, however, that what is objectionable about genetic engineering in general, if anything is—what is disrespectful of *telos*—is simply its departure from the natural; partly because this amounts to little more than a description of what the technology involves, as opposed to a criticism. What is so bad about departing from the natural? One thinks of species of farm animal—varieties of farm dog, geese, or sheep, let us say—whose history of domestication, at any rate, has in no obvious way seriously impaired their ability to function satisfactorily. And just as, across the whole range of cultivated plants and domesticated animals, there are varieties and breeds which prove less than satisfactory, so, too, there are many whose loss would be mourned as the passing of natural species would be mourned. Moreover, it might also be borne in mind that, as the case of feral animals shows, it is not necessary for domestic animals to recover their original "wild" natures in order for them to be able to establish viable and flourishing populations in the wild.

THE AVOIDABLE HARM PRINCIPLE

Perhaps the most obvious way of failing to respect the *telos* of an animal is through the infliction of avoidable harm. This will include any form of suffering or distress, but may also include various forms of injury, disease or incapacity—castration, for example—even if the undesirable nature of these conditions is not felt or experienced. Is it possible to argue that a genetically engineered animal, living an unsatisfactory life as a result of being genetically engineered, has thereby been harmed? There is a special difficulty here, recently brought to the fore by Attfield (1994): it is the difficulty of giving sense to the idea that it is even *possible* to harm a creature by *bringing it into existence*. For to harm something is surely to bring about some deterioration in its circumstances; and for this to happen it seems as if it must first exist in order for its circumstances then to deteriorate. Citing the work of Parfit, Attfield responds forcefully that one can indeed inflict harm in these circumstances, chiefly because of the concept of "a life that is not worth living," or even of one that is worth not living, which does enable one to give sense to the idea that to inflict such a life upon an animal by bringing it into existence would be to do it harm. (Compare recent court cases in the United States, where charges of "wrongful life" are beginning to be brought.) However, a difficulty does remain, which is hard to adjudicate, over how we should judge the case of lives which are *just* worth living. One view might be that, compared with non-existence, a life that is just worth living is clearly a benefit; so it clearly cannot be a harm. A contrasting view, however, would be that, insofar as a life that is just worth living might have been engineered to be better, then it is harmed insofar as it is not better.

But suppose now that what the genetic engineer proposes to do, when faced by a domestic animal leading an unsatisfactory life, whether this is due to its modified nature or due to the circumstances under which it is kept, e.g., in certain forms of intensive farming, is not to remove the cause of the suffering but to use the techniques of genetic manipulation to build into the animal's constitution an indifference to suffering. There is certainly no wish to overestimate the likelihood of this happening, or underestimate the difficulty of bringing it about (Seidel, 1986). It should be thought of more as an illustrative device, as a token for a type of development which could conceivably be entertained, for example, in response to welfare concerns over intensive farming. (Philosophers have a well known reluctance for dealing with specifics.)

Rollin is quite clear about the answer: "One cannot argue that because it is wrong to violate the various aspects of a certain animal's *telos*, given the *telos*, it is therefore wrong to change the *telos*" (Rollin 1986a). And he adds: "On my view . . . it is only wrong to change a *telos* if the individual animals of that sort are likely to be more unhappy or suffer more after the change than before." Clearly, then, he can find no fault with an application of genetic technology that builds into an animal's constitution an indifference to suffering; and it might seem at least that it would be implausible to describe such an animal as having been harmed. But does this mean that transgenic manipulations of this kind are free of objection?

It is important to notice straightaway that even if Rollin is correct to claim that the conclusion, "it is wrong to change the *telos*" does not follow from the premise, "it is wrong to violate the *telos*," it may still be correct—but on other grounds—that it is wrong to change the *telos*. The claim that a given statement does not follow from another is different from the claim that it is not true. Another point to notice is that, although there may be a number of ways in which an indifference to suffering could be expressed in a phenotype, it is hard to see how they would not involve a reduction of sensitivity or some kind of "reduced capacity." On this basis, it is possible to match Rollin's argument with an opposing one, equally worthy of consideration. Thus, there will presumably be general agreement that we ought to minimize suffering, other things being equal, where we can. But there is a clear and important distinction to be drawn between the experience of suffering and the capacity to suffer. Because of this distinction, one can say, answering Rollin, that from the fact that one ought to reduce suffering it does not follow that one ought, or even that it is not wrong, to reduce the capacity to suffer.

Or does it? Surely, it might be argued, one ought to reduce suffering in any way one can; and reducing the capacity to suffer is a most effective way of achieving this; so it cannot be wrong to reduce the capacity to suffer. However, this argument cannot be right; because an even more effective way of reducing suffering would be to eliminate the sufferer! Clearly, the first premise of this argument is unreliable: it simply is not true that one ought, or even that one is entitled, to reduce suffering in *any* way one can. So far the position seems to be something of a stalemate, with neither side having clearly the better of the argument. To resolve the position another principle needs to be examined.

THE REDUCTION OF CAPACITY PRINCIPLE

A further suggestion is that the very project of bringing about a class of animals with a reduced range of capacities is itself an objectionable one (Attfield, 1994; cf. Holland, 1990). One might claim that simply to bring about an animal with a reduced range of capacities constitutes a failure to respect its *telos*: "it is wrong to cause there to be a lower and more restricted quality of life than would otherwise have been the case," suggests Attfield; and "It is bad that there be crippled lives where there might have been less restricted ones."

There appear to be two problems, however, with this attempt to proscribe forms of genetic engineering which bring into existence animals with reduced capacities, in the same way in which we spoke of a life having been harmed simply by virtue of its having been brought into existence. One is that to talk of a reduced state or condition presupposes a comparison with some state or condition of existence which might otherwise have obtained. Recognizing this point, Attfield suggests that the loss lies in the fact that "where there might have been creatures with a higher or ampler level of capacities . . . there is instead a lower or reduced level." The snag with this suggestion, however, is that if life was not to be brought into existence on these terms, that is, in the reduced form whose legitimacy is in question, it may well be that it would not have been brought into existence at all. In that case, and if the life would not otherwise exist, how can its nature constitute any kind of a "reduction"? Certainly, and unlike the case of harm, its existence cannot be compared adversely with non-existence in this respect. One can agree that a creature can be better off never having existed, so that it is harmed by being brought into existence; but it surely cannot be agreed that it would be leading a less "reduced" form of existence, if it had not been brought into existence at all.

The other problem is the difficulty of identifying what exactly is bad about bringing into existence forms of life which are in some sense "reduced," in particular where they are not supposed to be supplanting any more complex form of life. Where is the harm in that? one might ask, since we are now talking of a situation which is precisely beyond the reach of the concept of harm. Attfield seems to avoid the difficulty somewhat when he appeals to the undesirability of causing there to be "a lower and more restricted quality of life." The difficulty would indeed be eased if talk of differing "qualities" were in order. But what is engineered is simply a "lower" and less complex life; some argument is needed to justify the insinuation that this is of inferior quality. It is the same with Attfield's retention of the epithet "crippled" to describe the new reduced form of life. To be sure, a crippled life is an undesirable one. But what is being examined here is what is undesirable about a case where the genetic engineer has redesigned what counts as a worthwhile life and in which the condition in question is, precisely, no longer "crippling," relative to the environment in which the animal lives its life.

One resolution of the former problem is to observe the distinction between what *would* otherwise have obtained if a particular animal had not been brought into existence, and what *might* otherwise have obtained. Although it may be true with respect to any engineered form of life that it *would* not have existed otherwise, there remains a sense in which it *might* have existed otherwise. And this is because, although genetic engineers may be clever, they cannot work miracles; in particular, they cannot create life *de novo*, but must manipulate something pre-existing. This is usually the zygotic stage of life—when the individual identity is already established—and if not, then the gametes, which still provide a sufficient basis upon which to found a notion of the individual that there might have been if no engineering had occurred or of the individual who might have been engineered otherwise, and therefore of the individual in comparison with which the engineered life constitutes a "reduced" life. (Reliance is placed here on a strong principle of identity, based on the work of the philosopher Saul Kripke [1980], to maintain that the point holds even for second and subsequent generations, i.e., that one can still give sense to the notion of the individuals that they might have been.)

Regarding the latter problem (What is bad about reduction?), it is admittedly difficult to identify anything very tangible—unless, perhaps, it is that there is something peculiarly "life-denying" about such a procedure, a sense in which the very life-forming processes themselves are being suborned. There does seem to be something different involved, even from engaging in procedures which manifestly threaten life, e.g., producing substances which are toxic to certain life-forms, and which might superficially seem to be *more* life-denying. For even to bring hostilities to bear on a particular life-form is, in a way, to recognize its independent standing, and thereby to show a sort of respect. In this way genetically induced reduction also appears somehow different from a *natural* evolutionary progression towards a simpler form of life, which although not common (McShea, 1991), does occur, and which, presumably, would not incur any opprobrium.

KANT REVISITED

But suppose, finally, and with the same reservations as before, that it is proposed to use the techniques of genetic manipulation to build in a resistance to disease or injury, or, at any rate to *some* disease and *some* forms of injury. Once again, there may be a number of different ways in which resistance to disease or injury could be expressed in the phenotype, but it would be less plausible, in this case, to construe them as involving some form of reduced capacity. It would indeed be far more plausible to construe them as life-enhancing.

Rollin's position, again, is clear: "Suppose that one can change a telos, T1, by breeding, natural selection, genetic engineering, or whatever to T2, so that animals which have the new telos are less prone than their predecessors to certain diseases or injuries . . . surely such a change would not in and of itself be wrong" (Rollin, 1986a).

However, it is suggested here that even this "positive" application of the genetic technology turns out to fall foul of something akin to Kant's proscription against treating rational natures, which are ends in themselves, as means—even as means which could be regarded as beneficial to the animal in question. It was on grounds just such as these that Kant condemned suicide, believing that it involved treating rational natures as subservient to the cause of the alleviation of suffering. In a similar vein it would seem that changing an animal's nature for the sake of rendering it less susceptible to disease is less than respectful of that animal's nature, since it would involve subordinating the whole nature to the cause of relief from disease. Essentially, it puts respect for the states of a subject above respect for the subject. We are approaching basic values here, and not everyone might agree with Kant's condemnation of suicide, or at any rate with the grounds of his condemnation. However, once one fully understands what changing an animal's *telos* does, and does not, mean, it is possible to see that such a procedure may not, after all, entirely escape the orbit of Kantian criticism. There would remain the charge that one was using an animal's nature as a means, and failing to respect its ends in the process.

CONCLUSION

By all accounts, our understanding of *how* to bring about various results by means of genetic manipulation is growing apace. It is not clear, however, that our understanding of *what* it is that we are doing is quite keeping up. This reflective exercise has been an attempt to contribute to that understanding. In conclusion, a brief valediction on "burden of proof" is offered. Every position carries its theoretical burden. In this chapter the author has attempted to lay bare for critical inspection the principles and presumptions of the "concerned doubter," conscious of Glover's admonition that "by easy stages, we could move to a world which none of us would choose if we could see it as a whole from the start" (Glover, 1984). But the genetic engineer, too, may be assumed to have his or her principles and presumptions; which raises the question, What is the philosophy, where are the principles, which inform and legitimate the actual and proposed applications of genetic engineering? It would greatly help matters if these too could be brought more clearly into the light and opened up for critical inspection.

NOTE

Reprinted with permission from A. Holland, "Artificial Lives: Philosophical Dimensions of Farm Animal Biotechnology," in T. B. Mepham, ed., *Issues in Agricultural Biotechnology* (Nottingham, U.K.: Nottingham University Press, 1995), 293–304.

REFERENCES

Aristotle. 1979. *The Physics*. Bks. 1–2. Edited by W. C. Charlton. Oxford: Clarendon.
Attfield, R. 1994. Genetic engineering: Can unnatural kinds be harmed? In *Animal Genetics; Of Pigs, Oncomice, and Men*. Edited by P. Wheale and R. McNally. London: Pluto.

Bostock, S.St.C. 1993. *Zoos and Animal Rights*. London: Routledge.

Brambell Committee. 1965. *Brambell Committee Report*. London: HMSO.

Colwell, R. K. 1989. Natural and unnatural history: Biological diversity and genetic engineering. In *Scientists and Their Responsibility*, edited by W. R. Shea and B. Sitter, 1–40. Canton: Watson Publishing International.

Darwin, C. 1901. *The Origin of Species*. London: John Murray.

———. 1899. *The Variation of Animals and Plants under Domestication*. London: John Murray.

Fox, M. 1990. Transgenic animals: Ethical and animal welfare concerns. In *The Bio-Revolution,* edited by P. Wheale and R. McNally, 31–45. London: Pluto.

Ghiselin, M. T. 1987. Species concepts, individuality, and objectivity. *Biology and Philosophy* 2: 127–43.

Glover, J. 1984. *What Sort of People Should There Be?* Harmondsworth, U.K.: Penguin.

Holland, A. 1990. The biotic community: A philosophical critique of genetic engineering. In *The Bio-Revolution,* edited by P. Wheale and R. McNally, 166–74. London: Pluto.

Kant, I. 1948. *Groundwork of the Metaphysic of Morals*. Translated by H. J. Paton. London: Macmillan.

Kripke, S. 1980. *Naming and Necessity*. Oxford: Blackwell.

Mayr, E. 1987. The ontological status of species: Scientific progress and philosophical terminology. *Biology and Philosophy* 6: 145–66.

McShea, D. W. 1991. Complexity and evolution: What everybody knows. *Biology and Philosophy* 6: 303–24.

Mill, J. S. 1968. Nature: From three essays on religion. In *Collected Works of J. S. Mill.* Vol. 10. London: Routledge.

Rachels, J. 1990. *Created from Animals*. Oxford: Oxford University Press.

Regan, T. 1979. Exploring the idea of animal rights. In *Animals' Rights: A Symposium*, edited by D. Paterson and R. D. Ryder, 73–86. London: Centaur.

Ridley, M. 1989. The cladistic solution to the species problem. *Biology and Philosophy* 4: 1–l6.

Rollin, B. E. 1986a. On *Telos* and Genetic Manipulation. *Between the Species* 2: 88–89.

———. 1986b. The Frankenstein thing. In *Genetic Engineering of Animals: An Agricultural Perspective,* edited by J. W. Evans and A. Hollaender, 285–97. New York: Plenum.

———. 1989. The *Unheeded Cry: Animal Consciousness, Animal Pain, and Science.* Oxford: Oxford University Press.

RSPCA. 1992. RSPCA *Science Review*. Horsham, U.K.: RSPCA.

Seidel, G. E. 1986. Characteristics of future agricultural animals. In *Genetic Engineering of Animals: An Agricultural Perspective*, edited by J. W. Evans and A. Hollaender, 299–310. New York: Plenum Press.

Verhoog, H. 1992. The concept of intrinsic value and transgenic animals. *Journal of Agricultural and Environmental Ethics* 2: 147–60.

Webster, J. 1990. Animal welfare and genetic engineering. In *The Bio-Revolution,* edited by P. Wheale and R. McNally, 24–30. London: Pluto.

21

Genetic Engineering as Animal Slavery

Andrew Linzey

This chapter rejects absolutely the idea that animals should be genetically manipulated to provide better meat-machines or laboratory tools. According to the perspective embraced by animal theology, to refashion animals genetically so that they become only means to human ends is morally equivalent to the institutionalization of human slavery. There is, therefore, something morally sinister in the untrammeled development of genetic science which admits of no moral limits save that of the advancement of the controlling species. Nothing less than the dismantling of this science as an institution can satisfy those who advocate moral justice for animals. We reach here the absolute limits of what any reputable creation theology can tolerate.

ANIMAL REVOLUTION

Imagine a place called Manor Farm. The farmer, Mr. Jones, has retired for the night. Quite an ordinary farm of its type with a wide variety of animals: cart-horses, cattle, sheep, hens, doves, pigs, pigeons, dogs, a donkey and a goat. The only difference with this farm is that the animals can talk to one another. And in the dead of night when the farmer is sound asleep, the Old Major, a prize Middle White Boar, addresses a secret meeting in the barn. He begins:

> Now, comrades, what is the nature of this life of ours? Let us face it: our lives are miserable, laborious, and short. We are born, we are given just so much food as will keep the breath in our bodies, and those of us who are capable of it are forced to work to the last atom of our strength; and the very instant that our usefulness has come to an end we are slaughtered with hideous cruelty. No animal in England

325

knows the meaning of happiness or leisure after he is a year old. The life of an animal is misery and slavery: that is the plain truth.

The Old Major continues his oration with increasing passion:

> But is this simply part of the order of nature? Is it because this land of ours is so poor that it cannot afford a decent life to those who dwell upon it? No, comrades, a thousand times no! The soil of England is fertile, its climate is good, it is capable of affording food in abundance to an enormously greater number of animals than now inhabit it . . . Why then do we continue in this miserable condition? Because nearly the whole of the produce of our labour is stolen from us by human beings. There, comrades, is the answer to all our problems. It is summed up in a single word—Man. Remove Man from the scene, and the root cause of hunger and overwork is abolished for ever.
>
> Man is the only creature that consumes without producing. He does not give milk, he does not lay eggs, he is too weak to pull the plough, he cannot run fast enough to catch rabbits. Yet he is lord of all the animals. He sets them to work, he gives back to them a bare minimum that will prevent them from starving and the rest he keeps for himself . . . and yet there is not one of us that owns more than his bare skin.

Finally the oration reaches its crescendo to gladden the animal hearts that hear it:

> What then must we do? Why, work night and day, body and soul, for the overthrow of the human race! That is my message to you comrades: Rebellion! I do not know when that Rebellion will come, it might be in a week or in a hundred years, but I know, as surely as I see this straw beneath my feet, that sooner or later justice will be done. Fix your eyes on that, comrades, throughout the short remainder of your lives! And above all, pass on this message of mine to those who come after you, so that future generations shall carry on the struggle until it is victorious.[1]

By now, of course, you will have guessed the location of Manor Farm—in the *Animal Farm* of George Orwell's imagining. We all know that Orwell intended his book not as a satire on the oppression of pigs and horses but on the oppression of working-class humans by their indolent and unproductive bosses. Nevertheless, it could not have escaped Orwell's attention, as it may not have escaped ours, that there is indeed a similarity between the arguments used (so brilliantly summarized and rebutted by the Old Major) for the justifying of oppression of humans and animals alike.

And if we see this similarity we shall also have grasped something historically quite significant.[2] For the two arguments, or rather assumptions, alluded to in the rousing polemic of the Old Major, namely, that one kind of creature belongs to another and exists to serve the other, have not been confined to the animal sphere. Earlier we have made reference to how Aristotle—typically or untypically—held that animals were made for human use. "If then nature makes nothing without some end in view," he argues, "nothing to no purpose, it must be that nature has made all of them [animals and plants] for the sake of man."[3] Notice that Aristotle is not claiming here that we may sometimes make use of animals when necessity demands it, rather he is asserting that it is in accordance with nature, indeed it is *by nature* that animals are humans' slaves. And if we ask how Aristotle knows that animals are by nature slaves the answer seems to be that if they were not they would "refuse" but since they do not, it obviously follows that it is natural to enslave them. It is crucial to appreciate, however, that this ingenious argument does not stand alone in Aristotle's *The Politics*. When Aristotle comes to considering the right ordering of society, based in turn on the pattern of nature, he uses the example of animal slaves to underline and justify the existence of *human* slaves as well:

> Therefore whenever there is the same wide discrepancy between human beings as there is between soul and body or between man and beast, then those whose condition is such that their function is the use of their bodies and nothing better can be expected of them, those, I say, are slaves by nature.[4]

In a notorious section, Aristotle describes human slaves as "tools" and none other than pieces of property. "A slave is not only his master's slave but belongs to him *tout court*, while the master is his slave's master but does not belong to him."[5] In short: Aristotle does not demur from using the same two arguments, namely that one creature belongs to another and one kind of creature exists to serve the other—to justify both animal and human slavery. As for women, incidentally, they appear to stand somewhere in between, possessing some soul, that is reason, but not as much as men, and having a kind of half status depending upon their rationality.[6]

BELONGING TO AND EXISTING FOR

Aristotle represents what we may call the "belong to and exist for" element within the Western intellectual tradition which Christianity in particular has taken over and developed to the detriment of slaves and women as well as animals. In Aquinas, for example, a few centuries on, we find this same argument repeating itself. "There is no sin in using a thing for the purpose for which it is," he argues.[7] Also with women, though to a lesser degree, we may observe a similar logic. Men, not men and women, are made in the image of God and thus only males possess full rationality. Women are halfway between men and the beasts. "In a secondary sense the image of God is found in man, and not in woman," argues Aquinas, "for man is the beginning and end of woman." Some of us may not fail to see an echo of Aristotle in these words.[8]

Now the simple point I want to make is this: the debate about slavery, human or animal, is not over. Let us take human slavery first. Most of us think that the battle about human slavery was fought and won two hundred or more years ago. If we think that we are simply mistaken. The Anti-Slavery Society exists to combat slavery which continues to exist in many parts of the world, albeit under different guises and in different forms.[9] But if we stay with the issue of slavery and the slave trade of less recent history and go back about two hundred or more years, we will find intelligent, respectable and conscientious Christians supporting almost without question the trade in slaves as inseparable from Christian civilization and human progress. The argument is not an exact repeat from Aristotle, but one that may owe something to his inspiration. Slavery, it is argued, was "progress"—"an integral link in the grand progressive evolution of human society" as William Henry Holcombe, writing in 1860, put it. Moreover, slavery was a natural means of "Christianization of the dark races."[10] Slavery was assumed to be one of the means whereby the natural debased life of the primitives could be civilized. And in this it may not be too far-fetched to see at least a touch of the logic of Aristotle, who defended human slavery on the basis that domestic tame animals were better off "to be ruled by men, because it secures their safety."[11] As David Brion Davis points out: it is often forgotten that Aristotle's famous defence of slavery is embedded within his discussion of human "progress" from the patriarchal village, where "the ox is the poor man's slave to the fully developed polls, where advances in the arts, sciences, and law support that perfect exercise of virtue which is the goal of the city state."[12]

If slavery then was frequently defended on the basis of progress, on which basis, we may ask, did its opposers oppose it? We know that individuals—like Shaftesbury, Wilberforce, Richard Baxter and Thomas Clarkson—opposed the trade in slaves because they regarded it as cruel, dehumanizing, and the source of all kinds of social ills. But one argument we find them using time and again, namely that "man" had no right to absolute dominion over other "men." According to Theodore Weld's influential definition, slavery usurped "the prerogative of God." It constituted "an invasion of the whole man—on his powers, rights, enjoyments, and hopes [which] annihilates his being as a MAN, to make room for the being of a THING."[13] In other words, humans cannot be owned like things or as property. This argument was not peculiar to Weld and Wilberforce and the other reformers in the eighteenth century. Nearly fourteen hundred years before Wilberforce was born, St. Gregory of Nyssa made the first theological attack on the institution of slavery itself. His argument is simple: man is beyond price. "Man belongs to God; he is the property of God"; he cannot therefore be bought or sold. St Gregory was arguably the first to break decisively with that "belong to and exist for" element within the Western tradition.[14]

And yet St Gregory's argument contains a twist in its tail. For Gregory argues that humans cannot have dominion over other humans and therefore possess them, because God gave humans dominion not over other humans but over the world and animals in particular. In other words, humans belong to God and are therefore beyond price but the animals, since they belong to humans, can be bought and sold like slaves![15] One kind of slavery is therefore opposed on the grounds that another is self-evident.

We are now in a position to confront the second kind of slavery I want to consider, namely the slavery of animals. When it comes to animals we find almost without exception the kinds of arguments used to justify human slavery also used to justify the slavery of animals. Animals, like human slaves, are thought to possess little or no reason. Animals, like human slaves, are thought to be "by nature" enslaveable. Animal slavery, like human slavery, is thought to be "progressive," even of "benefit" to the animals concerned. But two arguments are used repeatedly, and we have already discovered them: Animals belong to humans and they exist to serve human interests. Indeed Davis describes what is meant by a slave in a way that makes the similarity abundantly clear:

> The truly striking fact, given historical changes in polity, religion, technology, modes of production, family and kinship structures, and the very meaning of 'property', is the antiquity and almost universal acceptance of the concept of the slave as a human being who is legally owned, used, sold, or otherwise disposed of as if he or she were a domestic animal.[16]

It may be asked: what has all the foregoing to do with the issue of genetic engineering? The answer is this: Genetic engineering represents the concretization of the *absolute* claim that animals belong to us and exist for us. We have always used animals, of course either for food, fashion, or sport. It is not new that we are now using animals for farming, even in especially cruel ways. *What is new is that we are now employing the technological means of absolutely subjugating the nature of animals so that they become totally and completely human property.* "New animals ought to be patentable," argues Roger Schank, Professor of Computer Science and Psychology at Yale University, "for the same reason that new robots ought to be patentable: because they are both products of human ingenuity."[17] When technologists speak, as they do, of creating "super animals,"[18] what they have in mind is not super lives for animals so that they may be better fed, lead more environmentally satisfying lives, or that they may be more humanely slaughtered; rather what they have in mind is how animals can be originated and exist in ways that are completely subordinate to the demands of the human stomach. In other words, animals become like human slaves, namely "things," even more so in a sense since human masters never, to my knowledge, actually consumed human slaves. Biotechnology in animal farming represents the apotheosis of human domination. In one sense it was all inevitable. Failing to have respect for any proper limits in our treatment of animals always carried with it the danger that their very nature would become subject to a similar contempt. Now animals can be not only bought and sold, but *patented*, that is owned, as with human artefacts, like children's toys, cuddly bears, television sets, or other throwaway consumer items, dispensed with as soon as their utility is over.

Again we are not, even at this point, as far away from Aristotle as some might suppose. For in an uncanny, prophetic-like part of his work, Aristotle seems to anticipate a time when human slaves would be automated, being slaves of their own nature, rather than by Nature or the will of their masters. "For suppose," he muses, "that every tool we had could perform its task, either at our bidding or itself perceiving the need . . . then master-craftsmen would have no need of servants nor masters of slaves."[19] Some might argue that biotechnology has transformed this ancient dream into a present nightmare.

PATENTING AND CREATION DOCTRINE

The nightmare intensifies when we look further into the concept of patenting. In 1992, the European Patent Office in Munich actually granted a patent for the onco-mouse, the first European patent on an animal. The controversy over this has not unnaturally focused on the issue of suffering to animals and whether genetically engineered animals (in this case a mouse genetically designed to develop cancer) are likely to lead to an increased level of suffering among laboratory animals. This is an important consideration, but the one which requires even more attention is whether the granting of patents for genetically engineered animals is acceptable in principle. Such a step would, in my opinion, reduce their status to no more than human inventions, and signifies the effective abdication of that special God-given responsibility that all humans have towards the well-being and autonomy of sentient species. Animal patents should not be given; not now, not ever.

We should be clear what the full granting of a patent will mean. A patent confers the legal status of ownership. For the first time—in a European context at least—animals will be classed legally as property without any duty of care; animals will become human artefacts or inventions. If the application for this patent withstands opposition, it will mark the lowest status granted to animals in the history of European ethics. While historically animals have sometimes been thought of as "things"—beings without rights or value—the patenting of animals will mark their legal classification in these terms. I have, I hope, said enough in this and previous chapters to demonstrate that such a use and classification of animal life is not compatible with the Christian doctrine that animals are God's creatures. The Christian doctrine of creation requires us to grasp the fact that human estimations of our own worth and value cannot be the sole grounds for evaluating the worth of other creatures.

Allied to the debate about whether and in what ways humans may use animals is the debate about how far humans are justified in changing the nature of created beings, including their own. It would not be possible to argue that humans must not interfere with any part of nature as it now is, either animate or inanimate. According to traditional Christian belief, creation is "fallen," in a state of "bondage," and therefore—from the Creator's perspective—unfinished. It follows that there is scope for human development of nature, and the notion of dominion in particular presupposes an active role for humans in the care and management of the planet.

That accepted, such empowerment to "better" creation carries with it some strict limits. In the first place, the empowerment presupposed in Genesis is the power to do good in accordance with God's will. It is not a commission to do with creation willy nilly as one wishes. Second, in each and every case it must be shown that such "alteration" of nature is consistent with the designs of the Creator and not simply the pursuance of human avarice and self-advantage.

Here we reach the nub of the matter. Is the created nature of animals "bettered" by genetic engineering? It may be that there is some, albeit limited, case that can be made for such research if it seeks to genuinely enhance the welfare of the individual animals concerned. But in the case of the oncomouse we are dealing not with any bettering of nature either individually or generally but rather with a process that involves the deliberate and artificial creation of disease, suffering and premature death. What is more, the purpose of patenting is so to secure the legal rights to this "invention" that the patent holders concerned may uniquely secure any benefits that may flow from it, not least of all commercial gain. If successful, therefore, the granting of a patent would not only legitimize a morally questionable line of research, it would also financially reward those who carried it out.

While it cannot be claimed that all nature in every instance should be regarded as sacred and inviolable, it is a mistake to suppose that the pursuance of every conceivable human advantage no matter how indirect or trivial justifies each and every intervention in nature. While creation may be disordered, it does not follow that there is a total absence of integrity; maintaining and promoting the good that already exists is an essential task of stewardship. The artificial creation of disease in animals can hardly be claimed to be compatible with the designs of a holy, loving Creator.

Moreover, while it may not strictly speaking be an implication of creation doctrine, opposition to cruelty has been a long-standing feature of traditional moral theology. Whatever else may be said in favour of the oncomouse, it is difficult to see how it can pass any test of moral necessity. To show that something is necessary we have to show that it is essential, unavoidable or, arguably at its very weakest, that some higher good requires it and could not in any way be obtained without it. Even within those sub-traditions of Christendom which have been profoundly unreflective about animal welfare, there is a strong conviction that the infliction of pain can only be justified, if at all, on the most stringent criteria. Animals pose special problems in this regard. They cannot give consent to experimental procedures performed upon them; they cannot merit any infliction of pain, and moreover they cannot intellectually comprehend the meaning of the procedures to which they are subjected. These considerations always tell against the infliction of pain upon innocents, whether they be children, or the mentally handicapped, or animals.

In short: the innocence and defencelessness of animals, far from being considerations which should push animals away from our field of moral concern, are
precisely those which should make us exercise special care and enjoin extraordinary scrupulosity. Whatever views may be held generally about the use of animals
in scientific research, patenting represents the attempt to perpetuate, to institution-
alize, and to commercialize suffering to animals. It seems predicated on the
assumption that animal suffering is justifiable no matter how indirect the benefit
to humans.

The objection may be raised that the oncomouse is, after all, only a mouse.
According to many people, including Christians, a mouse has little value in comparison with a human being.

Proponents of patenting have certainly been clever in choosing a species which
apparently commands limited public sympathy. Because mice interact with
human environments in ways which are disadvantageous to us, they are frequently
classed as a pest species and often killed inhumanely. The limited sympathy this
species can invoke is, however, irrelevant to its moral status and therefore to the
issue of principle. Mice are intelligent, sentient, warm-blooded creatures. There
are no rational grounds on which we can include some sentient species while
excluding others from moral consideration.

It is important to remember that the patent for the oncomouse constitutes a test
case. The oncorabbit, the oncocow, the oncopig, the oncochimp will inevitably
follow. There is no limit to the species or the numbers which may be patented. If
the arguments in favour of patenting succeed in the case of the oncomouse, there
can be no rational grounds why they should not succeed in others. We shall wake
up and find that we have reduced innumerable species of animals to a class of
human inventions, tailor-made to laboratory needs and arguably unprotected in
law. If successful, the patenting of animals will represent the victory of short-term
utilitarianism over the constraints of Christian theology (however unimplemented
these may have been in the past). It may be no exaggeration to say that we stand
on the brink of a wholly new relationship to other creatures: no longer custodians
of our fellow creatures, but rather dealers in new commodities.

THE DISCREDITED THEOLOGY
OF GENETIC ENGINEERING

I was going to call this chapter "the discredited theology of genetic engineering"
but some individuals protested that it might be read as assuming that genetic engineering *had* a theology. In fact, as we have seen, it does and a strong and powerful
one at that. The Christian tradition, fed by powerful Aristotelian notions, has been
largely responsible for its propagation. For many centuries Christians have simply
read their scriptures as legitimizing the Aristotelian dicta: "existing for and
belonging to." The notion of "dominion" in Genesis has been interpreted as

licensed tyranny over the world, and animals in particular. God, it was supposed, cared only for humans within creation, and as for the rest, they simply existed for the human "goodies." According to this view, the whole world belongs to humans by divine right, and the only moral constraint as regards the use of animals was whether animal cruelty brutalizes humans or how we should treat them if they were other humans' property.[20] This god—not unfairly described as a "macho-god"—essentially masculine and despotical, who rules the world with fire and expects his human subjects to do the same, has trampled through years of Christian history, but his influence is now waning. There are many reasons for this and two in particular: First, most Christians do not believe in him any more. You will have to search high and low for any reputable theologian who defends the view that God is despotic in power and wants human creatures to be as well. Second, as we have seen, having re-examined their scriptures most theologians conclude that we misunderstand dominion if we think of it simply in terms of domination. What dominion now means, according to these scholars, is that humans have a divine-like responsibility to look after the world and to care for its creatures.[21] Indeed, for those who still hold to the "macho-god" version of divinity, I have some worrying news. Not only theologians (who tend in the nature of things to be either ahead of or behind the times) but also church people, even church leaders, have disposed of this old deity:

> The temptation is that we will usurp God's place as Creator and exercise a *tyrannical* dominion over creation . . . At the present time, when we are beginning to appreciate the wholeness and interrelatedness of all that is in the cosmos, preoccupation with humanity will seem distinctly parochial . . . too often our theology of creation, especially here in the so-called "developed" world, has been distorted by being too man-centered. We need to maintain the value, the preciousness of the human by affirming the preciousness of the nonhuman also—of all that is. For our concept of God forbids the idea of a *cheap* creation, of a throwaway universe in which everything is expendable save human existence. The whole universe is a work of love. And nothing which is made in love is cheap. The value, the worth of natural things is not found in Man's view of himself but in the goodness of God who made all things good and precious in his sight . . . As Barbara Ward used to say, "We have only one earth. Is it not worth our love?"[22]

These words come from a lecture given in 1988 by the then Archbishop of Canterbury, Robert Runcie. Notice how the earlier tradition is here confronted and corrected. God is a God of love. God's world is a manifestation of costly, self-sacrificial love. We humans are to love and reverence the world entrusted to us. And lest you should think that this is just "Anglican" theology which may at times tend to be a little fashionable, it is worth mentioning that an encyclical from that rather unfashionable, undoubtedly conservative Pope, John Paul II, specifically speaks of the need to respect "the nature of each being" within creation, and underlines the modern view that the "dominion granted to man . . . is not an absolute power, nor can one speak of a freedom to 'use and misuse,' or to dispose of things as one pleases."[23]

We have not yet brought our argument to its sharpest point, however. It is this: *No human being can be justified in claiming absolute ownership of animals for the simple reason that God alone owns creation. Animals do not simply exist for us nor belong to us. They exist primarily for God and belong to God. The human patenting of animals is nothing less than idolatrous.* The practice of genetic engineering implicitly involves the claim that animals are ours, to do with as we wish and to change their nature as we wish. The reason why it is wrong to use human beings as slaves is also precisely the reason why we should now oppose the whole biotech endeavour with animals as theologically erroneous. We have no right to misappropriate God's own.

FOUR OBJECTIONS

I anticipate four objections to this conclusion which I shall consider briefly in turn.

The first objection is as follows: We have always made animals our slaves. Our culture is based upon the use of animals. It is therefore absurd to suppose that we can change our ways.

I agree with the first part of this objection. It is true that human culture is based on the slavery of animals. I, for one, would like to see a root and branch cultural change. Christians may legitimately disagree about how far we can and should use animals. But one thing should be clear: we cannot own them, we should not treat them as property, and we should not pervert their nature for the sole purpose of human consumption. Genetic engineering—while part of this cultural abuse of animals—also represents its highest, or lowest, point. Because we have exploited animals in the past and now do so is no good reason for intensifying that enslavement and bringing the armoury of modern technology to bear in order to create and perpetuate a permanently enslaved species. The second objection is that the record of Christianity has been so terrible as regards the non-human that we must surely despair of any specifically theological attempt to defend animals.

I agree with the first part of this objection. Christianity has a terrible record on animals. But not only animals—also on slaves, gays, women, the mentally handicapped, and a sizeable number of other moral issues as well. I see no point in trying to disguise the poor record of Christianity, although I have to say that I do not quite share Voltaire's moral protest to the effect that "every sensible man, every honourable man, must hold the Christian sect in horror."[24] All traditions, religious or secular, have their good and bad points.

But to take one issue as an example: Recall my earlier point that if we go back in history two hundred years or so, we will find intelligent, conscientious, respectable Christians defending slavery as an institution. The quite staggering fact to grapple with is that this very same community which had in some ways provided the major ideological impetus for the defense of slavery came within a historically short period, one hundred, perhaps only fifty, years, to change its mind. The same tradition which helped keep slavery alive was the same community, one hundred or fifty years later, that helped end it. So successful indeed has this change been that I suppose that among Christians today we shall have difficulty in finding one slave trader, even one individual Christian who regards the practice as anything other than inimical to the moral demands of the Christian faith. While it is true that Christian churches have been and are frequently awful on the subject of animals, it is just possible, even plausible, that given say fifty or one hundred years we shall witness among this same community shifts of consciousness as we have witnessed on other moral issues, no less complex or controversial. In sum: Christian churches have been agents of slavery—I do not doubt—but they have also been, and can be now, forces for liberation.

The third objection is that genetic engineers are really good, honest, loving, generous, well-meaning people only trying to do their best for the sake of humanity, or at the very least they are no more awful than the rest of us. Why criticize what they are doing, when in point of fact *everybody* is in the difficult situation of moral compromise to a greater or lesser extent?[25]

Again, I agree with the first part of this objection. I have no reason for doubting the sincerity, the motivation, and the moral character of those who are actively engaged in biotech research. One of the really sad aspects of the campaign for the abolition of the slave trade was the way in which abolitionists tended, during the time of their ascendancy, to vilify their opponents by regarding them as the source of all evil. I have no desire to do the same. Indeed, what I want to suggest is that genetic engineers are really doing what they say they are doing, namely pursuing the cause of humanity according to their own lights. What I want to question, however, is whether a simple utilitarian humanist standard is sufficient to prevent great wrong. From the slave trader's perspective, it was only right and good to use slaves for the sake of the masters. From the genetic engineer's perspective it is only right and good to treat animals as utilities in order to benefit the human species. I doubt whether simple utilitarian calculation based on the interests of one's *own* class, or race, or species can lead otherwise than to the detriment of *another's* class, race, or species. Once we adopt this framework of thinking, there is no right, good or value that cannot be bargained away, at least in principle, in pursuit of a supposedly "higher" interest.

The fourth objection is to the analogy so far drawn between human and animal slavery. Animals are only animals, it is argued. Animals are not human.

This argument, which emphasizes a clear demarcation between animals and humans, whatever its merits in other spheres, is exceedingly problematic when applied to genetic engineering. After all, are not genetic engineers involved in the injection of *human* genes into nonhuman animals? According to a recent report, Vernon G. Pursel, a research scientist at the US Department of Agriculture's research faculty in Beltsville, responded to a recent move by various humane agencies and churches against genetic engineering by saying, "I don't know what they mean when they talk about species integrity." He went on to make a most revealing statement: "Much of all genetic material is the same, from worms to humans."[26] This statement is revealing precisely because it supposes what transgenic procedures must implicitly accept, namely that there is not a watertight distinction between humans and animals. Some may think that this is an argument in favour of treating animals in a more humane fashion, and so in a way it should be, but the argument is used to the practical detriment of nonhuman creatures. Here we have curious confirmation of the anxiety that besets bystanders like myself. For the question that must be asked is this: if the genetic material is much the same—from worms to humans—what is there logically to prevent us experimenting upon humans if we accept its legitimacy in the case of animals? Indeed genetic experiments on humans are not new. And neither is the view that there should be an eugenic programme for human beings. This view has received strong support from Christians at various times. One Christian writer in 1918 made clear that

> the man who is thoroughly fit to have children, and who either through love of comfort, or some indulgence of sentiment, refrains from marriage, defrauds not only himself and his nation, but human society and the Ruler of it . . . But the man or woman who knowing themselves unfit to have healthy children yet marry, are clearly guilty of an even more serious offence.[21]

This writer does not just advocate these moral imperatives as personal guidelines, rather he seeks to have them enshrined in law:

> The only kinds of legislation *for which the times are ripe* seem to be two. In the first place, marriage might be forbidden in the case of those mentally deficient, or suffering from certain hereditary diseases. And in the second place, much more might be done at present in the way of providing cottages in the country, and well-arranged dwellings in the towns, and by encouraging in every way the production of healthy children.[27]

This work by Percy Gardner was entitled *Evolution in Christian Ethics*. Gardner's view was straightforward: only those who are fit have the right to propagate the race. The well-being of the race was, as he saw it, threatened by the First World War because only the "weaker, and especially those whose vital organs are least sound, we retain at home to carry on the race."[28]

EUGENICS AND GENETIC ENGINEERING

Gardner's views had to wait another fifteen years before his ideas reached their fullest and most persuasive expression in the work of another writer, a political philosopher, of immense influence:

[The state] must see to it that only the healthy beget children; but there is only one disgrace: despite one's own sicknesses and deficiencies, to bring children into the world, and one's highest honour to renounce doing so. . . . [The state] must put the most modern medical means in the service of this knowledge. It must declare unfit for propagation all who are in any way visibly sick or who have inherited a disease.

And according to this view:

[The state's philosophy] of life must succeed in bringing about that nobler age in which men no longer are concerned with breeding dogs, horses, and cats, but in elevating man himself, an age in which the one knowingly and silently renounces, the other joyfully sacrifices and gives.[29]

These views are taken from the well-known work, *Mein Kampf*, and the author is, of course, Adolf Hitler.

Some may object that the analogy here breaks down. After all Hitler would hardly have approved of infecting Aryan blood with the genes of animals or, more accurately, allowing Aryan genes to be wasted on animals. He was hardly in favour of "hybrid humans"—as he called the children of mixed marriages—so he might well have had a certain disdain for the very idea of transgenic animals. And yet we cannot dismiss the fact that Hitler popularized, indeed did apparently much to develop, a medical science which aimed at "preserving the best humanity" as he saw it. And what is more his ideas of genetic control exercised through force, coercion, and legislation are by no means dead. Indeed the notion of creating a "super animal" is faintly reminiscent of the Hitler doctrine of creating a "superior" race.

Some may still feel that human eugenics and genetic engineering with animals are two quite separate things. Some may think that I am being simply alarmist. But *Mein Kampf* is, in my view, a much more important work of political philosophy than its detractors allow. But that is beside the point. What is the point is that I can find no good arguments for allowing genetic experiments on animals which do not also justify such experiments (or genetic programmes) in the case of human beings. I am alarmed by the way in which we have simply failed to recognize that animal experiments are often a precursor to experiments on human beings. Even in current established practice, animal experiments frequently precede the clinical trials on human subjects. We should not be oblivious to the fact that the century which has seen the most sustained and ruthless use of animals in scientific research is also the century that has seen experiments on human subjects as diverse as Jews, blacks, embryos, and prisoners of war. If "much of all genetic material is the same, from worms to humans," as Dr. Pursel maintains, what real difference does it make if the subjects are animals or humans?

If some of this still appears alarmist, it is perhaps worth emphasizing that one of the main planks of the case for anti-vivisection has always been that the experimental method, if morally valid, must logically extend to humans. C. S. Lewis based his critique on this very idea. His words deserve to be read in full:

But the most sinister thing about modern vivisection is this. If a mere sentiment justifies cruelty, why stop at a sentiment for the whole human race? There is also a sentiment for the white man against the black, for a *Herrenvolk* against the Non-Aryan,
for "civilized" or "progressive" peoples against "savage" or "backward" peoples.
Finally, for our own country, party, or class against others. Once the old Christian
idea of a total difference in kind between man and beast has been abandoned, then
no argument for experiments on animals can be found which is not also an argument
for experiments on inferior men. If we cut up beasts simply because they cannot
prevent us and because we are backing our own side in the struggle for existence, it
is only logical to cut up imbeciles, criminals, enemies, or capitalists for the same
reasons. Indeed experiments on men have already begun. We all hear that Nazi scientists have done them. We all suspect that our own scientists may begin to do so, in
secret, at any moment.[30]

Lewis was writing in 1947 and may be accused of hindsight. No such accusa-
tion, however, could be levelled at Lewis Carroll who seventy-two years earlier,
when vivisection was just beginning at Oxford, argued on precisely the same basis
but with even more vigour. Of the thirteen "Popular Fallacies about Vivisection,"
it was the thirteenth, "that the practice of vivisection shall never be extended so
as to include human subjects," that earned his greatest mockery:

That is, in other words, that while science arrogates to herself the right of torturing
at her pleasure the whole sentient creation up to man himself, some inscrutable
boundary line is there drawn, over which she will never venture to pass . . . And
when that day shall come, O my brother-man, you who claim for yourself and for
me so proud an ancestry—tracing our pedigree through the anthropomorphoid ape
up to the primeval zoophyte—what potent spell have *you* in store to win exception
from the common doom? Will you represent to that grim spectre, as he gloats over
you, scalpel in hand, the inalienable rights of man? He will tell you that it is merely
a question of relative expediency,—that, with so feeble a physique as yours, you
have only to be thankful that natural selection has spared you so long. Will you
reproach him with the needless torture he proposes to inflict upon you? He will smil-
ingly assure you that the *hyperaesthesia*, which he hopes to induce, is in itself a most
interesting phenomenon, deserving much patient study. Will you then, gathering up
all your strength for one last desperate appeal, plead with him as with a fellow-man,
and with an agonized cry for 'Mercy!' seek to rouse some dormant spark of pity in
that icy-breast? Ask it rather of the nether mill-stone.[31]

There is one important sense, however, in which Pursel was right. In addition
to the nature appropriate to each individual species, there is a nature which is
common to all human and non-human animals. But this realization alone should
make us think twice about genetic engineering. Animals, it is sometimes supposed, are simply "out there," external to ourselves like nature itself. Likewise, it
is thought, what we do to animals does not really affect *us*. In fact, however,
humans are not just tied to nature, they are *part of nature*, indeed inseparable from

nature. Because of this there is a profound sense in which we cannot abuse nature without abusing ourselves.[32] The genetic manipulation of animal nature is not just some small welfare problem of how we should treat some kinds of animal species, it is part of a much more disturbing theological question about "who do we think we are" in creation, and whether we can acknowledge moral limits to our awesome power, not only over animals, but also over our own species.

At the beginning of this chapter I invited you to imagine the Old Major addressing his fellow animal comrades and complaining that their state was none other than "misery and slavery." You may recall that a little provocatively the Old Major thought that the answer was the abolition of "Man." In one sense the Old Major was right. We need to abolish what St Paul calls the "old man" which is humanity in moral bondage or slavery to sin.[33] Demythologized a little, what St Paul might have said is that we must stop looking on God's beautiful world as though it was given to us so that we can devour, consume, and manipulate it without limit. I look forward to the final death of the "old man"—of which St Paul speaks—both in myself as well as in other human beings. Then, and only then, when we have surrendered our idolatrous power, which is nothing short of tyranny, over God's good creation, shall we be worthy to have that moral dominion over all which God has promised us.

NOTES

Reprinted from Andrew Linzey, *Animal Theology* (Champaign: University of Illinois Press, 1995), 138–55.

1. George Orwell, *Animal Farm,* Penguin edition.

2. For an illustration of some dramatic parallels between the two systems, see Marjorie Speigel, *The Dreaded Comparison,* esp. 37–57.

3. Aristotle *Politics* 1.8.

4. Aristotle 1.5.

5. Aristotle 1.4.

6. Aristotle 1.5.

7. See the discussion in chapter 1.

8. Aquinas *Summa Theologica* 1.93.4.

9. The Anti-Slavery Society Campaigns against chattel slavery, debt bondage, serfdom, child exploitation, and servile forms of marriage.

10. William Holcombe, cited in David Davis, *Slavery and Human Progress*, 23.

11. Aristotle 1.5.

12. Davis, *Slavery and Human Progress,* 25.

13. Theodore Weld, in Davis, *Slavery and Human Progress,* 146.

14. See Trevor Dennis, "Man beyond Price: Gregory of Nyssa and Slavery," in Andrew Linzey and Peter Wexler, eds., *Heaven and Earth,* 138.

15. Dennis, "Man Beyond Price," 137.

16. Davis, *Slavery and Human Progress,* 13.

17. Roberet Schank, cited in *Omni,* January 1988, and in *Agscene,* May 1988.

18. Cited in "Gene Splicing Aims for Super Animals," *St. Louis Post Dispatch,* December 8, 1986.

19. Aristotle 1.4.

20. For the Roman Catholic view that animals have the moral status identical to that of human property, see Henry Davis, *Moral and Pastoral Theology,* 2:258.

21. John Rogerson; Genesis 1–11.

22. Robert Runcie, "Address at the Global Forum of Parliamentary and Spiritual Leaders on Human Survival," 13–14.

23. Pope John Paul II, *Sollicitudo Rei Socialis,* par. 34.

24. Davis, 130.

25. See esp. Davis, *Slavery and Human Progress,* 129–53.

26. Vernon Pursel, cited in "Gene Splicing."

27. Percy Gardiner, *Evolution and Christian Ethics,* 188–89.

28. Gardiner, *Evolution,* 190.

29. Adolf Hitler, *Mein Kampf,* 367–69.

30. C. S. Lewis, *Vivesection,* 9–10.

31. Lewis Carroll, *Some Popular Fallacies about Vivesection,* 14–16.

32. For a theological discussion of nature which underlies this point, see Paulos Mar Gregorios, *The Human Presence: Ecological Spirituality and the Age of the Spirit.*

33. Romans 6:6.

22

Uncertainty in Xenotransplantation: Individual Benefit versus Collective Risk

F. H. Bach
L. A. Fishman
N. Daniels
J. Proimos
B. Anderson
C. B. Carpenter
L. Forrow
S. C. Robson
H. V. Fineberg

Clinical xenotransplantation, the transplantation of cells, tissues or organs from nonhumans to humans, crosses a species barrier that has evolved over millions of years. In doing so, it promises a great benefit to some patients, but it creates the possibility of new disease entering the human population. The ethical issues posed by this dramatic tradeoff of individual benefit against societal risk is the subject of this paper. These ethical issues demand an approach different from that usually taken in the evaluation of new medical technologies. We advocate a process of education in which public discussion and iterative evaluations are used to define the ethical concerns, the potential risks, and the benefits of clinical xenotransplantation at the societal level.

Clinical interest in xenotransplantation is prompted by the shortage of human donor organs for allotransplantation. Successful xenotransplantation would provide unlimited numbers of organs, making transplantation available to a greater number of patients. We focus on xenografts from pigs throughout, noting that grafts from evolutionarily more proximate species, such as non-human primates, likely pose even greater risks of infection.

The problems of rejection of a porcine organ by a primate are formidable. However, advances made in the past decade have allowed us to overcome some aspects of these problems and develop promising therapeutic approaches to others, making it conceivable that transplantation of porcine organs to humans will become a clinical reality.[1-9] Porcine cells are already being transplanted into the brain of patients with neurological diseases, and the Food and Drug Administration has tentatively approved "bridge transplant" protocols to perfuse the blood of patients with liver failure through porcine livers.

Four ethical considerations guide our thinking. First, a risk to the public requires a public mechanism for determining the acceptability of, and method of consent to, the risk. Second, since the risk to the public is not a "one time only" event, its assessment and regulation must be iterative. Third, the standard model of individual informed consent to medical interventions must be modified, since risks involve third parties, requiring that patients and close contacts be carefully monitored. Fourth, the possibility that a new infectious agent with altered pathogenicity will arise within the xenograft recipient may represent a danger to the pig population. Because of these four considerations, xenotransplantation requires a novel process of evaluation at the national level with novel institutional guidelines, responsibilities and resources.

Although a number of lengthy reports have been published dealing with xenotransplantation,[10-13] no organized method has been developed that addresses how to deal with the assessment and decision-making regarding this infectious risk to the public. A recent report of the U.S. National Research Council outlines an approach to assess and manage such risk-laden situations through an iterative process of analysis and deliberation involving public officials, scientists, and interested and affected parties.[14] We propose such an approach for xenotransplantation.

The Food and Drug Administration (FDA) has already established a broadly constituted advisory committee including both expert scientists and lay representatives to examine xenotransplantation. However, due to the unique aspects of public risk associated with xenotransplantation, initial discussions must be focused on the ethical issues. It is essential that the larger public interest, reflected in the ethical considerations we note above, be adequately aired and developed prior to developing a regulatory framework driven by technical considerations, and prior to making a commitment to proceed. A publicly constituted national advisory committee would be one vehicle to accomplish this type of broad public discussion.

Such a national committee should be comprised of individuals who are open-minded, sensible and broadly concerned citizens from many walks of life, thus representing a range of philosophical backgrounds and disciplines. Ethicists must be actively involved. Physicians and scientists familiar with technical aspects of the problem should also be included. As part of their own educational process, the committee would invite experts in relevant disciplines to answer questions and give advice. These should include, but not be limited to, those involved in the science of xenotransplantation, epidemiology, ethical aspects of the problem, animal welfare and rights, the medical profession, commercial efforts in transplantation, as well as the law and economics. Transplant recipients should be consulted. It will be important for the committee to have input from experts from the U.S. Food and Drug Administration and the Centers for Disease Control and Prevention.

Education of the members of the national committee is only a preliminary to their participating in decision-making about the future of xenotransplantation. The fundamental aim is to develop a consensus about the risks posed by present clinical trials, whether these efforts in xenotransplantation should be abandoned or expanded and if so, under what conditions. These efforts in the United States must be coordinated internationally with similar efforts, such as those of the Interim Regulatory Authority in the United Kingdom,[10] in countries and with organizations that are likely to become involved in xenotransplantation. Whatever safeguards are needed to avoid infectious spread to the population must be adhered to in all countries concerned.

There are some historical precedents to deal with concerns about risks to the public of infection from novel procedures, such as agents produced by genetic engineering. The Asilomar proposals set standards dealing with recombinant DNA research.[15,16] The fact that worst case scenarios that underlay that caution did not materialize is no reason to suspend caution in this case.

XENOSIS

The term xenosis was coined to describe the transfer of infections by transplantation of xenogeneic tissues or organs. Xenosis, or xenozoonosis, potentially poses unique epidemiological hazards due to the efficiency of transmission of pathogens, particularly viruses, with viable, cellular grafts. Transplantation in general enhances the risk of infection for a variety of reasons: (i) the graft itself serves as a nidus or "culture plate" from which organisms can spread in the human host, avoiding the need for a vector to achieve disease transmission; (ii) migration of cells from the graft throughout the host may carry cell associated infection, and (iii) administration of immunosuppressants leads to a diminished host response to infection, allowing infection to proceed in the absence of the usual manifestations of inflammation, often causing a delay in diagnosis.

The risks associated with xenotransplantation may be greater than those of allotransplantation for a variety of reasons. Whereas any type of organism can become a pathogen in the immunocompromised human host, viruses transferred with viable cellular grafts have been a major source of concern with both allotransplantation and xenotransplantation. Recent molecular data suggest that there is a family of closely related porcine endogenous retrovirus (PERV), some of which appear to be infective for human cells in vitro.[20] In vitro infection is not always predictive of in vivo infectivity and does not predict the ability of an agent to cause disease in the host. However, the possibility that a porcine retrovirus might be xenotropic, might have altered biological behavior in the human hosts, or might recombine with genetic material from the host (either in the germ line or from exogenous infection) raises the specter of a new disease entity developing. Such pathogens might spread undetected to the general population.

The level of risk for such infections to the recipient and the likelihood that such infection will spread to others are unknown.

A THREE-TIERED APPROACH TO POLICY DEVELOPMENT AND DECISIONS

The potential that xenotransplantation could introduce new infectious diseases into the population is the inverse of immunization. Immunization is intended to protect the population at the risk of having occasional individuals experience adverse reactions to the immunization. Xenotransplantation, on the other hand, offers potential benefit to the individual while putting the population at risk. Because the risk is societal and not merely individual, the decision whether to undertake the procedure involves more than ensuring the ability of the surgeon and the transplant team, the capacity of the institution, and the willingness of the patient. Where the risks are collective, the public must not only be educated about the risk but must also be involved in decision making. The first level of decision making must therefore occur at the level of social policy; the second at the level of the institutions performing the xenografts; and the third at the level of individual patients and physicians, affecting especially the processes of informed consent and of medical confidentiality.

Societal Level

Our focus on the risk of infection that can spread to the general population, while the most urgent ethical problem in xenotransplantation, should not diminish concerns about ethical issues related to the use of organs in humans from genetically manipulated donor animals, and the risk to pigs of infection, among others. All the major reports on xenotransplantation released to date have recommended comprehensive monitoring and surveillance of xenograft recipients because of the risk of xenosis. The legal and ethical problems associated with imposing such surveillance on recipients, and perhaps their sexual partners, for what will likely be many years or for life, as well as details about the nature and frequency of monitoring, require further discussion. It is not inconceivable that patients manifesting signs of a possible xenosis after transplantation would have to be quarantined. The maintenance of patient confidentiality, as in all areas of medicine, remains paramount and further complicates the need for adequate monitoring of recipients.

Neither the degree of risk nor the capacity of the medical community to deal with xenosis is known. Hence, perception of uncertain risk becomes key. In general, perceptions of the public about risk are quite different from those of experts. Public views are affected by the degree to which risk is familiar or mysterious, controllable or uncontrollable, and whether it evokes feelings of dread or suggests the potential for catastrophe. While news reports of breakthroughs in the use of animal tissue are appearing frequently, many in the lay community still fear the idea of organ farming and regard the exchange of body parts between animals and humans as macabre and the stuff of horror films. Fears and values play a key role in the way humans view risk.[14]

Experts tend to focus more directly on the degree of likelihood that mortality may result from the practice in question. With respect to the risk of transmission of infection from an animal donor to a human recipient there are essentially no data which could be used accurately to assess the level of such risk. The range of uncertainty is large, with the possibility of devastating cross species infection looming in the background.

Because of this uncertainty and because risk can mean very different things to different constituencies, the framing of statements about risk is particularly important.[14] First, the life-saving potential and enormous impact on the practice of clinically successful xenotransplantation would have to be made clear. Second, the public must be informed of any risk that could arise from xenotransplantation, whether or not the extent of that risk and the degree to which that risk is controllable can be precisely defined. Finally, it would be helpful for the public to have a better understanding of the process by which decisions are made in situations of uncertainty.

The problem cannot be dismissed by talking about education as if the experts have to eliminate ignorance and persuade the public; the public has its own concerns, rooted in quite diverse moral beliefs. There is a needed deliberative task: how do we coalesce the different ethical and factual considerations that merit consideration? A public deliberative process is key with the ultimate control over risk management vested in the public. Decisions as to how xenotransplantation will best proceed, and how risk will be handled at different stages of technological development, must emerge from such societal deliberations.

At the outset, there are two alternative, default positions. Either something has to happen to allow moving forward, or something has to happen to prevent moving forward. If the decision is to proceed, then the generally accepted policy in situations of uncertainty is to begin a limited series of experiments proceeding only to the point where risk reaches a certain pre-defined level. The committee would therefore serve to define the milestones for subsequent re-evaluations at each stage of the process.

Because it is impossible for all of the necessary evidence about risk to be available before xenotransplants begin to be performed, an appropriate approach might be to have a controlled initial trial involving a limited number of human recipients, and have those patients followed for a specified period of time. This approach has the advantage of allowing the refinement of regulatory and institutional mechanisms for evaluation of approaches to microbiological testing and other factors. This first phase would ideally last for as long as there is any risk of infectious transmission, which would mean waiting for some number of years. The length of this observation period is one appropriate issue for the committee to discuss and resolve. Society would then be presented with outcomes which could be used to revise the initial risk assessment, allowing reassessment of the safety and advantage of further experimentation. It may then be necessary to modify or extend the monitoring system in order to justify proceeding to the next phase of testing. Such a stepwise approach could be repeated and would require specifying in advance the points of reassessment with the assurance of regulatory controls as the process unfolds.

There will be great pressure on the committee to yield to the reality of "identified victims" who will die without an organ transplant. The central challenge to the committee will be to sustain in balance the uncertain societal risks against the palpable risks to individuals dying of organ failure.

In addition, the national committee must concern itself with at least two other problem areas: the financial commitments that must be made to allow monitoring for a period long enough to cover possible late occurring xenosis, and the use of transgenic animals (in this case transgenic donor animals) and risks to the pig population of infections arising from xenotransplantation.

The cost of monitoring patients and others for signs of infection would likely be very expensive and must be addressed prior to starting xenotransplantation. There must be discussion with national funding agencies, insurance companies, pharmaceutical and biotechnology companies interested in xenotransplantation and other health care financing institutions regarding a commitment to meet the required costs of the monitoring procedures for whatever period is deemed necessary.

Xenotransplantation has stimulated renewed interest in the use of animals in research. The use of pigs as a donor species has generally been viewed as reasonable. Some additional concerns have emerged as herds of swine are developed for possible use in xenotransplantation. First, pigs that are quarantined for the purpose of minimizing disease transmission must be assured of appropriate social interactions, which are important for general health and development.[17-19] Second, xenotransplantation presents the possibility that an infectious organism that did not previously exist might arise in humans that would be devastating to pigs. Transmission to a human recipient of a porcine retrovirus that did not cause diseases in pigs could result in modification of the nucleic acid sequences of that virus via recombination or mutation. Such a novel viral strain could cause disease in swine with devastating consequences for pig animal husbandry.

The report of a national advisory committee will provide guidelines for decisions that must be made at the institutional and patient-doctor levels. A supra-institutional public authority will need to be responsible for the regulation and management of xenotransplantation. In that role, the authority would define the conditions under which an institution would be authorized to proceed, and would fix the nature of the commitment that the patient and relevant contacts must make before xenotransplantation can take place. Regulations would likely be aimed at offering a uniform measure of protection to patients, to society and to animals.

As such, decisions at the institutional and individual levels should be guided by the deliberations at the societal level and should not be undertaken before the societal process has taken place. Because the societal discussions are likely to take some time, it is critical that these be started as soon as possible to avoid the unnecessary withholding of therapies from patients in need. Despite the fact that lives of patients needing transplantation may be lost with delay, we believe that the risks are sufficient to warrant refraining from human xenotransplantation until public deliberations on the ethical issues have occurred. Research in xenotransplantation should be strongly encouraged, including studies to define the potential risks of inter-species transplantation.

Institutional Policy

At the level of the hospital research center, institutions must be responsible for establishing and enforcing standards for quality of care, management of risk, monitoring of patients and their contacts, and evaluation of the effectiveness of the procedure in accordance with public guidelines and regulations. Institutions should avoid a situation in which individuals proceed with xenotransplantation in advance of adequate safeguards and should curtail clinical trials until societal guidelines are available.

Patient-Physician Interactions: Consent and Confidentiality

A new approach to informed consent as it relates to xenotransplantation is called for. A patient's agreement to participate in xenotransplantation must be premised on perceptions of both individual risk, as is the case with any experimental or extreme procedure, and the risk of new disease to family, friends, close contacts, and society at large. Because of the need for monitoring for signs of infection, the patient and others will have to commit to participate in such monitoring for a period of time that is considered to be longer than the potential time it might take for an infection to become manifest. Patients would not only have to agree to the risks attendant to a transplant procedure, but also to a contract binding the patient and others to carry out future obligations, including the patient's possible quarantine, as well as modification of the guarantees of confidentiality and surrender of the right to "drop out" of the study.[21] Whether such a contract could be enforced is an issue that the national committee will need to debate. In theory, xenotransplantation might not be allowed unless the patient and family members and sexual partners agree to some onerous conditions for monitoring. For instance, should there be a requirement that the same persons be informed of the fact that the patient is a xenotransplant recipient? Thus, the xenotransplant patient undertakes a social obligation to submit to close monitoring and frequent follow-up, even if this means relinquishing certain freedoms in order to gain the potential benefits of participation.

CONCLUDING COMMENT

We offer a strategy for handling the ethical issues related to xenotransplantation based on the optimistic perspective that xenotransplantation could become a clinically useful procedure, and our strong support of the science being performed. We propose a moratorium on xenotransplantation including those procedures that could be practiced at any time, such as using a pig organ as a temporary "bridge" until an allogenic organ is found, or as support for patients with hepatic failure. We have a feeling of urgency for a national review to be undertaken given the need of patients who might benefit from xenotransplantation, the impact on the field of the already-arrayed commercial interests, and the present and impending use of xenotransplantation procedures that could cause spread of disease. Past experience with problems involving uncertain public risks that are hard to quantify has shown that individuals and groups with various interests and concerns will have to work together in an interactive manner if appropriate decisions are to be made, and effective and responsible policy is to be generated.

The history of medical innovation has shown us unwilling to resist tangible individual benefit even in the face of unknown risks. It is incumbent upon us now to prepare for the moment when the decision to begin organ xenotransplantation will be well nigh irresistible.

NOTES

Reprinted by permission from *Nature Medicine* 4 (1998): 141–44.

1. F. H. Bach, M. Turman, G. M. Vercellotti, J. L. Platt, and A. P. Dalmasso, Accommodation: A working paradigm for progressing toward clinical discordant xenografting. *Transp. Proc.* 23 (1991): 205–7.

2. F. H. Bach et al., Delayed xenograft rejection, *Immunology Today* 17 (1996): 379–84.

3. F. H. Bach et al., Accommodation of vascularized xenografts, *Nat. Med.* 3 (1997): 196–204.

4. D. K. Cooper et al., Specific intravenous carbohydrate therapy, *Transplantation* 56 (1993): 769–77.

5. A. P. Dalmasso et al., Inhibition of complement mediated endothelial cell cytotoxicity by decay accelerating factor, *Transplantation* 52 (1991): 530–33.

6. A. P. Dalmasso et al., Expression of human regulators of complement activation on pig endothelial cells, *Xeno* 4 (1996): 55–57.

7. D. H. Sachs, Tolerance and xenograft survival, *Nat. Med.* 969 (1995): 1.

8. J. Platt, Xenotransplantation: Recent progress and current perspectives, *Current Opinion in Immunology* 6 (1996): 721–28.

9. T. Deacon et al., Histological evidence of fetal pig neural cell survival after transplantation into a patient with Parkinson's disease, *Nat. Med.* 3 (1997): 350–53.

10. *Animal Tissue to Humans* (London, 1996).

11. *Animal to Human Transplants: The Ethics of Xenotransplantation* (London, 1996).

12. Draft Public Health Service Guideline in Infectious Disease Issues in Xenotransplantation, 1996.

13. *Xenotransplantation: Science, Ethics, and Public Policy* (Washington, D.C.: Institute of Medicine, 1996).

14. P. C. Stern and H. V. Fineberg, eds., *Understanding Risk* (Washington, D.C., 1996).

15. P. Berg et al., Summary statement of the Asilomar conference, *Proceedings of the National Academy of Science* 72 (1975): 1981–84.

16. P. Berg et al., Asilomar conference on recombinant DNA molecules, *Science* 188 (1975): 991–94.

17. J. A. Fishman, Miniature swine as organ donors for man, *Xenotransplantation* 1 (1994): 47–57.

18. J. A. Fishman, Xenosis and xenotransplantation: Addressing the infectious risks posed by an emerging technology, *Kidney International—Supplement* 58 (1997): S41–S45.

19. L. E. Chapman et al., Xenotransplantation and xenogenic infections, *NEJM* 333 (1995): 1498–1501.

20. C. Patience, Y. Takeuchi, and R. Weiss, Infection of human cells by an endogenous retrovirus of pigs, *Nat. Med.* 3 (1997): 282–86.

21. A. S. Daar, Ethic of xenotransplantation—animal issues—consent, and likely transformation of transplant ethics, *World J. Surgery* 21, no. 9 (1997): 975–82.

23

Critical Ethical Issues in Clinical Trials with Xenotransplants

Harold Y. Vanderpool

Xenotransplants are widely regarded as the earliest foreseeable means of alleviating the dire shortage of transplantable organs from human donors. In anticipation of newly initiated clinical trials, councils and committees in the UK and the USA have explored scientific, ethical, and social issues surrounding xenotransplants: the US Public Health Service has published draft guidelines for animal to human transplants of cells, tissues, and organs, and the US Food and Drug Administration and National Institutes of Health have sponsored committee meetings and conferences.

WIDELY DISCUSSED ISSUES

The ethics of clinical trials with whole organ xenotransplants involve fascinating and serious issues. Four widely discussed issues include whether animal to human transplants violate the laws of nature; whether they wrongfully require animals to be sacrificed as organ sources for humans; whether they will cause social injustice through the lavishing of medical resources on a few people at the expense of basic care for the many; and whether they endanger public health. The first two issues are both controversial and philosophically intriguing, since they raise several questions. How should humans relate to the natural world? What is humankind's status relative to domestic animals and the higher apes? And should animals be granted moral rights? The airing of these issues has neither halted research on xenotransplantation nor undermined hopes for its medical promise.[1–3,5–10] Yet this discussion has revealed that modern interpretations of Judaism, Islam, and other

351

religions are not opposed to xenotransplants,[5] and has subdued initial fears about harmful psychological effects of animal organs in human recipients. The discussion has also elucidated why the use of primates as organ sources should be restricted, even if their infectious disease potential is excluded from consideration.[1–3]

The third issue—that xenotransplants will generate social injustice[11]—arises from assumptions that these transplants will absorb a large percentage of medical resources, and benefit only a few people. But xenografts and the scientific discoveries engendered by them are more likely to benefit large numbers of people in the future. The view that xenotransplants will cause social injustice also assumes that justice requires equal benefits to be given to everyone. This assumption disregards other conceptions of justice,[12] including the argument that unusual benefits should be given to people with unusual needs.

The fourth issue—that xenotransplants endanger public health—has received centre-stage attention in the light of discoveries relating to the initiating factors of the AIDS epidemic,[13] and of findings over the capacity of endogenous retroviruses in porcine tissue to infect human cells in vitro.[14,15] Concerns over the infectious disease potential of xenotransplants have led some groups and individuals to call for delays and even a moratorium[16] on human xenotransplantation. Other groups and individuals[17] view these fears as excessive, and are willing to approve of clinical trials if they are expertly reviewed and monitored with tissue testing and centralised registries and archives.

Wide discussion of these issues is understandable. The first two issues are philosophically interesting, culturally significant, and confined to xenotransplantation. The third issue centres on social ethics, and raises fund allocation questions that have been—and should be—debated whenever new medical technologies are developed and used.[18] In its focus on public health, the fourth issue is also collective and social.

Unfortunately, preoccupation with these collective concerns—particularly the last—has overshadowed four additional ethical issues that are critical and urgent. They are ethical prerequisites to clinical trials with whole organ and vasculated tissue xenotransplants, yet are being neglected despite the fact that cellular and tissue xenotransplant trials with human beings are already under way,[19–20] and the resumption of organ transplants is expected.

HARM-BENEFIT THRESHOLDS FOR INITIATION OF CLINICAL TRIALS

The first of these issues is the ethical imperative to identify permissible harm benefit thresholds for clinical trials with xenotransplants. What balance between expected harms and benefits should we have in mind as justification for these trials?

Contemporary suppositions about these risk benefit thresholds are variable and uncertain. Consider the following: in 1992, Polish surgeons justified the transplantation of a pig's heart into a human patient by arguing that it was the only way to extend the man's life in the absence of an available allogenic transplant. Although they surmised that the man could overcome hyperacute rejection,[21] he survived for only 23 h. In 1993, Pierson and colleagues asserted that a clinical definition of success must be established before cardiac or other xenografts are attempted. They proposed that "the median survival time" of weeks to months should "be established as the goal post"—marking attainment of a reasonable likelihood of clinical efficacy."[22] Platt has suggested that although hyperacute rejection can now be prevented, xenotransplants should be delayed until the barriers of acute vascular reaction and cellular and humoral reactions to donor antigens are understood and overcome.[23] Bach and colleagues argue that porcine xenotransplants with human beings should not be attempted until "documented long term function" is achieved with non-human primates.[24] Steele and Auchincloss propose that xenotransplants should not be undertaken until patients who are too sick to be candidates for allografts can be given xenografts that offer "at least the equivalent hope of success" as any allotransplant they might have received.[25]

These diverse proposals show the lack of systematic thinking over ethically sound risk benefit thresholds for the resumption of clinical trials with animal to human transplants. Only two investigators have specifically asked how a "successful" xeno organ transplant is defined.[26,27] Most of the brief discussions of risk benefit thresholds surface in departments of ethics focusing on the "scientific base" of xenotransplantation. The authors of these departments of ethics are sometimes unaware of the ethical underpinnings of a subject they view primarily in scientific and logistical terms.[25]

This issue—in which scientific, medical, and ethical questions are inextricably bound—is more complex, and fascinating, than it might at first appear. The issue encompasses three factors that must be played off against each other: scientific-medical feasibility, clinical urgency (i.e., the critical medical circumstances of patients with no therapeutic alternatives),[25,27] and the prospects of scientific discovery. Should xenotransplants be moved into clinical trials in ways similar to the trials, errors, and eventual successes of allotransplants?[27] This is one of the questions that needs to inform a search for greater coherence and consensus in relation to likely scenarios involving experimental xenotransplants with human patients.

Without contemplation of several realistic and imaginative scenarios, the vital prerequisite of predicating ethically acceptable clinical trials on a systematic analysis of foreseeable risks and harms may well be overlooked by local review committees,[21] and discredit researchers and clinicians. Deliberations on these scenarios will encourage critical thinking about the reasons and rationales used to justify clinical trials, and about the ethical assumptions contained within these reasons and rationales.[28]

INFORMED CONSENT FROM
THIRD-PARTY GROUPS

The second issue asks whether informed consent should be secured from third parties—in particular, from the close contacts of patients, and from medical personnel. This issue has been identified, but rarely explored. The draft guidelines of the US Public Health Service opt for education over consent for third parties. The guidelines propose that research patients should follow a detailed plan that will enable each person "to educate his/her close contacts regarding the possibility of the emergence of xenogenic infections."[4] Medical personnel must also be educated about risks of infection, and the precautions that must be taken.

This emphasis on education does not negate arguments in favour of informed consent for third parties. Consider three such arguments. First, clinical trials with xenotransplants directly affect the lives of the patients' close contacts who must act to assure the success and safety of the research. Such action includes abstention from sexual practices that could transmit infectious diseases, agreement to serum testing, and the reporting of any unexplained illnesses to medical authorities. Since informed consent is based on the ethical principle of respecting the views and choices of others as we would want ours to be respected, is there not a moral obligation to approach these parties personally, and to secure their consent for the parts they are expected to play?

Second, informed consent protects researchers and medical institutions from legal action that could arise from claims made by medical personnel and the close contacts of patients that they were never told about the risks they incurred.

Third, the personal privacy of medical staff must be considered. The guidelines of the US Public Health Service say that baseline serum samples should be drawn from all personnel who deal with research patients or who handle any of the tissues, cells, or organs—human or animal—related to xenotransplants.[4] These baseline serum samples are to be stored and subject to surveillance by two federal agencies. Should not the consent of these medical workers be requested after they are told who will have access to their test results, and what will happen should the tests reveal a difficult or embarrassing infection?

The arguments for and against informed consent for third parties should be explored in the light of these and other questions, and in relation to the ethical principles set down in *The Belmont Report*[12] and elsewhere.

INFORMED CONSENT FROM RESEARCH PATIENTS

Informed consent from the patients in clinical trials of xenotransplants must be thoroughly examined, because the challenges to fully informed consent in these trials may exceed those in any known research setting, including phase I of chemotherapy trials for cancer. The prospective participants in xenotransplant research are likely to be sick, or very sick, and desperate to have their lives extended. Nevertheless, if their autonomy is to be respected, they will need to hear, understand, weigh, and make decisions about a host of complex concerns.[10] These concerns include: the stage of xenotransplant research (incorporating discussions about data from previous experiments on mortality and quality of life); the likelihood of media attention and compromised confidentiality;[29] the risks and discomforts of opportunistic infections; information about the risk of development of animal mediated (zoonotic) and genetically innovative (xenogenetic) infections, and of transmission of these infections to others; the requirements of adherence to a schedule of frequent, long term, or lifelong medical surveillance; the granting of permission to public health agencies to examine private medical records; the granting of consent to a complete necropsy at time of death; the responsibility to educate close contacts about the risks and control of infections; and detailed information about the trying, sometimes traumatic psychological effects of immunosuppressive drugs.[30–32]

Out of concern for some of these complexities, and the possibility that overzealous surgeons will continue to underestimate the risks—and exaggerate the benefits—of xenotransplants the Nuffield Council in the UK recommended that participants' consent should be secured by "trained professionals who are independent of the xenotransplantation team."[1] This distrust driven policy, however, would keep surgical teams and their patients from openly and mutually dealing with the multi-faceted human dimensions of xenotransplants.

Essential features of the content and process of informed consent for patients and their proxy decision makers should be set out as guidelines based on a thorough understanding of moral autonomy, the circumstances of prospective recipients, the self-interests of researchers, for-profit economic pressures, and the advantages and disadvantages of various approaches. In addition to the specialists specified in the draft guidelines of the US Public Health Service,[4] every transplant team should include a psychological counsellor. And the process of consent should be spelt out in each protocol that is approved.

OVERSIGHT AND APPROVAL OF CLINICAL TRIALS

The fourth critical and urgent issue encompasses at least three sets of questions—the first of which has been dealt with at greater length than the others. First, who should be responsible for the oversight and approval of clinical trials? Should the governance of xenotransplant research fall to local ethics review committees, a national committee, or some combination of the two?[3] Although a solution to this question will involve turf battles over power and control, it should reflect a considered answer to the following ethical question: which group or groups will best foster concern, protection, and respect for patients, and a commitment to balance possible harms to the public with possible benefits to people in desperate need?

Good reasons support the view that a national body should establish mandatory rather than advisory guidelines for the make-up, resources, standard of approval, and monitoring of local review committees that would themselves carry out "hands on" reviews of xenotransplant protocols.

Second, what issues should be addressed in more developed guidelines? Although such guidelines should deal with the dangers of infectious disease, they should also address the critical issues discussed above. The US Public Health Service deserves praise for explicitly recommending the use of *The Belmont Report* in its draft guidelines.[4] Nevertheless, the interconnections between clinical trials involving xenotransplants, and the ethical principles and applications of these principles in *The Belmont Report*, are far from obvious and should not be left to chance.

Third, what should guidelines for clinical trials of xenotransplants require in relation to membership of oversight committees? Because the decisions of these committees will hinge on the understanding of the complex cognitive and emotional needs of prospective participants, membership should include a mental health professional. Furthermore, to be able to define and make reasonable assessments of the risks and benefits of xenotransplants to recipients, each oversight committee should review protocols in consultation with former transplant patients until sufficient published research emerges on patients' points of view. Former patients might thus have to become standing members of ethics review committees. The availability of consultants in research ethics and religious traditions should also be considered.

Why do the harm benefit assessments required of review committees call for input that is based on personal experience or social science (or both) from transplant patients? It is because every such assessment is derived from a particular frame of reference, the adequacy of which depends on the knowledge, experience, empathy, and personal agenda of every person involved in the assessment. Furthermore, the assessments of oversight committees need to be informed by patients, since it is they who will undergo the experimental procedures that are predicated on potential losses and gains.[33] As the best spokespeople, former transplant patients should have a say in what should count as risks and benefits, and what balance of the two is reasonable.

CONCLUSION

As pressures build for the initiation of new clinical trials with xenotransplanted organs and vasculated tissue, the four main issues discussed must be thoroughly explored and acted on. Attention currently focuses on measures to protect public health from the possibility of xenotransplant generated infectious diseases. This concern should not detract from the necessity of protecting the ethical integrity of clinical research involving xenotransplants. Without attending to the four main issues discussed in this paper, clinical trials with xenotransplants could begin to erode long and arduous efforts to uphold the rights of research patients, and to secure the public's mist and participation in medical research.

NOTES

Reprinted with permission from *Lancet,* May 2, 1998, 1347–50.

1. Nuffield Council on Bioethics UK, *Animal-to-human transplants* (London: Nuffield Council on Bioethics, 1966).

2. UK Advisory Group on the Ethics of Xenotransplantation, *The government response to "animal tissue into humans"* (London: Department of Health, 1997).

3. US Institute of Medicine Committee on Xenograft Transplantation, *Xenotransplantation: Science, ethics, and public policy* (Washington, D.C.: National Academy Press, 1996).

4. US Public Health Service, Draft public health service guidelines on infectious disease: Issues in xenotransplantation, *Federal Register* 1996: 61 49920–32.

5. R. M. Veatch, The Ethics of Xenografts, *Transplant Proc.* 18 (1986): 93–97.

6. J. L. Nelson, Moral sensibilities and moral standing: Caplan on xenograft donors, *Bioethics* 7 (1993): 314–22.

7. T. L. Beauchamp, The moral standing of animals in medical research, *Law, Med. Health Care* 20 (1992): 7–16.

8. S. G. Post, Baboon livers and the human good, *Arch. Surg.* 12S (1993): 131–33.

9. Al Caplan, Is xenografting morally wrong? *Transplant Proc.* 24 (1992): 722–27.

10. C. R. McCarthy, Ethical aspects of animal-to-human xenografts, *Institute of Laboratory Animal Resources* 37 (1995): 3–9.

11. R. C. Fox and J. P. Swazdy, Leaving the field, *Hastings Cent. Rep.* 22 (1992): 9–15.

12. The Belmont Report, *Federal Register* 44 (1979): 23192–97.

13. L. E. Chapman, T. M. Folks, D. R. Salomon, A. P. Panenon, T. E. Eggennan, and P. D. Noguchi, Xenotransplantation and xenogenic infections, *New Engl. Jour. Med.* 333 (1995): 1498–1501.

14. C. Patience, Y. Takeuchi, and R. A. Weis, Infection of human cells by an endogenous retrovirus of pigs, *Nat. Med.* 3 (1997): 282–86.

15. P. Le Tissier, J. P. Stove, Y. Takeuchi, C. Patience, and R. A. Weis, Two sets of human-tropic pig retrovirus, *Nature* 389 (1997): 681–82.

16. F. H. Bach, J. A. Fishman, N. Daniels et al. Uncertainty in xenotransplantation: Individual benefit versus collective risk, *Nat. Med.* 4 (1998): 141–44.

17. J. W. Ebert, L. E. Chapman, A. P. Patterson, Xenotransplantation and public health, *Current Issues in Public Health* 2 (1996): 215–19.

18. R. E. Bulger, E. M. Bobby, H. V. Fineberg, eds., *Society's Choices: Social and Ethical Decision Making in Biomedicine* (Washington, D.C.: National Academy Press, 1995).

19. T. Deacon, J. Schumacher, J. Dinimore et al., Histological evidence of fetal pig neural cell survival after transplantation into a patient with Parkinson's disease, *Nat. Med.* 3 (1997): 350–53.

20. R. P. Lanza, J. L. Hayes, and W. L. Click, Encapsulated cell technology, *Nat. Biotechnology* 14 (1996): 1107–11.

21. J. Cxaplicki, B. Blonska, Z. Riga, The lack of hyperacute xenogenic heart transplant refection in a human, *J. Heart Lung Transplant* 11 (1992): 393–97.

22. R. N. Pierson Jr., D. J. White, J. Wallwork, Ethical considerations in clinical cardiac xenografting, *J. Heart Lung Transplant* 12 (1993): 876–78.

23. J. L. Platt, Xenotransplantation: The need, the immunologic hurdles, and the prospects for success, *Institute for Laboratory Animal Resources* 37 (1995): 22–31.

24. F. H. Bach, C. Ferran, M. Scares et al., Modification of vascular responses in xenotransplantation: Inflammation and apoptosis, *Nat. Med.* 3 (1997): 944–48.

25. D. J. R. Steele, H. Auchincloss Jr., The application of xenotransplantation in humans: Reasons to delay, *Institute of Laboratory Animal Resources* 37 (1995): 13–15.

26. A. Hasollo, M. L. Hess, Heart xenografting: A route not yet trod, *J. Heart Lung Trans.* 12 (1993): 3–4.

27. R. E. Michler, Xenotransplantation: Risks, clinical potential, and future prospects *Emerg. Infect. Dis.* 2 (1996): 64–70.

28. H. Y. Vanderpool, Introduction and overview: Ethics, historical case studies, and the research enterprise, in *The Ethics of Research Involving Human Subjects: Facing the 21st Century* (Frederick, Md.: University Publishing Group, 1996), 1–11.

29. K. Reentsma, Ethical aspects of xenotransplantation, *Transplant Proc.* 22 (1990): 1042–43.

30. J. L. Craven, Cyclosporine-associated organic mental disorders in liver transplant recipients, *Psychosomatics* 32 (1991): 94–102.

31. P. De Groen and J. Craven, Organic brain syndromes in transplant patients, in *Psychiatric Aspects of Organ Transplantation*, ed. J. Craven and G. M. Rodin (Oxford: Oxford University Press, 1992), 67–88.

32. J. Soos, Psychotherapy and counselling with transplant patients, in *Psychiatric Aspects,* 89–107.

33. B. Freedman, The ethical analysis of clinical trials: New lessons for and from cancer research, in *The Ethics of Research Involving Human Subjects: Facing the 21st Century,* ed. H. Y. Vanderpool (Frederick, Md.: University Publishing Group, 1996), 327–31.

V

HUMAN GENETIC TESTING AND THERAPY

HUMAN TESTING AND SCREENING

Principles of Genetic Testing

The DNA chemistry and genetics we have already presented will help you understand the technology for testing the genetics of people. Remember that A's pair with T's and G's pair with C's, using hydrogen bonds to form the double strands of DNA. If these bonds are broken, the two strands separate into single strands. The process of two strands binding to form a double-stranded molecule is called hybridization. If two strands match well, they are said to be complementary, or homologous, to each other. Remember these terms because they occur frequently in reference to procedures of genetic testing. This binding of DNA in double strands is the basis for most genetic testing.

A number of different conditions cause DNA strands to bind or hybridize weakly or strongly to each other. The first obvious condition, as discussed earlier, is the extent of matching of the A's with the T's and the G's with the C's. It is possible that every single nucleotide in the DNA sequence matches perfectly between the two DNA strands. Yet these sequences may hybridize imperfectly due to multiple unmatched bases or due to only one mismatched nucleotide. An example of a mismatch is an A being opposite a G, even though the other adjacent nucleotides are perfectly matched. A percentage complementarity can be calculated to represent the extent of homology between the two strands. For example, two DNA strands might be said to be 99 percent homologous, which means that one out of every ninety-nine sequences, on the average, is a mismatch. Conversely, two sequences that are only 35 percent homologous are not very related. This theory is used in showing the relatedness between two sources of DNA, such as in a criminal investigation. DNA from tissue such as blood cells can be used to show the relatedness of DNA from a criminal suspect with DNA found at a crime scene.

359

The number of hydrogen bonds between the nucleotide pairs affects the strength at which two strands of DNA hybridize to each other. A series of G's opposite C's will bind more strongly than a series of A's opposite T's because the G-C base pairing has three hydrogen bonds and the A-T base pairing has two bonds. Therefore, strands of DNA that have a higher percentage of G's and C's will be more tightly hybridized. Another condition that affects the hybridizing of DNA is the temperature of the solution containing the DNA, referred to as the hybridization solution. If the temperature of the hybridization solution is very high, the DNA strands will separate or not connect to each other because the bonds between hydrogen molecules are very energetic and unstable and do not hold the hydrogen molecules together. Conversely, if the hybridization solution is cold, the bonds are not energetic and will be more stable to hold the strands together. Finally, if the concentration of salt in a solution is high, the hydrogen bonds will hold the DNA strands tightly together. Low salt concentrations cause weaker hydrogen bonds. People performing genetic testing procedures use these various principles of hybridization to manipulate the hybridization of DNA in their genetic testing assays.

Method of Southern Blot Hybridization for Genetic Testing

These principles of hybridization are involved in how the screening procedures are performed. We will describe two screening procedures. The first is called Southern blot hybridization. The term "Southern" honors Dr. E. M. Southern, who invented the procedure in 1975. First, the large chromosomal DNA is cut with restriction enzymes into smaller manageable pieces. Next, the cut samples of DNA are separated according to their size. Because DNA molecules have a slight electrical charge, they can be separated using an electrical current (figure 18). DNA migrates from a positive to a negative pole in an agarose gel, a gelatin-like material. The agarose gel can be thought of as a matrix or sieve. When the DNA fragments move by electrical current through the gel, the larger fragments are slowed down through the matrix more than the smaller fragments (figure 18). Consequently, the larger fragments are closer to the point of origin and the smaller fragments travel farther from the point of origin. The result is that fragments of the same size are located at a single place on the gel. This accumulation of same-size DNA fragments on the gel is important for good detection, as will be explained later. If the fragments of different sizes were mixed all together, proper analysis would not be possible.

After separation of fragments on the gel, the DNA must be transferred out of the gel onto nylon paper. The procedure cannot be performed using the gel. The position or location of the various bands must be maintained during the transfer from the gel to the nylon. The procedure for accomplishing this task is called Southern blot hybridization. There are many modifications of this procedure, but the results are basically the same when analyzing the DNA. Figure 19 shows an

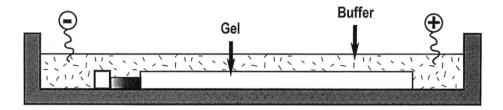

DNA, loaded into a well cut out of the gel

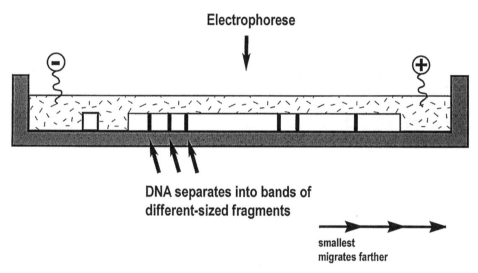

**DNA separates into bands of
different-sized fragments**

smallest
migrates farther

Figure 18.   A Cross-Section of Gel Electrophoresis

apparatus using a reservoir for holding transfer solution, the gels, the nylon paper or membrane, and paper absorbent pads for soaking up the transfer solution. The transfer solution moves up through the gel and nylon into the absorbent pads, like soaking up a spill on the kitchen floor with paper towels. The solution picks up the DNA as it passes through the gel and deposits it onto the nylon membrane, after which the solution continues to be soaked up into the absorbent pads. Because the solution moves in a straight line through the gel and nylon, the position of the DNA is maintained. This procedure prepares the DNA for hybridization.

Probes are known sequences of DNA that are labeled with a chemical or a radioactive molecule called an isotope. When the probe is labeled with an easily detected chemical or isotope, the presence of the probe can be easily detected. In essence, probes tell the scientist if a piece of DNA is related to the probe. DNA that is hybridized or connected to the probe of known sequence can be identified. The hybridization step involves soaking the nylon membrane in a hybridization

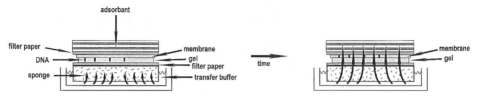

Figure 19. Southern Blot Hybridization

solution containing just the right amount of salt at the right temperature; remember that the hybridization is affected by salt and temperature. After the probe is allowed to hybridize to the DNA on the membrane, the unbound probe is washed free from the membrane so that only the hybridized probe remains on the membrane. If you want only the strongest probes to remain, the wash solution is at a higher temperature and lower salt concentration to remove the weakly bound probe. The point of this concept is that strictly related, or weakly related, sequences can be detected using the probe by adjustments to the salt and temperature during the washing step. The final step is to expose X-ray film to the nylon filter. For example, specific, complementary radioactive probes can hybridize or attach to DNA found on the nylon membrane (figure 20), and then the nylon membrane can be exposed to film that can detect the radioactivity. The developed film will have a dark spot at the position where the complementary DNA is located. The positions of all the other DNA will not have a dark spot because the radioactive probe would not have hybridized to any of these other DNA sequences (figure 20). The details of how probes are made are not important here, but a clear understanding of how labeled or tagged probes help detect hybridization of the probe to its target is important for understanding how probes work. Probes are labeled with radioactivity or a detectable chemical either at the terminal end of the probe or on nucleotides within the probe. Radioactive isotopes emit radiation that can expose film, more specifically X-ray film like that used to X-ray a broken bone. Detectable chemicals usually generate fluorescence or light, which can also expose film. The principle is that probes must be tagged with something that will allow detection of the probe if it is hybridized to its target sequence.

Method of Polymerase Chain Reaction (PCR) for Genetic Testing

The second method for doing genetic testing involves a procedure called the polymerase chain reaction (PCR). The essence of PCR is its capability to increase the numbers of a DNA molecule by a billionfold. To understand how PCR works, one must understand the role of primers in DNA replication, which is the synthesis of new complementary DNA molecules. Primers are small pieces of DNA that are homologous to a sequence of the DNA to be replicated, also known as the "tar-

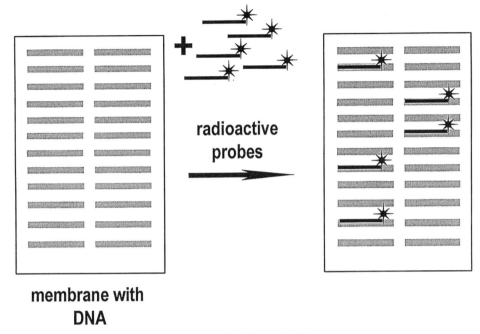

radioactive
probes

membrane with
DNA

Figure 20. Using Radioactive Probes to Identify DNA of Interest

get" DNA (figure 21). Binding primers to the target DNA is an essential step to start replication. If the primer is not homologous to the target DNA, it will not bind and DNA synthesis will not occur. The protein enzyme responsible for synthesis finds this primer, attaches to it, and then starts adding nucleotides at one end of the primer. This enzyme is called "DNA polymerase." If a G is on the target DNA to which the primer is bound, the DNA polymerase enzyme will add a C to the end of the primer because it is complementary to the G. As DNA extends out from the primer, the DNA polymerase will only add complementary nucleotides. The end product is a new strand of complementary DNA.

Another important concept for PCR is that the DNA polymerase adds nucleotides in a 5' to 3' direction with respect to the primer. Remember that the ends of the DNA primer are identified as having a 5' or 3' end, so the nucleotide is added to the 3' end of the primer. The new DNA only grows in one direction, from a 5' to 3' direction.

PCR involves the synthesis of new complementary DNA, but how does it increase the amount of DNA by a billionfold? It uses two primers in the synthesis reaction. The direction of synthesis of each primer points toward the other (figure 21). The direction of synthesis of one primer points toward the second primer, and the second primer faces toward the first. When the reaction is started, the DNA polymerase molecules attach to each primer and the synthesis is toward each primer, not away from each primer. After one reaction is completed, another syn-

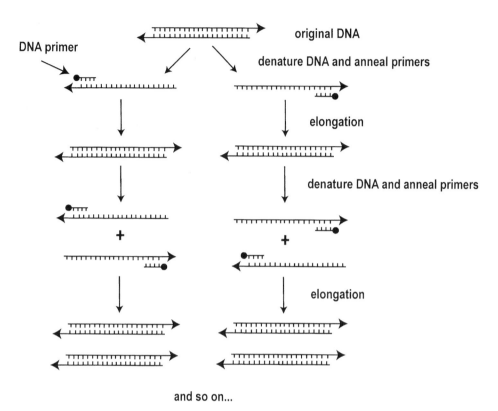

Figure 21. Polymerase Chain Reaction (PCR)

thesis reaction is done, except this time there are twice as many pieces of DNA from which to synthesize. The amount of target DNA has been doubled from the first to the second reaction. Each time the synthesis reaction is repeated, the amount of DNA is doubled. At first glance, this may not appear to increase the DNA levels by too much, but the reaction is repeated thirty times! After thirty cycles, DNA increases tremendously by about 1 billion as compared to the original starting amount.

Application of PCR for Genetic Testing

What features of the PCR make it valuable for DNA testing or screening? Often the amount of DNA that can be recovered from a crime scene, such as hair in the carpet or semen stains on clothing, is too small for the Southern blot hybridization procedure. PCR, however, needs only very small pieces of DNA for analysis because the PCR procedure can dramatically increase the amount of DNA. Therefore, when only small amounts of DNA are available, PCR is the method for analysis. Another feature that makes PCR valuable for DNA screening is the requirement for primers to be homologous to their targets. If primers are homologous to the target DNA, they will bind and synthesis will occur. If there are some mismatches, synthesis may not occur. So the synthesis of DNA from the PCR is indicative of a genetic match in a screening assay. Remember, when the Southern blot hybridization is employed for genetic screening, a panel of probes is used. The primers in a PCR reaction are analogous to the probes in a Southern blot hybridization. The ability of primers or probes to hybridize to target DNA sequences depends on their homology or complementarity with the target. A panel of primers or probes is used to increase the chances of accuracy. The larger the panel, the greater the accuracy. In this the PCR and the Southern blot hybridization are the same, but they differ in the amount of target DNA required to do the assay. Often both assays can be done on the same target DNA for even greater accuracy, even though large amounts of target DNA might be available for the assays.

Statistical Probabilities of Genetic Testing

Individual people obviously have different facial features, height, weight, bone structure, and hair color; we are all different and these differences are a result of our genetics. People have genetic sequences that are clearly the same; otherwise they would not look like human beings. If we wanted to identify the differences between individual people, we would not want to use a probe that can detect sequences common to each individual; rather, we would want to use probes that detect sequences found in some people and not other people. One certain probe might detect sequences in 10 percent of people, but it cannot distinguish between each individual in this group. Another certain probe recognizing a completely different sequence might recognize 2 percent of people. If both probes are used, only 0.2 percent of people's DNA hybridizes to both probes. When a panel of many probes is used, the percentage of the population hybridizing to all these probes becomes very small. This concept is used in criminal investigations to discover if a suspect and DNA found at a crime scene are reactive to the same panel of probes. One can then calculate the probability that the DNA at the crime scene is not the suspect's DNA. The important concept here is that the probabilities can be very low, that there is not a match, but the probability never reaches zero. In other words, we can be pretty sure that the person committed the crime, but not absolutely, unequivocally certain. The probability may approximate 100 percent that the person committed the crime—that the suspect's DNA matches the DNA at the crime—but it is never absolutely 100 percent. This raises the question of what probability is enough to convict a suspect in a court of law. Is a probability of one in ten that the DNA is an error good enough to convict a person? Probably not. A probability of one in a thousand that the conviction is an error may be good enough depending on other evidence, but one in a million may be convincing to a jury. By using enough probes in a panel, scientists can get pretty strong probabilities that a suspect is innocent or guilty, but as explained earlier, people have a difficult time comprehending the numbers. DNA evidence is being accepted more readily by the courts, but other evidence is still important in making a conviction or setting a person free.

Comparison of PCR with Southern Blot Hybridization

What are the disadvantages of the two methods? As already noted, Southern blot hybridization needs more DNA than the PCR does. It can also be more expensive and involved. But PCR has a major disadvantage if care is not taken to avoid it. Because PCR can identify very, very small trace amounts of DNA, it could inadvertently pick up complementary DNA from unwelcome sources such as the air or technicians. This presents a problem when a match is found because possible contamination could result in false positives: some question about the actual source of the target DNA. This problem is referred to as "PCR contamination," and is often considered the greatest disadvantage of PCR. The assay is so sensitive that it might accidentally pick up the wrong target DNA.

Let us suppose that a woman has been murdered, and the husband is a suspect. Hair is found at the crime scene that does not belong to the victim. Perhaps the victim was pulling the attacker's hair. The decision is made to do genetic testing on the hair using PCR. Identity is found with the husband's DNA. You are the defense attorney. What can you tell the jury that would place doubts in their minds as to the validity of the PCR results? You could explain in simple terms the principle of PCR and that it is very, very sensitive. You could find a scientist as an expert witness who would give the following illustration as to the sensitivity of the PCR reaction. If a drop of liquid containing DNA were put into an Olympic-size swimming pool and then the water was thoroughly mixed, PCR could still detect the presence of the DNA. Because the suspect and the victim were married, skin cells could easily have been in the carpet or on the wife's hands for the husband's DNA to be detected on the hair by the PCR even though the husband may not have committed the crime. If the Southern blot hybridization were used, however, to confirm identity of the hair at the crime scene with the hair of the husband, the evidence would be more convincing, because Southern blot hybridization cannot pick up small traces of contaminating DNA. It requires the larger amounts of DNA extracted from the hair.

Now let's change the scenario so that a woman was killed and the suspect had never known the family or the woman; no prior association could be made. It appeared to be a random killing. Then PCR evidence might be more convincing because the suspect had no prior contact with the woman that could cause PCR contamination. Clearly, some knowledge of the science can help sift through ethical and social issues of genetic testing.

HUMAN GENE THERAPY

Gene therapy uses beneficial genes as therapy against diseases, just as beneficial chemicals or drugs are used to treat illnesses. You now know enough about genes and DNA to understand that a DNA coding for a missing function of a protein could conceivably be inserted into a person's cells to provide therapeutic benefit. Theoretically, the person could be genetically engineered to contain the gene or separate cells could be engineered and the cells put into the individual. There are three main technical issues discussed in part 5: how DNA is put into the cells for gene therapy, what cells are targeted for therapy, and illnesses that are current candidates for gene therapy.

Germ Line Gene Therapy versus Somatic Gene Therapy

Gene therapy can be divided into two main categories, somatic or germ line gene therapy. A germ line cell is a sperm, an egg, or an early embryo—cells that are involved in reproduction. These are the cells from which future children or progeny are made. The genetic information contained in these germ line cells will be inherited by any offspring of these cells. Consequently, if a parent has a genetic illness or at least carries a disease gene, and the germ line cell contains the disease gene, the resulting baby will inherit the gene. In the same fashion, if the germ line cell is genetically engineered to contain the beneficial gene, then the progeny will inherit the beneficial gene. The beneficial gene, like any other gene, will be inherited through generations. This process of germ line therapy can theoretically correct the genetic disease in children of future generations.

Somatic gene therapy does not involve the genetic manipulation of germ line cells involved in reproduction, so the therapeutic gene is not inherited by future generations. If a parent has a genetic disease that is treated with somatic gene therapy, the genetic cure is not passed on to the children. Somatic cells are defined as any other cells besides germ line cells. They are lung, liver, brain, skin, or other such cells, but not germ line cells. The strategy for developing somatic gene therapy usually involves the genetic manipulation of the tissue having the disease. For example, cystic fibrosis is primarily, although not exclusively, a disease in the lung. Somatic gene therapy strategies for this disease have involved putting the corrective, beneficial gene directly into the lung cells. Diseases of the liver could involve putting corrective genes directly into liver cells. The same could be true for the brain, pancreas, or other tissues. When lung cells are manipulated for cystic fibrosis, the beneficial gene is not passed on to the patient's children, and they could still get cystic fibrosis. The beneficial gene dies when the cell or the person dies. Let us suppose that a cystic fibrosis patient does not want to pass the disease gene to her children, so she chooses to have germ line therapy to genetically engineer the germ line cells, not the lung cells. In this hypothetical example, the patient would not be directly treated for cystic fibrosis; only her children would receive the benefit. Germ line gene therapy treats the progeny, not the patient; somatic gene therapy treats the patient, not future generations.

Germ line gene therapy presents more complex ethical issues than does somatic gene therapy, and consequently far more clinical trials and research are directed toward somatic gene therapy. At the time of writing, no clinical trial of germ line gene therapy has been approved by the regulatory committee of the National Institutes of Health. Ironically, somatic gene therapy is more difficult theoretically than germ line gene therapy, based on animal studies. In animals, it is relatively easy to put DNAs of corrective genes into a single-cell embryo for implantation into a mother for fetal development as compared to putting genes into adult cells of diseased tissues. However, some of the minor problems encountered in generating transgenic animals would be major ethical problems if the same procedure were conducted in people. The procedure of microinjecting animal embryos with DNA illustrates the ethical issues that are encountered with other procedures. When the foreign DNA is injected into the nuclei of the single-cell embryo, the DNA is integrated randomly in the chromosome. Strictly speaking, this constitutes a genetic mutation. Some mutations caused by the integration are "silent" because they would not cause any disease in the animal. A certain number of animal embryos that develop after random integration of the foreign DNA, however, have physiological or developmental diseases. Sometimes they die in the uterus before delivery. The event goes almost unnoticed in an animal, but this would be a major undesirable event for human gene therapy. A human baby born with a genetic defect as a result of the procedure would be very unfortunate, especially since that baby had no choice in the matter of performing the therapy. In an attempt to prevent or cure one genetic disease, another accidental genetic disease could be caused, while the afflicted person has no choice in the matter. Two ethical issues arise with the use of germ line gene therapy in people: lack of consent by the child and the medical risk of the procedure.

Somatic gene therapy is potentially difficult because the genetically engineered cell may not last long enough in the tissue to give long-lasting benefit, and there may not be enough genetically engineered cells or protein produced in the tissue to make a noticeable difference. All the cells in the tissue may be in very large numbers as compared to the relatively small number of genetically engineered cells added to the tissue.

Methods of Somatic Gene Therapy

The enabling technology of somatic gene therapy is based on strategies for delivering genes. To do this, scientists have developed gene delivery vehicles called vectors, which encapsulate therapeutic genes for delivery to cells. Many of the vectors currently in use are based on attenuated or modified versions of viruses. Over billions of years of evolution, viruses have developed extraordinarily efficient ways of targeting cells and of delivering their genomes, which unfortunately leads to disease. The challenge is to remove the disease-causing components of the virus and insert recombinant genes that will be therapeutic to the patient. The modified viruses cannot replicate in the patient, but they retain the ability to efficiently deliver genetic material. Another strategy is based on synthetic vectors in which complexes of DNA, proteins, or lipids are formed in particles capable of efficiently transferring genes. The basic challenge for gene therapy is how to deliver the DNA into large numbers of cells of the body, and more specifically into the nucleus of the cells in a way that is efficient, specific for the right types of cells, and safe for the patient. We already mentioned that foreign DNA can be microinjected into each individual embryo for germ line gene therapy. This is not practically possible when doing somatic gene therapy because far too many cells need to receive the foreign DNA for enough therapeutic protein to be produced. Most strategies involve the use of viruses or artificial virus-like particles. Viruses have very efficient methods for getting into cells. The viruses are genetically modified so that they cannot replicate or reproduce, but they can still enter cells of the body. When the foreign DNA is put into the modified virus, the virus can infect body cells and release the foreign DNA into the cell; however, the virus will not reproduce so that it might be contained only for its intended use. Viruses have receptors that attach to specific molecules on the surface of cells. The process is analogous to traveling in a car from one country to the next. You are in the car that transports you, just as the foreign DNA is in a virus that can transport it into the cell. At the border, you are allowed to enter the next country with the suitable paperwork such as a passport, just as the virus attaches and enters the cell by using specific receptor molecules. Modified viruses that carry therapeutic genes could be called vehicles and more specifically "vectors."

ADVANTAGES AND DISADVANTAGES OF USING VIRUS VECTORS

What are the advantages and disadvantages of using virus vectors in somatic cell gene therapy? The virus vector is immunogenic, meaning that the body will develop an immune response to eliminate the virus from the body. Consequently, virus vectors may not be repeatedly administered. If large numbers of virus vectors are administered, sufficient numbers of cells may receive the therapeutic gene to have its beneficial effect. In clinical trials, these virus vectors have proven to be safe, except for one unfortunate case. Eighteen-year-old Jesse Gelsinger, a participant in the experimental gene therapy trial for ornithine transcarbamylase (OTC) deficiency, died from complications of a genetic therapy clinical trial. His OTC deficiency prevented the normal clearing of by-products of his immunological reaction to the virus vector. No other patients in the study had such side effects and the events could not have been foreseeable based on laboratory tests prior to initiation of the study. To date, the efficacy of gene therapy clinical trials has fallen short of expectations; much more research needs to be done to achieve sufficient success before the FDA approves gene therapy as safe for clinical use.

CURRENT STATUS OF SOMATIC GENE THERAPY

The following list includes diseases that are being considered for intervention by gene therapy. It illustrates the potential value for human gene therapy protocols. Certainly not all of these diseases will be treatable by gene therapy, but some of them will be: acute hepatic failure, acute myelogenous leukemia, adenocarcinoma of prostate, advanced cancer, advanced mesothelioma, amyotrophic lateral sclerosis, Alzheimer's disease, bladder cancer, brain tumors, breast cancer, Canavan disease, cervical cancer, chronic granulomatous disease, chronic lymphocytic leukemia, chronic myelogenous leukemia, colon carcinoma, colorectal cancer, critical limb ischemia, cubital tunnel syndrome, cystic fibrosis, cytomegalovirus diseases, diffuse coronary artery disease, diabetic insensate foot ulcer, Epstein Barr virus diseases, EBV-positive Hodgkin's disease, extraovarian cancer, Fabry disease, familial hypercholesterolemia, Fanconi's anemia, gaucher disease, glioblastoma multiform, gliomas, gyrate atrophy, head and neck squamous cell carcinoma, hip fractures, HIV, HIV lymphomas, leptomeningeal carcinomatosis, leukocyte adherence deficiency, limb girdle muscular dystrophy, liver cancers, lung cancers, lymphoid malignancies, melanoma, mild hunter syndrome, multiple sclerosis, myocardial angiogenesis, neuroblastoma, non-Hodgkin's lymphoma, non–small cell lung cancer (NSCLC), oral squamous cell carcinoma, ovarian cancer, partial ornithine transcarbamylase deficiency, pediatric brain tumors, pediatric malignant astrocytomas, peripheral artery disease, peripheral artery occlusive disease, peripheral vascular disease, prostate cancer, purine nucleoside phosphorylase deficiency, renal cell carcinoma, retinoblastoma, rheumatoid arthritis, sarcomas, severe combined immune deficiency, severe hemophilia A/B, small cell lung cancer, stale exertional angina, venous leg ulcer, and x-linked severe combined immune deficiency.

ETHICAL ISSUES

For the last generation much has been written about the issues posed by testing and screening for genetic abnormalities and diseases. As genetics has increased its power and predictive capabilities, the debate has been rejoined. Under what circumstances might it be reasonable to screen newborns for genetic abnormalities? How should society respond to developments in genetics and the genetic testing for disease? How should this knowledge be used to benefit persons equitably and responsibly? How can human freedom be protected in the new genetic world? Are privacy and confidentiality being adequately protected or are new rules and laws needed? What about our responsibilities for knowledge? Do we have a right not to know our genetic future?

The best place to begin thinking about these issues is a recent report by a committee from the National Academy of Sciences. The NAS identifies four key moral principles that should be used in considering screening and testing in genetic disease: autonomy, privacy, confidentiality, and equity.

Autonomy requires that accurate information be available to individuals about their genetic makeup and even, at a prior stage of the discussion, what testing may uncover. They need to know what testing is, how it works, and what it may lead to. This is the standard and formal consent law of medicine, but as the NAS points out, this model may not be adequate where one blood sample can be tested for dozens of genetic conditions. It also notes special problems in screening newborns for genetic problems, especially those that cannot be treated.

Knowledge may be power, but it can negatively affect people, especially when no treatment can be provided. What good is knowledge to a child? For future life decisions the testing can be done when the child is older and has the capacity to make informed decisions.

Confidentiality is a key value in medicine, and genetic medicine is no exception. Genetic information should only be known to the professional and the patient. Are there exceptions? Perhaps a few in cases: (1) the harms from not revealing that the genetic problem is great, for example a relative needs treatment, and (2) the tested person will not reveal the information voluntarily. What if third parties such as employers or insurers apply for such information? Perhaps individuals will be discriminated against in such cases and excluded from insurance.

Equity is at risk in genetic medicine because it reveals a world in which people are not created equal. Some people have genes that increase their risk of fatal or severe illness. Should these sorts of risks be shared or should the cost be borne by those who have them? Should medical insurance be rated across people in general or "risk related" so that those with high risk pay more? Should employers use genetic information to assign people to jobs? In general the NAS argues for strict rules of confidentiality and privacy and for the broadest possible rules against discrimination based on genetics. It believes that genetics is not something under the control of the individual, and therefore the only relevant reason for discrimination is lack of ability to perform a certain social task. Discrimination against people because they are different is unjust.

In chapter 25 Rhodes focuses less on autonomy and more on social bonds. She asks whether we have a right not to know about our genetic destiny or contribute to knowledge about the destiny of family members. She answers a qualified no to the problem of genetic autonomy. She believes that since all of our decisions will affect those closest to us, we have an obligation to know our genetic destiny. Failure to know it will result in harm to those who have close social ties to us. Furthermore she argues that in situations where we have close social ties we have an obligation to provide knowledge to our kinship groups. She believes that knowledge should be shared for the benefit of the social group and not just the individual and that individuals have responsibilities in genetic medicine to those closest to them. Surprisingly, to Rhodes these social bonds mean that the individual and the health professional are not the only people with responsibilities in the area of genetic medicine. Those with social ties have a network of rights and responsibilities as part of that network.

GENE THERAPY

Many human diseases have some basis in the genetic makeup of those afflicted. Our genes heavily influence our physical features and various genetic combinations may make some people more susceptible to some types of diseases. Genotype—one's genetic makeup—is known to play a role in depression, addiction, and cancer, for example. Some human diseases are more specifically and directly genetic. In those cases having a specific gene or cluster of genes means having a specific disease such as cystic fibrosis or Huntington's disease.

For the last decade gene therapy has been the most exciting area in research and treatment of diseases with a genetic origin. Gene therapy is used to attack the root of the problem, the genotype of the afflicted person. If you lack a gene that causes the development of a certain liver enzyme, for example, the best treatment is not taking pills every day to replace the enzyme. The best treatment would be causing the liver to make that missing enzyme. Once your liver had the missing gene, it could do the job. So why not supply the missing gene? For a number of diseases, researchers have been trying to do just that over the past decade. There are actually two ways to do it. One way is to use a vector to add the missing gene to the diseased organ of a specific human being; typically a virus is used as a vector. Vector in this case refers to a way to put the gene where it is needed. The new gene is attached to the virus and the virus is then injected into the organ that needs the gene. The hope is that the virus will invade enough cells to deliver the new gene to enough of the affected organ that it will produce the missing enzyme in sufficient quantity to be an effective treatment for the disease. Another way to perform gene therapy is to start at the beginning with germ line gene therapy, which does not work with adults or adolescents but with the fertilized egg. In this process a gene is added that covers or masks the genes that cause disease; thus the enzymes needed by a particular organ are provided right from the start. It is called germ line gene therapy because it attacks the whole genetic destiny of that individual.

From the description above it is obvious why the first gene therapy focused on somatic cell therapy for adults. It required the consent of a specific patient who was sick and needed treatment that would effectively combat his or her illness. Somatic cell therapy does not involve genetic changes that will alter the lives of future generations in ways that they could not and did not consent to. Second, somatic cell therapy did not lend itself to using gene therapy to enhance a person's physical features such as body build or eye color.

Germ line therapy involves making genetic changes for future generations without knowing what they might want. It also lends itself to being used to enhance genetic traits of individuals. The readings in part 5 offer a rich variety of views on gene therapy. Walters and Palmer discuss a variety of issues related to somatic cell therapy but they are also supporters of germ line interventions. The experts assembled by the AAAS doubt that germ line gene therapy will work, and they raise many questions about its moral propriety. They believe that it can be dangerously misused for enhancement schemes, which they strongly criticize. By contrast, Engelhardt assumes that we will use genetic therapy for enhancement and he sees little wrong with it. Engelhardt's argument is based on the idea that the profound moral pluralism of our day means that we do not know what the good life is and that we can no longer appeal to human nature to define it. Since we cannot appeal to human nature and since there are a variety of ways of living a good life, the only moral restriction we should put on gene therapy is that it not be used in an evil way. This would mean not using it to make someone worse off. On the other hand, people who want to try gene therapy to enhance their body build or change their eye color or hair color because they think that blond and blue-eyed people are better should not be prevented from doing this, in Engelhardt's view.

24

Social, Legal, and Ethical Implications of Genetic Testing

National Academy of Sciences

Each new genetic test that is developed raises serious issues for medicine, public health, and social policy regarding the circumstances under which the test should be used, how the test is implemented, and what uses are made of its results. Should people be allowed to choose or refuse the test, or should it be mandatory, as newborn screening is in some states? Should people be able to control access to the results of their tests? If test results are released to third parties such as employers or insurers, what protections should be in place to ensure that people are not treated unfairly because of their genotype?

The answers to these questions depend in part on the significance given to four important ethical and legal principles: autonomy, confidentiality, privacy, and equity. A review of the meaning of those concepts and how they are currently protected by the law provides a starting point for the development of recommendations on the degree of control people should have in deciding whether to undergo genetic testing and what uses should be made of the results. The task is a pressing one. In a 1992 national probability survey of the public, sponsored by the March of Dimes, 38 percent of respondents said that new types of genetic testing should be stopped altogether until the privacy issues are settled.[1] This chapter reviews some of the conflicts that will arise in the research and clinical settings, and suggests general principles that should be the starting point for policy analyses in this evolving field.

KEY DEFINITIONS

Autonomy

Ethical Analysis

Autonomy can be defined as self-determination, self-rule, or self-governance. Autonomous agents or actions presuppose some capacity of reasoning, deciding, and willing. Moral, social, and legal norms establish obligations to respect autonomous agents and their choices. Respect for personal autonomy implies that agents have the right or power to be self-governing and self-directing, without outside control. In the context of genetic testing and screening, respect for autonomy refers to the right of persons to make an informed, independent judgment about whether they wish to be tested and then whether they wish to know the details of the outcome of the testing. Autonomy is also the right of the individual to control his or her destiny, with or without reliance on genetic information, and to avoid interference by others with important life decisions, whether these are based on genetic information or other factors. Respect for autonomy also implies the right of persons to control the future use of genetic material submitted for analysis for a specific purpose (including when the genetic material itself and the information derived from that material may be stored for future analysis, such as in a DNA bank or registry file).

Even though respect for autonomy is centrally important in our society, it is not absolute. It can be overridden in some circumstances, for example, to prevent serious harm to others, as is the case in mandatory newborn screening for phenylketonuria (PKU) and hypothyroidism.

Legal Issues

The legal concept of autonomy serves as the basis for numerous decisions protecting a person's bodily integrity. In particular, cases have held that competent adults have the right to choose whether or not to undergo medical interventions.[2] Before people make such a choice, they have a right to be informed of facts that might be material to their decision,[3] such as the nature of their condition and its prognosis,[4] the potential risks and benefits of a proposed test or treatment,[5] and the alternatives to the proposed intervention.[6] In the genetics context, health care providers have been held liable for not providing the information that a genetic test is available.[7]

People also have a right to be informed about and to control the subsequent use of tissue that has been removed from their bodies.[8] There is some leeway under the federal regulations governing research involving human subjects for researchers to undertake subsequent research on blood samples provided for genetic tests (as in the newborn screening context) as long as the samples are anonymous and as long as the subsequent use was not anticipated at the time the sample was collected.[9] If the additional test was anticipated at the time the sample was collected, informed consent for that use should be obtained prior to the collection of the original sample. Such an approach is thought appropriate to avert conflicts of interest such as a physician/researcher suggesting that a patient undergo a particular test when the researcher actually wanted the tissue for the researcher's own additional use in a research or commercial project. In such a situation, the patient's autonomy is compromised even if the sample is used anonymously in the subsequent use. A report from the Office of Technology Assessment similarly stressed the importance of knowledge and consent:

> The consent of the patient is required to remove blood or tissue from his or her body, and also to perform tests, but it is important that the patient be informed of all the tests which are done and that a concern for the privacy of the patient extends to the control of tissues removed from his or her body.[10]

Privacy

Ethical Analysis

Among the various definitions of privacy, one broad definition captures its central element: privacy is "a state or condition of limited access to a person."[11] People have privacy if others lack or do not exercise access to them. They have privacy if they are left alone and do not suffer unauthorized intrusion by others. Once persons undergo genetic tests, privacy includes the right to make an informed, independent decision about whether—and which—others may know details of their genome (e.g., insurers, employers, educational institutions, spouses, and other family members, researchers, and social agencies).

Various justifications have been offered for rules of privacy; first, some philosophers argue that privacy rights are merely shorthand expressions for a cluster of personal and property rights, each of which can be explicated without any reference to the concept of privacy. In making this argument, Judith Jarvis Thompson holds that privacy rights simply reflect personal and property rights, such as the rights not to be looked at, not to be overheard, and not to be caused distress.[12] A second justification holds that rights to privacy are important instruments or means to other goods, including intimate relations such as trust and friendship. Being able to control access to themselves enables people to have various kinds of relationships with different people, rather than being equally accessible to all others. A third approach finds the basis for rights to privacy in respect for personal autonomy. Decisional privacy is often very close to personal autonomy. The language of personal autonomy reflects the idea of a domain or territory of self-rule, and thus overlaps with zones of decisional privacy.

Whatever their rationale or justification, rights of privacy are the subject of ongoing debate about their scope and weight. However, their scope is not unlimited, and they do not always override all other competing interests, such as the interests of others.

Legal Issues

In the legal sphere, the principle of privacy is an umbrella concept encompassing issues of both autonomy and confidentiality. The right to make choices about one's health care is protected, in part, by the right to privacy guaranteed by the U.S. Constitution, as well as state constitutions. This includes a right to make certain reproductive choices,[13] such as whether to use genetic testing.[14] It also includes a right to refuse treatment.

An entirely different standard of privacy protects personal information. A few court decisions find protection for such information under the constitutional doctrine of privacy,[15] but more commonly, privacy protection against disclosure of personal information is found under common law tort principles.[16] In addition, there is a federal privacy act,[17] as well as state statutes protecting privacy.

Confidentiality

Ethical Analysis

Confidentiality as a principle implies that some body of information is sensitive, and hence, access to it must be controlled and limited to parties authorized to have such access. The information provided within the relationship is given in confidence, with the expectation that it will not be disclosed to others or will be disclosed to others only within limits. The state or condition of nondisclosure or limited disclosure may be protected by moral, social, or legal principles and rules, which can be expressed in terms of rights or obligations. In health care and various other relationships, we grant others access to our bodies. They may touch, observe, listen, palpate, and even physically invade. They may examine our bodies as a whole or in parts, and parts, such as tissue, may be removed for further study, as in some forms of testing. Privacy is necessarily diminished when others have such access to us: rules of confidentiality authorize us to control and thus to limit further access to the information generated in that relationship. For example, rules of confidentiality may prohibit a physician from disclosing some information to an insurance company or an employer without the patient's authorization.

Rules of confidentiality appear in virtually every code or set of regulations for health care relationships. Their presence is not surprising, because such rules are often justified on the basis of their instrumental value: if prospective patients cannot count on health care professionals to maintain confidentiality, they will be reluctant to allow professionals the full and complete access necessary for diagnosis and treatment. Hence, rules of confidentiality are indispensable for patient and social welfare; without those rules, people who need medical, psychiatric, or other treatment will refrain from seeking or fully participating in it. Another justification for rules of confidentiality is based on the principles of respect for autonomy and privacy, above. Respecting persons involves respecting their zone of privacy and accepting their decisions to control access to information about them. When people grant health care professionals access to them, they should retain the right to determine who else has access to the information generated in that relationship. Hence, the arguments for respect for autonomy and privacy support rules of confidentiality. Finally, duties of confidentiality often derive from explicit or implicit promises in the relationship. For instance, if the professional's public oath or the profession's code of ethics promises confidentiality of information, and the particular professional does not specifically disavow it, then the patient has a right to expect that information generated in the relationship will be treated as confidential.[18]

There are at least two distinct types of infringements of rules of confidentiality. On the one hand, rules of confidentiality are sometimes infringed through deliberate breaches. On the other hand, rules of confidentiality are often infringed through carelessness, for example, when health care professionals do not take adequate precautions to protect the confidential information. Some commentators argue that both carelessness and modern practices of health care have rendered medical confidentiality a "decrepit concept," since it is compromised routinely in the provision of health care.[19]

It is widely recognized that the rules of confidentiality are limited in at least two senses: (1) some information may not be protected, and (2) the rules may sometimes be overridden to protect other values. First, not all information is deemed confidential, and patients do not have a right to expect that such information will be protected from disclosure to others. For example, laws frequently require that health care professionals report gunshot wounds, venereal diseases, and other communicable diseases such as tuberculosis. Second, health care professionals may also have a moral or legal right (and sometimes even an obligation) to infringe rules of confidentiality, for example, to prevent a serious harm from occurring. In such cases, rules of confidentiality protect the information, but they can be overridden in order to protect some other value. Judgments about such cases depend on the probability of serious harm occurring unless confidentiality is breached. Any justified infringements of rules of confidentiality should satisfy the conditions identified earlier in the discussion of justified infringements of the principle of respect for autonomy.

Legal Issues

The legal concept of confidentiality focuses on the information that people provide to their physicians. The protection of confidentiality is thought to serve an important public health goal in encouraging people to seek access to health care.

It is thought that the patient's interest can be served only in an atmosphere of total frankness and candor.[20] Without the promise of confidentiality, people might avoid seeking medical treatment, thus potentially harming themselves as well as the community. In fact, the first doctor-patient confidentiality statute was passed in 1828 in New York during the smallpox epidemic to encourage people to seek health care. Various legal decisions have protected confidentiality of health care information,[21] as have certain state and federal statutes. Confidentiality of health care information is also protected because disclosure of a person's medical condition can cause harm to him or her. An alternative set of legal principles—those penalizing discrimination (see below)—protects people against unfair uses of certain information.

Equity

Ethical Analysis

Issues of justice, fairness, and equity crop up in several actions, practices, and policies relating to genetic testing. It is now commonplace to distinguish formal justice from substantive justice. Formal justice requires treating similar cases in a similar way. Standards of substantive or material justice establish the identity of the relevant similarities and differences and the appropriate responses to those similarities and differences. For instance, a society has to determine whether to distribute a scarce resource such as health care according to persons' differences in need, social worth, or ability to pay.

One crucial question is whether genetic disorders or predispositions provide a basis for blocking access to certain social goods, such as employment or health insurance. Most conceptions of justice dictate that employment be based on the ability to perform particular tasks effectively and safely. For these conceptions, it is unjust to deny employment to someone who meets the relevant qualifications but also has a genetic disease. Frequently these questions of employment overlap with questions of health insurance. Practices of medical underwriting in health insurance reflect what is often called "actuarial fairness"—that is, grouping those with similar risks together so insurers can accurately predict costs and set fair and sufficient premium rates. Although actuarial fairness may be intuitively appealing, critics argue that it does not express moral or social fairness. According to Norman Daniels, there is "a clear mismatch between standard underwriting practices and the social function of health insurance" in providing individuals with resources for access to health care.[22]

The fundamental argument for excluding genetic discrimination in health insurance amounts to an argument for establishing a right to health care. One of the central issues in debates about the distribution of health care is one's view of the "natural lottery," in particular, a "genetic lottery."[23] The metaphor of a lottery suggests that health needs result largely from an impersonal natural lottery and are thus undeserved. But even if health needs are largely undeserved because of the role of chance, society's response to those needs may vary, as H. Tristram Engelhardt notes, depending on whether it views those needs as unfair or as unfortunate.[24] If health needs are unfortunate, but not unfair, they may be the object of individual or social compassion. Other individuals, voluntary associations, and even society may be motivated by compassion to try to meet those needs If, however, the needs are viewed as unfair as well as unfortunate, society might have a duty of justice to try to meet those needs. One prominent argument for the societal provision of a decent minimum of health care is that, generally, health needs are randomly distributed and unpredictable, as well as overwhelming when health crises occur.[25] Because of these features of health needs, many argue that it is inappropriate to distribute health care according to merit, societal contribution, or even ability to pay. Another version of the argument from fairness holds that health needs represent departures from normal species functioning and deprive people of fair equality of opportunity. Thus, fairness requires the provision of health care to "maintain, restore, or compensate for the loss of normal functioning" in order to ensure fair equality of opportunity.[26] Several committee members expressed concerns that these stated arguments are somewhat weakened by the fact that a number of diseases are not the result of random events, but are brought on or exacerbated by dispensable habits such as cigarette smoking and excessive alcohol ingestion. While our and other societies attempt to discourage such habits by education and taxation, there is general agreement that access to full health care must be ensured once illness develops. If a tendency to abuse alcohol, for example, were to have a genetic predisposition, an additional argument could be made for providing the same level of health care to everyone since a person does not choose his or her genetic propensities. The argument that society should guarantee or provide a decent minimum of health care for all citizens and residents points toward a direction for health policy, but it does not determine exactly how much health care the society should provide relative to other goods it also seeks. And within the health care budget there will be difficult allocation questions, including how much should be used for particular illnesses and for particular treatments for those illnesses. Questions of allocation cannot be resolved in the abstract. In democratic societies, they should be resolved through political processes that express the public's will in specifying and implementing a conception of a decent minimum, an adequate level, or a fair share of health care in the context of scarce resources. As the President's Commission noted in 1983, it is reasonable for a society to turn to fair democratic political procedures to choose among alternative conceptions of adequate health care, and in view of "the great imprecision in the notion of adequate health care, it is especially important that the procedures used to define that level be—and be perceived to be—fair."[27]

Legal Issues

The concept of equity serves as the underpinning for a variety of legal doctrines and statutes. Certain needy people are provided health care, including some genetics services, under government programs such as Medicaid. In addition, some legislative efforts have been made to prohibit discrimination based on genotype. For example, some states have statutes prohibiting discrimination in employment based on one's genotype.[28] And nearly all people over age 65 are deemed to have a right to care (under Medicare).

CURRENT PRACTICE OF PROTECTION IN GENETICS

The development of genetic testing has raised numerous concerns about autonomy, confidentiality, privacy, and equity that are exacerbated by the range of contexts in which such tests are undertaken, the sheer volume of tests that could be offered, the many uses that can be made of test results, and the variety of institutions that store genetic information. To date, most genetic testing has been done in the reproductive context or with newborns, to identify serious disorders that currently or soon will affect the fetus or infant. However, the types of genetic conditions or predispositions that can potentially be tested for are much broader than those signaling serious, imminent diseases. These include characteristics (such as sex or height) that are not diseases, potential susceptibility to diseases if the person comes into contact with particular environmental stimuli, and indications that a currently asymptomatic person will suffer later in life from a debilitating disease such as Huntington disease. The genetic anomalies that can be tested for range widely in their manifestations, their severity, their treatability, and their social significance. People's ability to define themselves, to manage their destiny and self-concept, will depend in large measure on the control they have over whether they and others come to know their genetic characteristics. Most medical testing is done within a physician-patient relationship. With genetic testing, however, the potential range of contexts in which it can be undertaken is large. Already, in the public health context, more than 4 million newborns are tested annually for metabolic disorders so that effective treatment can be started in a few hundred. Researchers are inviting people to participate in family studies and undergo genetic testing, including collection of DNA samples for present or future analyses. There are a growing number of nonmedical applications of genetic testing as well. In the law enforcement context, DNA testing is undertaken to attempt to identify criminal offenders. At least 17 states have DNA fingerprint programs for felons.[29] The armed services are collecting DNA samples from all members of the military, the primary purpose of which is to identify bodies of deceased soldiers. Employers and insurers may require people to undergo testing for genetic disorders for exclusionary purposes. One challenge for policy posed by this wide array of testing settings is that many of the existing legal precedents about autonomy, confidentiality, and privacy apply only to the traditional doctor-patient relationship. For example, some state statutes governing confidentiality deal only with information provided to physicians and might not cover information provided to Ph.D. researchers or employers.

There seems to be great variation among institutions and among providers in the amount of attention paid to autonomy, confidentiality, and privacy. For example, some obstetricians recognize the patient's autonomy by providing them the information about maternal serum alpha-fetoprotein (MSAFP) screening but acknowledging the patient's right to decide whether or not to undergo the test. Other obstetricians run the test on blood gathered from the woman for other purposes, if the woman does not even know she has been the subject of the test unless the obstetrician delivers the bad news that she has had an abnormal result. Geneticists differ with respect to the emphasis they place on the confidentiality of the results of genetic testing. In a survey by Dorothy Wenz and John Fleicher,[30] numerous geneticists suggested that there were at least four situations in which they would breach confidentiality and disclose genetic information without the patient's permission, even over the patient's refusal: (1) 54 percent said they would disclose to a relative the risk of Huntington disease; (2) 53 percent said they would disclose the risk of hemophilia A; (3) 24 percent said they would disclose genetic information to a patient's employer; and (4) 12 percent said they would disclose such information to the patient's insurer. Primary care physicians may be even more likely to disclose such information.[31] Health care providers should explain their policies for disclosure in advance, including for disclosure to relatives.

Institutions that store DNA samples[32] or store the results of genetic tests also differ in the amount of respect they give to autonomy, confidentiality, and privacy.[33] Some institutions do additional tests on DNA samples without the permission of the person who provided the sample. Some share samples with other institutions. Some store samples or information with identifiers attached, rather than anonymously. Indeed, storage conditions themselves differ widely. Some newborn screening programs store filter papers in a temperature-controlled, secure setting; others merely pile them in a file cabinet or storage closet. Programs also differ in the length of time the sample or the test results are maintained. Once DNA material has been submitted, there are few safeguards concerning other present or future uses that may be made of the material. DNA from the blood spots collected for newborn screening can now be extracted for further testing.[34] No standards or safeguards currently exist to govern the appropriate use of DNA analysis and storage from newborn screening tests. These possibilities raise questions about the need to obtain consent for additional and subsequent uses (particularly since consent is almost never obtained initially in newborn screening), as well as questions about the duty to warn if disorders are detected in the blood by using the new DNA extraction testing techniques.

The issue of confidentiality of genetic information will be underscored with the introduction of "optical memory cards," a credit card–sized device that stores medical information.[35] These cards have already been introduced for use in Houston city health clinics. There is sufficient computer memory on the cards to include genetic information about the person and, in the future, to include a person's entire genome.

Congressional legislation has been introduced that would require all patients to use optical memory cards This bill, the Medical and Health Insurance Information Reform Act of 1992, would mandate a totally electronic system of communication between health care providers and insurers. Such a system would be based either on the optical memory card (with a microchip capable of storing data) or on a card similar to an Automated Teller Card (which simply provides access to data stored elsewhere).

APPLYING THE PRINCIPLES TO GENETIC TESTING

The principles of autonomy, privacy, confidentiality, and equity place great weight on individuals' rights to make personal decisions without interference. This is due, in part, to the importance placed on individuals in our culture and our legal system. However, individual rights are not without bound, and the area of genetics raises important questions of where individual rights end and where responsibilities to a group—such as one's family or the larger society—begin. Medicine is generally practiced within this culture of individual rights (with provisions for patients' right to refuse treatment and right to control the dissemination of medical information about themselves), but there have been circumstances in which the medical model has been supplanted by the public health model, which encourages the prevention of disease—for example, by requiring that certain medical intervention (such as vaccinations) be undertaken and by warning individuals of health risks (e.g., through educational campaigns against smoking or through contact tracing with respect to venereal diseases). Some commentators have suggested that the public health model be applied to genetics,[36] with mandatory genetic screening and even mandatory abortion of seriously affected fetuses. A related measure might be warning people of their risk of genetic disorders.

There are several difficulties with applying the public health model to genetics, however. Certain infectious diseases potentially put society as a whole at immediate risk since the diseases can be transmitted to a large number of people in a short time. The potential victims are existing human beings who may be total strangers to the affected individual. In contrast to infectious disease, the transmission of genetic diseases does not present an immediate threat to society. Whereas infectious disease can cause rapid devastation to a community, the transmission of genetic disorders to offspring does not necessarily have an immediate detrimental effect, but rather creates a potential risk for a future generation in society.[37] U.S. Supreme Court cases dealing with fundamental rights have held that harm in the future is not as compelling a state interest as immediate harm.[38]

Moreover, the very concept of "prevention" docs not readily fit most genetic diseases. In the case of newborn screening for PKU, treatment can prevent mental retardation. However, with many genetic diseases today, the genetic disease itself is not being prevented, but rather the birth of a particular individual with the disease is prevented (e.g., when a couple, each of whom is heterozygous for a serious recessive disorder, chooses not to conceive or chooses to terminate the pregnancy of a fetus who is homozygous for the disorder). This sort of prevention cannot be viewed in the same way as preventing measles or syphilis, for example. There is a great variation among people in their view of disability and what constitutes a disorder to be "prevented." Many people will welcome a child with Down syndrome or cystic fibrosis into their family. In addition, some individuals have religious or other personal moral objections to abortion; even mandatory carrier status screening or prenatal screening without mandatory abortion may be objected to because people who object to abortion are concerned that the abortion rate will rise among those in the general population who learn of genetic risks to their fetus. Furthermore, some people with a particular disability or genetic risk may view mandatory genetic testing for that risk or disability as an attempt to eradicate their kind, as a disavowal of their worth.

Mandatory genetic testing might also have devastating effects on the individuals who are tested. Unlike infectious disease (which can be viewed as external to the person), genetic disease may be viewed by people as an intractable part of their nature. Persons who learn, against their will, that they carry a defective gene may view themselves as defective. This harm is compounded if they did not choose to learn the information voluntarily. This assault on personal identity is less likely with infectious diseases, although AIDS and genital herpes (for example) can also have a negative impact on self-image. Moreover, most genetic defects, unlike most infectious diseases, generally cannot now be corrected.[39] Thus, the unasked-for revelation that occurs through mandatory genetic testing can haunt the person throughout his or her life and can have widespread reverberations in the family, including others who may be at risk or related as partners. The information can serve as the basis for discrimination against the individual. Additionally, policy concerns raised by attempts to stop the transmission of genetic diseases differ from those addressed to infectious diseases because genetic diseases may differentially affect people of different races or ethnic backgrounds. For that reason, some commentators contest the applicability of the infectious disease model to government actions regarding genetic disorders. Catherine Damme notes that "unlike infectious disease which generally knows no ethnic, racial, or gender boundaries, genetic disease is the result of heredity"—leaving open the possibility for discriminatory governmental actions.[40]

The government has discretion with respect to which infectious diseases it tackles. For example, it can decide to require screening for syphilis but not chlamydia, or to require vaccinations for smallpox but not for diphtheria. Government action with respect to genetic diseases is likely to be regarded much differently, especially with respect to disorders for which an effective treatment does not exist and, consequently, the only medical procedure available is the abortion of an affected fetus. Minority groups who have been discriminated against in the past may view a screening program that targets only disorders that occur within their racial or ethnic group as an additional attack, and may view abstention from reproduction or the abortion of offspring based on genetic information as a form of genocide.[41]

Those commentators who argue that the infectious disease precedents justify mandatory genetic screening fail to recognize that even in the case of infectious disease, very few medical procedures are mandated for adults. Adults are not forced to seek medical diagnosis and treatment even if they have a treatable infectious disease. Laws that required compulsory infectious disease screening prior to marriage (e.g., for venereal disease) are being repealed. For example, New York abolished its requirements for premarital gonorrhea and syphilis testing. One of the reasons for the abolition of the requirements was that they were not the most appropriate way to reach the population at risk.[42]

Mandating diagnosis and treatment for genetic disorders is particularly problematic when the concept of disease is so flexible. Arno Motulsky has noted that "the precise definition of 'disease' regardless of etiology, is difficult."[43] He notes that maladies such as high blood pressure and mental retardation are based on arbitrary cutoff levels. David Brock similarly noted that most disorders lie between the extremes of Tay-Sachs disease and alkaptonuria: what a physician advises "depends as much on the physician's ethical preconceptions as his medical experience."[44]

Despite the fact that the public health model does not fit the situation of genetics, the individual rights model should not be seen as absolute. There are certain situations in which the values of autonomy, privacy, confidentiality, and equity should give way to prevent serious harm to others. Determining the exceptions to these general principles is no easy matter, however. There may be instances in which harm can be prevented by violating one of these principles, but in which the value of upholding the principles will nonetheless outweigh the chance of averting harm. In each instance, it will be necessary to assess several factors: How serious is the harm to be averted? Is violating one of the principles the best way to avert the harm? What will be the medical, psychological, and other risks of violating the principle? What will be the financial costs of violating the principle? The following section addresses the issues raised by the application of these principles—autonomy, privacy, confidentiality, and equity—in the contexts of clinical genetics, other medical practice, genetics research, and so forth. It also provides guidance for determining the appropriate circumstances for exceptions to these principles. The chapter concludes with the committee's recommendations on these issues.

ISSUES IN GENETIC TESTING

Autonomy

One important way to ensure autonomy with respect to genetic testing is to provide adequate information upon which a person can make a decision whether or not to undergo testing. A proper informed consent in medicine generally involves the presentation of information about the risks, benefits, efficacy, and alternatives to the procedure being undertaken. In addition, recent cases and statutes have recognized the importance of disclosures of any potential conflicts of interest that the health care professional recommending the test may have, such as a financial interest in the facility to which the patient is being referred. In the genetics context, this would include disclosure about equity holdings or ownership of the laboratory, dependence on test reimbursement to cover the costs of counseling, patents, and so forth. It would also include disclosure of any planned subsequent uses of the tissue samples, even if such uses are to be anonymous.

Various kinds of information are relevant to people who are attempting to exercise their autonomy by deciding whether or not to undergo genetic testing. This includes information about the severity, potential variability, and treatability of the disorder being tested for. If, for example, carrier status testing is being proposed for a pregnant woman or prenatal testing is being proposed for her fetus, she should be told whether the disorder at issue can be prevented or treated, or whether she will be faced with a decision about whether or not to abort. The proposed informed consent guidelines for research involving genetic testing suggested by the Alliance of Genetic Support Groups provide an excellent starting point for the development of informed consent policies in the genetics area. The potential development of multiplex testing adds another wrinkle to the issue of informed consent for genetic testing. If 100 disorders are tested from the same blood sample, it may be difficult to apply the current model of informed consent in which a health care provider gives information about each disorder and the efficacy of each test to the patient in advance of the testing. The difficulty in applying the traditional mechanisms for achieving informed consent does not provide an excuse for failing to respect a patient's autonomy and need for information, however. New mechanisms may have to be developed to protect these rights. It will be possible to have results reported back to the physician and patient only about those tests (or types of tests) the patient chooses. The choices can be made by the patient, based, for example, on the patient learning through a computer program about the various disorders and the various tests. Or the choices can be made according to general categories—for example, the patient might choose to have multiplex testing but choose not to be informed of the results of testing for untreatable or unpreventable disorders.[45] In addition to the recognition that people are entitled to information before they make decisions, a second application of the autonomy principle comes with the recognition that the decision to participate in genetic testing and other genetics services must be voluntary. Voluntariness has been a recognized principle in past recommendations and practices involving genetics. This is in keeping with the recognized right of competent adults to refuse medical intervention, as well as the right to refuse even the presentation of medical information in the informed consent context.[46] If, for example, it becomes possible to accurately screen fetal cells isolated from a pregnant woman's blood in order to determine the genetic status of the fetus, state public health departments might be interested in requiring the test on the grounds that it is a minimally invasive procedure that can provide information to the woman (perhaps leading her to abort an affected fetus and saving the state money for care of that infant). Mandating such a test, however, would show insufficient respect for the woman's autonomy and would violate her right to make reproductive decisions.

Special Issues in the Screening and Testing of Children

The expansion of available tests fostered by the Human Genome Project will present complicated issues with respect to the testing of newborns and other children. Although there are clear legal precedents stating that adults are free to refuse even potentially beneficial testing and treatment, legal precedents provide that children can be treated without their consent (and over their parents' refusal) to prevent serious imminent harm. The U.S. Supreme Court has said that, while parents are free to make martyrs of themselves, they are not free to make martyrs of their children.[47] Medical intervention over parents' objection has been allowed in situations in which a child's life was in imminent danger and the treatment posed little risk of danger in itself.[48] Blood transfusions have been ordered for the children of Jehovah's Witnesses when the child's life was imminently endangered.[49] All states have programs to screen newborns for certain inborn errors of metabolism for which early intervention with treatment provides a clear medical benefit to the child, such as phenylketonuria. Currently, the statutes of at least two jurisdictions (the District of Columbia and Maryland) clearly provide that newborn screening is voluntary.[50] In at least two states (Montana and West Virginia), screening is mandatory and there is no legal provision for parental objection or refusal based on religious grounds.[51] In the rest of the states, there are grounds for parental refusal for religious or other reasons. However, although the majority of states allow objection to screening on some grounds, very few statutes require that the parents or guardians of an infant either be sufficiently informed that they can choose whether or not their infant should submit to the screening or be told they have the right to object. Two states (Missouri and South Carolina) have criminal penalties for parents who refuse newborn screening of their children.[52]

The idea behind mandatory newborn screening is a benevolent one—to try to ensure that all children get the benefits of screening for PKU and hypothyroidism, for which early treatment can make a dramatic difference in the child's well-being by preventing mental retardation. Yet there is little evidence that it is necessary to make a newborn screening program mandatory to ensure that children are screened under the program. Recent studies show that the few states with voluntary newborn screening programs screen a higher percentage of newborns than some states with mandatory newborn screening programs; for 1990 voluntary programs reported reaching 100 percent of newborns in their states, states with mandatory programs reported reaching 98 percent, and some even less than 96 percent.[53] Relevant research has suggested that even when a newborn screening program is completely voluntary and parents may refuse for any reason, the actual refusal rate is quite low, about 0.05 percent (27 of 50,000 mothers). In that study, most nurses reported that it required only one to five minutes to inform a mother about newborn screening.[54]

Newborn screening for PKU—like a necessary blood transfusion for a child over the parents' refusal—has been justified on the basis of the legal doctrine of *parens patriae,* where the state steps in to order an intervention to protect a child from substantial, imminent harm. In the era of the Human Genome Project, when additional tests are being developed, some people are promoting newborn screening in part for less immediate and less clear benefits. Proposed guidelines have suggested that another benefit of newborn screening "might take the form of inscription in registries for later reproductive counseling (material PKU) or of surveillance of phenotypes (congenital adrenal hyperplasia)."[55] To achieve such an outcome, the resulting children would need to be followed until the age when reproductive counseling was appropriate—or when symptoms manifest—a daunting task in this age of mobility.

The first newborn screening programs were for disorders in which early treatment of the newborn was effective. Increasingly, however, testing is suggested for untreatable disorders. In such instances, the justification is not the benefit to the newborn but the benefit to the parents for future reproductive plans. For such reasons, several countries—and some states in the United States (e.g., Pennsylvania)—screen newborns for Duchenne muscular dystrophy. This medical intervention has no immediate medical benefit for the newborn, and earlier screening of the parents could be obtained through other methods, even when (as in the case of Duchenne muscular dystrophy and some other conditions) they may not realize they are at risk.

Moreover, screening newborns for genes for untreatable disorders or carrier status may have disadvantages. The children may be provided with information that, at the age of consent, they would rather not have. Parents might treat them differently if the results are positive. Parents may stigmatize or reject children with the abnormal genes, or may be less willing to devote financial resources to education or other benefits for such children. In addition, release of the test results might cause them to be uninsurable, unemployable, and unmanageable. There are additional benefits from voluntariness in newborn screening. Informing parents about newborn screening in advance of testing allows quality assurance: parents can check to see if the sample was actually drawn. As children are being released from the hospital increasingly early, due to insurance pressures, they might receive a false negative result because blood levels of phenylalanine have not yet risen sufficiently to be detected if elevated. Informed motivated parents may need to bring their babies to be screened after release from the hospital in order to ensure an accurate test result. The recommended informed consent process can provide the necessary education and motivation that will be required to make the return trip far better than mandatory programs.

In the postgenome era, people will be facing the possibility of undergoing many more genetic tests in their lifetimes, and will need to master a wealth of genetic information that is relevant to their health, their reproductive plans, and the choices they make about what to eat, where to live, and what jobs to take. The more settings in which they can be informed about genetics, the more able they will be to make these decisions. In addition, when newborn screening programs are voluntary, there is a greater chance that parents will be provided with material in advance about the disorder and have their questions answered, thus presenting the possibility that they will view it more seriously and will make a greater effort to ensure that the child receives proper treatment if a condition is detected. The disclosure of information to parents about newborn screening prior to newborn screening can be an important tool for public education about genetics.

Mandatory newborn screening should only be undertaken if there is strong evidence of benefit to the newborn from effective treatment at the earliest possible age (e.g., PKU and congenital hypothyroidism). Under this principle, screening for Duchenne muscular dystrophy would not be justified. In addition, mandatory newborn screening for cystic fibrosis would currently not be justified.[56] A prospective double-blind study in Wisconsin (the only controlled study on the subject) has not found benefits of early detection in newborn screening for CF; the treatment of children could be initiated with just as successful results based on the occurrence of symptoms. In addition to its lack of clear benefit, newborn screening for CF has a clear downside. Screening by its nature is overly broad; in newborn screening for cystic fibrosis, for example, "only 6. 1 percent of infants with positive first tests (in the Colorado and Wyoming program) were ultimately found to have cystic fibrosis on sweat chloride testing."[57] Yet one-fifth of parents with false positives on newborn screening for cystic fibrosis "had lingering anxiety about their children's health."[58] Of the parents whose infants had initial, later disproven positive reports of CF in the Wisconsin study, 5 percent still believed a year later that their child might have CF.[59] Such a reaction may influence how parents relate to their child. A report on the Wisconsin newborn screening for CF stated that of the 104 families with false positives, 8 percent planned to change their reproductive plans and an additional 22 percent were not sure whether they would change their reproductive plans.[60] In fact, in France, the newborn screening program for cystic fibrosis was terminated at the request of parents who objected to the high number of false positives.[61] Denmark stopped screening for alpha 1 amitrypsin deficiency because of negative long-term effects on the mother-child interactions associated with identifying the infant's alpha 1 amitrypsin deficiency.[62]

Even in cases where a treatment is available for a disorder detectable through newborn screening, it may not be of unequivocal benefit if started after symptoms appear. Treatment of children identified through screening for maple syrup urine disease may have only limited effectiveness at best, and parents may face a quandary about whether or not to treat. Even if hypothetical benefits exist newborn screening programs need close scrutiny to determine if the necessary treatments are actually provided to the children. In states that support screening but not treatment, families may be unable to afford treatment and thus children may not benefit from screening. Many children with sickle cell anemia, for example, do not get their necessary penicillin prophylaxis.[63] Although most states provide education about diet and nutrition to parents of infants with PKU, not all states provide the expensive essential diet or other food assistance.

Beyond the issue of the testing of newborns in state-sponsored programs, there are more general issues regarding the genetic testing of children in clinical settings. Some technologies designed to identify affected individuals will also provide information about carrier status. If an infant is tested for sickle cell anemia, for example, the test will reveal whether the infant is a carrier. In that case, the carrier status information is a by-product of the test for sickle cell anemia since obtaining information on carrier status is not the primary purpose of the testing. Questions arise as to whether that information should be reported to the infant's parents. One advantage of reporting the information is that it is relevant to the parents' future reproductive plans. If the infant is a carrier, at least one of the parents is a carrier. If both are carriers, then they are at 25 percent risk of having an affected child. On the other hand, there are disadvantages to the reporting of such information to parents. Unless education and counseling are available, they may erroneously worry that the child will be affected with a disease related to the carrier status. They may stigmatize the child or otherwise treat the child as different. In addition, the disclosure of the child's carrier status may result in disruption to the family if neither of the social parents is a carrier (which most often indicates that another man fathered the child).

Since numerous tests can be added in a newborn screening program using the initial filter paper spot, the pressure to add new tests may be difficult to resist. Under the American Society of Human Genetics (ASHG) guidelines, however, before tests are added, a rigorous analysis should be made about who will benefit, who will be harmed, and who consents. In state programs for newborn screening, subsequent anonymous uses of samples for research may be undertaken.

Voluntariness of Subsequent Uses

Many state newborn screening programs, as well as research and clinical facilities, store the filter paper spots or other DNA samples for long periods after (their initial use in genetic testing. Some states use newborn screening spots to experiment with new tests, and this would seem permissible as long as the samples are not identified and the uses were not anticipated prior to the initial test.[64] If the samples are identified, the person's permission would be required. However, researchers constitute just one group that might want access to the newborn screening spots. Such spots are of interest to law enforcement officials; in one case, police contacted a newborn screening laboratory when they were trying to identify a young murder victim.

The American Society of Human Genetics issued a statement on DNA banking and DNA data banking in 1990.[65] ASHG recommended the purposes for which samples are acquired for DNA analysis be defined in advance:

> Later access to DNA samples or to the profiles for other purposes should be permitted only when (a) a court orders the information to be released, (b) the data are to be anonymously studied, or (c) the individual from whom the sample was obtained provides written permission. In general, regardless of the purpose for which it was compiled, this information should be accorded at least the confidentiality that is accorded to medical records.[66]

Confidentiality

Confidentiality is meant to encourage the free flow of information between patient and physician so that the patient's sickness may be adequately treated. The protection of confidentiality is also justified as a public health matter, since ill people may not seek medical services in the first place if confidentiality is not protected. As a legal matter, confidentiality is generally protected in the doctor–patient relationship. However, genetic testing may not always occur within a doctor-patient relationship: a non-M.D. scientist may undertake the testing, or screening may occur in the employment setting. Moreover, it is not just the result of the test that raises concern about confidentiality. The sample itself may be stored (as in DNA banking or family linkage studies) for future use.

Genetic information is unlike other medical information. It reveals not only potential disease or other risks to the patient, but also information about potential risks to the person's children and blood relatives. The fact that geneticists may wish to protect third parties from harm by breaching confidentiality and disclosing risks to relatives is evidenced in the study by Wertz and Fletcher, cited earlier. in which half of the geneticists surveyed would disclose information to relatives over a patient's refusal. The geneticist's desire to disclose is based on the idea that the information will help the relative avoid harm. Yet this study indicated that about the same number of geneticists would disclose to the relative when the disorder was untreatable as when the disorder was treatable (53 percent would contact a relative about the risk of Huntington disease; 54 percent about the risk of hemophilia A), since most people at risk for Huntington disease have not chosen testing to see if they have the genetic marker for the disorder.[67] Geneticists may be overestimating the relative's desire for genetic information and infringing upon the relative's right not to know. They may be causing psychological harm if they provide surprising yet unwanted information for which there is no beneficial action the relative can take.[68]

In the legal realm there is an exception to confidentiality: A physician in certain instances may breach confidentiality in order to protect third parties from harm, for example, when the patient might transmit a contagious disease or commit violence against an identifiable individual.[69] In a landmark California case, for example, a psychiatrist was found to have a duty to warn the potential victim that his patient planned to kill her.[70]

The principle of protecting third parties from serious harm might also be used to allow disclosure to an employer when an employee's medical condition could create a risk to the public. In one case, the results of an employee's blood test for alcohol were given to his employer.[71] The court held the disclosure was not actionable because the state did not have a statute protecting confidentiality, but the court also noted that public policy would favor disclosure in this instance since the plaintiff was an engineer who controlled a railroad passenger train.

An argument could be made that health care professionals working in the medical genetics field have disclosure obligations similar to those of the physician whose patient suffers from an infectious disease or a psychotherapist with a potentially violent patient. Because of the heritable nature of genetic diseases, a health professional who—through research, counseling, examination, testing, or treatment—gains knowledge about an individual's genetic status often has information that would be of value not only to the patient, but to his or her spouse or relatives, as well as to insurers, employers, and others. A counterargument could be made, however, that since the health professional is not in a professional relationship with the relative and the patient will not be harming the relative (unlike in the case of violence or infectious diseases), there should be no duty to warn. The claims of the third parties to information, in breach of the fundamental principle of confidentiality, need to be analyzed, as indicated earlier, by assessing how serious the potential harm is, whether disclosure is the best way to avoid the harm, and what the risk of disclosure might be.

Disclosing Genetic Information to Spouses

The genetic testing of a spouse can give rise to information that is of interest to the other spouse. In the vast majority of situations, the tested individual will share that information with the other spouse. In rare instances, the information will not be disclosed and the health care provider will be faced with the issue of whether to breach confidentiality. When a married individual is diagnosed as having the allele for a serious recessive disorder, the spouse might claim that the health care provider has a duty to share that information with him or her to facilitate reproductive decision making.[72] A few court cases have allowed physicians to disclose medical information about an individual in order to protect a spouse or potential spouse.[73] The foundation for this approach is laid by cases allowing disclosure of communicable diseases.[74] In situations such as disclosure of information about venereal disease or AIDS, the argument is made that sacrificing confidentiality, by notifying spouses and lovers, is necessary for public health and welfare, and is essential as a warning to seriously endangered third parties where the risk of transmission is high.

Since genetic disorders are not communicable to the spouse, a counter-argument could be made that there is no legitimate reason for disclosing them. However, the spouse might have a great interest in the genetic information because he or she would like to protect any potential children from risk. Consider the case of a doctor who learns that a young man will later suffer from Huntington disease. The wife would appear to have at least some claim to that information since, if she and her husband have children, there will be a 50 percent chance that each child would inherit the disease. Similarly, each spouse would seem to have a claim to the information that the other was a carrier of a single gene for a recessive defect. Because of the importance of reproductive decisions, such information is crucial to the spouse.

Another instance in which genetic risk information arises in the marriage context is through prenatal screening. A fetus may be found to have an autosomal recessive disorder, which occurs only if both parents transmit the particular gene. If, in the course of prenatal diagnosis, it is learned that the mother is a carrier of the gene but her husband is not, the health care professional has knowledge that the husband is almost certainly not the father of the child. A claim could be made that the health care professional has a duty, or at least a right, to advise the husband of his misattributed paternity, so that he will know that any future children he has will not be at risk for that particular disorder.

On the other hand, an argument could be made that spouses should not be entitled to genetic risk information about a patient, even if it is arguably relevant to their future reproductive plans.[75] The right of reproductive decision making is viewed as the right of the individual.[76] The U.S. Supreme Court has held that a woman can abort without her husband's consent even if this will interfere with her husband's reproductive plans.[77] More recently, the U.S. Supreme Court held that a husband was not even entitled to notice that his wife intends to abort.[78] The court expressed concerns that the husband might react to the disclosure with violence, with threats to withhold economic support, or with psychological coercion.[79] Similar reactions could occur with information about misattributed paternity, particularly because the primary purpose of the testing was not to get paternity information.

Disclosing Genetic Information to Relatives

Blood relatives of the patient may have a more convincing claim than spouses for requiring that health care providers breach confidentiality. They could argue that the information about genetic risks or the availability to genetic testing may be relevant to their own future health care.[80] The strongest case for warning would exist when there is a high likelihood that the relative has the genetic defect, the defect presents a serious risk to the relative, and there is reason to believe that the disclosure is necessary to prevent serious harm (e.g., by allowing for treatment or by warning the person to avoid harmful environmental stimuli). Malignant hyperthermia is an autosomal dominant genetic condition causing a fatal reaction to common anesthesia. Prompt warning of families can literally save lives, especially from death due to minor surgeries such as setting broken bones in children.

If the patient does not want to inform relatives, however, questions arise as to whether the health care provider or counselor should contact the relative over the patient's refusal. The President's Commission for the Study of Ethical Problems in Medicine and Biomedical and Behavioral Research (1983) recommended that disclosure be made only if (1) reasonable attempts to elicit voluntary disclosure are unsuccessful; (2) there is a high probability of serious (e.g., irreversible or fatal) harm to an identifiable relative; (3) there is reason to believe that disclosure of the information will prevent harm to the relative; and (4) the disclosure is limited to the information necessary for diagnosis or treatment of the relative.[81]

Even in the more compelling situation of disclosure to relatives, the health care provider is not in a professional relationship with the relative, and previous legal cases regarding a duty to provide genetic information have all involved a health care provider in a professional relationship with the person to be informed. Although infectious disease cases provide a precedent for warning strangers about potential risks,[82] genetic diseases are simply different from infectious diseases. The only potential argument that the health care professional could make for contacting the relative is that through diagnosis of the patient, the health care professional has reason to believe that the relative is at higher risk than the general population of being affected by a genetic disorder. If disorders are highly likely and are treatable or preventable, many medical geneticists would overrule a patient's refusal to disclose and would inform a relative. Although there may be no legal obligation to single out relatives as creating a special duty for physicians, the knowledge that a defined, unknowing relative is at high risk for a serious or life-threatening, treatable disease may allow rare exceptions to the principle of confidentiality.

Confidentiality and Discrimination When Third Parties Seek Genetic Information

Many entities may have an interest in learning about people's genetic information. Insurers, employers, bankers, mortgage companies, educational loan officers, providers of medical services, and others have an interest in knowing about a person's future health status. Already, people have been denied insurance, employment, and loans based on their genotype. Such discrimination has occurred both when the information has been obtained through genetic testing and when the information has been obtained in other ways (e.g., inadvertent release of a relative's medical record or disclosure of payment for medical service for a child).[83]

In the future, third parties may want access to genetic information or may wish to mandate genetic testing. In child custody cases, one spouse may claim that the other spouse should not get custody because of his or her genetic profile, for example, when the latter person has the gene for a serious, untreatable late-onset disorder. Professional schools (such as medical schools or law schools) may wish to deny admission to someone with such a disorder on the theory that such a person will have a shortened practice span.

Insurers underwriting individual health insurance currently use medical information to determine whether coverage should be granted and to determine how to price a particular policy. According to the Office of Technology Assessment, each year about 164,000 applicants are denied individual health insurance.[84] Far more Americans are covered by group plans—85 to 90 percent—with about 68 percent[85] covered by employment-based group plans rather than by individual plans. Although medical underwriting is not generally done as part of large employers' group policies, medical information is sometimes used against people in other ways in that context. People with medical problems or whose family members have medical problems have been refused jobs because employers do not want their insurance premiums increased due to payments for the care of the employee or the employee's family members.

In addition, employers that self-insure may choose to restrict coverage under their insurance plans so as not to pay for care for existing employees. One major airline already permanently excludes coverage for preexisting conditions for new employees.[86] Other employers have curtailed plan benefits once an employee has been diagnosed as having a particular disorder. In *McGann v. H. A. H. Music Co.,* for example, a man was covered by employer-provided commercial insurance that had a million dollar medical benefit maximum.[87] Once the employee was diagnosed as having AIDS, however, the employer switched to self-insurance and established a $5,000 limitation for AIDS, while keeping the million dollar cap for other disorders. The court held that an employer who is self-insured could modify its plan in this way—an ominous decision when one considers that at least 65 percent of all companies and 82 percent of companies with more than 5,000 employees are self-insured.[88] The U.S. Supreme Court decided not to hear the case and let stand the lower court's decision. Employees who are covered by their employers' self-insurance are thus in a precarious position, akin to having no insurance at all:

> When one considers that many employees contribute substantial amounts of money to purchase this "coverage," that many of them forego purchasing other insurance products in reliance on this coverage, and that few of them understand the precise nature of the self-insurance system, the entire system verges on fraud.[89]

This is particularly true, given that many people choose jobs because of ill health benefits.[90] The Equal Employment Opportunity Commission is reportedly endeavoring to use the Americans with Disabilities Act to challenge companies' practices of setting caps on health insurance payouts for employees with AIDS.[91] The advent of genetic testing, as well as the increasing identification of genetic diseases, makes genetic information, like other medical information, available for use as a basis for medical underwriting in health insurance. The danger, according to one study, is that "genetic testing made possible as we continue to map the human genome may result in many more individuals being denied private insurance coverage than ever before."[92] Genetic tests are not necessary to find out genetic information on applicants. Insurers already obtain genetic information from medically underwritten applicants through family histories and laboratory tests (e.g., cholesterol levels). This was of as much concern to the committee as the use of genetic information from other sources. Although insurers generally do not perform or require genetic tests when doing medical underwriting, they may seek to learn the results of any genetic tests from which an applicant may have information. This could deter people from seeking these tests. The existence of medical underwriting can lead people to avoid needed medical services:

> If people worry that their use of health services may disqualify them from future insurance coverage, they may limit their use of needed services, fail to submit claims for covered expenses, or pressure physicians to record diagnoses that are less likely to attract an underwriter's attention. The last two actions add error to data bases used for health care research and monitoring.[93]

A survey of insurers undertaken by the Office of Technology Assessment (OTA) of the U.S. Congress found that insurers see a role for genetic information in medical underwriting. OTA surveyed commercial insurers, Blue Cross and Blue Shield companies, and large health maintenance organizations, which offered individual and medically underwritten small-group health insurance coverage. Data were gathered on underwriting practices, including requirements for diagnostic tests or physical examinations before an insurance policy can be issued. Data on reimbursement practices, as well as general attitudes toward genetic testing, were also obtained.

Insurers generally believed that it was fair for them to use genetic tests to identify those at increased risk of disease; slightly more than one-fourth of medical directors indicated that they disagreed somewhat that such use was fair. Three-quarters of the responding companies said they thought "an insurer should have the option of determining how to use genetic information in determining risk."[94]

OTA's survey of insurers found that genetic information is not viewed as a special type of information.[95] What seems important to insurers when making insurability and rating decisions is the particular condition, not that the condition is genetically based, OTA found. But the majority of insurers did not anticipate using specific genetic tests in the future. However, a majority of medical directors from commercial insurers agreed with the statement that "it's fair for insurers to use genetic tests to identify individuals with increased risk of disease." In a comparison survey, OTA found that 14 percent of responding genetic counselors reported that they had clients who had experienced difficulties obtaining or retaining health care coverage as a result of genetic testing.

Surveys by Paul Billings and colleagues,[96] as well as by the Office of Technology Assessment,[97] uncovered specific examples of people being denied health insurance coverage based on their genotype. These incidents include cases in which a person with a positive test for a genetic disorder had his or her insurance canceled or "rated up" as a result;[98] where genetic disorders such as alpha-antitrypsin were defined as preexisting conditions, thus excluding payment for therapy; where a particular genetic condition resulted in exclusion from maternity coverage;[99] and where the birth of a child affected with a serious recessive disorder led to the inability of the parents and unaffected siblings to obtain insurance.[100] Genetic information provides serious challenges to the traditional operation of insurance. Health insurance in this country is premised on the notion that risks can be predicted on a population-wide basis, but not well on an individual basis; thus insurance becomes a mechanism for spreading risks. If, through genetic testing or the use of genetic information acquired by other means, insurers can learn of people's actual future health risks (e.g., the risk of a serious late-onset disorder), the benefit of risk spreading will be lost; the individual will be charged an amount equal to future medical costs, which may in some cases make insurance prohibitively expensive.

Currently it is permissible in most states to do medical underwriting based on genetic information. However, the expansion of genetic testing presents a serious challenge to medical underwriting and could lead to an alternative policy approaching which medical underwriting is eliminated altogether. Originally, health insurance was based on health risks for entire communities, known as community rating, rather than on individual rating of health risks or conditions. Insurers gradually began to offer lower rates to employers based on the generally better health and lower risks of employed persons, and competition ensued among insurers to insure the "best" (i.e., lowest) risks. This has led to many of the problems in our current health insurance system in which some people have become permanently uninsurable.[101] In a system of community rating,

> there would be no place for the use of the results of genetic testing since applicants
> would not be rated according to their individual health risks and conditions.[102]

Rochester, New York, has had a successful system of community rating; a key factor in its success has been the belief of large employers who would normally self-insure that their participation in a system that emphasizes risk sharing and collective strategies to contain costs results in a system that will keep costs lower over the long term than they would be in a segmented, risk rated competitive health insurance market.[103] The states of Maine and New York have recently passed legislation requiring health insurers offering policies in their states to return to community rating by 1993.[104] Several other states have introduced legislation to protect people from discrimination based on their genotype. In addition, more general antidiscrimination laws may provide some remedy for people who are discriminated against because of their genotype.

Much of this legislation has been a direct response to the debacle in the early 1970s with respect to sickle cell screening. When mandatory sickle cell screening laws were adopted, some insurers and employers began making decisions about insurance coverage and employment opportunities based on the results of the testing. In particular, carriers of sickle cell trait were denied jobs and charged higher insurance rates without evidence that possession of the trait placed a person at a higher risk of illness or death.[105] As a result, some states have adopted laws protecting people with sickle cell trait. At least two states prohibit denying an individual life insurance[106] or disability insurance,[107] or charging a higher premium,[108] solely because the individual has sickle cell trait. A few states have similarly adopted statutes to prohibit mandatory sickle cell screening as a condition of employment,[109] to prohibit discrimination in employment against people with sickle cell trait,[110] and to prohibit discrimination by unions against people with sickle cell trait.[111]

More recently, some states have adopted laws with a broader scope. A California statute prohibits discrimination by insurance companies against people who carry a gene that has no adverse effects on the carrier, but may affect his or her offspring.[112] Under a Wisconsin law,[113] insurers are prohibited from requiring that applicants undergo DNA testing to determine the presence of a genetic disease or disorder, or the individual's predisposition for a particular disease or disorder. Nor may insurers ask whether the individual has had a DNA test or what the results of the test were. Insurers are also prohibited from using DNA test results to determine rates or other aspects of coverage. However, insurance discrimination based on genetic information not obtained through DNA testing is not forbidden by the law.

There is also much concern about the use of genetic information in the employment context. The Council of Ethical and Judicial Affairs of the American Medical Association has taken the position that it is inappropriate for employers to perform genetic tests to exclude workers from jobs.[114] The opinion acknowledges that the protection of public safety is an important rationale for medical tests of employees. However, the opinion states: Genetic tests are not only generally inaccurate when used for public safety purposes, but also unnecessary. A more effective approach to protecting the public safety would be routine testing of a worker's actual capacity to function in a job that is safety-sensitive.[115]

The opinion points out that capacity testing is more appropriate because it would not cause discrimination against someone who has the gene for a disorder but who is totally asymptomatic, yet it would "detect those whose incapacity would not be detected by genetic tests, either because of a false-negative test result or because the incapacity is caused by something other than the disease being tested for."[116] In the employment context, a New Jersey law prohibits employment discrimination based on an "atypical hereditary cellular or blood trait."[117] In New

York, a statute prohibits genetic discrimination based on sickle cell trait, Tay-Sachs trait, or Cooley anemia (beta-thalassemia) trait.[118] In Oregon, Wisconsin, and Iowa, even more comprehensive laws prohibit genetic screening as a condition of employment.[119]

At the federal level, it is still an open question whether the Americans with Disabilities Act (ADA)[120] will provide adequate protection against genetic discrimination. There are three definitions of persons considered to have a disability and, therefore, protected under the statute. Individuals currently with a disability comprise the first group, persons with a history of a disability comprise the second group, and persons who have the appearance of being disabled constitute the third. This latter category should protect carriers of genetic disease who are themselves healthy but could be refused employment because they have a high risk of giving birth to a child with a genetic disorder that might be expensive in insurance or health care costs to the employer. This third category for those with the appearance of disability should also protect persons with an increased risk of disease due to genetic susceptibility to breast cancer, or who have a gene for a late-onset disorder such as Huntington disease.

The NIH-DOE Joint Working Group on Ethical, Legal, and Social Implications (ELSI) of the Human Genome Project petitioned the Equal Employment Opportunity Commission (EEOC), which is responsible for implementing the law. ELSI requested that the EEOC broaden its proposed rulemaking to include these protections related to genetic testing and genetic disorders, or susceptibility to a genetic disorder.

However, according to an interpretation by the EEOC, the act does not protect carriers of genetic diseases who are themselves healthy but could be refused employment because they have a 25 percent risk of giving birth to a child with a genetic disorder. Also, the EEOC does not view a person with an increased risk of disease due to genetic factors, or who has the gene for a late-onset disorder such as Huntington disease, as having a disability and thus being protected by the law. Legislation has been introduced to extend the definition of disability to a "genetic or medically identified potential of, or predisposition toward, a physical or mental impairment that substantially limits a major life activity."[121]

Another limitation of the ADA is that it allows employers to request any type of medical testing on an employee after a conditional offer of employment is made. In contrast statutes in 11 states limit such testing to that which is job related.[122]

There may in fact be a narrow set of circumstances in which genetic testing may be appropriate to determine a person's ability to undertake a particular job. For example, a person with an active seizure disorder might be excluded from a job in which he or she could cause serious harm. Such a possibility would seem to be appropriate only if the potential harm were serious and screening were the most appropriate way to avert the harm. The committee was concerned, however, that employers might confuse having the gene for, or a genetic predisposition to, a particular disorder with currently being symptomatic. The possibility that someone, later in life, might become incapable of doing a job does not provide a sufficient rationale for not letting him or her undertake the job at the current time. Consequently, in most situations, periodic medical screening for symptoms rather than genetic screening will be a more appropriate means of determining whether an employee presents a serious risk of harm to third parties.[123]

FINDINGS AND RECOMMENDATIONS

Overall Principles

The committee recommends that vigorous protection be given to autonomy, privacy, confidentiality, and equity. These principles should be breached only in rare instances and only when the following conditions are met: (1) the action must be aimed at an important goal—such as the protection of others from serious harm—that outweighs the value of autonomy, privacy, confidentiality, or equity in the particular instance; (2) it must have a high probability of realizing that goal; (3) there must be no acceptable alternatives that can also realize the goal without breach of these principles; and (4) the degree of infringement of the principle must be the minimum necessary to realize the goal.

The committee recommends that regardless of the institutional structures of the entity offering genetic testing or other genetics services, there be a mechanism for advance review of the new genetic testing or other genetics services not only to assess scientific merit and efficacy, but also to ensure that adequate protections are in place for autonomy, privacy, confidentiality, and equity. The usual standards for review of research should be applied no matter what the setting. In particular, an institutional review board (IRB) should review the scientific and ethical issues related to new tests and services in academic research centers, state public health departments, and commercial enterprises. These reviews should include any proposed investigational use of genetic, as well as more extensive pilot studies. In all instances the review body should include people from inside and outside the institution, including community representatives, preferably consumers of genetic services. In the clinical practice setting, professional societies should be encouraged to review studies and issue guidelines, thereby supplementing the guidance provided by IRBs.

The committee also recommends that the National Institutes of Health (NIH) Office of Protection from Research Risks provide guidance and training on how review bodies should scrutinize the risks to human subjects of genetic testing. IRBs may also need technical advice from a local advisory group on genetics (see Chapter 1). To the extent that a National Advisory Committee on Genetic Testing and its Working Group on Genetic Testing are established these bodies should be consulted by IRBs and the NIH Office of Protection from Research Risks. All laboratories offering genetic testing are included under the Clinical Laboratory Improvement Amendments of 1988 (CLIA88), and the committee recommends that the Health Care Financing Administration expand its existing lists of covered laboratory tests to include the full range of genetic tests now in use.

New tests, not validated elsewhere, that are added to the battery of tests should be considered investigational if they are used to make a clinical decision. The committee recommends that IRB approval be obtained in universities, commercial concerns, and other settings where new tests for additional disorders are being undertaken, even if the tests rely on existing technologies. IRB approval should be obtained before new tests are added to newborn screening.

AUTONOMY

Informed Consent

The committee recommends that for a proper informed consent to be obtained from a person who is considering whether to undergo genetic testing, the person should be given information about the risks, benefits, efficacy, and alternatives to the testing; information about the severity, potential variability, and treatability of the disorder being tested for; and information about the subsequent decisions that will be likely if the test is positive (e.g., whether the person will have to make a decision about abortion). Information should also be disclosed about any potential conflicts of interest of the person or institution offering the test (e.g., equity holdings or ownership of the laboratory performing the test, dependence on test reimbursement to cover the costs of counseling, patents). The difficulty in applying the traditional mechanisms for achieving informed consent should not be considered an excuse for failing to respect a patient's autonomy and need for information.

The committee recommends that research be undertaken to determine what patients want to know in order to make a decision about whether or not to undergo a genetic test. People may have less interest in information about the label for the disorder and its mechanisms of action than they have in information about how certainly the test predicts the disorder, what effects the disorder has on physical and mental functioning, and how intrusive, difficult, or effective any existing treatment protocol would be. Research is also necessary to determine advantages and disadvantages of various means of conveying that information (e.g., through specialized genetic counselors, primary care providers, single-disorder counselors, brochures, videos, audiotapes, and computer programs). People also need to know about potential losses of insurability or employability or social consequences that may result from knowledge about the disorder for which the testing is being discussed.

Multiplex Testing

Performing multiple genetic tests on a single sample of genetic material—often using techniques of automation—has been called multiplex testing. The committee recommends that informed consent be gained in advance of such multiplex testing. New means (such as interactive or other types of computer programs, videotapes, and brochures) should be developed to provide people—in advance of testing—with the information described in the previous recommendations, such as descriptions of the nature of tests that are included in multiplex testing and the nature of the disorders being tested for. A health care provider or counselor should also provide information about each of the tests, or if that is not possible because of the number of tests being grouped together, the provider or counselor should supply information about the categories of disorders so that the person will be able to make an informed decision about whether to undergo the testing.

The committee identified the area of multiplex testing as one in which more research is needed to develop ways to ensure that patient autonomy is recognized. The more general research the committee has advocated on determining what information should be conveyed and how it should be conveyed should be supplemented with additional research dealing with the unique case of multiplex testing where many disorders could be tested for at once, and those disorders may have differing characteristics. In multiplexing, tests should be grouped so that tests requiring similar demands for informed consent and education and counseling may be offered together. Only certain types of tests should be multiplexed; some tests should only be offered individually, especially tests for untreatable fatal disorders (e.g., Huntington disease).

The committee also recommends that research be undertaken to make decisions about which tests to group together in multiplex testing, based on the type of information the tests provide. The committee believes strongly that tests for untreatable disorders should not be multiplexed with tests for disorders that can be cured or prevented by treatment or by avoidance of particular environmental stimuli.

Voluntariness

The committee reaffirms that voluntariness should be the cornerstone of any genetic testing program. The committee found no justification for a state-sponsored mandatory public health program involving genetic testing of adults, or for unconsented-to genetic testing of patients in the clinical setting.

Screening and Testing of Children

The committee recommends that newborn screening programs be voluntary. The decision to make screening mandatory should require evidence that—without mandatory screening—newborns will not be screened for treatable illnesses in time to institute effective treatment (e.g., in PKU or congenital hypothyroidism). The committee bases its recommendation and preference for voluntariness on evidence from studies of existing mandated and voluntary programs that demonstrate that the best interests of the child can be served without abrogating the principle of voluntariness. Voluntary programs have delivered services as well or better than mandated programs. There is no evidence that a serious harm will result if autonomy is recognized, just as there is no evidence that mandating newborn screening is necessary to ensure that the vast majority of newborns are screened.

The committee recommends that newborn screening should not be undertaken in state programs unless there is a clear, immediate benefit to the particular infant being screened. In particular, screening should not be undertaken if presymptomatic identification of the infant and early intervention make no difference, if necessary and effective treatment is not available, or if the disorder is untreatable and screening is being done to provide information merely to aid the parents' (or the infant's) future reproductive plans. The committee recommends that states that screen newborns have an obligation to ensure treatment of those detected with the disorder under state programs, without regard to ability to pay for treatment.

The committee recommends that in the clinical setting, children generally be tested only for disorders for which a curative or preventive treatment exists and should be instituted at that early stage. Childhood screening is not appropriate for carrier status, unbeatable childhood diseases, and late-onset diseases that cannot be prevented or forestalled by early treatment. Because only certain types of genetic testing are appropriate for children, tests specifically directed to obtaining information about carrier status, unbeatable childhood diseases, or late-onset diseases should not be included in the multiplex tests offered to children. Research should be undertaken to determine the appropriate age for testing and screening for genetic disorders in order to maximize the benefits of therapeutic intervention and to avoid the possibility that genetic information will be generated about a child when there is no likely benefit to the child in the immediate future.

The majority of the committee recommends that carrier status of newborns and other children be reported to parents only after the parents have been informed of the potential benefits and harms of knowing the carrier status of their children. Because of the risk of stigma for the newborn, such pretest information should be provided to parents when they are informed about newborn screening. Provision should be made for answering any questions the parents may have; these questions are best answered in the context of genetic counseling. The decisions of the parents about whether to receive such information should always be respected. Where such information is not disclosed after parents are given the option to get such information and then knowingly refuse the information, the courts should take this policy analysis and the recommendation of this committee into consideration and not find liability if parents sue because the carrier status of their child was not disclosed and they subsequently give birth to an affected child. Research is needed on the consequences of revealing carrier status in newborns to identify both harms and benefits from disclosing such information in the future.

Subsequent Uses

The committee recommends that before genetic information is obtained from individuals (or before a sample is obtained for genetic testing), they (or, in the case of minors, their parents) be told what specific uses will be made of the information or sample; how—and for how long—the information or sample will be stored; whether personal identifiers will be stored; and who will have access to the information or sample, and under what conditions. They should also be informed of future anticipated uses for the sample, asked permission for those uses, and told what procedures will be followed if the possibility for currently unanticipated uses develops. The individuals should have a right to consent or to object to particular uses of the sample or information. Subsequent anonymous use of samples for research is permissible, including in state newborn screening programs. Except for such anonymous use, the newborn specimen should not be used for additional tests without informed consent of the parents or guardian.

If genetic test samples are collected for family linkage studies or clinical purposes, they should not be used for law enforcement purposes (except for body identification). If samples are collected for law enforcement purposes, they should not be accessible for other nonclinical uses such as testing for health insurance purposes.

Confidentiality

Disclosure to Spouses and Relatives

As a matter of general principle, the committee believes that patients should be encouraged and aided in sharing appropriate genetic information with spouses. Mechanisms should be developed to aid a tested individual in informing his or her spouse and relatives about the individual's genetic status and informing relatives about genetic risks. These mechanisms would include the use of written materials, referrals for counseling, and so forth. On balance, the committee recommends that health care providers not reveal genetic information about a patient's carrier status to the patient's spouse without the patient's permission. Furthermore, information about misattributed paternity should be revealed to the mother but should not be volunteered to the woman's partner.

Although confidentiality may be breached to prevent harm to third parties, the harm envisioned by the cases generally has been substantial and imminent.[124] The spouse's claim of future harm due to the possibility of later conceiving a child with a genetic disorder would not be a sufficient reason to breach confidentiality. The committee found no evidence of a trend on the part of people to mislead their spouses about their carrier status. Moreover, since most people do tell their spouses about genetic risks, breaching of confidentiality would be needed only rarely. The committee believes that patients should share genetic information with their relatives so that the relatives may avert risks or seek treatment.

Health care providers should discuss with patients the benefits of sharing information with relatives about genetic conditions that are treatable or preventable or that involve important reproductive decision making. The committee believes that the disadvantages of informing relatives over the patient's refusal generally outweigh the advantages, except in the rare instances described above. The committee recommends that confidentiality be breached and relatives informed about genetic risks only when attempts to elicit voluntary disclosure fail, there is a high probability of irreversible or fatal harm to the relative, the disclosure of the information will prevent harm, the disclosure is limited to the information necessary for diagnosis or treatment of the relative, and there is no other reasonable way to avert the harm. When disclosure is to be attempted over the patient's refusal, the burden should be on the person who wishes to disclose to justify to the patient, to an ethics committee, and perhaps in court that the disclosure was necessary and met the committee's test. If there are any circumstances in which the geneticist or other health care professional could breach confidentiality and disclose information to a spouse, relative or other third party—for example, to an employer—those circumstances should be explained in advance of testing; and, if the patient wishes, the patient should be given the opportunity to be referred to a health care provider who will protect confidentiality.

On a broader scale, the committee recommends that:

- all forms of genetic information be considered confidential and not be disclosed without the individual's consent (except as required by law), including genetic information that is obtained through specific genetic testing of a person as well as genetic information about a person that is obtained in other ways (e.g., physical examination, knowledge of past treatment, or knowledge of a relative's genetic status);
- confidentiality of genetic information should be protected no matter who obtains or maintains that information, including genetic information collected or maintained by health care professionals, health care institutions, researchers, employers, insurance companies, laboratory personnel, and law enforcement officials; and
- to the extent that current statutes do not ensure such confidentiality, they should be amended so that disclosure of genetic information is not required.

The committee recommends that codes of ethics of those professionals providing genetics services (such as those of the National Society of Genetic Counselors [NSGC], or of geneticists, physicians, and nurses) contain specific provisions to protect autonomy, privacy, and confidentiality. The committee endorses the NSGC statement of a guiding principle on confidentiality of test results:

> The NSGC supports individual confidentiality regarding results of genetic testing. It is the right and responsibility of the individual to determine who shall have access to medical information, particularly results of testing for genetic conditions.[125]

The committee also endorses the principles on DNA banking and DNA data banking contained in the 1990 ASHG statement.

To further protect confidentiality, the committee recommends that

- patient's consent be obtained before the patient's name is provided to a genetic disease registry and that consent be obtained before information is redisclosed;
- each entity that receives or maintains genetic information or samples have procedures in place to protect confidentiality, including procedures limiting access on a need to-know basis, identifying an individual who has responsibility for overseeing security procedures and safeguards, providing written information to each employee or agent regarding the need to maintain confidentiality and taking no punitive action against employees for bringing evidence of confidentiality breaches to light;
- any entity that releases genetic information about an individual to someone other than that individual ensure that the recipient of the genetic information has procedures in place to protect the confidentiality of the individual;
- any entity that collects or maintains genetic information or samples separate them from personal identifiers and instead link the information or sample to the individual's name through some form of anonymous surrogate identifiers;
- the person have control over what parts of his or her medical record are available to which people; if an optical memory card is used, this could be accomplished through a partitioning-off of data on the card; and
- any individual be allowed access to his or her genetic information in the context of appropriate education and counseling, except in the early research phases during the development of genetic testing when an overall decision has been made that results based on the experimental procedure will not be released and the subjects of the research have been informed of that restriction prior to participation.

Discrimination in Insurance and Employment

In general, the committee recommends that principles of autonomy, privacy, confidentiality, and equity be maintained, and the disclosure of genetic information and the taking of genetic tests should not be mandated. Such a position, however, is in conflict with some current practices in insurance and employment.

Although more than half the U.S. population (approximately 156 million people) is covered by some kind of life insurance,[126] the use of genetic information in medical underwriting[127] decisions about life insurance appears to raise different and somewhat lesser concerns than the use of genetic information in health insurance underwriting. More of life insurance has historically been medically underwritten. Complaints of genetic discrimination in life insurance have been made.[128] Apparently, fewer Americans believe that life insurance is a basic right. In contrast, the Canadian Privacy Commission believes life insurance is a basic right and recommends that Canadians be permitted to purchase up to $100,000 in basic life insurance without genetic or other restrictions; underwriting for larger amounts of life insurance could be subject to a variety of lifestyle and health restrictions, including the use of genetic information.[129] Most of the committee agrees with the spirit of the Canadian Commission's recommendation that a limited amount of life insurance be available to everyone without regard to health or genetic status. However, health insurance was considered a much more pressing ethical, legal, and social issue.

The committee recommends that legislation be adopted so that medical risks, including genetic risks, not be taken into account in decisions on whether to issue or how to price health care insurance. Because health insurance differs significantly from other types of insurance in that it regulates access to health care, an important social good, risk-based health insurance should be eliminated and a means of access to health care should be available to every American without regard to the individual's present health status or condition, including genetic makeup. Any health insurance reform proposals need to be evaluated to determine their effect on genetic testing and the use of genetic information in health insurance.

The committee recommends that the unfair practices highlighted by the McGann case be prevented. Such situations could be eliminated by Congress in three ways. First, the antidiscrimination provision of the Employee Retirement Income Security Act (ERISA), section 510, could be amended to prohibit various types of employer conduct. For example, the legislation could prohibit: (1) the alteration of benefits or the alteration of benefits without a certain notice period; (2) the reduction of coverage for only a single medical condition; (3) the reduction of benefits after a claim for benefits already had been submitted, and so forth. At the very least, the committee recommends that an amendment be adopted making those practices illegal.

A second way of legislatively preventing McGann-type situations would be to amend the ERISA preemption provision, section 514. By amending this section to limit the preemptive effect of ERISA (e.g., that permits ERISA provisions to override state insurance regulations) or to eliminate ERISA preemption entirely, the result would be to allow the states to regulate self-insured employer benefits in the same way that state insurance commissions regulate commercial health insurance benefits. Although state regulation may be preferable to no regulation, it could lead to the burdensome multiplicity of state regulations that ERISA was intended to eliminate. For this reason, the committee believes that federal prohibition of the type of conduct in the McGann case would be preferable.

A third way to eliminate discrimination in employee health benefits by self-insured employers would be to amend section 501 of the ADA. The ADA is essentially neutral on the issue of health benefits; clauses on preexisting conditions, medical underwriting, and other actuarially based practices, to the extent permitted by state law, do not violate the ADA. Thus, the ADA could be amended to prohibit differences in health benefits that result in discrimination against individuals with disabilities. Amending the ADA in this manner would, in effect, mandate uniform coverage (although it is not clear what conditions would be covered) at community rates for employees. If Congress wanted to mandate that all employers offer a package of health benefits, a good argument could be made that it ought to do so separately and not via amendments to the ADA. The committee recommends that legislation be adopted so that genetic information cannot be collected on prospective or current employees unless it is clearly job related. Sometimes employers will have employees submit to medical exams to see if they are capable of performing particular job tasks. The committee recommends that if an individual consents to the release of genetic information to an employer or potential employer, the releasing entity should not release specific information, but instead answer only yes or no regarding whether the individual was fit to perform the job at issue.

The committee recommends that the EEOC recognize that the language of the Americans with Disabilities Act provides protection for presymptomatic people with a genetic profile for late-onset disorders, unaffected carriers of disorders that might affect their children, and people with genetic profiles indicating the possibility of increased risk of a multifactorial disorder. The committee also recommends that state legislatures adopt laws to protect people from genetic discrimination in employment. In addition the committee recommends an amendment to the ADA (and adoption of similar state statutes) limiting the type of medical testing employers can request or the medical information they can collect to that which is job related.

Ultimately, new laws on a variety of other topics may also be necessary to protect autonomy, privacy, and confidentiality in the genetics field, and to protect people from inappropriate decisions based on their genotypes.[130] The ability of genetics to predict health risks for asymptomatic individuals and their potential offspring presents challenges in the ethical and social spheres. The committee recommends that careful consideration be given to the development of policies for the implementation of genetic testing and the handling of genetic test results.

NOTES

1. March of Dimes Birth Defects Foundation, *Genetic Testing and Gene Therapy National Findings Survey* (New York, 1992).

2. See, e.g., *Satz v. Perlmutter*, 362 So. 2d 160 (Fla. App. 1978), *aff'd.*, 379 So. 2d 359 (Fla. 1980).

3. See, e.g., *Salgo v. Leland Stanford Jr. Univ. Bd. of Trustees*, 154 Cal. App. 2d 560; 317 P.2d 170 (1957); *Canterbury v. Spence*, 464 F.2d 772 (D.C. Or.), *cert. denied*, 409 U.S. 1064 (1972).

4. *Gales v. Jensen*, 92 Wash. 2d 246; 595 P.2d 919 (1979).

5. *Salgo v. Leland Stanford Jr. Univ. Bd. of Trustees*, 154 Cal. App. 2d 560; 317 P.2d 170 (1957).

6. *Kogan v. Holy Family Hospital*, 95 Wash. 2d 306; 622 P.2d 1246 (1980).

7. See, e.g., *Becker v Schwartz*, 46 N.Y. 2d 401; 386 N.E. 2d 807; 413 N.Y. S.2d 895 (1978). For a review of relevant cases, see Lori B. Andrews, "Tons and the Double Helix: Liability for Failure to Disclose Genetic Risks," 29 U, *Houston L Rev* 143 (1992).

8. L. Andrews, "My Body, My Property," 16(5) *Hastings Center Report* 28 (1986).

9. The federal regulations governing informed consent in the context of human experimentation provide that informed consent is not necessary for research on pathological or diagnostic specimens "if these sources are publicly available or if the information is recorded by the investigator in such a manner that subjects cannot be identified, directly or through identifiers linked to the subjects" 45 C.F.R. 46, 101(b)(5) (1991). Similarly, some state human experimentation laws do not seem to extend their coverage to research on removed parts. In Virginia, for example, the law does not cover "the conduct of biological studies exclusively utilizing tissue or fluids after their removal or withdrawal from a human subject in the course of standard medical practice" Va Code Ann § 37 I 234(1) (1984). Under the New York law, human research is defined to exclude "studies exclusively utilizing tissue or fluids after their removal or withdrawal from a human subject in the course of standard medical practice." N V Public Health Law § 2441(2) (McKinney 1977).

10. Office of Technology Assessment (OTA), U.S. Congress, *Human Gene Therapy* 72 (Washington, D.C.: U.S. Government Printing Office, 1984).

11. Ferdinand D. Schoeman, "Privacy: Philosophical Dimensions of the Literature," in *Philosophical Dimensions of Privacy: An Anthology*, ed. Ferdinand D. Schoeman (New York: Cambridge University Press, 1984).

12. Secretary Judith Jarvis Thomson, "The Right to Privacy," *Philosophy and Public Affairs* 4 (1975): 315–33.

13. See, e.g., *Griswold v. Connecticut*, 381 U.S. 479 (1965); *Roe v. Wade*, 410 U.S. IU (1973); *Planned Parenthood v. Casey*, 505 U.S. 833 112 S. Ct. 2791 (1992).

14. *Lifchez v. Hartigan*, 735 F. Supp. 1361 (ND III 1991), *aff'd.* without opinion *sub nom; Scholber v. Lifchez*, 914 F.2d 260 (7th Cir 1990), *cert. denied,* 111 S. Ct. 787 (1991).

15. *Carter v. Broadlawn Medical Center*, 667 F. Supp. 1269, 1282 (SD Iowa 1987). In that case the court held that the privacy of patient records in a county hospital is protected by the Fourteenth Amendment's concept of personal liberty. See also *Whalen v. Roe*, 429 U.S. 589, 599 n 23 (1977).

16. *Home v. Paiton*, 291 Ala. 701, 287 So. 2d 824 (1973). *MacDonald v. Clinger*, 84 A D 2d 482. 444 N.Y. S.2d 801 (1982).

17. Privacy Act of 1974, 5 U.S.C. § 552a (1991).

18. For further discussion of these arguments (and others in this section), see Tom Beauchamp and James F. Childress, *Principles of Biomedical Ethics*, 3d ed. (New York: Oxford University Press, 1989), chap. 7.

19. Mark Siegler, "Confidentiality in Medicine: A Decrepit Concept," *N Eng J Med* 307 (1983): 1518–21.

20. Research has found that people who are told that their answers would not be confidential provide less intimate information. Woods and McNamara, "Confidentiality: Its Effect on Interviewee Behavior," *Prof. Psychology* 11 (1980): 714–19.

21. See, e.g., *Home v. Ration*, 291 Ala. 701, 287 So. 2d 824 (1973); *MacDonald v. Clinger*, AD 2d 482. 444N. Y-S. 2d 801 (1982). See W. Prosser and W. P. Keeton, *Prosser and Keeton on Torts*, 856–863 (1984). See also S. Newman, "Privacy in Personal Medical Information: A Diagnosis," 33 *U Ha L Rev* 33 (1981): 394–424. According to the latter article, the tort of invasion of privacy has been rejected only in Rhode Island, Nebraska, and Wisconsin. *Id.* 403. However, two of these states now recognize it by statute. R.I. Gen. Laws 5 9-1-28 (1984); Wisc Stat § 895, 50 (1985–86).

22. Norman Daniels, "Insurability and the HIV Epidemic: Ethical Issues in Underwriting," *Milbank Quarterly* 68 (1990): 497–515. See also Mark A. Rothstein, "The Use of Genetic Information in Health and Life Insurance," in *Molecular Genetic Medicine*, ed. T. Friendman (New York: Academic Press, 1993).

23. On the natural lottery, see H. Tristram Engelhardt Jr., *Foundations of Bioethics* (Oxford University Press, 1986).

24. *Id.*

25. See Gene Outka, "Social Justice and Equal Access to Health Care," *Journal of Religious Ethics* 2 (1974). see also President's Commission for the Study of Ethical Problems in Medicine and Behavioral Research, *Securing Access to Health Care*, vol. 1 (Washington, D.C.: U.S. Government Printing Office, 1983).

26. Norman Daniels, "Equity of Access to Health Care: Some Conceptual and Ethical Issues," *Milbank Memorial Fund Quarterly/Health and Society* 60 (1982). See also Norman Daniels, *Just Health Care* (New York: Cambridge University Press, 1985).

27. President's Commission for the Study of Ethical Problems in Biomedical and Behavioral Research, *Securing Access to Health Care*, Vol. 1 Report (Washington, D.C.: U.S. Government Printing Office, 1983), 42.

28. Rothstein, "Genetic Discrimination in Employment and the Americans with Disabilities Act," 29 *Houston L Rev*. 23 (1992): 31. See also the discussion infra in this chapter.

29. Office of Technology Assessment (OTA), U.S. Congress, *Genetic Witness: Forensic Uses of DNA Tests* (Washington, D.C.: U.S. Government Printing Office, 1990).

30. D. C. Wertz and J. C. Fletcher, *Ethics and Human Genetics: A Cross Cultural Perspective* (New York: Springer-Verlag, 1989); D. C. Wertz and J. C. Fletcher, "An International Survey of Attitudes of Medical Geneticists toward Mass Screening and Access to Results," 104 *Public Health Reports*, 1989, 35–44.

31. G. Geller, E. Tambor, G. Chase, K. Hofman, R. Faden, and N. Holtzman, "How Will Primary Care Physicians Incorporate Genetic Testing Directiveness in Communication?" *Medical Care* 31 (1993): 625–31.

32. There are several reasons why institutions store DNA samples, rather than just information from samples. The first is for future potential clinical benefit, such as when a new test may be developed that could provide a more accurate diagnosis. The second is for litigation purposes, so that the sample can be retested if the results are challenged. The third is for research purposes, to use the DNA for the development of additional tests.

33. Philip Reilly, Presentation to Ethical, Legal, and Social Implications Program Committee, January 1991.

34. Yoichi Maisubara, Kuniaki Narisawa, Keiya Tada, Hiroyuki Ikeda, Yao Ye-Qi, David M. Danks, Anne Green, Edward R. B. McCabe, "Prevalence of K329E Mutation in Medium-Chain AcyiCoA Dehydrogenase Gene Determined from Guthrie Cards," *Lancet* 338 (1991): 552–53.

35. Joe Abernathy, "City Health Clinics Unveil Controversial 'Smart Card,'" *Houston Chronicle*, October 11, 1992, A1.

36. See, e.g., M. Shaw, "Conditional Prospective Rights of the Fetus," *Jour. Legal Med.* 5 (1989): 63.

37. Moreover, it should be noted that this risk (transmission of genetic disease to offspring) is one that society has always lived with and seems to have flourished despite that risk.

38. For example, in *The New York Times v. United States*, 403 U.S. 713 (1971), the U.S. Supreme Court, in a *per curiam* opinion, held that the government had not met its "heavy burden" of proving that national security required that the Pentagon Papers be suppressed. The logic of the case was explained further in the concurrences: the right of free speech is to be infringed by a prior restraint only when disclosure "will surely result in direct, immediate, and irreparable damage to our Nation or its people." *Id.* at 730 (Stewart, J., concurring). Or when there is "governmental allegation and proof that publication must inevitably, directly, and immediately cause the occurrence of an event kindred to imperiling the safety of a transport already at sea" during wartime. *Id.* at 726–27 (Brennan, J., concurring). The standard of irreparability for granting an injunction against protected speech is an absolute not a comparative standard. Even if the speech could cause great harm, that would not be sufficient. As Justice White pointed out in his concurrence in *New York Times,* it is not sufficient that there may be "substantial damage to public interests," *Id.* at 731 (White, J., concurring). Similarly, Justice Stewart said, "I am convinced that the Executive is correct with respect to some of the documents involved [i.e., they should not, in the national interest, be published]. But I cannot say that disclosure of any of them will surely result in direct, immediate, and irreparable harm to our Nation or its people. That being so, there can under the First Amendment be but one Judicial resolution of the issue before us." *Id.* N 730 (White, J., concurring).

Even if irreparable harm were a possibility, *New York Times* indicates that an injunction should not be issued against the press unless such harm would come about directly and immediately. The term "immediately" is easy enough to understand; it requires a present, not future, harm. The term "directly" relates to the lack of intervening influences during that time period. The irreparable harm would not occur directly if another important influence would or could intervene. Another way of expressing the immediacy and directness that is necessary is by saying the harm inevitable—it will occur within a short period of time during which nothing will or could change or stop it.

Even when a prior restraint is not at issue, high standards are required for showing a compelling state interest when a fundamental right is at issue. Also in the First Amendment area, speech that is not false should not be the basis for subsequent punishment unless it provided an immediate threat of real harm (see *Bridges v. California,* 314 U.S. 252, 263 (1941).

39. AIDS, of course, is an infectious disease that cannot be cured and that is strongly identified with certain minority groups (e.g., homosexuals). It is interesting to note that for many of the same reasons that are applicable to mandatory genetic screening, manda-

tory AIDS screening has not been adopted. Instead, anonymous voluntary screening has been the model.

40. C. Damme, "Controlling Genetic Disease Through Law," 15 *U. Cal Davis L Rev* 15 (1982): 801, 807.

41. Such charges were leveled by blacks when mandatory sickle cell carrier screening was put in place.

42. L. Andrews, "Medical Genetics: A Legal Frontier," 233 (1987).

43. A. G. Motulsky, The Significance of Genetic Disease," 59, 61 in B. Hilton, D. Callahan, M. Harris, P. Coodliffe, B. Berkley, eds., *Ethical Issues in Human Genetics: Genetic Counseling and the Use of Genetic Knowledge,* Fogarty International Proceedings, no. 13, 1973.

44. Statement of D. Brock in B. Hilton et al., eds., *id.* at 90.

45. The idea of choosing by category was discussed by Alta Charo, J.D., at the committee's June 1992 workshop.

46. E.g., *Cobbs v. Grant,* Cal. 3d 829; 104 Cal. Rptr. 505, 502 P.2d 1, 12 (1972). Some states' informed consent statutes explicitly recognize a right to refuse information. Alaska Stat. § 09 55, 556<bM2) (1983); Del. Code Ann. tit. IS. i 6852(bX2) (Supp. 1984); N H Rev Stat Ann § 507-C:2(HXbX3) (1983); N.Y. Public Health Law 2805-d(4Kb) (McKinney 1985). Or Rev. Stat. § 677, 097(2) (1985). Utah Code Ann. 78-14-5 (2)(c)(l977); Vi. Stat. Ann. in 12. § 1909(c)(2) (Supp. 1991); Wash. Rev. Code Am. 17. 70. 060(2) (Supp. 1991).

47. *Prince v. Massachusetts,* 321 U.S. 158, 170 (1944).

48. *In re Green,* 448 Pa. 338, 292 A.2d 37, 392 (1972). See also Brown and Truit, "The Right of Minors to Medical Treatment," *DePaul L Rev.* 28 (1979): 289, 299.

49. *Jehovah's Witnesses v. King County Hospital,* 278 F. Supp. 488 (W D Wash. 1967). 390 U.S. 598 (1968), *denied,* 391 U.S. 961 (1968).

50. D.C. Code Ann. 16-314(3) (1989); Md. Health-Gen. Code Ann. §§ 13 102(10) and 109(c)(f) (1982).

51. Iowa Admin Code 641-4, 1 (136A) (1992); Mich. Comp Laws Ann. § 333 5431(1) (WCM 1992); Mont. Code Ann. 50-19-203(1) (1991); W. Va. Code Ann. § 16 22 3 (Supp. 1992).

52. Mo. Ann. Stat. 191. 33l(5) (Vernon 1990); S.C. Code § 44 37 30(B) (1991).

53. Council of Regional Networks for Genetic Services (CORN), *Newborn Screening* 1990, Final Report, February 1992.

54. R. R. Faden, A. J. Chalow, N. A. Holtzman, and S. Horn, "A Survey to Evaluate Parental Consent as Public Policy for Neonatal Screening," 72 *Am J Pub. Health* 1347 (1982).

55. Consensus Statement Proposed for Routine Newborn Genetic Screening, Based on October 1989 Conference in Quebec, Canada. Reported in Barta Maria Knoppers and Claude M LaBerge, eds., "Genetic Screening: From Newborns to Data Typing," *Excerpta Medico* 382 (1990).

56. Currently, Colorado and Wyoming include cystic fibrosis testing as a part of their mandatory newborn screening program; Wisconsin includes cystic fibrosis in newborn screening as part of an experimental research protocol.

57. Neil A. Holtzman, "What Drives Neonatal Screening Programs?" *New Engl J Med* 325 (1991): 802, 809, referring to K. B. Hammond, S. H. Abman, R. J. Sokol, F. J.

Accurso, "Efficacy of Statewide Newborn Screening for Cystic Fibrosis by Assay of Tryp-sinogen Concentration," 325 *New Eng J Med* 325 (1991): 769–74.

58. *Id.*

59. *Id.,* citing P. Farrell, personal communication.

60. Norm Fost presentation, June 1992 committee workshop.

61. Statement of Claude LaBerge at June 1992 committee meeting.

62. See T. McNeil, B. Hany, T. Thelin, E. Aspergren-Jansson, T. Sveger, "Identifying Children at High Somatic Risk: Long-Term Effects on Mother Child Interactions," *Acta Psychiatrica Scandinavia* 74, no. 6: 555–62 (1986).

63. Ellen Wright Clayton, "Screening and Treatment of Newborns," 29 *Houston L Rev* 29 (1992): 85.

64. The federal regulations governing informed consent in the context of human experimentation provide that informed consent is not necessary for research on pathological or diagnostic specimens "if these sources are publicly available or if the information is recorded by the investigator in such a manner that subjects cannot be identified, directly or through identifiers linked to the subjects." 45 C.F.R. 46, 101(bX4) (1991).

Similarly, some state human experimentation laws do not seem to extend their coverage to research on removed parts. In Virginia for example, the law does not cover "the conduct of biological studies exclusively utilizing tissue or fluids after their removal or withdrawal from a human subject in the course of standard medical practice." Va. Code Ann. 37, 1-234(I) (1990). Under the New York law, human research is defined to exclude "studies exclusively utilizing tissue or fluids after removal or withdrawal from a human subject in the course of medical practice." Some researchers argue that blood and urine left over after a patient's tests are done should be available without requiring the patient's informed consent. However, the issue is not as straightforward as it might seem. Research surreptitiously done on a patient's blood might generate information that could be damaging to the patient. If a cystic fibrosis test were being developed with excess blood from a PKU test and the blood tested positive for cystic fibrosis, it could be argued that, depending on how reliable the test seemed, the infant's parents should be informed. But if the result ultimately turned out to be a false positive, the family may nave been harmed by unnecessary worry. With respect to research on leftover blood from adults, if testing is being developed for a potentially stigmatizing disorder (such as AIDS) or a disorder that might influence employment or other opportunities for the person (such as Huntington disease), the risks to the patient if confidentiality were compromised might be so high that it would seem unethical not to solicit the individual's consent before the research is undertaken.

65. ASHG Ad Hoc Committee on Individual Identification by DNA Analysis, "Individual Identification by DNA Analysis: Points to Consider," *Am J Hum Genet* 46 (1990): 631–34.

66. *Id.* at 632.

67. Nancy Wexler, "The Tiresias Complex: Huntington's Disease as a Paradigm of Testing for Late-Onset Disorders," *FASEB Journal* 6 (1990): 2820–25.

68. See, e.g., *Skillings v. Allen*, 143 Minn. 323, 173 N.W. 663 (1919); *Davis v. Rodman*, 147 Ark. 385, 227 S. W. 612 (1921). For a more recent case regarding a duty to warn third parties of communicable diseases, see *Gammill v. U.S.*, 727 F.2d 950 (10th Cir. 1984).

69. *Tarasoff v. Regents of the University of California*, 131 Cal. Rptr. 14, 17 Cal. App. 3d 425, 551 P.2d 334 (1976).

70. *Id.*

71. *Collins v. Howard*, 156 F. Supp. 322, 325 (S. D. Ga. 1957) (dicta).

72. The individual generally asks that his or her spouse be informed as well. Statement of J. Lejeune in B. Hihon, D. Callahan, M. Harrii, P. Condliffe, B. Berkeley, eds., *Ethical Issues in Human Genetics: Genetic Counseling and the Use of Genetic Knowledge*, Fogarty International Proceedings, no. 13, 1973, 70. In a March of Dimes–sponsored national public opinion survey, 71 percent of respondents said that if a doctor of a woman who plans to have children finds through testing that her children might inherit a serious or fatal genetic disease, the doctor has an obligation to tell her husband. March of Dimes Birth Defects Foundation, *Genetic Testing and Gene Therapy National Survey Findings*, September 1992, 7. However, in some instances, an individual may or may not want personal genetic information disclosed to his or her spouse.

73. *Berry v. Moensch*, 8 Utah 2d 191, 331 P.2d 814 (1958); *Curry v. Corn*, 52 Misc. 2d 1035, 277 N Y. S.2d 470 (1966) (during marriage, each has the right to know the existence of any disease that may have bearing on the marital relation).

74. See, e.g., *Simonsen v. Swenson*, 104 Neb. 224. 177 N.W. 831 (1920).

75. The man whose nonpaternity is shown through prenatal screening might argue that, in addition to its relevance to his future childbearing plans, the information that he is not the father of his wife's child has an immediate financial implication since he might not wish to support the child. However, state paternity statutes preserve that a child born during a marriage is the husband's child and require him to support the child, even if it could conceivably be shown through genetic testing that he was not the biological father of the child. The logic behind such cases is that there is a societal interest in the integrity of the family.

76. *Eisenstadt v. Baird*, 405 U.S. 438 (1972).

77. *Planned Parenthood v. Danforth*, 428 U.S. 52 (1976).

78. *Planned Parenthood v. Casey*, 505 U.S. 833, 112 S. Ct. 2791 (1992).

79. *Id.*

80. If a patient is the carrier of the gene for a serious autosomal recessive disorder, his or her relatives might also argue that they would be harmed by not knowing that they too are at risk of having children with that disorder. However, the risk to future offspring may be too remote to warrant breaching confidentiality, as it is in the case of a spouse.

81. The President's Commission for the Study of Ethical Problems in Medicine and Biomedical and Behavioral Research, *Screening and Counseling for Genetic Conditions* 6 (1983).

82. See, e.g., *Simonsen v. Swenson*, 104 Neb. 224, 177 N.W. 831 (1920).

83. In one instance, for example, a woman was denied disability insurance when her father's medical records were released to the insurer.

84. Office of Technology Assessment, U.S. Congress, *Medical Testing and Health Insurance* 73 (1988).

85. S. Rep. 100-360, 100th Cong., 1st Sess. 20 (1988).

86. Report of Committee on Employer-Based Health Benefits, citing Seeman (1993).

87. *McCann v. H. & H. Music Co.*, 946 F.2d 401 (5th Cir. 1991), *cert. denied sub nom. Greenberg v. H. & H. Music Company*, 506 U.S. 179 112 S. Ct. 1556 (1992).

88. Eric Zicklin, "More Employers Self-Insure Their Medical Plans, Survey Finds," *Business and Health*, April 1992, 74.

89. Mark Rothstein, "The Use of Genetic Information in Health and Life Insurance," in *Molecular Genetic Medicine*, ed. Ted Friedman (New York: Academic Press, 1993).

90. According to American Healthline, Briefing on Health Insurance, November 17, 1992. More than half of Americans would not accept a job that did not provide health insurance.

91. "EEOC Said Ready to Fast Track Complaints of Insurance Caps under Title 1 of the ADA,"*AIDS Policy and Law,* October 2, 1992, 1.

92. N. Kass, "Insurance for the Insurers," *Hastings Center Report*, November-December 1992, 611.

93. Institute of Medicine (IOM), National Academy of Sciences, M. Field and D. H. Shapiro, eds., *Employment and Health Benefits: A Connection at Risk*, Committee on Employer Based Health Insurance (Washington, D.C.: National Academy Press, 1993).

94. OTA, 1992a, 80.

95. OTA, 1992b.

96. See, e.g., P. R. Billings, M. A Kohn, M de Cuevas, J. Beckwith, J. S. Alper, M. R. Natowicz, "Discrimination as a Consequence of Genetic Testing," *Am J Hum Genet* 50 (1992): 476, 482.

97. U.S. Congress, Office of Technology Assessment, *Cystic Fibrosis and DNA Tests*: *Implications of Carrier Screening* (Washington, D.C.: U.S. Government Printing Office, 1992).

98. The disorders included adult polycystic kidney disease, Huntington disease, neurofibromatosis, Marfan syndrome, Down syndrome, Fabray disease.

99. The conditions included a balanced translocation.

100. The disorder was cystic fibrosis.

101. Kass, 1992.

102. *Id.* at 10.

103. Report of Institute of Medicine Committee on Employer-Based Health Benefits.

104. See Kass, 1992 for descriptions of these plans.

105. P. Reilly, *Genetics, Law, and Social Policy* (Cambridge: Harvard University Press, 1977), 62–86.

106. Fla. Stat. Ann. f 626. 9706(1) (West 1984); La. Rev. Stat. Ann. 22:652. 1(D) (West Supp. 1992).

107. 22 Fla. Stat. Ann. 1626. 9707(1) (West 1984). La. Rev. Stat. Ann. 22:652. 1 (D) (West Supp. 1992).

108. Fla. Stat. Ann. 626. 9706(2) (West 1984) (life insurance). 626 9707(2) (West 1984) (disability insurance); La. Rev. Stat. Ann. 22:652. 1(0) (West Supp. 1992).

109. This same law appears in three places in the Florida statutes: Fla. Stat. Ann. 448. 076 (West 1981); i 228. 201 (West Supp. 1989); and 163. 043 (West 1985).

110. Fla. Stat. Ann. 1448. 075 (West 1981); N.C. Gen. Stat. 95-28. 1 (1989); La. Rev. Stat Ann. 23:1002 (West 1985).

111. La. Rev. Stat. Ann. 23:1002(C)(1) (West 1985).

112. Cal. Ins. Code 10143 (West Supp. 1992).

113. 1991 Wisc. Act 269, codified as Wisc. Stat. Ann. 1631. 89.

114. Council on Ethical and Judicial Affairs, "Use of Genetic Testing by Employers," *JAMA* 266 (1991): 1827.

115. *Id.* at 1828.

116. *Id.* at 1828.

117. N.J. Stat. Ann. 10-5-12(a) (West Supp. 1992).

118. N.Y. Civ. Rights Law 48 (McKinney 1992).

119. 1992 Iowa Legis. Serv. 93 (West); Or. Rev. Stat. 1659. 227 (1991); 1991 Wis. Laws 117.

120. 104 Stat. 327 (1991). For sections of the Americans with Disabilities Act relating to employment, see 42 U.S. C. A. 12101-12117 (Supp. 1992).

121. L. Gostin and W. Roper, "Update: The Americans with Disabilities Act," *Health Affairs* 11, no. 3: 248–58.

122. Alaska Stat. 18. 80. 220(a)(l)(1991); Cal. Gov't. Code 1 12940(d) (West Supp. 1991); Colo. Rev. Stat. 24-34-402(1)(d) (1988); Kan. Stat. Ann. 44-1009(a)(3) (Supp. 1991); Mich. Comp. Laws Ann. 37. 1206(2) (West 1985); Minn. Stat. Ann. 363. 02(I)(8)(i) (West Supp. 1991); Mo. Ann. Stat. 213. 055. 1(1) (3) (Vernon Supp. 1992); Ohio Rev. Code Ann. 4112. 02(E)(1) (Anderson 1991); 43 Pa. Cons. Stat. Ann. 955(b)(I) (1991); R. I. Gen. Laws 28-5-7(4)(A) (Supp. 1991); Utah Code Ann. 34-35-6 I(d) (Supp. 1991).

123. Office of Technology Assessment (OTA), U.S. Congress, *Genetic Screening in the Workplace*, OTA-BA-456 (Washington, D.C.: U.S. Government Printing Office, 1990).

124. See, e.g., *Simonsen v. Swenson,* 104 Neb. 224. 177 N.W. 831 (1920); *Tarasof v. the Regents of the University of California*, 131 Cal. Rptr. 14. 17 Cal. App. 3d 425, 551 P.2d 334 (1976).

125. National Society of Genetic Counselors (NSGC), Guiding Principles, Perspectives on Genetic Counseling, October 1991.

126. American Council on Life Insurance, *Life Insurance Fact Book* 19 (1992).

127. Medical underwriting is the evaluation of a person's insurability, usually assessed through a combination of answers to a written questionnaire and physical examination to identify certain conditions determined by medical underwriters (and underwriting manuals) to reduce life expectancy. Actuarial norms, standards for medical underwriting, vary substantially by insurance company, and underwriting decisions are considered crucial business decisions by insurers and are thus considered "trade secrets."

128. Paul Billings, "Testimony before Human Resources and Intergovernmental Relations Subcommittee of the Committee on Government Operations," U.S. House of Representatives, 102d Congress, July 23, 1992.

129. Canadian Privacy Commission, *Genetic Testing and Privacy*.

130. Neil A. Holtzman and Mark A. Rothstein, "Invited Editorial: Eugenics and Genetic Discrimination," 50 *Am J Hum Genet* (1992): 457–59.

25

Genetic Links, Family Ties, and Social Bonds: Rights and Responsibilities in the Face of Genetic Knowledge

Rosamond Rhodes

I. INTRODUCTION

An old adage tells us that blood is thicker than water. The message of the saying is that blood ties carry moral weight. A thoughtful person might wonder how to act on that advice. It had been assumed that hereditary similarities gave those of the same blood common characteristics and, hence, special responsibilities to each other. So, before modern genetics we might have considered which of our familial relations count as blood. How thick are our responsibilities to blood? And, how thin are our responsibilities to water?

Knowledge of genetics, however, complicates the message that we might previously have drawn from the saying. Since learning that all living organisms have 80% of their DNA in common, that apes differ from humans in only 2% of their DNA, and that humans differ from one another by less than .1% of their DNA, the concept of blood ties becomes much more complicated. We still need to learn which social bonds involve thick responsibilities and which ties demand less of us. As we get a clearer picture of our genetic similarities and differences, we also need to find out more about the bearing of our genetic links on our ethical obligations.

Currently, some of the most significant moral issues involving genetic links relate to genetic knowledge. Some of these issues relate to responsibilities of the genetics community for informing and guarding confidentiality of patients, clients, and the general population (Annas, 1993). Some relate to the special case of determining a child's best interest with respect to genetic knowledge (The American Society of Human Genetics Board of Directors and The American College of Medical Genetics Board of Directors, 1995; Hoffmann and Wulfsberg, 1995; Wilfond, 1995). Some relate to the uses of genetic knowledge in personal decision making and in the public domain where there are potential problems of genetic discrimination in appointments, employment, and insurance (ASHG Ad Hoc Committee on Insurance Issues in Genetic Testing, 1995; Editorial, 1995; Gostin, 1994; Juengst, 1991; Hudson, 1995). Other questions concern genetic knowledge itself: when are the tests sufficiently sensitive; when are the results adequately interpreted; what standards of predictability, sensitivity, and reliability do tests have to meet before they are made available to individuals and to the population at large (American Society of Clinical Oncology, 1996)? These issues have received a good deal of attention in the bioethics and genetics literature.

The particular genetic knowledge questions that I shall address turn attention to another set of issues. Instead of looking at the responsibilities professionals or institutions have to individuals, I will take up the question of what responsibilities individuals have to one another. In other words, I want to examine a different set of questions about genetic knowledge. I am interested in whether individuals have a moral right to pursue their own goals without contributing to society's knowledge of population genetics, without adding to their family's genetic history, and without discovering genetic information about themselves and their offspring. These questions are not dramatically distinct from other personal decisions. The point of this discussion is rather to demonstrate the similarity that may be obscured by dazzling new technology and to point a way toward structuring our thinking about these novel circumstances in terms of traditional moral frameworks.

To approach these questions of personal decision making in the face of new genetic technology we will have to examine the presumed right to genetic ignorance and its relation to rights of privacy. We will also need an account of why the various social bonds have different strengths and why more or less demanding obligations attach to them. Certainly an analysis of these issues will have implications for the mainstream discussion of professional and institutional responsibilities with respect to genetic knowledge. I leave those extrapolations for other discussions. I offer four cases to help us focus on the issues I have identified, the cases of Tom, Dick, Harry, and Harriette.

II. CLINICAL CASES

A. Case 1: Tom. Human DNA contains 6 billion base pairs. An average gene contains about 2,000 base pairs. Our approximately 100,000 genes contain about 200,000,000 base pairs of DNA. DNA is a double helix molecule of strands of just four bases, adenosine, thymine, guanine, and cytosine, referred to by their abbreviations, ATCG. The genetic code is carried by triplets of these nucleotides, sequences of three bases. Only a small portion of the genetic sequences of DNA is related to our particular recognizable characteristics.

By studying families with a history of Huntington disease, however, geneticists have learned that 40 or more CAG repeats identify a person who will develop the disease. Even though the typical small number of repeats is not associated with any identifiable trait, or phenotype, an expanded number of repeats is associated with this adult onset degenerative nervous system disorder. When a parent has the disease there is a 50% chance of a child inheriting it. And when the genetic marker is inherited, if the person does not die early from some other cause, Huntington will eventually develop.

Population studies of Huntington disease have not been undertaken. Without studying the general population no one knows whether there are individuals who have long CAG repeats and no familial history of the disease. No one knows whether long CAG repeats without a family history are indications of the disease. As the technology for genetic blueprints approaches our grasp, it becomes easy to imagine that population information about the phenotypic implications of long CAG repeats could be significant to people who might have them.

Tom's family has no history of Huntington disease. If researchers were to undertake a population study to learn more about the genetics of Huntington disease should Tom volunteer to be a research subject and allow his cheek to be swabbed for collection of genetic material? Does he owe this debt to his fellows? Tom is reluctant to learn anything about himself that might give him cause to worry. Could his reluctance justify refusing to participate?

B. Case 2: Dick. Dick has been diagnosed with Marfan syndrome, an inherited disorder of the connective tissue. People with this syndrome are very tall and typically have heart and eye defects. Dick's cousin, Martha, is tall but it is not clear whether she too has Marfan syndrome. Martha understands that Marfan syndrome is a dominant genetic disorder, so that she would have a 50% chance of having a child with this disorder if she has the mutation causing Marfan syndrome.

She and her husband have consulted a genetic counselor because they want to have children but would want to avoid having children with this problem. The counselor explains that while the gene for Marfan syndrome has been fully mapped and cloned, each family has its own specific familial mutation. Searching for the specific mutation in the DNA of Martha's family will involve a lengthy and expensive process. A better alternative would be a linkage study to discover the pattern of this family's co-inheritance of variations in neighboring genes with the Marfan mutation. Genes that are close to one another on a chromosome tend to be inherited together. By collecting genetic samples from close relatives and comparing the patterns of linked genes, geneticists are able to identify a familial pattern and associate it with the genetic defect. Future fetuses can then be screened for the identified genetic pattern.

Martha's mother has average stature and so does her only sister. But her father, who died in a car accident, and his brother were both tall. Martha's uncle Henry has agreed to provide a blood sample for the linkage study. Martha has also asked Dick, Uncle Henry's only child, to participate in the linkage study. Does Dick have a moral responsibility to allow himself to be tested?

C. Case 3: Harry. After three years of deterioration and suffering, Harry's father died of Huntington disease at age 49. When his father was diagnosed all of the immediate family were invited to participate in a genetic counseling session. Harry, who was 22 years old at the time, learned that this degenerative disease of the basal ganglia is physically and mentally disabling and that he has a one in two chance of developing it. Because the number of CAG repeats tends to increase in the offspring of affected males, and because more CAG repeats indicate earlier onset of the disease, if he has inherited the genotype Harry is likely to be afflicted earlier than his father was. Although Harry's brother was tested for the genetic marker and found to be unaffected, Harry has refused the test saying that he does not want to know.

Harry and Sally have fallen in love. They want to marry and have a family. Does Harry have a moral right not to know whether he has a long CAG repeat?

D. Case 4: Harriette. Harriette's sister had a child who died of Tay-Sachs disease after a brief and agonizing life. Harriette's family as well as her husband's are part of a community that is known to have a significant number of Tay-Sachs disease carriers. Harriette wants to have children and she knows that Tay-Sachs is a recessive inherited disease. She knows that she and her husband could be tested and learn whether their offspring might have the disease. She and her husband have discussed the situation and decided not to be tested. Harriette says that she does not want the information to affect her choice. She has decided to take whatever she gets.

Does Harriette have the moral right to act in ignorance? Is it ethically permissible for her to take this chance for her child? If she should have one affected child, would it be her moral right to have another?

III. A PRESUMED RIGHT TO GENETIC IGNORANCE FROM RESPECT FOR AUTONOMY

The several societies of the genetics community and most of the authors writing on issues of genetic knowledge have embraced a policy of nondirective, value neutral counseling.[1] Historically, this univocal commitment can be understood as a moral stance to disassociate modern genetics from the eugenics movements of the first half of this century which directed and coerced in light of their political, economic, and social agendas (Paul, 1995; Alien, 1986; Pernick, 1996). Philosophically, non-directive counseling also reflects the centrality of respect for autonomy in modern bioethics. Appropriately, the genetics community has adopted the view that value neutral, or "non-directive," counseling implies that counselors and physicians should allow their clients and patients to decide whether or not to pursue genetic testing.[2] Less fortuitously, in their conversations, if not explicitly in their writing, they have come to speak of their patients' right not to know (Quaid, 1996; Shaw, 1987). This supposed right, in turn, has become another crucial precept in the ethics of genetic counseling. Our natural psychological aversion to being coerced, the historical commitment to personal liberty in the United States, and bioethics' attachment to the principle of autonomy can all account for the genetics community's comfort with the right not to know. Yet, inclinations do not add up to an argument. This crucial moral assumption may represent an unsupported and misleading philosophical leap of faith.[3]

IV. RIGHTS AND DUTIES

Before launching into a fuller discussion of the right not to know, and to insure that we have a shared understanding of the crucial terms, allow me a digression to explain a philosopher's understanding of rights and duties.[4] To have a right is to have a freedom to do or not do, while to have a duty is to have no moral freedom. Having a duty is being morally obliged, constrained, or bound to do one thing or another. We also say that rights are correlative with duties.[5] This means that rights and duties express a relationship that can be described from either perspective, like two sides of the same coin. When someone has a right, someone else has a duty, and when someone has a duty, someone else has a right. For example, when one person has a right to speak, another has the duty or obligation to allow her to speak. The same situation can be described by saying that one person is bound to hold his peace while the other has the right to speak or not speak.

   In the case of genetic knowledge this use of the terms "right" and "duty" makes it clear that if someone has a right to genetic ignorance, he has no duty to pursue genetic knowledge, and if someone has an obligation to pursue genetic knowledge, she has no right to preserve her genetic ignorance. Furthermore, using the correlativity model, we can appreciate the implications for human relationships with respect to genetic knowledge. When someone has a right to genetic ignorance, others have the responsibility to allow her that ignorance and to respect her choice not to know. When someone has a right to genetic information, someone else has a responsibility with respect to trying to provide that information; when one person has a duty to pursue genetic knowledge, someone else has a right with respect to that information.[6] With this sense of rights in mind we can return to the analysis of a person's right not to know genetic information about himself.[7]

V. A RIGHT TO GENETIC IGNORANCE, CONTINUED

When we hear the phrase, "a right not to know," it does not sound like something that we should obviously challenge. Hearing it applied in the medical context might even sound acceptable. Imagine the patient saying to his doctor, "If you find something terrible, do what you have to but I do not want to know." It is not hard to imagine the doctor justifying going along with the patient's request by maintaining that "the patient has a right not to know." However, when we begin to examine the supposed right not to know we cannot fail to notice a stark contrast to our ordinary moral thinking. This discrepancy invites us to challenge the ethical status of the right not to know.

A television commercial shows a blindfolded man test driving a car so that he can fairly assess the smoothness of its ride. That man behind the wheel might later decide that he prefers driving blindfolded. When he does not see what is in front of him he does not have to take those obstacles into account: his decision making is simplified and he is freed from a host of troubling concerns. Even though we might sympathize with his motives, we would, nevertheless, find the prospect of his driving blindfolded totally ridiculous and morally unacceptable to such an extreme as to be not worth considering. Obviously, driving a car without being able to see where you are going puts the property and lives of others in danger. No one has the right to do that. In other words, anyone who gets behind the wheel of a car is obliged to pay careful attention to his surroundings. If he is obliged to know his situation when he is driving, he has no right to choose to be in ignorance when he drives.[8]

This is the line of reasoning that supports negligence law. Even when people can honestly assert that they were ignorant of the hazards posed by their action or some possibly dangerous circumstance involving their property, they are still held liable for consequent harms. They are liable because they have an obligation to know: they have no right not to know.

The eighteenth century philosopher Immanuel Kant is the most acknowledged source of our appreciation of the ethical significance of respect for autonomy.[9] It seems fitting and prudent, therefore, to turn to his writings for support in what may appear, at first glance, as an attack on autonomy. We can appreciate the general form of the argument against a right not to be informed by looking at a short essay by Kant. The essay I have in mind is "On a supposed right to tell lies from benevolent motives," published in 1797 (Kant, 1967, pp. 361–365). There Kant recounts the situation of someone planning a murder asking you if your friend, his intended victim, is in your house when you had previously observed your friend entering. According to my reading of this controversial essay, Kant argues that benevolence cannot justify lying to the murderer because you cannot know whether your lie will accomplish good or actually harm the intended victim who might have already left the house unobserved. On the other hand, you will know with certainty that your lying will undermine our general reliance on veracity and that would do "wrong to men in general" (Kant, p. 362). In other words, those who believe that benevolence justifies this lie are mistaken because, according to Kant's argument, concern for benevolence justifies not lying. Regardless of whether you are willing to accept Kant's conclusion in his case, the form of the argument, that benevolence cannot justify lying because benevolence is the ground for not lying, is compelling. As I see it, the argument is actually more persuasive when applied to the supposed right to genetic ignorance.

Those who have argued for a right not to know genetic information about themselves have grounded their argument on the right to have one's autonomous choices respected. Following Kant's model, I want to argue that respect for autonomy actually leads to the opposite conclusion, the obligation to pursue genetic knowledge.

Through the active history of the American bioethics movement, the principle of respect for autonomy has been a cornerstone for bioethics argument. It has supported the consensus for a strong commitment to confidentiality, sustained arguments for informed consent in research, and upheld arguments for truth telling and patient self determination, particularly in decisions about withholding and withdrawing treatment. Respect for autonomy entails regarding others as capable of making choices for themselves that reflect their own values and commitments. It is, in part, because we recognize that people could have good reasons for not wanting to share personal medical information with everyone that we have to respect patient confidentiality. It is because we recognize that a host of personal commitments and a personal attitude toward risk can impact on judgments about whether to participate in a research protocol that we require informed consent in human experimentation. It is because people can have good reasons for making different choices about their medical treatments that patients need to have information about their medical condition and the alternative treatments (Pelias, 1991). Clearly, having information has been recognized as essential for making choices. Any piece of information might change a person's choice and lead her to pursue a different course than she might have without the information or with different information. So, because it allows others to make choices according to their own lights, health care providers are ethically obligated to make available the information that may be relevant to the patient. The reason for providing information in the typical medical context is that the patient is presumed to be an autonomous agent. Without the relevant information, the patient cannot make autonomous choices. From my point of view as an individual autonomous agent (as opposed to the point of view of some professional inside or policy-maker outside of the medical context who might be inclined to limit my autonomy) when I choose to remain ignorant of relevant information, I am choosing to leave whatever happens to chance. I am following a path without autonomy. Now, if autonomy is the ground for my right to determine my own course, it cannot also be the ground for not determining my own course. If autonomy justifies my right to knowledge, it cannot also justify my refusing to be informed. I may not be aware of the moral implications of ceding autonomy by insisting on genetic ignorance, but the ramifications are there, nevertheless.

From a Kantian perspective, autonomy is the essence of what morality requires of me. The core content of my duty is self determination. To say this in another way, I need to appreciate that my ethical obligation is to rule myself, that is, to be a just ruler over my own actions. As sovereign over myself I am obligated to make thoughtful and informed decisions without being swayed by irrational emotions, including my fear of knowing significant genetic facts about myself. When I recognize that I am ethically required to be autonomous, I must also see that, since autonomous action requires being informed of what a reasonable person would want to know under the circumstance, I am ethically required to be informed. So, if I have an obligation to learn what I can when genetic information is likely to make a significant difference in my decisions and when the relevant information is obtainable with reasonable effort, I have no right to remain ignorant. From the recognition of my own autonomy, I have a duty to be informed. I have no right to remain ignorant.[10]

Another example from Kant can be explicated to make the point about genetic ignorance in a different way. In this example from the *Foundations of the Metaphysics of Morals* (Kant, 1959, p. 40) Kant proves that it is immoral to make a promise with the intention of breaking it. According to his case someone who is in need of money procures a loan by promising to repay it, while knowing that he will not be able to return the money. Again Kant uses a consistency proof to show that making a promise without intending to keep it is immoral. The borrower relies upon the institution of promising to secure his loan, but unless people fulfilled their promises no one would trust another to meet a pledge of future performance.

Taking a broad view of promising, any commitment that we undertake with respect to another is some sort of promise that carries a moral responsibility to do one's part. My commitment to look after my sister's children while she goes off on a business trip certainly counts as a promise, but so does getting behind the wheel to drive a car, and developing a friendship. Drawing on this inclusive view of promises, we can see that I would have been immoral to have said that I would baby-sit for my niece and nephew if I had no intention of ever spending time with them. I would have unethically treated others who use the roads if I had started off in my car with no intention of keeping my eyes wide open and focused on what was in front of me. I would have failed in my obligations to my friend if I had accepted the benefits of her friendship without ever being willing to do my part in return. From this Kantian perspective we recognize that human interactions are crammed with undertakings that implicitly require performance by those who take on the responsibilities that define those undertakings.

Sometimes being able to satisfy a responsibility turns on some bit of knowledge that I could easily know. I would be morally culpable to undertake the obligation without first attempting to ascertain whether I had a reasonable chance of meeting that obligation. If I borrowed some cash and gave you a check in return without looking up my balance at the back of the same checkbook, I should be counted as blameworthy for my failure to repay my debt by the promised date. It is a small step to recognize that the same reasoning could apply to genetic knowledge. When I believe that meeting an obligation to another turns on genetic knowledge that is easy to obtain, I should not undertake the obligation until I have first learned that I am likely to be able to do what I have committed myself to doing.

I need to support the institution of promise keeping because I also rely upon others to keep their commitments to me. I should also keep my promises because I have given others a reason to rely on me. Unless I try to obtain pivotal information before undertaking commitments, I am likely to find myself having made promises that I cannot keep. Had it not been for my reluctance to be informed, I would not have been in the situation of violating the moral command against breaking promises.

VI. TOM, DICK, HARRY, AND HARRIETTE

In the same way, when we face a choice between alternative actions and when the choice turns on genetic knowledge, preserving our autonomy requires that we pursue the information. And, when we have to decide whether to assume an obligation and when the decision even in part turns on genetic knowledge, we have a moral duty to pursue that information. These strong conclusions have implications for Tom, Dick, Harry, and Harriette. None of them can claim a right to genetic ignorance. Because there is no right to genetic ignorance Tom cannot use it as a ground for refusing to participate in the population study and Dick cannot use it as a reason for not helping his cousin Martha. And as for Harry and Harriette, because they are on the brink of making important decisions that involve commitments to others, there is at least a prima facie reason for maintaining that they are ethically obligated to learn crucial genetic facts about themselves.

Harry is considering marriage and fatherhood but he knows that he has a 50% chance of developing Huntington disease. Because marriage and fatherhood are essentially commitments to others, he has the duty to learn whether he is likely to be able to meet these responsibilities to his future wife and children. If he worried that he might develop the disease and seriously tried to take his duties into account while still refusing genetic tests, he would be likely to make different decisions than he would have made in the face of knowledge. For example, knowing that he was not going to develop Huntington disease might free him to take a job in Australia that would allow him better opportunities for career advancement and the development of his talents.[11] Knowing that he was going to develop the disease might lead him to accept a less promising job that would allow him to remain geographically close to Sally's family. Taking a farsighted view of his situation allows us to imagine that his trying to second guess what he ought to do while considering two dramatically different scenarios could lead him to do what he might otherwise have seen as wrong. His informed assessment could lead him to a different conclusion.

Harriette is considering motherhood. She knows that she may have a chance as high as one in four of having a child with Tay-Sachs disease. She and her husband have agreed to forgo testing and she expresses a willingness to undertake responsibility for any children they should have. If neither she nor her husband carry the trait, or even if only one of them carries the trait, they are worrying for nothing and making choices in light of that concern that may be inappropriate for their actual situation. On the other hand, if they were to have the genetic testing and learn that they were both carriers they could consider their reproductive choices in that light.

The analysis of the presumed right to genetic ignorance has taken us some way in discerning the rights and responsibilities of genetic knowledge. Exploring the thickness of social ties may take us some steps farther in understanding more about what the individuals in our cases owe to others.

VII. SOCIAL BONDS

We need to recall the popular beliefs that blood counts and that family comes first. We need to acknowledge that in our world the social bonds between family members, the descendants of the Mayflower, the Irish, and the Mormons do matter. Yet we also need to notice that it is because of the DNA we share with yeast, drosophila (fruit flies), and mice that we can use them to learn about our own genetic makeup and about how to use genetic knowledge for human good. There is an important sense in which we share genetic links with them. But our shared DNA does not seem to be a source of ethical duties or moral responsibilities to yeast, or drosophila, or mice. Commonality of DNA seems to be an important component of blood ties and blood does seem to count somewhat in moral relationships, but neither DNA nor social history fully explains our ethical responsibilities to one another.

Looking at DNA does allow us dramatically to see how much alike we all are and which individuals are most similar. If genetic similarity were the source of our moral relationships, genetic maps could identify our most similar sibling, or even some distant DNA matching stranger, as the one to whom we owed the most. But if anyone maintained that we had different degrees of responsibilities to different siblings it is not likely that they would attribute that distinction to our degrees of genetic matching. More likely reasons would be related to the intimacy and dependency of our previous relationship, or the strength of our feelings, or the history of our interactions, or something about our relative wherewithal and neediness. These reasons, however, point to social relations, not blood or genes, as relevant features for moral responsibility. Furthermore, the genetically closest matching stranger has less of a moral claim on us than many others because of the absence of a previous relationship and the lack of history of interacting. Blood alone does not tell the story of our moral responsibility to one another. The bonds that have moral weight and give us thick responsibilities to one another typically include a social component.

Although many philosophers have taken a simple view of human relationships and morality and argued that we owe the same to everyone,[12] the cases we are considering suggest a need for a richer account of moral responsibility.[13] For some insight into the complexities of social bonds, allow me to go back farther into the history of ethics. Aristotle discusses justice and friendship as an apportionment of rights, responsibilities, and things between individuals. From what he and others have written on the subject it is not completely clear whether we should behave as if these features had to be considered because they are necessary for social inter-action, or because our intuition, or reflective endorsement, or sense that our action should stand the test of interpersonal justification requires us to take them into account. That ethical intuitionism, rationalism, contractarianism, and feminism take the particularity of relationships seriously gives us reason to look seriously at what Aristotle has to say about these considerations.

Writing on friendship in his *Nicomachean Ethics*, Aristotle maintains that

> the duties of parents to children and those of brothers to each other are not the same. nor those of comrades and those of fellow citizens, and so too with other kinds of friendship. There is a difference, therefore, also between the acts that are unjust towards each of these classes of associates, and the injustice increases by being exhibited towards those who are friends in a fuller sense; e.g. it is a more terrible thing to defraud a comrade than a fellow citizen, more terrible not to help a brother than a stranger, and more terrible to wound a father than anyone else. (Aristotle, viii. 9. pp. 207–208)

In short, and without picking over the nits, Aristotle takes the position that out of friendship, we owe something to everyone and more to those who are biologically or socially closer to us. For him, friendship is akin to what we now call brotherly love. He argues that justice binds us to express the caring compassion of friendship in our actions toward others to varying degrees depending on the relationship.

Then, after he reminds us that "[a]ll such questions are hard . . . to decide with precision," Aristotle goes on to explain

> that we should not give the preference in all things to the same person is plain enough; and we must for the most part return benefits rather than oblige friends, as we must pay back a loan to a creditor rather than make one to a friend. But perhaps even this is not always true: e.g. should a man who has been ransomed out of the hands of brigands ransom his ransomer in return, . . . or should he ransom his father? It would seem that he should ransom his father even in preference to himself. (Aristotle, ix. 2. p. 223)

Here Aristotle suggests that just distributions turn on assessment of an array of features. He recognizes that the history of a relationship must be taken into account and may make more of a moral difference than social or even biological relationships. He is also making the point that the particularity of situations and relationships can make a significant difference in prioritizing our duties to others.

Aristotle concludes this discussion of friendship (which, we should recall, encompasses all human relationships, even those with strangers) by reminding his readers that "it is characteristic of friendship to render services" (Aristotle, ix. 11. p. 246). He goes on to say that "friendship is a partnership, and as a man is to himself, so is he to his friend" (Aristotle, ix. 12. p. 246). This last expression of a golden rule instructs us to do for our friends and to help them further their own ends. Taken together these Aristotelian conclusions can provide some additional guidance for resolving our cases. First, they express the moral stance that we owe something to everyone. Second, they outline a collection of distinct considerations that we need to accord some weight and take into account in the assessment of our duties to others.

(1) Family relationships count.
(2) Social relationships count.
(3) The history of a relationship counts.
(4) The particulars of the relationship and situation count.

With this much guidance in hand, let us return to the cases and see if we can make more headway in identifying the ethical duties of Tom, Dick, Harry, and Harriette. These considerations should be seen as prima facie duties that must be taken into account in moral deliberation. They could however, be out-weighed by other considerations that may be relevant to the particular case.

VIII. CASE DISCUSSION

Because we have duties to our fellows, Tom has a responsibility to participate in the population study. What is being asked of him requires little effort, minuscule discomfort, and no physical risk. Yet the information to be gained by the study could have a significant impact on the well-being and decisions of others. Because the information can be a significant good to them and because we are morally required to render service to our brethren, Tom has a duty to participate in the study.[14]

Similarly, Dick is obligated to provide a blood sample for a family linkage study.[15] His familial relationship with Martha is closer than his bond with a stranger, so it would be more terrible to ignore her need than it would be to ignore a stranger's. Compassionate concern for Martha who wants to have a child who is unaffected by Marfan syndrome should incline him to cooperate. If Dick had been particularly close lifelong friends with his cousin, or if Martha had been his sister, he might have even stronger moral reasons for trying to be of service. On the other hand, if the cousins had minimal contact in their childhood, and if they had not even spoken to one another in the past decade, his responsibility might be somewhat diminished.

Dick's case is particularly interesting because it is only his blood ties to Martha that give him an obligation to participate in the linkage study. No one else can do it for her and no one else could take his place. Dick's case makes the point that we have some of our responsibilities because of our unique ability to help, others only because of our biological ties. So morality is not entirely constructed out of socially created links. Adoption would be another relevant example. A birth mother can relinquish all of her social rights and duties to her child and the adopting parents can take them all on. But the birth mother cannot transfer her biological responsibilities. Perhaps this insight calls for collecting a blood sample from the birth mother when the child is given up for adoption so that it could provide the genetic information that her child might need from her later in life.

We have already discussed some of the issues relating to *Harry's* obligation to pursue genetic knowledge. In addition, his "fuller friendship" with Sally gives her a special claim on him. Their social relationship gives them mutual responsibilities to treat one another as partners. To use the language of feminist philosopher Patricia Mann, they have a mutual responsibility to recognize and support one another's "social agency," which, according to Mann, is associated with "motivation, responsibility, and expectation of recognition or reward" (Mann, 1994, p. 14). For Harry this involves acknowledging Sally's motivation with respect to entering a marriage, the responsibilities it would entail for her, and the rewards she would anticipate and how all of that might be impacted by his developing Huntington disease. Recognizing her agency would require him to inform her of his chance of developing Huntington disease, to discuss the alternatives relating to marriage and reproduction with her as a genuine partner, and to support her agency in whatever choice she should make for herself with respect to their future.[16]

Harry's standing with respect to his future children raises another set of concerns. What duties does he undertake when he chooses to become a father? Is there some way that he can meet his obligations to his child if he becomes disabled and dies before the child matures?

And as for *Harriette's* case, a lurking issue of family ties involves informing other family members about their risk for having an affected child. Was Harriette's sister obligated to inform her siblings that she had an affected child? If she were estranged from the others and lived on the other side of the country would she be bound to tell? We know that families do keep secrets. Yet, because the impact of this disease is so devastating to the affected child and the entire household, and because family relations seem to carry some moral weight, and because we should care about the projects of close friends as well as strangers, there seem to be sufficient reasons for claiming that Hariette's sister should inform her siblings regardless of their present social relationship.

The most troubling issue of family bonds related to this case, however, turns on whether there are some children who are better off not being born. A Tay-Sachs baby appears normal for the first few months. Then, between the fourth and eighth month of life, the nerves begin to be affected by a fatty substance that accumulates in the cells. The child becomes blind, deaf, unable to swallow. The muscles atrophy, response to the environment dwindles, and the relentless deterioration continues until the child is totally debilitated and develops uncontrollable seizures. Death from pneumonia or infection usually occurs between age five and eight. Are there some lives that parents have no right to create?[17]

IX. CONCLUSION

The cases considered in this paper have led to some troubling unanswered questions about genetic links, family ties, and social bonds. When confronting the most difficult cases, it may be useful to recall Aristotle's caution that it is difficult to decide these matters with precision. Perhaps we should sometimes be satisfied with just ruling out the worst alternatives and allowing some room for differences of opinion.

The arguments we have considered have led to some strong and perhaps surprising conclusions. The clearest conclusion is that no one has a moral right to genetic ignorance. The other noteworthy conclusion is that moral responsibility depends on a variety of factors including blood ties, social relationships, the history of an interaction, and particular features of the situation and the individuals involved.

These conclusions have obvious and forceful implications for the individuals who may have reasons for considering genetic services. They are also significant for geneticists and genetic counselors in that they present good reasons for taking a certain view on what would be right for their patients and clients to choose. However, I need to caution that my line of argument does not go so far as to over-ride the genetics community's well-modulated commitment to non-directive counseling. A vivid imagination can allow us to see that in spite of the arguments against a right to genetic ignorance, in some unusual situation a rational person could have good reasons for making another choice.[18] Beyond that, we know that people are not ideally rational and idiosyncratic psychological facts could make a significant difference. Furthermore, the general consequences of maintaining a non-directive counseling policy could far out-weigh the alternative by encouraging people to enter the counseling setting.[19]

The surprising conclusion of the argument against the presumed right to genetic ignorance has another significant implication. While the genetics community deserves praise for their concerted efforts to address the ethical issues raised by their new technology, they also need to be warned. Ethics draws on analogy and examples, and then requires careful argument and careful analysis that turns on careful use of technical terms. At each step it is easy to stumble and mistakes are easily compounded. It is easy to reach a wrong conclusion when you start with an inappropriate analogy, misuse a technical term, or claim to have provided ethical analysis without understanding what is involved. Misunderstandings about the nature and moral force of autonomy have led some in the genetics community to a false conclusion about genetic ignorance. That error in turn has led to extreme positions on non-directive counseling. The moral of this story is that we all have our limitations. Some are in our genes, some are not, and no one is eager to hear bad news.

NOTES

Reprinted by permission from *Journal of Philosophy and Medicine* 23 (1998): 10–30.

An earlier version of this paper was read on September 27, 1996, at the Ninth Annual Fredrick Womble Speas Memorial Colloquium, "Redesigning Ourselves, Engineering Our Children: The Ethical, Legal and Social Implications of the Human Genome Project," Davidson College, Charlotte, N.C. I am grateful to Rosemarie Tong and Lance Stell for inviting me to consider this subject. This paper has benefited from genetics counseling by Randi Zinberg and comments by Richard Epstein, Joe Fitschen, Kurt Hirschhorn, John Robertson, and Lance Stell.

1. See, for example, Ad Hoc Committee on Genetic Counseling, 1975; Andrew et al., 1993; Caplan, 1993; Chadwick, 1993; Kessler, 1992; Fine, 1993; Gervais, 1993; Kopinsky, 1992; National Society of Genetic Counselors, 1991; Singer, 1996; Sorenson, 1993; Wilfond and Baker, 1995; Yarborough et al., 1989.

2. The argument here has limited implications. The thrust of this discussion is merely to urge that we recognize that genetic information and genetic ties have some moral weight. They are factors that need to be taken into account in moral deliberation. Prescribing social policy and policy for the professions is beyond the scope of this paper.

3. Again, what follows is not an argument for informing patients/clients who choose not to know genetic information about themselves or their offspring. The argument below has the much more limited scope of merely showing that there is no patient/client "right not to know." There may certainly be excellent policy considerations and personal reasons for not imposing information on someone who does not want it.

4. For the purposes of this discussion it is not necessary to make further Hohfeldian distinctions.

5. An argument in support of the correlativity thesis is beyond the scope of this discussion.

6. The context of consideration determines the bounds of the right or responsibility.

7. In what follows I will be using "the right not to know" as the semantic equivalent of "the right to genetic ignorance." I have framed the right not to know as a negative right, simply a right to be left alone in ignorance, what Hohfeld (1919) would have counted as a "privilege." Showing that there is no such right does not imply that anyone else has a duty to inform the person who wants to be ignorant. An additional argument would be required to make that case.

From the use of the phrase "the right not to know" and the assertion that it is derived from autonomy rights, I have assumed that it should be treated as a privilege. Of course, someone might assert that there is a "claim" right to genetic ignorance. A claim would impose strong duties on others not to interfere. A new and different sort of argument is needed in order to assert that position.

8. This is an example of a voluntary undertaking of a special obligation. The point of this comparison is to show that distinctive moral responsibilities arise from such undertakings and that there is a model for considering them.

9. Other non-Kantian conceptions of autonomy have certainly been invoked, particularly in the bioethics literature. Without entering the debate about the appropriate or proper analysis of the term, my point is only that any concept of autonomy that can support a right to be informed cannot also be the ground for the opposite.

Furthermore, I choose to rely on the Kantian conception of autonomy because one aim of this paper is to show that bioethical issues involving new technology can be addressed with the tools of traditional moral philosophy.

10. The need for this clarification was pointed out by Lance Stell. The importance of introducing the metaphor of the just ruler was offered by Richard Epstein.

11. Kant maintains that we have duties to ourselves, e.g., the duty to develop our talents.

12. Utilitarian Jeremy Bentham is clearly a case in point. Many would also argue that Immanuel Kant and John Rawls should be counted in this group.

13. Bernard Gert makes a similar point about the inadequacy of simple accounts of moral judgments and he too recommends looking to Aristotle for a more comprehensive framework (1996, p. 33).

14. A contractarian argument could also conclude in the assertion that Tom has an obligation to participate. Tom can easily imagine that at some point he or a loved one might need some genetic information that could only be learned from a population study. Since he would want others to cooperate with that study, he too should be willing to participate. Assuming this kind of rational foresight, Tom could be taken to have given his implicit consent to cooperate in population studies so long as others also do their part. This kind of hypothetical reciprocal agreement is the basis for contractarian ethics. The model does not require personal, face-to-face promises or even explicit individual commitment. It only requires that a reasonable person in a similar situation should agree, that is, would have no good reason to refuse to participate in the practice.

15. Lamport takes this position (1987, pp. 307–314).

16. If Sally should choose to continue with the relationship and accept Harry's not learning his genetic status with respect to HD, then their situation would be very much like the situation of Harriette and her husband. We would have another case of two people wearing blindfolds and sharing the driving.

17. Others have discussed this and related issues, e.g., Alien, 1986, 1986b; *Dietrich v. Inhabitants of Northampton,* 1884; Engelhardt, 1975; Kevles, 1985; Paul, 1995; Pernick, 1996; Shaw, 1987.

18. We need to notice the significant difference between the high standard for counting one's own action as autonomous and the much lower standard that we should invoke for respecting the autonomy of others. Respect for the choices of others requires us to presume that their decisions flow from self-governance. Our ethical stance toward them commands us to regard people with the supposition that they are acting from the moral rules

that they have embraced and that they have good reasons for what they decide to do. Only in the face of strong evidence to the contrary are we entitled to question whether they have the mental capacity to make the decision before them. This is why we accept a striking asymmetry in medical practice: a patient's capacity to accept recommended treatment is rarely challenged, while a patient's refusal of highly beneficial treatment calls for a more thorough examination of the patient's capacity.

19. The history of directive genetic counseling for sickle-cell disease makes this point. The directive approach had a dramatic impact on the African American community, the group most likely to include carriers of this trait. Directiveness seems to have fostered militant opposition to genetic screening among this population.

BIBLIOGRAPHY

Ad Hoc Committee on Genetic Counseling. American Society of Human Genetics. 1975. "Genetic counseling." *American Journal of Human Genetics* 27: 240–42.

Allen, G. E. 1986. "The eugenics record office at Cold Spring Harbor, 1910–1940: An essay in institutional history." *Osiris,* 2d ser., 2: 225–26.

Allen, G. E. 1986b. "Eugenics and American social history, 1880–1950." *Genome* 31: 885–89.

American Society of Clinical Oncology. 1996. "Statement of the American Society of Clinical Oncology: Genetic testing for cancer susceptibility." *Journal of Clinical Oncology* 14, no. 5: 1730–36.

American Society of Human Genetics Board of Directors and the American College of Medical Genetics Board of Directors. 1995. "ASHG/ACMG report, points to consider: Ethical, legal, and psychosocial implications of genetic testing in children and adolescents." *American Journal of Human Genetics* 57: 1233–41.

Annas, G. J. 1993. "Privacy rules for DNA databanks," *JAMA* 270: 2346–50.

Andrew, L. B., J. E. Fullarton, N. A. Holtzman, A. G. Motulsky, eds. 1993. *Assessing Genetic Risks: Implications for Health and Social Policy.* Washington, D.C.: Institute of Medicine/National Academy Press.

Aristotle. 1971. *The Nicomachean Ethics of Aristotle.* Translated by Sir David Ross. London: Oxford University Press.

ASHG Ad Hoc Committee on Insurance Issues in Genetic Testing. 1995. "Background statement: Genetic testing and insurance." *American Journal of Human Genetics* 56: 327–31.

Caplan, A. L. 1993. "Neutrality is not morality: The ethics of genetic counseling." In *Prescribing Our Future,* edited by D. M. Bartels, B. S. LeRoy, A. L. Caplan, 149–65. New York: Aldine De Gruyter.

Chadwick, R. F. 1993. "What counts as success in genetics counseling?" *Journal of Medical Ethics* 19: 43–46.

Dietrich v. Inhabitants of Northampton, 138 Mass. 14. 52 *Am Rep* 242 (1884).

Engelhardt, H. T., Jr. 1975. "Ethical issues in aiding the death of young children." In *Beneficent Euthanasia,* edited by M. Kohl. Buffalo, N.Y.: Prometheus.

Fine, B. 1993. "The evolution of nondirectiveness in genetic counseling and implications for the human genome project." In *Prescribing Our Future,* 101–18.

"Genetic prophecy and genetic privacy: Can we prevent the dream from becoming a night-mare?" 1995. *American Journal of Public Health* 85, no. 9: 1196.

Gert, B. 1996. "Moral theory and the human genome project." In *Morality and the New Genetics,* edited by Bernard Gert et al., 29–55. Sudbury, Mass.: Jones & Bartlett.

Gervais, K. G. 1993. "Objectivity, value neutrality, and nondirectiveness in genetic counseling." In *Prescribing Our Future,* 119–30.

Gostin, L. 1994. "Genetic discrimination: The use of genetically based diagnostic and prognostic tests by employers and insurers." In *Genes and Human Self-Knowledge: Historical and Philosophical Reflections on Modern Genetics*, edited by R. F. Weir, S. C. Lawrence, and E. Fales, 122–63. Iowa City: University of Iowa Press.

Hoffmann, D. E., and E. A. Wulfsberg. 1995. "Testing children for genetic predispositions: Is it in their best interest?" *Journal of Law, Medicine, and Ethics* 23: 331–44.

Hohfeld, W. N. 1919. *Fundamental Legal Conceptions,* 35–64. New Haven: Yale University Press.

Hudson, K. L., et al. 1995. "Genetic discrimination and health insurance: An urgent need for reform." *Science* 270: 391–93.

Juengst, E. T. 1991. "Priorities in professional ethics and social policy for human genetics." *JAMA* 266, no. 13: 1835–36.

Kant, I. 1959. *Foundations of the Metaphysics of Morals*. Translated by Lewis White Beck. New York: Library of Liberal Arts Press.

Kant, I. 1967. "On a supposed right to tell lies from benevolent motives." Rosenkranz, vol. 7:295. Anthologized in *Kant's Critique of Practical Reason and Other Works on the Theory of Ethics*, 361–65. 6th ed. Translated by Thomas Kingsmill Abbott. London: Longman, Green.

Kessler, S. 1992. "Psychological aspects of genetic counseling: Thoughts on directiveness." *Journal of Genetic Counseling* 1, no. 1: 9–17.

Kevles, D. J. 1985. *In the Name of Eugenics.* New York: Alfred A. Knopf.

Kopinsky, S. M. 1992. "'Value-based directiveness' in genetic counseling." Letter to the editor. *Journal of Genetic Counseling* 1, no. 4: 345–48.

Lamport, A. T. 1987. "Presymptomatic testing for Huntington chorea: Ethical and legal issues." *American Journal of Medical Genetics* 26: 307–14.

Mann, P. 1994. *Micro-Politics: Agency in a Postfeminist Era.* Minneapolis: University of Minnesota Press.

National Society of Genetic Counselors. 1991. *Code of Ethics.*

Paul, D. 1995. *Controlling Human Heredity, 1865–Present.* New Jersey: Humanities.

Pelias, M. Z. 1991. "Duty to disclose in medical genetics: A legal perspective." *American Journal of Medical Genetics* 39: 347–54.

Pernick, M. S. 1996. *The Black Stork.* New York: Oxford University Press.

Quaid, K. 1996. *The Scientific, Ethical, and Social Challenges of Contemporary Genetic Technology.* Tacoma, Wash.: NEH/NSF Institute/University of Puget Sound.

Shaw, M. W. 1987. "Testing for the Huntington gene: A right to know, a right not to know, or a duty to know." *American Journal of Medical Genetics* 26: 243–46.

Singer, G. H. S. 1996. "Clarifying the duties and goals of genetic counselors: Implications for nondirectiveness." In *Morality and the New Genetics*, 125–45.

Sorenson, J. R. 1993. "Genetic counseling: Values that have mattered." In *Prescribing Our Future*, 3–14.

Wilfond, B. S. 1995. "Screening policy for cystic fibrosis: The role of evidence." *Hastings Center Report* 25, no. 3: S21–S23.

Wilfond, B. S., and D. Baker. 1995. "Genetic counseling, non-directiveness, and clients' values: Is what clients say, what they mean?" *Journal of Clinical Ethics* 6, no. 2: 180–82.

Yarborough, M., J. A. Scott, and L. K. Dixon. 1989. "The role of beneficence in clinical genetics: Non-directive counseling reconsidered." *Theoretical Medicine* 10: 139–49.

26

Privacy and the Control of Genetic Information

Madison Powers

The prospects of extensive collection, storage, and transmission of genetic information raise new and troubling issues of privacy. Many of these concerns can be traced to the vast array of potential uses for genetic information and to the fear that the proliferation of such information will result in social stigma, loss of economic and social opportunities, and loss of highly valued freedoms. Because the potential uses of genetic information are numerous and diverse, the inventory of potential benefits and harms will vary depending on context, and the best strategies for the protection and promotion of privacy will be a function of those contextual differences. The aim of this chapter is to present one philosophical approach to the resolution of such issues.

The first section outlines a general framework for the moral analysis of privacy issues. Any discussion of privacy must start from the fact that privacy theories and their applications are matters of considerable academic and public controversy. For example, critics have raised doubts about privacy, either as a legal doctrine or, more fundamentally, as a coherent philosophical category for understanding the moral issues at stake in practical controversies (Parent 1983). Other critics have argued that social policies and legal doctrines designed to ensure individual privacy can impede medical and scientific research, prevent the implementation of comprehensive strategies to protect the health of the community, and ignore the importance of individual responsibility to act for the protection of others when a slavish adherence to rights of the individual puts others at risk of harm (Black 1992).

In light of such objections, no discussion of privacy can proceed without some preliminary account of its meaning and moral importance. A definition of privacy is discussed, as well as the underlying interests that make it morally significant, and how those interests are related to privacy rights is analyzed.

applies the analytical framework to a discussion of how
ought to be dealt with in the development of public policy.
three distinct kinds of privacy rights is debated: Two rights
its of informational self determination reflect the individual's
controlling, first, what information is generated and, second, what
is disclosed to others. A third kind of privacy right makes no refer-
liberty right or right of control vested in the individual. Instead, it is
as a right against the government for protection against the loss of pri-
id it is of primary importance when privacy protection through the exercise
dividual control over information is ineffective, infeasible, or undesirable.
ally, it is argued that what often matters most is that social and economic insti-
tions are organized in ways that reduce the harmful consequences of increased
access to personal information.

PRIVACY: A FRAMEWORK FOR MORAL ANALYSIS

Privacy Definitions

Some privacy definitions are too broad for our purposes, such as the one frequently attributed to Samuel Warren and Louis Brandeis. Their view of privacy is
often described as the right to be let alone (Warren and Brandeis 1890). However,
as critics have noted, there are many ways of failing to let someone alone (e.g.,
hitting someone with a baseball bat), and most do not involve a loss of privacy
(Thomson 1984).

Other privacy theorists have defined privacy as a condition of limited access to
some aspect of the person (Alien 1987). A loss of privacy is said to occur when
there is increased access to some aspect of the person. The specific kind of privacy
loss individuals experience will depend upon the particular aspect of the person
made more accessible. For example, limited access definitions might count as a
privacy loss any one or more of the following aspects of the person for which
access is gained: personal information (informational privacy), the person's body
(physical privacy), the physical proximity of the person (solitude or seclusion)
(Gavison 1984), a person's relationships with others (relational privacy), or a person's sphere of decision-making (decisional privacy) (Tribe 1978).

Multidimensional privacy definitions, or ones that count inaccessibility to more
than one aspect of a person as forms of privacy, are the subject of considerable
debate (Schoeman 1984). However, this chapter focuses primarily on informational privacy. Informational privacy can be understood as a condition or state of
affairs in which access to information about a person's physical and mental condition, biological and genetic makeup, psychological states, dispositions, habits, and
activities is limited.

Several implications of this definition, as well as some additional assumptions about the value of privacy, should be noted at the outset. First, privacy is always a matter of degree, and concern with its protection must always be tempered by the knowledge that it is not possible to ensure complete privacy.

Second, privacy merely describes a state of affairs; it is morally neutral. Privacy or its loss may be viewed as good, bad, or a matter of indifference. Its moral importance, if there is any, lies elsewhere. Thus, privacy is not assumed to be a fundamental category of moral thought. Its value is derivative from the underlying interests that may be promoted or protected when a condition of limited access to information exists.

Third, to the extent that privacy matters morally, what matters most in some instances is a certain kind of desired outcome, namely that access to information is restricted. It is an outcome that is desirable, independent of whether it is achieved by the individual's retaining control over personal information. In other instances, what matters most is that the individual retain and exercise control over access to information, even if doing so results in greater privacy losses than otherwise would occur under different social arrangements.

Fourth, the proposed definition of privacy comprehends limited access to information, not only by others, but by oneself as well. Thus, privacy losses include instances in which the individual gains access to previously unknown information about himself or herself.

Fifth, a loss of privacy is a morally significant event only when there is a reasonable expectation that access to information about a person will remain restricted.

Underlying Interests

What considerations form the basis of a reasonable expectation that access to personal information should be limited? This question takes on added urgency if privacy is not seen as a fundamental category of moral thought and if its value is viewed as wholly derivative of some underlying interests. One prominent approach to answering this question seeks to identify the *one* interest that explains why privacy matters in all instances (Reiman 1976). A number of possibilities have been suggested by various privacy theorists.

First, a person's well-being, or life prospects, may depend on limiting others' access to certain information, which if revealed can have adverse psychological, social, or economic consequences. Second, the ability to make personal choices without substantial interference by others may be compromised by access to information by those who may unduly influence personal decisions (Benn 1971). Third, access to information about a person may inhibit his or her ability to form intimate relationships with others where a selective sharing of information is often an essential element of what distinguishes deep personal commitments from the variety of other personal relations people share (Fried 1968). Fourth, intrusions into decisions about what information individuals will have access to have the power to affect profoundly personal self-concepts and the capacity for maintaining a sense of psychic stability (Reiman 1976). Fifth, dissemination of highly sensitive personal information may result in emotional distress, social stigma, embarrassment, and loss of self esteem and the respect of others.

However significant any one interest may be in a wide range of cases, there is no compelling rationale for supposing that the moral significance of privacy in all instances is reducible to one underlying justification. A more plausible hypothesis is that there is not just one, but many interests that may matter morally; and often it is a cluster or constellation of interests that justifies concern for privacy protection.

Consideration of the importance of privacy in the context of genetics adds support to the cluster theory of privacy in forests. Access to genetic information has the potential to adversely affect each of the interests identified above; in many instances, access to such information can create risks to social, psychological, and economic well being, as well as threats to the exercise of autonomy and to the development of personal relationships.

The Nature of Genetic Information

The focus upon the cluster of interests at stake directs attention to the following questions: Is the information obtained from genetic testing and screening qualitatively different from other medical information? Is there anything about the nature of genetic information, or its potential uses, that makes it more deserving of privacy protection than other kinds of medical information? Several differentiating characteristics of genetic information have often been noted.

1. Genetics can reveal much more personal information about an individual than other types of medical testing and evaluation.
2. Genetic testing offers the often unwelcome prospect of revealing to individuals detailed predictions of their medical futures.
3. Generic disorders generally affect people throughout their lives, and thus knowledge of genetic information may have a greater impact upon individuals than knowledge of other kinds of medical information.
4. The development of genetic information has potentially serious adverse financial, emotional, and social consequences.
5. The analysis of DNA samples may reveal information in the future not contemplated at the time of initial consent for testing or sample collection.
6. Genetic information has the potential to be misleading or incorrect in particular cases, in that it often relies upon probabilities for whole populations, whereas other personal and environmental factors can influence health outcomes.
7. Genetic testing or research into familial patterns of genetic inheritance may

reveal information about other family members as well as information about a single individual.

None of these claims is sufficient to distinguish genetic information from a variety of other medical information routinely gathered. Many other medical tests and routine medical histories reveal much about other family members and about an individual's probable medical future. Information gathered from sources other than genetics may have adverse social and economic consequences, and it may be misleading or false. Other medical tests may reveal conditions that might affect persons throughout their lives, and future researchers and practitioners may learn things about the subject or patient not contemplated at the time of the initial encounter.

All these concerns taken together, however, illuminate the larger set of concerns attendant to the growth of scientific knowledge, of which genetics is but one very conspicuous reminder. What seems to magnify concerns about individual privacy is the increased potential for comprehensive, systematic, and efficient collection of an abundance of medical information about a person. Accordingly, concern for the privacy of genetic information should be enhanced, even if there is nothing clearly unique about this particular form of medical information.

The Foundation of Privacy Rights

The discussion so far has proceeded in terms of an inquiry into the interests that give informational privacy in general, and the privacy of genetic information in particular, its moral significance. However, debates about privacy typically are cast in terms of privacy rights, and indeed, one rhetorical function of rights language is the way it highlights the need for enhanced privacy protection. Thus, the connection between a cluster of morally significant interests and privacy rights—apart from the rhetorical one—needs specification.

First, if the moral significance of privacy depends upon the cluster of underlying interests at stake, then the moral significance of those interests in a particular context will depend upon the kinds of harms threatened by a loss of privacy. Those harms will depend upon a variety of factors, including the vulnerabilities of persons under any given set of economic and social arrangements, the existence of prejudice and impediments to an individual's ability to establish relationships with others within one's community, and the cultural traditions and customs that shape the individual's sense of dignity and self worth.

Second, rights need not be seen as unconditional; they represent generalizations about the kinds of interests that are of such importance as to justify the imposition of certain duties on others (Raz 1986). However, current privacy rights and the duties they entail, in any particular context, are a function of the contingent character of a person's interests, the specific threats to those interests, the available alternatives for their protection, and the extent to which each alternative addresses the full range of interests at stake.

Third, although the particular kinds of informational privacy that might be important to protect will vary, the persistent fact of human vulnerability will ensure a continuing value for privacy. However, to the extent that scientific, technological, and cultural developments transform the context in which human vulnerabilities exist, then the kinds of information that ought to be protected from access will change, as well as the kinds of protection needed.

Fourth, a commitment to privacy rights does not entail a commitment to absolute rights. In many instances, there is simply no reasonable expectation that access to some information should be limited; and even where there is a reasonable expectation of limited access, there may be countervailing reasons favoring disclosure. The task is to determine when the position that Alan Westin calls in this book "privacy fundamentalism" should be adopted and when the position he calls privacy pragmatism should be embraced (Westin 1993). The former view represents a strong commitment to privacy protection at the expense of other social goals, while the latter is more willing to make tradeoffs on a case by case basis.

A morally acceptable public policy with respect to genetic privacy will be a mixed strategy or one in which elements of both the fundamentalist and the pragmatist view are incorporated, depending on the particular balance of competing interests at stake. In some instances, the privacy fundamentalist position should be adopted, and accordingly, the existence of stringent rights to informational privacy must be recognized. Although there may be relatively few instances in which individuals have a reasonable expectation of exercising an absolute veto over what information about them is made available to themselves or others, the point of discussing informational privacy in terms of rights is a more modest one. It is the claim that informational privacy reflects the existence of a variety of morally significant interests of such importance that any proposed intrusion must overcome a substantial threshold presumption against it for it to be justified.

THE APPLICATION OF PRIVACY RIGHTS TO GENETIC INFORMATION

Rights of Informational Self-Determination

One interest often at the root of concern for protecting the privacy of genetic information is autonomy, or the ability of individuals to control their own destinies. Rights associated with autonomy interests are classified as liberty rights, or the rights of individuals to make their own choices and decisions substantially free from the interference of others. In the context of genetic information, such liberty rights include the right of the individual to control access to highly personal information. In large part, the importance of retaining control over information reflects a deeper concern for the protection of more general autonomy interests. For example, rights of informational self-determination are especially valuable when a loss of control over information can result in a loss of ability or freedom to make other important life choices.

Autonomy interests provide the principal foundation for two rights that can be understood as species of what a German court has labeled the "right of informational self-determination" (Flaherty 1989). Such rights may be exercised in either of two ways. Individuals may assert their rights to control what information is collected or generated, or they may exercise rights to control aspects of further disclosures of information, including the purposes of further disclosures and the identities of the recipients of information.

Control over What Information Is Generated

Some rights of informational self-determination are based upon the assumption that as autonomous agents, individuals are entitled to decide for themselves (at least to some extent) what they, or others, may know about them. However, rights to control the uses of information, or to control the identities of its recipients, may not be adequate to ensure autonomy. In some instances, the only (or most) effective way to secure the individual's ability to make important life decisions free from the undue interference of others is through the exercise of control over the initial production or creation of some kinds of information.

One illustration of the importance of the right to exercise control over the generation of genetic information involves calculations of social and economic risks. In some instances, individuals can plausibly claim that they want to weigh for themselves a variety of medical and nonmedical risks against medical and other benefits of knowing information about their genetic makeup. As long as there are substantial risks of social stigma or potential loss of insurance or employment, and no available medical intervention once test results are known, it is not unreasonable that an individual may rationally conclude that the benefits of testing and the information it produces do not outweigh the risks that might flow from unwanted or improper disclosure (Faden et al. 1991).

Additionally, a right of control over what information gets generated has implications for the moral duties of those who collect and store genetic information. Adequate respect for individual autonomy and the right of informational self-determination make it incumbent upon health professionals to discuss the economic and social risks of unwanted disclosure and to present an accurate and realistic account of the legal and other privacy protections as an integral part of the process of obtaining informed consent for testing (Powers 1991).

A second risk individuals arguably would want to weigh for themselves is that, with further scientific developments, the predictive information generated might turn out to be incorrect. Announcement of the discovery of some new genetic marker is an increasingly common occurrence, but some of the reports that purport to identify genes responsible for certain predispositions to disease may turn out to be false or misleading. One risk is that there may be a short term scientific consensus on some markers for genetic predisposition that may, after great harm is done to those tested and labeled, turn out to be not well founded. The risks of false labeling include unjustified discrimination in employment or insurance, emotional distress, or pressure to undertake ill advised actions pursuant to medical recommendations for risk reduction, e.g., prophylactic mastectomies or hysterectomies designed to reduce cancer risks. These are matters that ought reasonably be left for individuals to weigh against the claimed benefits of testing.

A third set of examples includes what is often called a right not to know (Shaw 1987). The emphasis of informational privacy is on ensuring that certain information is not made available to those the individual does not want to have access. This can include oneself as well as others. It is a right of privacy because it is one of the most important aspects of the person for which there is a reasonable expectation of limited access. The reasonableness of this expectation is supported by its close connection with what any conception of what autonomy minimally requires, together with considerations of the extreme sensitivity of certain types of information, and the potential for such information to produce profound psychological harm or interference with the exercise of autonomy.

Such considerations point to the kinds of risks one should reasonably expect to weigh in the decision about what information gets generated. The frequently discussed example of Huntington's disease is illustrative. Each individual may want to weigh for himself or herself the benefits of knowledge, which may aid in planning for the future, against the psychological burdens of emotional distress, chronic anxiety, and anticipation of development of a debilitating late onset condition.

Other examples highlight the close connection between risk of error and the potential for severe adverse psychological consequences. One example involves the sensitivity of certain tests, such as those used in carrier screening. If there is a substantial risk that couples identified as at-risk through population screening may be falsely labeled, there is the potential for needlessly increased levels of anxiety and for unwarranted interference with reproductive plans. Similarly, the risk of psychological harm from learning information about genetic predisposition to disease—which may never develop or may occur with only moderate severity—is the kind of risk that ought to be left to individuals to weigh for themselves.

The Generation of Information in Research Contexts

There are other instances in which arguments for a strong right of informational self-determination appear less well supported. Special problems of privacy protection arise in the context of pedigree studies generated from genetic registries. Such registries are designed to enable researchers to learn more about the natural history of that condition and the patterns of genetic inheritance. Patients identified as having symptoms or presymptomatic genetic predisposition for an autosomal dominant disorder (e.g., Huntington's disease, polycystic kidney disease, etc.) may be recruited for participation in a genetic registry. Each participant provides a detailed medical and family history, including the names of others in the family who may be affected by, or are at risk for, the disorder. Researchers use this information to develop a pedigree plot, which permits them to study the patterns of inheritance within a family. Family histories, as well as individual medical histories and perhaps even certain demographic information, are generally included in such registries.

Although the primary source for identifying persons to be included in a registry may be through referral by health care providers, and their inclusion in the registry comes only after informed consent has been obtained, a registry is likely to contain information about numerous other members of families in which the disorder has been prevalent. Typically, an individual gives consent for inclusion of his or her own name and history in the registry, as well as the names of and pertinent data about other family members. Thus, important information regarding nonparticipants is generated without the opportunity for nonparticipants to refuse.

Should nonparticipants be given the absolute right of control over what information is generated in this context? An answer depends upon a number of considerations. First, if the inclusion of nonparticipants poses a substantial risk of their learning of their own at-risk status or of learning that they are affected, then the case for a right to control what information is generated would be strengthened. In most of these cases, however, this is a minimally important consideration. Such persons are likely to have some awareness of their at-risk status as a result of their experiences with other family members. For example, with autosomal dominant conditions both sexes inherit the disease with equal frequency, and at-risk offspring of an affected individual have a 50 percent chance of inheriting the disease. In such cases, it seems unlikely that many individuals will be put at risk of learning new and unwanted information about their own conditions or risks.

Moreover, unless the confidentiality protections of the registry are especially poorly designed, there is a comparatively small risk of either the nonparticipant or others gaining access to information that could cause psychological, social, or economic harm. Researchers can take a variety of steps to ensure privacy that result in considerably less risk of adverse consequences of unwanted disclosure than with medical information in ordinary medical records. First, they should refuse to release data to participants regarding other family members without the latter's permission.

Second, researchers should segregate the data collected in the registry from all patient care records so that no individual will be adversely affected in insurance or employment by virtue of its inclusion in records accessible to employers and insurers. Any data obtained from the registry should be incorporated into medical records only with informed consent of the person who may be affected by possible disclosures.

Third, such registries should be protected from access by those having unrelated purposes. This can be achieved through the process of obtaining a Certificate of Confidentiality for research supported by the Public Health Service (PHS) as well as for nonfederally funded research related to alcohol abuse and alcoholism, drug abuse, mental health, and other health research. (1) The effect would be to shield such information from access by others, including law enforcement officers who might assert a need to know for purposes other than research. The certificate is not currently available for research not directly supported by the PHS, but the protection it affords should be extended to cover it.

If such safeguards are adopted, then the argument for a right of nonparticipants to control what information is generated is weakened considerably. In research contexts having adequate safeguards in place, there are reasonably persuasive arguments for thinking that the privacy pragmatist position is better justified than the privacy fundamentalist view.

However, the substantial costs of building a wall of separation between research activities and institutions delivering medical services must be acknowledged. The costs are justified in striking a reasonable balance between the need to conduct valuable scientific research for the common good and the need for individual privacy protection.

Control over Disclosures

The second form of the right of informational self-determination is related to the interest in controlling further disclosures. The right presupposes that information has been generated and that once it is generated and revealed to designated persons for specific purposes, individuals have a continuing interest in controlling what happens to it. This is an important element of privacy and confidentiality protection notably absent from most laws on medical information (Powers 1991). Control over the identity of the recipient, the purposes for which the information is used, and the period of time within which future disclosures may be made is too often lost once an initial consent to disclosure is given.

One illustration of these kinds of concerns in the genetic context involves the use of DNA databanks and databases. Although most scientific researchers ordinarily think of DNA banks established in conjunction with specific genetic registries, such as the Huntington's Disease Research Roster and DNA Data Bank at Indiana University, there is no established definition of what counts as a DNA bank. Databanks may be established for commercial, military, law enforcement, or a variety of other purposes. Even something as simple as the storage of a relatively small number of blood samples could have the potential for great harm if no procedures are in place to ensure privacy protection. Some current repositories of blood samples collected in conjunction with specific newborn screening programs (e.g., Guthrie cards) could become the basis for establishing databanks for new and previously unimagined purposes. In short, any collection of blood samples on filter paper—if well preserved, capable of being linked to identifiable persons, and organized in a searchable manner—can be transformed into a databank with an enormous impact on privacy. The important point about the vagueness of the notion of a databank is that any repository of genetic information, regardless of the purposes for which DNA material was initially obtained, can be an attractive source for mining information for other purposes.

Nonetheless, the most troubling issues are raised by certain kinds of DNA databanks. Unlike academic databanks established for specific scientific purposes, some large-scale databanks can be far more comprehensive in the information they might contain, less well defined in their purposes and objectives, and developed on a mass scale.

One statewide plan to take blood samples of convicted felons illustrates the problems of databanks created for purposes other than research (de Gorgey 1990). For all persons convicted of felonies, a blood sample is taken and labeled with a numeric code. In a second file are personal identifiers and personal information including prison medical examination information, sentencing reports, etc. A third file consists of a barcode, or digitalized equivalent of the actual blood sample. In a fourth file is an integrated database containing all the barcodes. However, in most instances where someone submits to particular medical tests, that person may anticipate that a particular bit of information will be obtained and that it will reveal information of an anticipated sort for which consent for testing has been given. But storage of DNA information in blood samples—and perhaps barcodes, depending upon what is included—provides a ready source for discovering new information about which the individual did not give consent and which may not have been possible to anticipate. Those in possession of DNA samples have the ability to ascertain further information that vastly outstrips the scope of the purposes which provided the initial justification for collection.

If the right of informational self-determination is to be taken seriously, one public policy option would be to require specific consent for every contemplated redisclosure not comprehended in the initial plan. Although there are strong arguments to be made against any implementation of databanks in prisons and the military, at minimum, the individual should have the right to control the subsequent release of information for new purposes.

Disclosure Issues in Research Contexts

Although it has been argued that multipurpose and large-scale databanks are likely to pose the greatest threats to privacy, there are some additional privacy issues that emerge in research databanks. For example, academic genetic databanks are an attractive resource for others who assert a need to know for reasons unrelated to research, such as law enforcement. However, such arguments should be resisted. The ready access to medical and research data by law enforcement officials involves an unacceptable blurring of institutional roles and purposes. Health care providers and researchers should not be expected to bear the burdens of effectuating other arguably important social goals, however great those needs may be. To demand that health care and medical researchers assume the functions of law enforcement agencies compromises the ability of health care and medical research institutions to fulfill their primary missions. Without a social agreement that recognizes the need for a moral division of labor among its institutions, important social institutions cannot be expected to maintain the integrity and independence necessary to function effectively.

However, the justification for limiting further access for other health-related purposes is less compelling. Particularly vexing is the issue of how to strike a balance between the needs of epidemiologists and the needs of subjects for privacy protections. The controversy over the appropriateness of individualized consent for access to personal data for epidemiological research has been especially acute in European countries where members of the European Data Protection Commission have issued a draft report, already ratified by some countries. The draft report prevents access to and processing of data revealing ethnic or racial origin and data concerning health or sexual life without express written consent of each person about whom information is obtained. Critics argue that such directives will "legislate epidemiology right out of existence" (Editorial 1992). The concern is that such absolute rights of informational self-determination would prevent transmission of individual data to regional cancer registries, exclude use of data from persons who are dead or untraceable, and make retrospective studies of health records impossible (Knox 1992). Even if some of these complaints are overstated, it seems more reasonable to prefer privacy protection measures that are less disruptive of scientific advancement than individualized and explicit consent requirements.

Consider once again the rights of privacy in the context of pedigree studies. The position that individuals should exercise an absolute veto over their inclusion in such registries, subject to the important qualification that stringent privacy protections be in place, has been argued. In part, this argument was premised on the claim that these registries and databanks are distinguishable from the more general kinds of databanks we have just considered. In the nonresearch databanks, individuals are less likely to be adequately protected from unwanted disclosures. Additionally, genetic registries differ in the limited purposes for which they are created, in the kinds of protection that can be expected from exercise of rigorous review by Institutional Review Boards, in the use of Certificates of Confidentiality that can shield them from inquiries for unrelated purposes, and, one hopes, by the special sensitivities to the particular kinds of privacy risks associated with specific heritable disorders for which the registry was created.

Even if research and nonresearch registries and databanks differ in the ways previously described, there are some circumstances in which subjects should have the right to exercise some control over subsequent disclosures to other researchers. At minimum, participants should have the benefit of an informed consent procedure that adequately informs them of the kinds of risks posed by disclosures to other researchers and by the subsequent publication of their findings. Moreover, it would seem reasonable to require explicit consent for disclosures to researchers when the new uses of information exceed or differ from the purpose for which original consent was given. It would be presumptuous to assume that consent to disclosure for some research purposes is sufficient for all research purposes, especially in light of the fact that future research may focus on different conditions or genetic predispositions.

In most cases of epidemiological research not involving the release of the names of the persons, the risk of harm from disclosures to other researchers is small. However, the distinction between participants and nonparticipants in genetic registries is relevant. If nonparticipants have no right to veto inclusion of data regarding their conditions, then there is a need to ensure that institutional privacy protections are extended to external researchers who have access to identifiable data. When the names of those in the registry are shared with others, additional risks of unwanted disclosure are greatly increased. Before data are provided to someone not under direct institutional supervision and control, members of the Institutional Review Board need to approve a comprehensive and detailed privacy protection plan. No less stringent safeguards can be justified when nonparticipants are involved.

Pedigree Studies and the Protection of Anonymity

When the data are themselves the pedigree plot of the family, publication becomes problematic (Powers 1993). The risk is that publication of such data may result in inferential identification of individual family members. The risk of inferential identification may be increased when the publication of data reflects the geographic region from which the subjects were recruited or when the data describe a comparatively rare disorder.

One approach to the problem would be to require specific consent to the publication of pedigree data. Each family member would have the opportunity to personally review the proposed material in order to evaluate the likelihood of unwanted inferential identification. By contrast, an argument for a more pragmatic approach is that studies that are based upon incomplete family pedigrees, i.e., where individuals can prevent inclusion in the registry, undermine the scientific usefulness of pedigree studies. However, the fact that the publication of such data can potentially reveal the identity of nonparticipants who have not given informed consent (and who may not have had the opportunity to prevent the inclusion of their name in the registry) supplies a strong counterargument in support of the more stringent approach.

One can look for additional guidance on this question by examining the standards for protecting against inferential identification when the data reported in journals are photographs or case reports. The Statement from the International Committee of Medical Journal Editors offers some insight to the prevailing norms in these contexts (International Committee 1991). In the case of photographs, masking (e.g., covering a subject's face) may be adequate to prevent physical identification. Case reports are more problematic. They provide a richer, more detailed body of data that increases the possibility of inferential identification of specific persons. Where substantial risk to anonymity remains, the guidelines encourage specific informed consent as well as a clear statement in the article that such consent was obtained. The reasoning appears especially relevant to issues surrounding publication of pedigree studies. Pedigree studies can further compromise anonymity in that the base of information from which inferences may be drawn often is larger and the number of persons potentially affected is greater. By extension, it would seem appropriate to expect application of the same standard to the publication of pedigree studies. Hence, in spite of the practical problems of obtaining informed consent, it seems reasonable and consistent with established privacy norms among medical journal editors and researchers to start with a strong presumption in favor of informed specific consent from all potentially affected.

Particular attention should be given to the following considerations in any decision whether or not to publish. First, there are differences in the probability of inferential discovery. The greater the probability of loss of anonymity, the more important individual consent becomes. Second, there are differences in magnitude of expected harms. For example, some matters may be especially sensitive, and the expected magnitude of harm will be greater. These would include studies of families with a high prevalence of psychiatric conditions such as bipolar affective disorder. Third, there are differences in expected social benefit. Where the contribution to the growth of scientific knowledge is substantial, the case for publication even without explicit consent is strengthened. Fourth, the foreseeability of false labeling or having unaffected family members perceived by others as at greater than actual risk or affected by the disorder is especially important when stigma and prejudice are associated with certain conditions. Fifth, cases for which there is no informed consent to inclusion in a registry or, more importantly, where some have explicitly refused to participate increase the threshold presumption against publication of the data in their complete form.

What alternatives are available if risks of inferential identification are significant and the prospects of obtaining informed consent from all the family members are poor? Two obvious possibilities exist. First, the decision might be made to scramble or disguise the data, such as by deleting a branch of the family tree or by altering birth order or gender. However, to the extent data are materially altered, this would be an unacceptable research practice, not only for reasons of scientific integrity, but because it undermines the risk-benefit calculus essential to the initial decision whether to publish. The undermining of the scientific validity and the ability of others to evaluate and replicate results destroys the expected benefit that provides a rationale for publication (Nylenna 1991). A second alternative is to publish results using some representative or hypothetical pedigree, coupled with the retention of the raw data by the journal under safeguards designed to limit inspection to bona fide researchers. Such suggestions present a variety of logistic difficulties, but the search for a new approach to communication with professional peers would be preferable to publication in the original form.

Information Disclosed for the Protection of Others

The issue of disclosure of genetic information to third parties in order to protect them from harm is one of the most complex issues to be addressed. Numerous persons have some claim of a legitimate need to know genetic information about particular persons for their own protection. Spouses and other reproductive partners may assert a need to know information that would affect reproductive decision making (Wertz and Fletcher 1991). Fellow employees and unions may assert a need to know information that is relevant to assessing shared occupational risks and to protecting coworkers (Andrews 1991). Employers and consumers of services may seek information relevant to assessing fitness and suitability to perform certain kinds of jobs, such as whether airline pilots are likely to suffer the debilitating effects of diminished motor skills and judgment in the early stages of Huntington's disease.

However, not all potential interests that may be adversely affected justify disclosure of genetic information. The judicial system has not resolved these questions, and it is somewhat premature to predict the outcome of such cases (Robertson 1992). The recommendations of the President's Commission for the Study of Ethical Problems in Medicine and Biomedical and Behavioral Research (1983) are relevant both to the ethical and legal analysis of these issues. The authors of the report argue that patient confidentiality may be overridden when the following four conditions are met:

1. Reasonable efforts to elicit voluntary consent to disclosure have failed;
2. There is a high probability that harm will occur if the information is withheld, and the disclosed information will actually be used to avert harm;
3. The harm that would result to identifiable individuals would be serious; and
4. Appropriate precautions are taken to ensure that only the genetic information needed for diagnosis and/or treatment of the disease in question is disclosed.

There are several important questions to be considered in the application of those criteria to the genetic context. For example, how does one compare the risks of harm associated with failure to disclose genetic information with the risks associated with the failure to disclose psychiatric information to third parties in imminent danger of physical harm, or risks associated with the failure to disclose HIV status to sexual partners of infected patients? How do we analyze the conflict in cases such as those in which a person with the diagnosis of Huntington's disease refuses to disclose the diagnosis to relatives at risk of developing the disorder or of passing it on to their children? How is the future prospect of disability handled, as in the example of the airline pilot? Are the obligations of medical researchers different from those of medical practitioners (Applebaum and Rosenbaum 1989)? Are the obligations of medical researchers and health care practitioners to particular people (such as spouses) more stringent than their obligations to the public at large? How important is it that the harm that would result is to identifiable individuals as opposed to the public at large, as in the airline pilot case?

In most instances, those who are potentially harmed by lack of genetic information are not personally at risk in quite the same way as in the cases of psychiatric and HIV disclosure. One may not be convinced that the model used in these instances is appropriate in the genetic context. Additional questions might include the following: Are there further steps that others could reasonably take in the genetic context to avoid potential harms to the next generation, including initiation of alternative forms of testing for themselves? Is there the same immediacy of harm that only disclosure by physicians and researchers could foreseeably prevent? Moreover, apart from the harm to the next generation, often there is nothing that can be done to avert harm to existing relatives, and even if so, it is important to consider whether they could take steps for their own protection. Is it likely in cases such as that of the airline pilot that current procedures for detecting disability and fitness to perform such a demanding and responsible job are inadequate to protect the public such that the responsibility for doing so justifiably falls upon health professionals?

Until questions of these sorts are resolved, it is advisable to proceed cautiously with the extension of legal duties to protect in the genetics area, especially with extension to researchers who are not involved with patient care and counseling. In addition, the further moral importance in preserving the trust and privacy of the physician-patient relationship is of such importance that we should not rush to extend exceptions to the usual rule of confidentiality to these newer contexts, absent a compelling demonstration that the prevention of harm could not be achieved in a less intrusive manner.

When Control Is Not What Matters Most

There are some circumstances in which an individual has an interest in information being inaccessible to others or perhaps even to oneself; but in such circumstances, that interest may not be well served by means of individual control over access to information. The point is not that the efforts of others to protect privacy can be an adequate substitute for the privacy protection that one might otherwise achieve for oneself through one's own control. The point is that, in some instances, it is preferable that others act on one's behalf. The existence of such circumstances suggests that, in addition to liberty rights, there are cases in which the promotion and protection of an individual's privacy interests justify the existence of rights to be protected by others against certain harms. Two types of examples are illustrative.

First, individual control is not sufficient for protection against genetic privacy losses when genetic testing of one person reveals information about another. There are a number of examples of this phenomenon. Carrier screening for cystic fibrosis (CF) reveals information about the status of siblings as potential CF carriers. The use of genetic linkage studies in Huntington's disease produces results for other family members.

The heart of the problem is that the mere liberty right of each individual to exercise control over his or her own genetic information may not be sufficient for privacy protection when genetic information about one person can be obtained by third parties from another person. Hence, the privacy of each individual simply cannot be assured through a system of individual liberty rights where each exercises control over his or her own information; hence, the family member's dilemma.

What are the potential solutions to the family member's dilemma? One possibility would be to view the rights of control as collective rights, or ones vested in groups rather than in individuals. One could prohibit any testing of one person unless all others potentially implicated by such tests consented. The result would be a system of privacy protection that would prevent access by third parties in a way that a scheme of individual rights would not.

Although the performance of some genetic tests requires at least some cooperation by some other family members, the collective rights approach has severe drawbacks. One is the obvious problem of delineating the domain of all those affected by a decision to be tested (or to participate in linkage or pedigree studies). A more significant problem is that it would unreasonably deprive some persons of valuable benefits that may be derived from testing. Persons tested for CF could use the information obtained to make important reproductive decisions, and persons tested for susceptibility to Huntington's disease could make use of the information obtained for the development of other significant life plans. It would seem difficult to justify the denial of such benefits to one family member in order to preserve the privacy of another family member.

An alternative involves a right of the individual to be protected from third party (e.g., employer and insurer) access to information about one individual *via* access to a relative's medical file or other health data record. In short, the rationally preferred strategy would be a system of rights, which includes rights that those who are in a superior position to secure privacy protection act on the behalf of those who cannot protect their own interests through their own choices and actions. In most cases, these rights will be rights against the government for the kinds of protection that neither the individual nor nongovernmental entities are able to guarantee.

A second example in which individual control over genetic information is not sufficient to protect privacy focuses on another inadequacy of a liberty rights approach. In these cases, individuals would rationally prefer some sort of mandated restriction prohibiting their own disclosure of genetic information. The result would be that they could not voluntarily relinquish their own rights to privacy, i.e., the Ulysses strategy. Like the mythical figure from which its name derives, it may be rational to make present choices that have the result of relinquishing one's freedom to make choices in the future (Elster 1979).

The rationale for adopting Ulysses's strategy can be seen in the context of a market for health insurance. A system of liberty rights that vests in each individual the freedom to decide what information is given to insurers or employers may be an inadequate approach to privacy protection. When individuals lack equality of bargaining power, and the consequence of failure to disclose genetic information is an inability to obtain insurance or employment, then it is rational to prefer institutional restrictions that eliminate the liberty of personal control over such information.

Although the examples, in which privacy protection without control over information matters most, may be few in number, they are nonetheless among the most significant elements of a comprehensive genetic privacy protection strategy. Such rights to have one's privacy interests protected by others (typically by government regulation) reflect the limits of a liberty rights approach to privacy. Developments in genetics reveal contexts in which reliance on individual informational control is either impossible or rationally undesirable.

The Mitigation of Adverse Consequences

A paradoxical feature of rights is that it may be more rational to prefer a system with fewer rights to a system with more rights. This paradox is a consequence of the fact that rights have their moral importance in the face of threats to individual interests. Thus, the elimination of the conditions that threaten those interests makes a system of rights less important. The paradox of rights has important implications for arguments about the need for privacy rights in genetics.

Many of the motivations for seeking greater privacy protections are predicated upon assumptions about the nature of the threats to our interests under a particular set of social and economic arrangements. However, it may be preferable to modify existing institutional arrangements in order to mitigate the adverse consequences that flow from unwanted disclosures.

Among the most important consequences of privacy losses in contemporary American society are economic ones, such as the loss of insurability. For example, insurers (and self-insured employers) might have fewer incentives to gain access to personal information if reforms in the market for health insurance were enacted into law. If the costs and eligibility requirements for health insurance were no longer dependent upon individual medical status or on the risk-based characteristics of a pool of insureds, then one important source of privacy concerns would diminish. However, it must be noted that any attempt to mitigate privacy concerns through the elimination of adverse consequences of disclosures raises fundamental questions of distributive justice in an explicit way. Any discussion of the relative weight of privacy in competition with other individual interests and social goals implicitly involves some view of just distribution. Consideration of policy options designed to make the need for privacy protection less pressing simply makes it clear that discussions of privacy rights cannot be pried apart from the hard questions of distributive justice.

Policy Implications and Conclusions

Reflection on the need for privacy protections in the context of the proliferation of genetic information raises some additional policy questions: Should there be special policies and laws for confidentiality and privacy protection with respect to genetic information? Or can existing laws governing medical information be relied on?

There are three reasons to think a separate genetic information policy, by itself, would be inadequate. First, to the extent that genetic testing becomes integrated into routine medical practice, the information will become a part of all medical records and thus open to the wide audiences to which medical records are now available. Second, because blood samples from any medical context can yield DNA information, there are serious difficulties with any proposal that attempts to define a databank and to specially regulate it separate and apart from the rest of medical practice. Third, to the extent feasible, public policy should leave it to individuals themselves to decide what kinds of information they are most concerned about keeping private and confidential. Thus, laws are needed that are adequate for the protection of medical and health related information generally, and one should be reluctant to presume what kinds of information are most important to different persons.

Even if some special rules are desirable for pedigree studies and genetic databanks, the likelihood that genetic information will be increasingly incorporated into routine medical records suggests the need to review the adequacy of present medical privacy protections generally. Although a comprehensive assessment is beyond the scope of this discussion, there are three primary problem areas that should be mentioned (Powers 1991).

First, there are problems of regulation. Current laws contain numerous gaps. There are simply too many individuals and institutions that are not under a clear legal duty to maintain confidentiality. Moreover, even for those under a clear legal duty to protect confidentiality, too often they may be required to disclose information to a large number of persons and institutions for a wide variety of reasons having nothing directly to do with the delivery of patient care.

Second are problems of legal remedy. There are numerous impediments to effective civil suits for breach of privacy. Often breaches occur without the knowledge of those affected. Lawsuits are costly, both in financial and emotional terms.

Lawsuits may be counterproductive inasmuch as the substance of the original grievance is the fact that highly personal or sensitive information has been revealed without one's permission and damage has thereby been caused, but the successful prosecution of a civil case requires the further airing of that information in an open public forum. Criminal prosecution may be even less effective. Successful criminal prosecution requires a higher burden of proof than civil trials, and in some instances victims may encounter prosecutorial reluctance to pursue powerful individuals and institutions. Discrimination suits designed to compensate for adverse consequences associated with unwanted disclosure may be ineffective for many of the same reasons that civil suits for breach of confidentiality are problematic. They present similar proof problems, particularly when intent is an issue in the case. Moreover, they may not offer an effective deterrent, especially if potential benefits of largely undetected discrimination outweigh risks of a few successful lawsuits.

Third are problems of jurisdiction. Medical information is collected, stored, and analyzed in a variety of states with different laws. A consent to medical tests in one state (under its laws) provides no assurance that similar legal protections will be available in other states where the information ultimately reaches.

Most current policies for privacy and confidentiality protection are designed to deter improper disclosures and to provide compensation for adverse consequences when the laws have failed to deter. The privacy interests outlined suggest a need for very different kinds of public policy options than the ones traditionally recognized in American law. There is a need for greater individual control over the generation and flow of information in some instances, for greater restrictions on the sharing of information than the currently accepted doctrines of a bona fide need to know now recognize, and for the elimination of certain predictable adverse consequences of information dissemination.

The challenges to personal privacy posed by the advances in genetic knowledge offer a welcome opportunity to rethink public policies with respect to medical and other health-related information in general.

NOTES

Reprinted by permission of the American Association for the Advancement of Science from J. Tisch and M. Powers, eds., *The Genetic Frontier: Ethics Law and Policy* (Washington, D.C.: AAAS, 1993).

1. Under Section 301 (d) of the PHS Act, a Certificate of Confidentiality may be applied for through the following agencies: for alcohol abuse and alcoholism research, contact the National Institute on Alcohol Abuse and Alcoholism, 14-C-20 Parklawn Building, 5600 Fishers Lane, Rockville, MD 20857; for drug abuse research, contact the National Institute on Drug Abuse, 10–42 Parklawn Building (same address as above); for mental health research, contact the National Institute of Mental Health, 9–97 Parklawn Building; and for all other areas of health research, contact the Office of Health Planning

and Evaluation, PHS, 740G Humphrey Building, U.S. Department of Health and Human Services, Washington, D.C. 20201.

REFERENCES

Allen, A. 1987. *Uneasy Access: Privacy for Women in a Free Society.* Totowa, N.J.: Rowman & Allenheld.

Andrews, L. 1991. "Confidentiality of Genetic Information in the Workplace." *American Journal of Law and Medicine* 17: 75–108.

Applebaum, P., and A. Rosenbaum. 1989. "Tarasof and the Researcher: Does the Duty to Protect Apply in the Research Setting?" *American Psychologist* 44: 885–94.

Benn, S. 1971. "Privacy, Freedom, and Respect for Persons." In J. Pennock and J. Chapman, eds., *Nomos XIIJ: Privacy.* New York: Atherton.

Black, D. 1992. "Personal Health Records." *Journal of Medical Ethics* 18: 5–6.

De Gorgey, A. 1990. "The Advent of DNA Databanks: Implications for Informational Privacy." *American Journal of Law and Medicine* 16: 381–98.

Elster, J. 1979. *Ulysses and the Sirens.* Cambridge: Cambridge University Press.

Faden, R., G. Gelier, and M. Powers. 1991. "HIV Infection, Pregnant Women, and Newborns: A Policy Proposal for Information and Testing." In R. Faden, G. Geller, and M. Powers, eds., *AIDS, Women and the Next Generation.* New York: Oxford University Press.

Flaherty, D. H. 1989. *Protecting Privacy in Surveillance Societies.* Chapel Hill: University of North Carolina Press.

Fried, C. 1968. "Privacy: A Rational Context." *Yale Law Journal* 77: 475–93.

Gavison, R. 1984. "Privacy and the Limits of Law." In F. Schoeman, ed., *Philosophical Dimensions of Privacy.* Cambridge: Cambridge University Press.

International Committee of Medical Journal Editors. 1991. "Statement from the International Committee of Medical Journal Editors." *Journal of the American Medical Association* 265: 2697–98.

Knox, E. J. 1992. "Confidential Medical Records and Epidemiological Research: Wrongheaded European Directive on the Way." *British Medical Journal* 304: 727–28.

Parent, W. A. 1983. "Recent Work on the Concept of Privacy." *American Philosophical Quarterly* 20: 343.

Powers, M. 1993. "Publication-Related Risks to Privacy: The Ethical Implications of Pedigree Studies." In *IRB: A Review of Human Subjects Research.*

Powers, M. 1991. "Legal Protections of Confidential Medical Information and the Need for Anti-Discrimination Laws." In R. Faden, G. Geller, and M. Powers, eds., *AIDS, Women, and the Next Generation,* 221–55. New York: Oxford University Press.

President's Commission for the Study of Ethical Problems in Medicine and Biomedical and Behavioral Research. 1983. *Screening and Counseling for Genetic Conditions.* Washington, D.C.: The Commission.

"Protecting Individuals; Preserving Data." 1992. Editorial. *Lancet* 339: 3.

Raz, J. 1986. *The Morality of Freedom.* Oxford: Oxford University Press.

Reiman, J. 1976. "Privacy, Intimacy, and Personhood." *Philosophy and Public Affairs* 6: 26–44.

Riis, P., and M. Nylenna. 1991. "Patients Have a Right to Anonymity in Medical Publication." *Journal of the American Medical Association* 265: 2720.

Robertson, J. 1992. "Legal Issues in Genetic Testing." In American Association for the Advancement of Science, *The Genome, Ethics, and the Law: Issues in Genetic Testing,* 79–110. AAAS Publication no. 92-115. Washington, D.C.: American Association for the Advancement of Science.

Schoeman, F. 1984. "Privacy: Philosophical Dimensions of the Literature." In F. Schoeman, ed., *Philosophical Dimensions of Privacy: An Anthology*. Cambridge: Cambridge University Press.

Shaw, M. 1987. "Testing for Huntington's Disease: A Right to Know, a Right Not to Know, or a Duty to Know?" *American Journal of Medical Genetics* 26:243–246.

Thomson, J. 1984. "The Right to Privacy." In *Philosophical Dimensions of Privacy.*

Tribe, L. 1978. *American Constitutional Law.* Mineola, N.Y.: Foundation Press.

Warren, S., and L. Brandeis. 1990. "The Right to Privacy." *Harvard Law Review* 4: 193–220.

Wertz, D., and J. Fletcher. 1991. "Privacy and Disclosure in Medical Genetics Examined in an Ethics of Care." *Bioethics* 5: 212–32.

27

The Ethics of Somatic Cell Gene Therapy

Leroy Walters and Julie Palmer

SCIENTIFIC ISSUES

Gene Therapy for ADA Deficiency

On September 14, 1990, the first officially sanctioned human somatic-cell gene-therapy experiment began. A 4-year-old girl was the first of two children to receive a dose of her own cells in which a functioning counterpart of her malfunctioning gene had been previously inserted. As early as December of the same year, and continuing through the end of the first phase of the experiment in July 1991, newspaper headlines dramatically cheered news of the experiment's success.[1]

The subjects of this successful gene therapy experiment were children suffering from a rare genetic disease called *adenosine deaminase* (ADA) *deficiency*. ADA deficiency arises from the malfunction of a gene expressed in bone marrow *stem cells*, long-lived cells that differentiate into multiple types of blood cells, including infection-fighting T and B cells (varieties of white blood cells). The malfunctioning ADA genes in these cells fail to produce functional versions of adenosine deaminase, an enzyme that normally metabolizes a compound called *deoxyadenosine*. Absent ADA, toxic levels of *deoxyadenosine* accumulate. T cells suffer the most from this buildup. These important infection fighting cells are devastated, leaving the patient prey to infections which healthy individuals could easily withstand. Ultimately, most patients suffering from this rare disease die from infection unless a matched bone marrow donor can be found.[2] Unfortunately, matched donors are available for only about 30% of patients.[3] ADA-deficiency patients usually die before they reach the age of 2.[4]

ADA deficiency was an appropriate initial candidate for gene therapy for several reasons. First, researchers had long been convinced that even a small level of

461

ADA production would be sufficient to operate as a cure for an ADA-deficient patient. They therefore did not have to achieve high expression levels in order to test the therapy. Additionally, the ADA gene is subject to relatively simple regulation, which made the achievement of appropriate ADA gene expression less complex than for other genes. Most significant was the hope that T cells armed with properly functioning ADA genes might have a selective advantage in the human body, allowing them to persist and multiply while noncorrected, ADA-deficient T cells expired.[5] Indeed, research had shown that human T cells, to which the ADA gene was added, survived in the bloodstreams of laboratory mice for months, while the cells with defective ADA genes died rapidly. The mouse experiments were evidence that T cells containing an inserted, functioning ADA gene were likely to thrive in patients suffering from ADA deficiency.[6]

In addition, numerous other animal experiments, in mice and in monkeys, had demonstrated that T cells could be successfully grown in vitro (literally, "in glass," meaning in a test tube), genetically altered, and reintroduced into animals where they would function appropriately as immune cells.[7] Although researchers had initially hoped to insert the functioning ADA gene into bone marrow stem cells, where the gene could be expected to persist indefinitely, they were, until recently, unable to purify the stem cells in the laboratory. Nevertheless, the T-cell results in animals were good enough to warrant a human experiment of the therapy.[8]

A team of investigators, led by National Institutes of Health doctors W. French Anderson, R. Michael Blaese, Kenneth W. Culver, and Steven Rosenberg, designed the first human gene therapy experiments in accordance with these extensive, preliminary animal experiments. During the human gene therapy experiment, doctors obtained blood from each ADA-deficient patient. The research team isolated T cells from these blood samples in the laboratory and grew the T cells in vitro. Then they inserted properly functioning ADA genes into the T cell using a process known as transduction in which an engineered RNA virus called a retrovirus carried the genes into the T cells (both retroviruses and transduction will be described below). Finally, doctors reinfused into each patient her (both are female) own altered T cells. The patients each received approximately six doses of transduced cells during the several-month first phase of the experiment.[9]

In conjunction with the gene therapy, the patients received a drug called *polyethyleneglycol* (PEG)-ADA, a version of the ADA enzyme itself. PEG-ADA had previously proven somewhat effective, but not effective enough to allow patients to fully regain immune function. Although able to decrease extracellular levels of deoxyadenosine, PEG-ADA cannot gain entrance to the T cells. Hence, PEG-ADA treatment by itself left the patients' T cells disabled. The investigators hoped that gene therapy would correct the T cells themselves and decrease intracellular levels of deoxyadenosine.[10]

The first scientific report on this initial gene therapy study was published in *Science* in October 1995.[11] Both young patients continue to receive PEG-ADA, although the dose has been reduced by more than half.[12] In laboratory measures of immune system function, both patients have shown improvement since the initiation of gene therapy in late 1990 and early 1991, In fact, the T cell count for the younger of the two patients rose from below normal (about 500) to the normal range (1,200–1,800), and her blood levels for the enzyme ADA rose to about 50% of the levels in heterozygotes who carry one normal and one nonfunctioning ADA gene. The researchers estimate that about half of the T cells in the circulation of this younger patient carry the ADA gene.[13] In contrast, the older patient has much lower levels of ADA, and only 0.1–1.0% of her circulating T cells carry the ADA gene.[14]

More important than the laboratory values of these two patients is the effect of gene therapy plus PEG-ADA on their daily lives. Here is the researchers' summary of their clinical progress.

> The effects of this treatment on the clinical well being of these patients is more difficult to quantitate. Patient 1 (the younger patient), who had been kept in relative isolation in her home for her first 4 years, was enrolled in public kindergarten after 1 year on the protocol and has missed no more school because of infectious disease than her classmates or siblings. She has grown normally in height and weight and is considered to be normal by her parents. Patient 2 was regularly attending public school while receiving PEG-ADA treatment alone and has continued to do well clinically. Chronic sinusitis and headaches, which had been a recurring problem for several years, cleared completely a few months after initiation of the protocol.[15]

The same issue of *Science* in which the U.S. ADA study was reported also included a report by Italian researchers on their study of gene therapy in two young children with ADA deficiency. The Italian research team was led by Claudio Bordignon of the Scientific Research Institute H. S. Raffaele in Milan. Both Italian patients had been started on enzyme therapy with PEG-ADA at the age of 2 years, and both had "failed"—that is, their immune systems were not adequately protecting them against infection—after treatment with PEG-ADA alone.[16] In the Italian study the two patients received both genetically modified T cells and genetically modified bone marrow cells plus ongoing support with PEG-ADA. Three years after the start of gene therapy in patient 1 (at age 5) and 2½ years after the start of treatment in patient 2 (at age 4), both patients are doing well clinically. The Italian researchers report that the genetically modified T cells have gradually died off after remaining in the children's circulation for 6–12 months but that the percentage of new T cells produced by genetically modified bone marrow progenitor cells is gradually increasing.[17]

The news media have greeted these two gene therapy studies with cautious optimism. For example, the headline of the story announcing the *Science* article on the U.S. study read, "Gene Therapy's First Success Is Claimed, But Doubts Remain."[18] These doubts have to do mainly with two features of the studies performed to date. First, for understandable reasons the researchers have been reluctant to withdraw PEG-ADA from children who are doing well on a combination of PEG-ADA and (past) gene therapy. Second, the experimental step of removing, selecting, and expanding the number of T cells may have a beneficial effect for patients, in addition to the beneficial effects of the genetic modification itself. Further research will help to resolve these remaining questions.[19]

Important Features of the First U.S. Gene Therapy Study

The disease selected for the initial research. During the late 1970s and early 1980s considerable discussion was devoted to deciding which diseases would be the best targets for early gene therapy attempts. For example, in 1977 molecular biologist David Baltimore sketched a scenario for gene therapy using sickle cell anemia as his example.[20] However, in the 1980s diseases in which a single enzyme was missing in patients came to the fore, in part because the genes involved in the gene therapy would not need to be so precisely regulated as they would be in the treatment of a hemoglobin disorder like sickle cell anemia. In other words, new genes could be introduced into a patient's body and, even if the genes produced more of the missing enzyme than the patient's body needed, the patient would not be harmed. The excess enzyme would simply be eliminated from the patient's body like any other waste matter.

The target cells selected. Before the first gene therapy study was undertaken, there was considerable discussion of the target cells into which new genes could be introduced. The assumption was always that the cells would be removed from the patient's body, genetically modified in the laboratory, and then returned to the patient. Skin cells, or fibroblasts, were sometimes mentioned as possible targets for gene therapy. In a long preliminary proposal submitted for review in early 1987, W. French Andersen and his colleagues suggested that bone marrow cells might be the best targets.[21] However, the harvesting of bone marrow is a relatively invasive and painful procedure for the donor; it involves the introduction of a large needle into the hip bones, from which the marrow is removed. When Anderson, Blaese, Rosenberg, and Culver sat down to design their first actual gene therapy protocol, they decided that white blood cells, specifically the T cells, would be better targets. The harvesting of these cells would be less traumatic for the patient, and there was at least a hint in studies with mice that T cells that had the ADA gene survived longer than cells that lacked the gene.

As their research continued, the Blaese-Anderson group decided to aim their vectors (delivery vehicles) at another kind of cell that is closely related to the T cell. In fact, stem cells are the "parents" that give rise to several different types of blood cells, including T cells. The researchers reasoned that if they could introduce the ADA gene into even a few of these stem cells, all of the offspring of these parental cells would contain the previously missing gene. In the study with young children, Blaese and his colleagues attempted to perform stem cell therapy with one of the two female patients—without apparent success. The companion study in newborns modified only stem cells derived from the umbilical cords of the infants.

It is important to note that all of the proposed target cells for the earliest studies of gene therapy have been somatic, or nonreproductive, cells. The new genes introduced into those cells are not likely to escape and "spread" to the sperm or egg cells of the recipient. Thus, the new genes will probably not be passed on to the descendants of the patient who receives this experimental treatment.

The vector employed. The ADA gene therapy experiment involved the addition of genes to cells which had been removed from a patient's body and grown in vitro. The genes were delivered to the cells in vitro by a clever manipulation of the naturally occurring RNA viruses called retroviruses. Retroviruses are so named because upon entering cells, they use an enzyme called reverse transcriptase to transcribe their genomes from RNA to DNA, a kind of "backward" transcription.[22] These resourceful viruses have the ability to integrate into the host genome, thereby tricking the host cell into treating them as if they were part of the host cell's own DNA.

Retroviruses can be virulent, multiplying and destroying their host cells. In fact human immunodeficiency virus (HIV), the virus responsible for acquired immune deficiency syndrome (AIDS), is one example of a deadly retrovirus. However, retroviruses can be engineered so that their harmful sequences are deleted and the sequences that allow them to integrate into a host genome are preserved. By replacing the deleted sequences with genes that a researcher wants to deliver to cells, he or she can exploit the retrovirus's ability to efficiently enter cells and deliver genes to host chromosomes. There is a sort of poetic justice in using these viruses that are themselves so manipulative of their hosts' machinery.

Retroviruses are comprised of RNA and protein. The fully assembled retroviral package (for any virus, called a virion) contains two identical strands of RNA, structural proteins, and reverse transcriptase. All of these core components are surrounded by an envelope, or coat, consisting of glycoprotein (a protein that contains small amounts of carbohydrate[23]), which is responsible for attaching to and allowing the virus to penetrate cells.

The retroviral genome (that is, one complete set of genetic information contained in the retroviral RNA) consists of three genes and a regulatory sequence.[24] The three genes are *gag, pol,* and *env*. Gag encodes several structural proteins derived from an initially expressed product that is cleaved after translation. Pol encodes reverse transcriptase as well as other enzymes. Env encodes the envelope glycoprotein. The regulatory sequence, which is called a long terminal repeat (LTR), contains sequences vital for regulation of RNA synthesis, transcript processing, and integration of the virus into the host genome.[25] In addition, the genome contains a nucleotide sequence called psi that must be present on a strand of RNA in order for that strand to be packaged as a virus.

Its normal life cycle makes the retrovirus a good medium for delivering genes into cells. The virion binds to membrane receptors on a host cell and the retrovirus enters the cell, leaving behind its glycoprotein envelope. Once inside the cell, the retrovirus uses reverse transcriptase to transcribe a single-stranded complementary DNA (cDNA) molecule from its own single strand of RNA. The host cell's enzymes cannot discriminate the viral cDNA from host DNA, and they synthesize from the cDNA template a double-stranded DNA molecule with an LTR at each end. This DNA molecule integrates into the host genome. becoming what is called provirus. The provirus is transmitted along with the rest of the host genome to daughter cells when the host divides. At another stage in its life cycle, the retrovirus may be reproduced and repackaged separately from the host genome, thereby reasserting its virulence.[26]

Molecular biologists use recombinant DNA techniques to obtain the proviral DNA form of the retroviral genome and to replace gag, pol, and env with the nonviral gene they desire to use for gene therapy. They preserve the LTR sequences which are necessary for integration, but without gag, pol, and env, the retrovirus cannot package itself into a virion after transcription. With psi remaining, however, these recombinant retroviruses can be packaged by researchers, who transcribe RNA genomes from the altered proviral DNA and supply an independent source of gag, pol, and env gene products.[27]

After disabling and packaging retroviruses for use as vectors, researchers use them to transduce cells in vitro, that is, to deliver the inserted gene to recipient cells, where the vector integrates but is unable to reproduce itself. The integrated provirus is passed along to daughter cells when the cell divides, resulting in a whole colony of cells that contain the added gene.[28] In the ADA gene therapy experiment, this process was used to insert ADA genes into T cells or stem cells, which were then infused into the patient.

Gene addition as the basic strategy. The gene therapy study that seeks to treat patients with severe combined immune deficiency (SCID) adds a functioning copy of the ADA gene to as many T cells or stem cells as possible. With current techniques it is possible to achieve only random integration of the vector and the new gene into the genome of each cell. Thus, except for the offspring of stem cells and the daughter cells of T cells, every integration site is likely to be unique. In the future, researchers hope to be able to achieve genuine gene replacement or gene repair, but at present gene addition is the most that can be accomplished.

The involvement of industry. The retroviral vector used in the initial gene therapy study had been developed by a company called Genetic Therapy, Incorporated (GTI). However, all of the clinical research in the study has been conducted either at the NIH Clinical Center or at hospitals in California. In 1990, when the study was initiated, the four primary researchers, W. French Anderson, R. Michael Blaese, Kenneth Culver, and Steven A. Rosenberg, held full-time appointments at the National Institutes of Health (NIH). Blaese and Rosenberg remain at NIH, while Anderson and Culver have moved to a faculty position at the University of Southern California and a research position with OncorPharm, respectively. The initial study has been funded primarily by NIH; GTI has made a substantial investment in developing and certifying the vector.[29]

Developments in Gene Therapy since 1990

Five years after the treatment of the first gene therapy patient in an approved protocol there are 100 protocols that have either been approved by the NIH Recombinant DNA Advisory Committee or that are under review at the U.S. Food and Drug Administration. The target diseases in these 100 studies are the following (numbers refer to number of studies):

Cancers of various types: 63 (63%)
HIV infection/AIDS: 12 (12%)
Genetic diseases: 22 (22%)
Other diseases: 3 (3%)
Rheumatoid arthritis: 1
Peripheral artery disease: 1
Arterial restenosis: 1

The category "Genetic Disease?" can be further subdivided among the following disorders (numbers refer to number of studies):

Cystic fibrosis: 12
Gaucher disease (type I): 3
SCID due to adenosine deaminase (ADA) deficiency: 1
Familial hypercholesterolemia: 1
Alpha 1-antitrypsin deficiency: 1
Fanconi anemia: 1
Hunter syndrome (mild form): 1
Chronic granulomatous disease: 1
Purine nucleoside phosphorylase (PNP) deficiency: 1

Generalizations about the Later Gene Therapy Studies

Target cells. As we have noted, Blaese, Anderson, and associates had sought initially to add the gene for adenosine deaminase (ADA) to the type of white blood cell called the T cell. Later they added the gene to stem cells derived from the circulatory system of one of their female patients and from the umbilical cords of their three newborn patients. Other researchers took aim at a wide variety of somatic, or nonreproductive, cells. Many of the cystic fibrosis protocols tried to modify the cells on the surface of the lungs. Other protocols have targeted liver cells, tumor infiltrating white blood cells, blood vessel cells, and cancer cells. Approximately two-thirds of the first 100 gene therapy studies treated the cells outside the recipient patient's body and then introduced the genetically modified cells into the patient through one of several means. In these studies the cells were said to have been treated ex vivo. In the remaining one-third of the protocols, the vector and gene were joined outside the patient's body but only were united with the patient's cells inside the patient's body. For example, in the cystic fibrosis studies, the vector gene combination is sometimes introduced into the lungs through an aerosol spray. These cells are treated in vivo.

Vectors. As one might expect, different target cells sometimes require different kinds of vectors. The first changes from the traditional retroviral vectors came with the early cystic fibrosis protocols. For these studies researchers initially chose a virus that causes flu-like respiratory infections, the adenovirus. Again the virus was domesticated to reduce the likelihood that it would make the recipient ill or spread infection to the people around the recipient. Other studies have used the adeno-associated virus, or AAV. Some researchers have forsworn viral vectors and used other methods to place new genes into target cells. One of the most used methods encapsulates the new DNA in fatty spheres called Uposomes and uses injections to move these spheres close to the target cells.

Researchers are generally agreed that there is currently no one ideal vector for all situations. Each of the available vectors has some advantages and some disadvantages. New vectors are on the drawing board in many laboratories. One researcher, Francis Collins, has even proposed the construction of a human artificial chromosome as a vector for the future.[30] Presumably such a synthetic chromosome would take its place beside the 46 already in the patient's cells and would include all of the control mechanisms that are present in the original 46.

Gene addition. In gene therapy protocols 2–100, gene addition continues to be the only approach. Gene replacement or gene repair methods are not sufficiently advanced to be proposed for human studies.

Industry involvement. Between 1990 and 1995 private industry became increasingly involved in gene therapy at two levels. As in the original ADA study, commercial firms have continued to develop vectors and to supply them to researchers. About two-thirds of the first 100 studies received their vectors from companies; the remaining one-third secured vectors from academic institutions. In addition, several companies have become industrial sponsors of human gene therapy trials, especially the studies involving larger numbers of patients. Among these companies are Genetic Therapy, Inc., Genzyme, Viagene, and Vical.

One other mode of industry participation in gene therapy deserves at least brief mention. In March of 1995 NIH was granted a broad patent on ex vivo gene therapy—that is, on gene therapy that modifies human cells outside the patient's body.[31] NIH had agreed in advance that this patent, when issued, would be licensed exclusively to Genetic Therapy, Inc. (GTI). Any other researcher or company that wishes to use the ex vivo method of modifying cells to develop a commercial product will need to purchase a sublicense from GTI. How this patent will affect the future of gene therapy remains to be seen.

Detailed Information about the Later Studies

Familial hypercholesterolemia: in vitro. Familial hypercholesterolemia (FH) is a genetic disease associated with astronomically high cholesterol levels in its sufferers. Patients who have the most severe form of FH can have cholesterol concentrations four or five times as high as healthy individuals. Most FH patients die prematurely from heart attacks.[32]

The abnormally high levels of cholesterol in FH patients are the result of malfunctioning genes responsible for directing the production of cellular receptors that capture and remove low-density lipoprotein (LDL) cholesterol from the bloodstream. Without the help of the LDL receptors normally present on the surface of liver cells, FH patients are unable to clear LDL cholesterol from their bloodstreams.

James M. Wilson, chief of Molecular Medicine and Genetics at the University of Pennsylvania, first worked with ADA-deficiency and Lesch-Nyhan syndrome patients but became interested in pursuing gene therapy for FH after reading about a well-known FH patient named Stormy Jones who was near death before being cured by a combined liver-heart transplant. Wilson realized that if he could devise a way to introduce the LDL receptor gene into the liver cells of FH patients without out a full-scale liver transplant, he could increase the number of FH patients able to receive treatment (matched livers are not available for every FH patient) and reduce the invasiveness of treatment (liver transplants require immunosuppression while gene therapy does not).[33]

Wilson's gene therapy experiment was successfully tested first in *Watanabe rabbits*, animals which suffer from the equivalent of human FH. The human experiment began in June 1992. It involved several steps. The patients were placed under anesthesia before surgery. Fifteen to fifty percent of the patients' livers were removed and a catheter was placed into the portal vein, which supplies the liver with blood. In the laboratory, liver cells called hepatocytes were isolated and cultured in vitro. Forty eight hours later, these cells were exposed to retroviral vectors containing properly functioning LDL receptor genes. The exposure lasted 12–18 hours, after which the cells were washed, harvested, and reinfused into each patient's liver via the catheter previously inserted. Finally, the catheter was removed.[34]

Eighteen months after treatment of a 28-year-old woman, the first patient under the FH gene therapy protocol, Wilson and his colleagues reported that the patient had significantly reduced LDL levels and that the study had demonstrated "the feasibility, safety and potential efficacy of ex vivo liver directed gene therapy in humans." At the time of their published report, however, it was still unclear whether the patient's improved LDL levels would result in an improved clinical outcome. The researchers were satisfied at least that the patient's coronary artery disease had not progressed during the 18 months since the gene therapy procedure was performed. In addition, they announced that this FH gene therapy represented the first example of a stable gene-therapy-derived correction of a "therapeutic endpoint," in contrast to the ADA gene therapy trials, which required repeated administration of genetically modified cells.[35]

Like Anderson's ADA-deficiency gene therapy, Wilson's FH experiment involves in vitro gene transfer and requires that patients' cells be removed from their bodies before the functioning genes are administered. Wilson believes that gene therapy will not achieve real medical relevance until genes can be bottled, distributed, and then administered in vivo by ordinary medical practitioners. In accordance with this vision, Wilson spends most of his time working on the development of what he calls *injectable genes*.[36]

Still working with FH and genes expressed in the liver, Wilson and his colleague George Y. Wu, of the University of Connecticut, have developed an artificial DNA virus. Their constructed virus consists of DNA, including a functional LDL receptor gene, coated with protein that is recognizable by liver cells. Within minutes after this complex is infused into the ear vein of Watanabe rabbits, liver cells scoop up the protein from circulation and pick up the DNA along with it. The Watanabe rabbits' cholesterol levels fall, but only transiently because the injected gene is soon degraded. Wilson and his colleagues are pursuing a strategy for accomplishing prolonged or permanent LDL receptor gene activity by following the gene injection with stimulation of liver cell growth. When genes are taken up by cells during cell division, the genes are more likely to integrate into the host genome and result in stable expression.[37]

Wilson's injectable genes are not yet ready for human trials because of a number of limitations, including the risk that some of the injected genes may be randomly picked up by cells other than liver cells, possibly resulting in LDL expression in inappropriate cells. Nevertheless, when perfected, in vivo gene therapy will have several advantages over gene therapy mediated by gene transfer into cells in vitro. It will allow the delivery of genes to appropriate cells without invasive procedures, and will reach a greater number of cells than those which can be removed, thereby holding out the promise of more complete amelioration of disease symptoms.[38]

Hereditary emphysema. Hereditary emphysema is a lung disease caused by a mutation in the gene that encodes an enzyme called alpha 1-antitrypsin (AAT). Normally, AAT protects the lungs from damage by neutrophil elastase, a powerful enzyme which is capable of destroying lung tissue. Without functioning AAT, a person's lungs are easy prey for neutrophil elastase. Current treatments for this disease, which causes progressive respiratory impairment and a shortened lifespan, are effective but extremely expensive.

Ronald G. Crystal of New York Hospital–Cornell Medical Center and his collaborators have experimented with two different gene therapy alternatives that might be used to treat AAT deficiency.[39] Crystal's early AAT experiments involved inserting the AAT gene into mouse T cells and injecting the T cells into mice. Subsequent injections of substances designed to stimulate the immune system caused the T cells to proliferate. Crystal's group theorized that they would be able to induce the same proliferation of modified T cells in patients and, using aerosolized antigens, draw the AAT-expressing T cells to the patient's lungs, where AAT was required.[40]

In more recent experiments, Crystal's team aerosolized viral vectors carrying the AAT gene and sprayed them directly into the lungs of rats, where they transduced epithelial cells that lined the bronchial tubes. After 6 weeks, the AAT gene still persisted in the rats—promising evidence that the therapy could work in human AAT-deficiency patients. Crystal's team used recombinant adenoviruses. Adenoviruses were advantageous for many reasons, including the fact that they normally infect human respiratory cells and have a proven safety record as human vaccine viruses.[41]

Cystic fibrosis. Gene therapy holds promise for another common inherited lung disease, cystic fibrosis. Cystic fibrosis (CF) is one of the most common genetic diseases.[42] The gene responsible for CF, the cystic fibrosis transmembrane conductance regulator gene (CFTR), normally encodes a protein that helps transport ions (electric charge) across cell membranes. When this gene is defective, mucus-producing cells cannot access normal quantities of water. As a consequence, these cells produce thick, sticky mucus that chokes the lungs and other organs, harboring bacteria and leading to life-threatening infections.[43]

Delivering aerosolized genes directly to the lungs of CF patients may prove to be a nonintrusive and effective therapy.[44] Several research teams, including Crystal's, are working on experiments that use the adenovirus which naturally infects the tissue lining the lung (the *lung epithelium*) to deliver normal copies of the defective gene responsible for CF. Other groups have already succeeded in correcting the genetic defect in vitro in cells taken from CF patients.[45] In addition, using adenoviral vectors, researchers have applied the normal CFTR gene to the nasal epithelia of CF patients, thereby correcting their defective ion channel without any serious inflammatory response and without replication of the adenovirus.[46]

Researchers would like to be able to use AAV as a vector for gene therapy in CF patients. AAV naturally infects the epithelial cells of the respiratory tract and the gastrointestinal tract. It can be produced in high concentrations in solution and it is nonpathogenic in its normal form. It also usually integrates into a specific, preferred chromosomal location, lowering the risk of insertional mutagenesis. Researchers have conducted animal trials and have submitted a protocol for a human trial to assess the usefulness of the AAV vector as a method for delivering the CFTR gene.

Additional delivery systems under scrutiny for CF gene therapy include retroviral vectors, liposomes, and receptor-mediated endocytosis (a method which involves attaching DNA to a molecule that binds to a specific cellular receptor, where it is internalized by the host cell).[47]

As of late 1995, 12 human CF gene therapy protocols have been submitted to regulatory bodies for approval, and 10 human trials are in progress or completed.

Based on all of the current research, one reviewer has predicted that an effective gene therapy for CF will be available by 1999.[48]

Coronary artery disease. Sufferers of coronary artery disease, whether genetically derived or not, might be helped by the addition of genes encoding proteins that dissolve artery-clogging substances. A few different groups have developed techniques for adding such genes to the cells lining the inside of artery walls.

Judith L. Swain and colleagues at Duke University Medical Center have successfully inserted genes directly into the femoral arteries of the legs of several dogs, using a catheter to deliver the gene-fortified solution. For their experiment, they used luciferase genes, which encode an enzyme normally responsible for giving fireflies their glow. After a few days, the glow of the luciferase enzyme in surgically removed tissue allowed the researchers to easily detect that the added genes had not only entered the arterial cells, but also turned on and expressed.[49]

Elizabeth Nabel, Gregory Plautz, and Gary Nabel, of the University of Michigan Medical School, achieved similar results using a different gene in the arteries of pigs. They had success using both in vitro and in vivo techniques. In 1989, the Nabel team showed that endothelial cells could be genetically altered in vitro and then stably implanted on the arterial walls of Yucatan minipigs, where they expressed a recombinant gene.[50]

In September 1990, the Nabel group reported the results of an experiment in which they used a double-balloon catheter to deliver a viral vector containing a marker gene encoding a stainable protein directly into pig arteries. The two balloons blocked the artery on both sides of a small space into which the vector was flushed, thereby containing the vector in a protected space, where it transduced arterial endothelial cells. Even as long as 21 weeks later, the pigs' arterial cells still produced the stainable gene product.[51] The Nabel group achieved similar results using liposome transfection, demonstrating that, whether by transduction or transfection, they could deliver a recombinant gene to a specific site in vivo.[52]

Using methods for transducing or transfecting endothelial cells, researchers hope one day to be able to deliver to patients not just marker genes, but helpful genes that produce, for example, clot-preventing proteins. The same techniques could be used also for treatment of other illnesses in addition to artery disease. For instance, the addition of functioning genes to artery endothelial cells might be a workable, future mechanism for delivering clotting factors to the blood of hemophilia patients or for dispensing insulin to diabetics.[53]

Cancer. Cancer is a generic name for a "class of diseases characterized by uncontrolled cellular growth."[54] It is now the "leading cause of death for women in the United States and, if trends continue, will be the overall leading cause of death in the United States by the year 2000."[55] Scientists have identified several environmental and genetic causes of cancer, including tobacco (the primary cause). Because many of the causes of cancer are associated with lifestyle choices, prevention research contributes importantly to the fight against cancer.[56] Complementary research in cancer treatment has led to the development of a variety of different theories which involve the use of gene therapy as a cancer treatment. All of these include attempts to specifically target cancer cells for destruction or arrest. More than 50 clinical trials of gene therapy for the treatment of cancer are currently under way.

The first cancer gene therapy proposal actually became one of the first gene therapy experiments attempted in humans (shortly after the ADA experiment). The experiment, led by Dr. Steven A. Rosenberg, chief of surgery at the National Cancer Institute, involved an attempt to genetically engineer cancer patients' own cells to dispense products that destroyed their tumors. Rosenberg removed naturally occurring white blood cells called tumor infiltrating lymphocytes (TILs) from patients suffering from malignant melanoma, a deadly skin cancer. These TILs have the ability to specifically home in on tumors growing in the patients' bodies. Using retroviral vectors in vitro, Rosenberg's team inserted into the TILs a gene that encoded tumor necrosis factor (TNF), a protein shown to vigorously fight tumors in mice. The TILs were then returned to the patients' bodies. It was hoped the engineered TILs would deliver the inserted TNF genes specifically to malignant tissue, where the genes would direct the production of TNF and destroy the patients' tumors. The human trial, which began on January 29, 1991, remained inconclusive in July 1991, when it was reported that Rosenberg and his colleagues had experienced difficulty in achieving persistent in vivo gene expression. Although Rosenberg said it was too early to tell whether the patients were benefiting from the gene therapy experiment, he did announce that none of the patients had been harmed.[57]

The second human cancer gene therapy experiment involved genetic alteration of patients' cancer cells that was expected to induce cancer sufferers' own immune systems to mount an attack on their tumors. This line of human research was based in large part on the compelling work of Drew Pardoll, of Johns Hopkins University. Working in mice, Pardoll and his team inserted genes for an immune system chemical called *interleukin* 4 (IL-4) into kidney cancer cells that had been removed from the mice's own spontaneously arising tumors. They reintroduced the engineered tumor cells into the mice and found that in addition to rejecting and destroying the engineered tumor cells, the mice immune systems mounted attacks which destroyed the parent tumors.[58]

Rosenberg and his co-workers later tested this new gene therapy approach to cancer in humans, again working with melanoma patients. They removed tumor cells from the patients and, after growing them in laboratory cultures, inserted into them genes for TNF or interleukin-2. Interleukin-2, like TNF, is an immune system chemical that battles tumors. After readministering the engineered cells to the patients, the Rosenberg team anticipated that the tumor cells containing the added genes would stimulate an elevated immune system response in vivo before dying off.[59]

A similar human experiment, using liposomes to transfect tumor cells in vivo, was proposed in October 1991. A group led by Gary Nabel (University of Michigan) proposed to directly inject cancer patients' tumors with liposomes containing a gene that encodes HLA-B7, a protein that stimulates tissue rejection. The Nabel group hopes that tumor cells will take up and express the HLA-B7 gene, causing those cells to be recognized as foreign by the patients' immune systems.[60]

A third line of inquiry arises from basic research into the molecular genetics of cancer. During the 1980s and into the 1990s, accumulating evidence has suggested that the normal, controlled growth of cells is regulated by two contrary types of genes. One type, genes that promote the growth of cells, are called *proto-oncogenes*, or when they go awry, *oncogenes*. Certain mutations of proto-oncogenes activate them, turning them into oncogenes and fostering uncontrolled cell growth. Activated oncogenes are almost always dominant, switching on unregulated cell growth even when present in heterozygous form.[61]

Later research has unearthed counterposing genes, descriptively named *tumor suppressor genes*. Genetic mutations that inactivate tumor suppressor genes destroy the limitations they impose on cell growth and allow the unsuppressed growth of cancer cells. Such mutations may be inherited, but the great majority of them appear to be somatic.[62] Tumor-suppressor genes act recessively; both alleles must be knocked out by mutation before a cell becomes cancerous.[63] The multiplying discoveries of and about tumor-suppressor genes have generated research and speculation on the possibility of harnessing them for cancer gene therapy.[64]

Early research results have been promising. The *retinoblastoma* (RB) gene encodes a tumor suppressor protein that has been found inactivated in several types of cancer, including breast, lung, and prostate cancers. Robert Bookstein and his colleagues at the University of California at San Diego have shown that adding functioning RB genes to human prostate cancer cells suppresses the ability of those cells to participate in tumor formation when the cells are implanted in mice. Human prostate cancer cells without the added genes readily form tumors in the mice. Bookstein has postulated that at some future time, a therapy for cancer will evolve that is based on the addition of functioning RB genes or their RB protein products to cancer cells.[65]

Perhaps even more hopeful are the results obtained in experiments involving the tumor suppressor gene p53. Mutations in the p53 gene are the most frequently observed genetic alterations in human cancers.[66] Various p53 mutations have been found in colon cancers, breast cancers, lymphomas, leukemias, lung cancers, and esophageal cancers, among others.[67] Surprisingly, the p53 gene has the capacity to act as an oncogene as well as a failed repressor when it undergoes certain mutations. In other words, it sometimes actually switches from suppression to stimulation of cell division.[68] Researchers do not have a complete understanding of the p53 gene and its activities, but at least one interesting model has been suggested: While normal p53 works transiently in conjunction with a complex of other molecules, certain mutants of p53 may attach too tightly to the other molecules, sequestering them from their usual activities and leaving the cell without any of the active complexes that it needs for negative growth regulation.[69]

A team of researchers led by Bert Vogelstein, of The Johns Hopkins University School of Medicine, has demonstrated that the normal human p53 gene can specifically repress the growth of human colorectal cancer cells in vitro.[70] Another team, at Temple University, has found that the normal p53 gene can halt the growth of brain cancer cells in vitro.[71] It is hoped that continuing research will result in a mechanism for using functioning p53 genes to stop the growth of cancer cells in vivo in the many patients whose uncontrolled cell growth is a result of mutant p53. At least four human-cancer gene-therapy proposals aimed at modifying the expression of tumor-suppressor genes have been submitted to regulatory bodies; several have received approval as of this writing.

A group of researchers led by Edward H. Oldfield, a neurosurgeon at the National Institutes of Health, has begun an experiment with people who have incurable, malignant brain tumors. The gene therapy strategy involves injecting the brain tumors with cells that have been genetically engineered to produce a vector containing a drug-sensitivity gene. It is hoped that this gene will be transferred into the tumor cells and make them sensitive to the drug ganciclovir (GCV). GCV is nontoxic to normal tissues, but will kill cells that express the gene in question.[72]

Other researchers have applied an inverse theory in designing human-cancer gene therapy trials. They have transferred drug resistance genes into human subjects' noncancerous cells for the purpose of protecting these normal cells from the effects of chemotherapy.[73]

Researchers are also testing in humans a gene therapy strategy called antisense gene therapy. The antisense strategy involves administering to cancer sufferers vectors that express RNA molecules which specifically block the function of certain genes that promote cancer growth. The antisense RNA molecules block gene expression by binding with the complementary base pairs of the targeted gene's transcript. Researchers hope that vectors expressing antisense RNA will be able to stop the growth and spread of cancer cells in patients.[74]

AIDS. AIDS is a somatic cell genetic disease that, in some respects, resembles ADA deficiency. AIDS is associated with a person's infection by HIV, which causes HIV-infected patients to suffer immune system impairment, resulting in increased infections and an increased risk of developing cancer. Several gene therapy strategies for treating HIV infection are being studied in animal models or have been proposed for human clinical trials.

One strategy that has already entered second generation human clinical trials was originally proposed in 1993. This study involves administering retroviral vectors containing a gene coding for an HIV envelope protein to HIV-infected individuals. Scientists hope that the vector will trigger immune responses in HIV-infected patients. Specifically, it is hoped that their bodies will produce T cells which target and kill cells infected with viruses.[75]

Philip Greenberg, of the University of Washington, along with his research team, proposed in October 1991 to use gene therapy techniques to augment a bone marrow transplant treatment for HIV infection and one of its associated cancers, lymphoma. The treatment first entailed removing some of a patient's T cells— those cells with a specific ability to fight HIV infection. The harvested T cells were transduced in vitro with a gene (the HyTK gene) that caused those T cells to self-destruct if the patient were later exposed to particular drugs. Then doctors irradiated the patient's bone marrow in order to destroy both the patient's lymphoma and HIV-infected cells. Next, the patient received a bone marrow transplant from a matched donor along with a dose of his own engineered T cells. It was hoped that the T cells would help the patient resist HIV infection of his new bone marrow. If, instead, the engineered T cells caused damage to the patient, they would be destroyed by administering a drug that would cause them to self destruct. The HyTK gene also enabled researchers to detect whether the T cells were surviving in vivo.[76] This strategy can be classified as a cell therapy rather than as a gene therapy. The genetic engineering really serves as a safety device. The clinical trial of this approach began in February 1993 and is still ongoing.[77]

A different plausible AIDS gene therapy is derived from the infection mechanism of HIV. HIV acts first by binding to certain sites on the surface of cells. These sites are called CD-4 receptors and are encoded by CD-4 genes. Researchers hope to find a mechanism that can deliver CD-4 genes, engineered into decoy cells, into patients, thereby luring HIV away from the important immune system cells it usually invades.[78]

Additional approaches involve transducing cells in vitro with genes that could enhance a patient's ability to fight an HIV infection or interfere with the HIV life cycle. When the transduced cells are placed into the patient's body, they should secrete a factor into his or her circulation that will bolster the body's defenses or directly affect HIV. Several strategies designed to directly kill HIV-infected cells or to interfere with HIV's life cycle by inhibiting HIV RNA or HIV proteins are also being tested.[79]

As French Anderson has commented: "HIV is a very clever adversary. It is highly unlikely that any single tactic will work. A combination of several types of attack simultaneously and/or sequentially will probably be needed."[80]

Other Candidate Diseases for the Future

Hemoglobin diseases. Sickle cell anemia and beta-thalassemia are two of several genetic diseases characterized by abnormalities in human hemoglobin, the blood component that is responsible for transporting oxygen. Individuals afflicted with these diseases suffer severe symptoms such as anemia, painful crises, and sometimes death.[81] Early in the development of gene therapy, long before the first human experiment was launched, hemoglobin diseases were considered good candidates for initial gene therapy trials. As knowledge accumulated, however, it became apparent that gene therapy for hemoglobin diseases such as beta-thalassemia and sickle cell disease would be difficult to achieve, because the production of the hemoglobin molecule depends on extraordinarily complex gene regulation.[82] The hemoglobin molecule is comprised of several subunits that must be present in blood cells in delicately balanced amounts. Not only are these subunits produced by several separate genes, but the genes are grouped on two distinct chromosomes.[83] That the healthy human body coordinates the expression of these several genes is nothing short of miraculous. For scientists to understand and mimic that coordination will be no easy matter. And if one added gene's expression is not coordinated with the expression of the other genes that produce subunits, the resulting imbalance itself constitutes a disease state.[84] Therefore, although sickle cell anemia and beta-thalassemia are still posited as potential gene therapy candidates by some, researchers have, for the most part, moved on to other diseases as early candidates for gene therapy. It is possible that recent research results involving retroviral vectors expressing hemoglobin genes in vitro will revive interest in the use of gene therapy for hemoglobin disease treatment.[85]

Muscular dystrophy. Muscular dystrophy is a disease characterized by progressive muscle deterioration, which starts in the lower extremities and steals up the legs to the hips, the respiratory muscles, and the heart. Muscular dystrophy almost always leads to death by the age of 30. Duchenne muscular dystrophy (DMD), the most common form of this cruel disease, primarily afflicts boys. DMD is caused by a mutation in the dystrophin gene, most often a deletion. For unknown reasons, if the dystrophin protein does not function properly, muscle cell death results. The key behind gene therapy for DMD is the delivery of an error-free copy of the dystrophin gene into the muscle cells of DMD patients.[86]

The dystrophin gene, the largest human gene known, is too large to be delivered to cells via any of the commonly used viral vectors.[87] Similarly, the dystrophin protein is large, complex, and not directly deliverable to cells in its active form. Nevertheless, Dr. Peter Law, formerly of the University of Tennessee,[88] and others have developed a cell-grafting type of method which may prove effective for indirectly delivering properly functioning dystrophin genes into patients' cells. This method takes advantage of the peculiar ability of muscle cells to fuse and share proteins.

The proposed treatment, which has worked in mice and is currently being tested in humans, involves injecting healthy muscle cells into the muscles of DMD patients. The healthy cells are obtained from the fathers of the patients and grown in culture in the laboratory before they are administered. It is hoped that these healthy cells will contribute normal dystrophin genes and the normal dystrophin protein to the patients' muscles, thereby preventing their deterioration.[89]

Researchers would like to deliver functioning dystrophin genes directly into the muscle cells of DMD patients. But so far, no one has devised a method for delivering the dystrophin gene into all the muscle of the body.

ETHICAL ISSUES

An Initial Question: Is This Kind of Treatment Different?

In the ethical discussions that began in the late 1960s, commentators on human gene therapy sometimes seemed to assume that this technique was qualitatively different from other types of therapeutic interventions. However, as the ethical discussion of gene therapy has progressed, somatic cell gene therapy has increasingly been viewed as a natural and logical extension of current techniques for treating disease. Which of these views is correct?

On balance, the gene-therapy-as-extension view seems to be the more appropriate one. There are several reasons for adopting this view. First, because somatic cell gene therapy affects only nonreproductive cells, none of the genetic changes produced by somatic cell gene therapy will be passed on to the patient's children. Second, in some cases the products of the genetically modified somatic cells are similar to medications that patients can take as an alternative treatment. For example, there are enzyme therapies currently available for both ADA deficiency and Gaucher disease, but both enzyme therapies are very expensive and must be administered frequently. Third, some of the techniques currently used in somatic cell gene therapy closely resemble other widely used medical interventions—especially the transplantation of organs or tissues.

Several examples noted in the earlier part of this chapter illustrate how similar at least some somatic cell approaches are to transplantation. In the protocol for treating ADA deficiency, some of the patients' T cells were removed from their bodies and had the missing ADA gene added to them. The genetically modified cells were then returned to the patients' bodies, where they began to produce the missing enzyme, ADA. If the patients involved in this early gene therapy study had had healthy siblings whose cells closely matched their own, the alternative to gene therapy would have been a bone marrow transplant. In effect, the T cells of the healthy sibling (or more technically the stem cells that produce the T cells) would have replaced the patients' own ADA-deficient T cells.

The case is similar with cystic fibrosis (CF). Increasingly, lung transplants are being employed in the treatment of this disease. The cells in the transplanted lung will not have the genetic defect that causes CF and will therefore be able to function normally in the recipient. However, such transplants are expensive and highly invasive procedures, and there is a perpetual shortage of healthy organs for transplantation. Further, because transplanted organs never match the recipient's genotype exactly, except in the case of identical twins, the recipient will likely have to take drugs indefinitely to prevent his or her immune system from rejecting the transplant as foreign tissue. Somatic cell gene therapy seems to many observers to be a less invasive approach than the transplantation of a major organ. In addition, because it is the patient's own cells that are being genetically modified, there is a much lower probability that the cells will be rejected as foreign.

Major Ethical Questions Concerning Gene Therapy Research

In the review of proposals to perform gene therapy research with human beings seven questions are central:

1. What is the disease to be treated?
2. What alternative interventions are available for the treatment of this disease?
3. What is the anticipated or potential harm of the experimental gene therapy procedure?
4. What is the anticipated or potential benefit of the experimental gene therapy procedure?
5. What procedure will be followed to ensure fairness in the selection of patient-subjects?
6. What steps will be taken to ensure that patients, or their parents or guardians, give informed and voluntary consent to participating in the research?
7. How will privacy of patients and the confidentiality of their medical information be protected?

Taken together, questions 1–4 constitute a kind of first hurdle, or initial threshold, that gene therapy research proposals must clear. If these questions are not satisfactorily answered, questions 5–7 will not even need to be asked. However, if the first four questions are satisfactorily answered, and the risk-benefit ratio for the proposed research seems appropriate, questions 5–7 remain as a second hurdle—a set of important procedural safeguards for prospective subjects in the research.

What is the disease to be treated? Question 1 asks both a simple and a more profound question. It asks simply for the name of a condition or a disorder that is regarded by reasonable people to be a malfunction in the human body. Thus, "cystic fibrosis" would be an acceptable answer to the first question, while "average height" would not. At a more profound level, question 1 asks whether the disease or condition put forward as an early candidate for somatic cell gene therapy is sufficiently serious or life-threatening to merit being treated with a highly experimental technique. As the list of disorders treated in the first 100 gene therapy studies indicates, the conditions proposed for possible treatment by means of gene therapy do gravely compromise the quality and duration of human life.

It must be acknowledged that two of the disorders included in the above list do not qualify as life-threatening: rheumatoid arthritis and peripheral artery disease. Rheumatoid arthritis is a chronic condition that often causes severe pain to the person who is suffering from it, but this disease alone generally does not cause a patient's death. Similarly, peripheral artery disease in, for example, the lower leg and ankle of a person with diabetes will not cause the patient's death. However, this condition can be limb-threatening. That is, a limb that does not receive a sufficient flow of oxygen may need to be amputated in order to prevent gangrene from developing in the limb. In the future, gene therapy protocols may also be submitted that seek to preserve sight by taking new approaches to the treatment of currently untreatable eye diseases.

Philosophers have argued at considerable length about the precise definitions of health and disease.[90] For our analysis we have adopted a rather standard definition of health, "species-typical functioning."[91] It seems clear that all of the conditions to be treated by the first 100 gene therapy protocols represent significant deviations from the physiological norm of species-typical functioning and therefore qualify as bona fide diseases. But are there reasonable limits to the notion of disease? How far would we be willing to extend the concept? At the far extremes exemplified in discriminatory social programs both past and present, we would never want to see human characteristics like gender, ethnicity, or skin color regarded as diseases. However, serious mental illness would be included within the scope of our definition. We would not consider mild obesity or a crooked nose or larger-than-average ears to be significant deviations from species-typical functioning. However, serious obesity that threatened to shorten life might well qualify as a disease. In each case a condition will need to be evaluated in the light of species-typical functioning and a judgment will have to be made about the extent to which the condition compromises such functioning.[92]

What alternative interventions are available for the treatment of this disease? Question 2 asks about alternative therapies. If available modes of treatment provide relief from the most serious consequences of a disease without major side effects and at reasonable cost, the disease may not be a good candidate for early clinical trials of gene therapy in humans. For example, phenylketonuria (PKU) is a hereditary disorder that can be detected in newborns through a simple blood test. Dietary therapy suffices to prevent the brain damage that would otherwise occur in children afflicted with the disorder. Therefore, PKU is probably not a good early candidate for gene therapy. Similarly, the harmful effects of diabetes can be controlled quite well in most patients through the use of insulin produced by recombinant DNA techniques. Thus, diabetes may be a later rather than an earlier candidate for gene therapy research.

The determination that an alternative therapy is sufficiently effective is always a judgment call. In the review process for the first ADA deficiency study, it was noted that bone marrow transplantation was an effective treatment for children who had genetically matched siblings and that a synthetic form of ADA was available for use in ADA-deficient patients. The synthetic form of ADA was derived from the ADA produced by cattle and was linked, or conjugated, with the chemical PEG. The proponents of gene therapy for ADA deficiency pointed out that PEG-ADA could stimulate a hostile response by the human immune system because the synthetic compound was derived from cattle and thus might be perceived as foreign. The proponents also noted that, while most ADA-deficient patients benefited somewhat from treatment with PEG-ADA, they were still susceptible to many infections. Further, the high cost of PEG-ADA, about $250,000 per year, put this synthetic treatment out of the reach of most families. Thus, the reviewers of the initial ADA gene therapy protocol ultimately concluded that in families where children had no matched sibling donors, the alternative therapy of PEG-ADA was not wholly satisfactory. There was therefore space or justification for the development of gene therapy as a possibly superior approach to the treatment of this life threatening pediatric disease.

As the field of gene therapy matures, the requirement that there be no effective alternative therapy may need to be relaxed. At some point there will need to be well controlled studies that compare the gene therapy approach with alternative approaches to treatment of the same disease. However, given the novelty of gene therapy in 1990 and the uncertainty about its potential benefits and harms, it seems to us to have been appropriate to limit the earliest gene therapy trials to diseases and groups of patients for whom no alternative therapies were available.

What is the anticipated or potential harm of the experimental gene therapy procedure? Question 3 concerns the anticipated or possible harm of somatic cell gene therapy. In responding to this question, researchers are asked to base their statements on the best available data from preclinical studies in vitro or in laboratory animal models like mice or monkeys. The use of domesticated (technically, replication deficient) viruses as vectors in many somatic cell gene therapy studies raises one important safety question: How certain can researchers be that the domesticated viruses will not regain the genes that have been removed from them and thus regain the capacity to reproduce and cause an infection in the patient? A second kind of safety question arises from the "unguided missile" quality of retroviral vectors. As noted earlier in this chapter, researchers cannot predict where a retroviral vector with its attached gene and marker will "land" within the nucleus of a target cell. It is possible that the vector will integrate into the middle of a gene that is essential to the functioning of the cell and will therefore kill the cell. A further concern is the theoretical possibility that a retroviral vector might integrate beside a quiescent oncogene (cancer-causing gene) and stimulate it into becoming active. If so, a previously healthy cell might begin to divide uncontrollably and even start a cancer in a particular site. Because of these concerns about risk and safety, researchers are asked to provide data about preclinical studies in animals that, insofar as possible, exactly duplicate the gene therapy studies that they propose to do in human subjects.

What is the anticipated or potential benefit of the experimental gene therapy procedure? The fourth question is in many ways the mirror image of the third. It asks researchers once again to provide data from preclinical studies, but in this case the data should indicate that there is a reasonable expectation of benefit to human patients from their participation in the gene therapy study. One step in the review process for the first approved human gene therapy study in 1990 illustrates the importance of this point. Drs. Blaese and Anderson had provided impressive safety data to the RAC (the Recombinant DNA Advisory Committee—established in 1974 by NIH) based on their long term studies in mice and monkeys. However, RAC members were not convinced that the genetic modification of ADA-deficient

T cells would be beneficial to human patients. What if the T cells died off quickly, or what if they were overwhelmed by the patients' own ADA-deficient T cells? Fortunately there was a researcher in Milan, Italy, Dr. Claudio Bordignon, who was willing to speak to the RAC about his own research in mice that had an immune deficiency similar to the one that afflicts ADA-deficient human patients. In these animals Dr. Bordignon was able to show that *human* T cells that carry a functioning ADA gene survive longer than ADA-deficient T cells. This was the information that the RAC was seeking, and Dr. Bordignon's report helped to ensure the approval of the Blaese-Anderson proposal.

There has been considerable debate about the appropriate relationship between question 3 and question 4—or between anticipated harm and anticipated benefit of a gene therapy study. Some researchers have argued that if a gene therapy study is not likely to make patients worse off, it should be approved even if the probability of benefit is very low. This rationale may have been the basis for the controversial approval of a gene therapy protocol by former NIH director Bernadine Healy in December 1992. The proposal to treat a single cancer patient had not gone through regular view by the NIH RAC, and there was, in the opinion of most experts, little probability that it would benefit the terminally ill patient. Nonetheless, Dr. Healy approved the protocol on a "compassionate use" basis.[93] Since late 1992 the RAC has further discussed the requisite harm-benefit ratio in gene therapy studies. A majority of the committee members seem to have adopted the following view: Even if a gene therapy protocol provides a satisfactory answer to question 3 (about harm), it must also offer at least a low probability of benefit to the patients who are invited to enter the protocol, and it must have an excellent scientific design, so that the information gathered from studying the early patients will be useful to later patients and to the entire field of gene therapy research. That is, there must also be a satisfactory answer to question 4.

If these first four questions are satisfactorily answered, researchers have cleared a first hurdle or crossed an important threshold. They have demonstrated that the ratio of probable benefit to probable harm in a proposed study is sufficiently positive to justify proceeding to research in human beings. The remaining three questions ask how the research will be done, or, in other words, outline procedural safeguards for the patients who will be invited to take part in the gene therapy studies.

What procedure will be followed to ensure fairness in the selection of patient subjects? The first of the procedural questions, and the fifth question overall, asks how patients will be selected in a fair manner. With very rare disorders, like familial hypercholesterolemia and ADA deficiency, fairness in the selection process was relatively unproblematic. Virtually every patient with the condition who was not too ill to participate was considered a candidate for gene therapy and invited to enroll in the studies. However, when somatic cell gene therapy began to be used for more prevalent conditions like brain tumors, the selection process became much more difficult. For example, the first study of gene therapy for brain tumors was led by Edward Oldfield and Kenneth Culver at the National Institutes of Health. Their proposal was initially approved for the study of 20 patients. Within the first year of their study the offices of Drs. Oldfield and Culver received more than 1,000 inquiries by patients, their family members and friends, and even governors and legislators. It was therefore quite important that a fair procedure, like first-come first-served, be in place to use in selecting among the many candidates for treatment.

In the review of the first 100 gene therapy protocols two other questions of fairness in selecting subjects have also arisen. The first question is whether children should be included in the initial studies, assuming that some children do survive to adulthood with the disease. The first gene therapy study, for severe combined immune deficiency (SCID), involved children precisely because, until now, almost no children with this disorder live to complete their teenage years.

However, some patients with familial hypercholesterolemia do live to be 30, as do an increasing number of cystic fibrosis (CF) patients. There are two opposing ethical perspectives on the involvement of children in the early stages of clinical trials. The classical position, formulated in the early 1970s in the United States, was that clinical trials should be completed in adults first, before children are exposed to the potential harms of such trials.[94] The revisionist position of the late 1980s and 1990s is that participants in clinical trials are carefully monitored and that they often are the first people in a society to have access to new and possibly effective treatments. Therefore no class of individuals, whether women or members of ethnic minorities or children, should be excluded from the potential benefits of timely participation in clinical trials.[95]

A second question is closely related to the alternative therapy question discussed above. It concerns the stage of disease at which people with serious disease should receive gene therapy. Already in 1987, during the review of the "Preclinical Data Document," there had been vigorous debate about whether gene therapy should be regarded as last ditch therapy. In the early studies of gene therapy, patients had generally not been helped by alternative therapies, or no alternative therapies were available. As the field has matured, and as researchers have gained more confidence in the safety of gene therapy (if not in its effectiveness), the question has increasingly been raised, "Why not employ gene therapy at earlier stages in the disease process where it might have a better chance to prevent the deterioration of the patient's condition?"

What steps will be taken to ensure that patients, or their parents or guardians, give informed and voluntary consent to participating in the research? The next procedural question, and the sixth question overall, concerns the voluntary and informed consent of patients or, in the case of minors, of their parents or guardians. It is always a challenge for researchers to convey to potential patient-subjects the important facts about their disease or condition, the major alternative treatments, and the precise procedures to be followed in the research. With a cutting-edge technology like human gene therapy this generic problem becomes even more difficult. For gene therapy studies, patients or their parents or guardians will frequently need a short course on how recombinant DNA research is conducted, how vectors are constructed, what cell types are targeted, how genes are inserted into cells, and how genes function in cells. In the case of specific gene therapy studies, additional modules may have to be added to the short course. For example, in the case of gene therapy for SCID, patients or their parents or guardians will need additional information on the human immune system and how specific kinds of cells, like T cells, function. While this educational responsibility may initially seem daunting, the question about voluntary, informed consent simply points to the importance of an extensive and ongoing dialogue between researchers and subjects, rather than a momentary act of signing a multiple-page consent form.

The response of researchers to question 6 has varied considerably. Some gene therapy proposals have included detailed information, including charts that outline the sequence and timing of proposed procedures, for patients. Other consent forms have been woefully incomplete or have included ambiguous wording about who pays for procedures required by the research or about how the sponsoring institution deals with patients who are accidentally injured when they participate in research. Further, no external observers are present to monitor the quality of the consent process as patients are invited to take part in therapy studies. All that can be said with certainty is that the RAC has provided detailed guidance to researchers about the major points that RAC members think should be included in consent forms and that the consent forms themselves should be reviewed in a public forum by peers who are not employees of the sponsoring institutions.

How will the privacy of patients and the confidentiality of their medical information be protected? The third procedural question, and the seventh question overall, concerns privacy and confidentiality. There is no single "correct" answer to this question. It merely asks researchers to think through in advance how they and the subjects participating in gene therapy studies will deal with inquiries from the press and the general public. A particular concern was that patients have sufficient privacy and rest following their treatment to allow "space" to benefit from this experimental approach to treatment. Different researchers and families have adopted varying policies regarding privacy and confidentiality. In the case of the ADA-deficiency protocol, the first two young women treated remained relatively anonymous until the second anniversary (in September 1992) of the first child's initial treatment. Less than a year later, the two children were featured, named, and pictured in a *Time* magazine story.[96] In contrast the parents of two newborns treated at birth (in a modification of the same protocol) by a different technique using stem cells from umbilical cord blood allowed their names and pictures, and the names and pictures of their infants, to be disclosed almost immediately.[97] Similarly, the first patient in the study that aimed to treat peripheral artery disease was interviewed by a reporter for the *New York Times* before he received his first gene transfer.[98]

This final question was included in the "Points to Consider" not to prescribe or proscribe any particular actions by patients or researchers but simply to encourage all parties involved in gene therapy studies to think through, in advance, a strategy for dealing with the press and the media. There had, in fact, been veritable media circuses when several earlier biomedical technologies had first been introduced. One thinks, for example, of the earliest heart transplant recipients, of Barney Clark and his artificial heart, of Baby Fae and her baboon heart, and of Louise Brown, the first "test tube baby." In general, the introduction of gene therapy has been, in our view, more respectful of patients and their families.

Other Major Questions about Gene Therapy

In addition to the seven questions raised about gene therapy by the "Points to Consider," there are important questions about the current system for conducting and overseeing gene therapy, both in the United States and abroad. The following section of this chapter attempts to step back from the specifics of the first 100 U.S. gene therapy protocols to examine some of these broader issues.

Was gene therapy attempted in patients too soon? This first question asks whether more laboratory research in cell cultures and in animal models should have preceded the first clinical studies with gene therapy. Some critics of current practice have noted that the available vectors for delivering genes to cells are relatively primitive. Other critics argue that basic research on how stem cells work and how they can be modified may facilitate new approaches to gene therapy. A third line of criticism is that, while a few protocols submitted by researchers for review by the NIH RAC have represented excellent science, many of the protocols have been unoriginal and probably would not have been funded by NIH if the protocols had been reviewed according to the stringent standards of the usual NIH grant application process.

What can be said in response to these criticisms? First, the first somatic-cell gene therapy studies in human beings were preceded by almost 20 years of public discussion and debate.[99] Even the initial protocol, approved in 1990, had been debated in near-final form from 1987 through early 1990. Thus, the introduction of gene therapy as a new biomedical technology has been more deliberate and more cautious than the first use of several other important techniques—for example, genetic screening for PKU in newborns, heart transplantation, and test tube (in vitro) fertilization. Second, even though the currently published evidence is sparse, there is at least a suggestion of clinical benefit in the initial gene therapy study for patients with SCID. At the same time, however, it must be acknowledged that the level of success in the first 60–70 gene therapy studies in the United States has been disappointing even to the technique's strongest advocates. In part because of this disappointment, the NIH director, Harold Varmus, recently asked an expert committee to review the NIH investment in gene therapy.[100]

Which diseases are the best early candidates for gene therapy? We have seen in this chapter what the target diseases are in the first 100 U.S. gene therapy protocols. Are various types of cancers sufficiently important to merit having more than 60% of the protocols directed toward cancer? Is 22 of the first 100 studies the appropriate share for the thousands of genetic diseases? And do HIV infection and AIDS deserve more or less than 12% of the research effort?

The answers to these questions, like the answers to many resource allocation questions, stretch human knowledge and judgment to their limits. One approach to answering the questions is to ask: What is the burden of disease, in terms of premature death, suffering, lost work, and disability, caused by each disorder? The answer to this question provides at least one dimension of the answer to the larger question, although it should be noted that the comparison of death, permanent disability, pain and suffering, and temporary disability involves a series of value judgments.[101] However, the burden caused by a certain disease is only part of the picture. There must also be a genuine research opportunity. That is, enough progress in understanding and perhaps even in treating a disease must have been made to allow a next step to have a reasonable hope of succeeding. In the case of SCID, a very rare genetic disorder, one feature that made the disease a good early candidate for gene therapy studies was that it had been cured through bone marrow transplants from matched sibling donors. Thus, researchers had good leads in their quest to treat the disease in children who lacked such donors. Further, the gene for ADA had been isolated in the laboratory and was available for clinical use. One other safety feature for SCID was that the gene did not have to be carefully regulated. An overproduction of ADA by a turned on gene would not, it seemed, harm the recipient patient. For these reasons SCID appeared to be an excellent early candidate for gene therapy, even though it is a rare disorder with a rather small national disease burden.

For the future the expert committee appointed by the NIH director may have some suggestions about an overall strategy for setting priorities among diseases, particularly if one or two diseases seem to have received inordinate attention thus far. On the other hand, centralized planning and long-term strategies may also have their pitfalls. There is much to be said for allowing considerable latitude to researchers in choosing the diseases that they consider to be plausible early candidates for gene therapy and in developing creative new approaches to the experimental treatment of those diseases. One of the most intriguing aspects of science is the factor of serendipity. One never knows in advance which scientists, which laboratory, which topic, and which approach will produce the next breakthrough, with unexpected benefits in a wide variety of seemingly unrelated fields.

To what extent should commercial considerations drive the selection of target diseases? This question is closely related to the preceding question. It is not surprising that, with the increasing involvement of private industry in somatic cell gene therapy there has also been an increasing focus on diseases that are prevalent in the United States. The largest trials to date, in terms of the numbers of patients, are studies directed toward HIV infection and AIDS. As noted earlier, cancers of various types are the targets in almost two-thirds of the first 100 protocols. And among genetic diseases cystic fibrosis, the most prevalent genetic disorder among Caucasians, commands the most attention.

We should not be surprised that commercial firms would look, first and foremost, at larger rather than smaller markets. If gene therapy proves to be a successful strategy, these firms will one day provide the bridge from the laboratory to large numbers of patients. At the same time, however, there are the so-called *orphan diseases*, which strike people in numbers too small to provide a strong incentive for an investment in research. There are, again, no easy solutions to this general problem in the U.S. health care system. Public investment in research is no doubt one part of the solution: NIH researchers and researchers supported by federal funds have the option of targeting diseases that affect relatively few people. Success in treating rare diseases may, in turn, provide clues that will assist in the treatment of more prevalent diseases. However, by far the most important part of the solution to the problem of gene therapy for rare disorders will have to await major reform in the health care system.

How is national oversight for gene therapy working? Virtually all U.S. gene therapy proposals go through a national public review process at the National Institutes of Health. Under a new consolidated review process with the Food and Drug Administration (FDA) only proposals that raise novel issues are reviewed publicly at quarterly RAC meetings. For the other proposals, basic information about the researchers, the title of the protocol, the target cell, the vectors, and the sponsors is made available in a public summary. All serious adverse events experienced by patients are also reported at each quarterly RAC meeting. In addition, at periodic intervals—annually in the future—a public report on virtually all U.S. gene therapy studies is compiled and presented at a public RAC meeting. This report allows any interested person to learn the most important facts, except for the effectiveness of treatment, about ongoing and already closed clinical trials of gene therapy. For example, one can find out how many patients are enrolled in each trial and how many adverse effects have occurred. If the results of a study have been published, the publication is cited.

One of the authors of this book has been much too closely involved with the public review process for gene therapy to be objective in judging it. However, several comments are in order. The level of public accountability that has existed for the early stages of gene therapy research is unprecedented in the history of clinical research. In fact, the NIH RAC has functioned as a kind of national research ethics committee (or institutional review board) for one new biomedical technology, gene therapy. Further, the RAC review process has provided a model for several other countries as they have established their own national review committees for gene therapy. In most cases the parallel committees in other nations do not meet in public, but their members often attend RAC meetings and sometimes report that RAC deliberations are helpful to them in evaluating protocols that come before them in their confidential review process.

Some U.S. commercial firms and some university researchers have found dual review to be burdensome and have campaigned to have the FDA provide the sole review for gene therapy studies. They argue that, even if the RAC was needed in the early 1990s for the earliest gene therapy studies, it has now outlived its usefulness. Our own view as authors is that the public review of selected protocols and the periodic reports on the current status of gene therapy are still important to the U.S. public, to the press and the media, and to members of the United States Congress. Public review and regular monitoring are essential to public accountability, in our view.

To what extent are researchers in other countries involved in gene therapy research? We have focused primarily on gene therapy in the United States because more studies have been initiated here than in any other nation. However, gene therapy research is proceeding in the United Kingdom, France, the Netherlands, Germany, Italy, Japan, and China—and in perhaps other nations as well. For the most part, the studies that have been made public parallel those that are being conducted in the United States. For example, SCID is the target disease in several countries. A study for which there is no parallel is the Chinese protocol that seeks to treat hemophilia B.[102]

How useful will somatic cell gene therapy be in the long run? The honest answer to this final question is, "It's too early to know the answer to this question with any degree of certainty." Some critics of gene therapy have argued that the future does not look bright for this approach to treatment. They note that, using current techniques, gene therapy probably costs at least $100,000 per year per patient, with ongoing retreatment and monitoring being required for all patients. They also assert, quite rightly, that alternative treatments, like drug treatment for people with cystic fibrosis, are also improving and may ultimately make gene therapy unnecessary.

So long as gene therapy requires repeated treatments by very specialized laboratories that can introduce genes into target cells by means of engineered viral vectors, gene therapy will remain a relatively expensive approach to the treatment of disease. It will, under these conditions, be of very limited utility in the war on human disease. There are visionaries, however, who foresee at least the possibility that the gene therapy approach may become as routine and pervasive as current techniques like the use of immunizations or antibiotics. W. French Anderson, for example, dreams of a day when a "magic bullet" will be available that would "enable healing genes to enter the bloodstream and go directly to the cell that needs help."[103] In the words of Anderson, "I'd like to go to Africa with 10,000 vials and inject the gene to cure sickle-cell anemia."[104]

NOTES

Reprinted by permission of Oxford University Press, Leroy Walters and Julie Palmer, *The Ethics of Human Gene Therapy* (1996).

1. Larry Thompson, "Human Gene Therapy Test Working," *Washington Post,* December 16, 1990, A6; Natalie Angier, "Doctors Have Success Treating a Blood Disease by Altering Genes," *New York Times,* July 28, 1991, I20.

2. Experimental protocol submitted to the Human Gene Therapy Subcommittee on the Recombinant DNA Advisory Committee, entitled "Treatment of Severe Combined Immunodeficiency Disease (SCID) due to Adenosine Deaminase (ADA) Deficiency with Autologous Lymphocytes Transduced with a Human ADA Gene," 2–3.

3. *Ibid.,* 3.

4. Eve K. Nichols, *Human Gene Therapy* (Cambridge: Harvard University Press, 1988), 217.

5. Response to Points to Consider for ADA Deficiency Protocol, 15.

6. Thomas D. Gelehrter and Francis S. Collins, *Principles of Medical Genetics* (Baltimore: Williams & Wilkins, 1990), 295; Natalie Angier, "New Genetic Treatment Given Vote of Confidence," *New York Times,* June 3, 1990, I25.

7. Response to Points to Consider, 13 (22).

8. W. French Anderson, "Prospects for Human Gene Therapy," *Science* 226 (1984): 402; October 26, 1984; John E. Dick, "Retrovirus-Mediated Gene Transfer into Hematopoietic Stem Cells," *Annals of the New York Academy of Sciences* 507 (1987): 242–51. Shortly after the human ADA gene therapy experiment achieved success, the U.S. Patent and Trademark Office granted a patent to Stanford University molecular biologist Irving Weissman and colleagues at SyStemix, a biotechnology company, for their process for isolating stem cells and for the isolated stem cells themselves. See Beverly Merz, "Researchers Find Stem Cell; Clinical Possibilities Touted," *American Medical News,* November 25, 1991, 2.

9. Angier, "Doctors Have Success" (note 1); experimental protocol (note 2), p. 24 (59).

10. Response to Points to Consider (note 5), p. 12 (21).

11. R. Michael Blaese et al., "T Lymphocyte Directed Gene Therapy for ADA SCID: Initial Trial Results after 4 Years," *Science* 270 (1995): 475–80; October 20, 1995.

12. *Ibid.,* 479.

13. *Ibid.,* 475–76.

14. *Ibid.,* 477.

15. *Ibid.,* 478. For more detailed accounts of this initial gene therapy study in the United States, see Larry Thompson, *Correcting the Code: Inventing the Genetic Cure for the Human Body* (New York: Simon & Schuster, 1994); Jeff Lyon and Peter Corner, *Altered Fates: Gene Therapy and the Retooling of Human Life* (New York: W. W. Norton, 1995). For an excellent survey of the status of somatic cell gene therapy in 1995, see Ronald G. Crystal, "Transfer of Genes to Humans: Early Lessons and Obstacles to Success," *Science* 270 (1995): 404–10; October 20, 1995.

16. Claudio Bordignon et al., "Gene Therapy in Peripheral Blood Lymphocytes and Bone Marrow for ADA-immunodeficient Patients," *Science* 270 (1995): 470–75; October 20, 1995.

17. *Ibid.,* 473–74.

18. Gina Kolata, *New York Times,* October 20, 1995, A22.

19. A few weeks before the two *Science* articles on gene therapy were published, a newborn substudy of the Blaese-Anderson protocol reported initial results in the treatment

of three newborn infants. According to the authors, the genetic modification of CD34 + cells in the umbilical cord blood of three ADA-deficient newborns had resulted in the persistence of the introduced gene in leukocytes (white blood cells) from bone marrow and peripheral blood for 18 months after the initial treatment. All three patients are also receiving PEG-ADA, although the dosages have been reduced. No clinical effects of the genetic modification were expected or observed at the time the article was published. See Donald B Kohn et al., "Engraftment of Gene Modified Umbilical Cord Blood Cells in Neonates with Adenosine Deaminase Deficiency," *Nature Medicine* 1, no. 10: 1017–23; October 1995.

20. David Baltimore, "Case Analysis 5. Genetic Engineering: The Future-Potential Uses," in National Academy of Sciences, *Research with Recombinant DNA: An Academy Forum*, March 7–9, 1977 (Washington, D.C.: The Academy, 1977), 237–40.

21. W. French Anderson et al., *Human Gene Therapy: Preclinical Data Document* submitted to Human Gene Therapy Subcommittee, Recombinant DNA Advisory Committee, April 24, 1987.

22. Robert C. King and William D. Stansfield, *A Dictionary of Genetics,* 4th ed. (New York: Oxford University Press, 1990), 276; Thomas D. Gelehrter and Francis S. Collins, *Principles of Medical Genetics* (Baltimore: Williams & Wilkins, 1990), 88.

23. King and Stansfield, *Dictionary,* 133.

24. The genomic structure described applies primarily to the type C leukemia viruses. Other retroviruses, such as the lentiviruses (HIV, for example), have a more complicated genomic structure.

25. H. von Melchner and K. Hoffken, "Retrovirus Mediated Gene Transfer into Hemopoietic Cells," *Blut* 57, no. 1: 1–5; July 1988; D. A. William, "Gene Transfer and Prospects for Somatic Gene Therapy," *Hematology/Oncology Clinics of North America* 2(2): 277–87; June 1988; Dick, *op. cit.* (note 8); and Maxine Singer and Paul Berg, *Genes and Genomes: A Changing Perspective* (Mill Valley, Calif.: University Science Books, 1991), 310–11.

26. Singer and Berg, *op. cit.*

27. Singer and Berg, *op. cit.*

28. Singer and Berg, *op. cit.*

29. Personal communication from James Barren, Genetic Therapy, Inc., May 24, 1996.

30. Oral presentation to the NIH Committee on the Investment in Human Gene Therapy, May 15, 1995.

31. Michael Waldholz, "Genetic Therapy Wins Patent for Use of Gene Treatment," *Wall Street Journal,* March 22, 1995, B6; and Teresa Riordan, "A Biotech Company Is Granted Broad Patent and Stock Jumps," *New York Times*, March 22, 1995, D1.

32. Carol Ezzell, "Gene Therapy for Rare Cholesterol Disorder," *Science News* 140, no. 15: 230; October 12, 1991.

33. Interview with James M. Wilson, October 29, 1991.

34. Interview with James M. Wilson, October 29, 1991; Mariann Grossman et al., "Successful ex vivo Gene Therapy Directed to Liver in a Patient with Familial Hypercholesterolemia," *Nature Genetics* 6, no. 4: 335–41; April 1994. See the comments on this paper in *Nature Genetics* 7, no. 3: 349–50; July 1994.

35. Grossman et al., *op. cit.*

36. Interview with James S. Wilson, October 29, 1991.

37. Interview with James S. Wilson, October 29, 1991. See also the following report from Wilson's research group: Karen F. Kozarsky et al., "In Vivo Correction of Low Density Lipoprotein Receptor Deficiency in the Watanabe Heritable Hyperlipidemic Rabbit with Recombinant Adenoviruses," *Journal of Biological Chemistry* 269, no. 18: 13695–702; May 6, 1994.

38. *Ibid.*

39. Melissa A. Rosenfeld el al., "Adenovirus-Mediated Transfer of a Recombinant Alpha 1-Antitrypsin Gene to the Lung Epithelium in Vivo," *Science* 252 (1991): 431–34; April 19, 1991; Richard C. Hubbard, "Anti Neutrophil Elastase Defenses of the Lower Respiratory Tract in Alpha 1 Antitrypsin Deficiency Directly Augmented with an Aerosol of Alpha l Antitrypsin," *Annals of Internal Medicine* 111, no. 3: 206–12; August 1, 1989; Ronald G. Crystal et al., "The Alpha 1–Antitrypsin Gene and Its Mutations: Clinical Consequences and Strategies for Therapy," *Chest* 95, no. 1 (1989): 196–208; January 1989.

40. Hubbard et al., *op. cit.*; Crystal et al., "The Alpha 1–Antitrypsin Gene."

41. Rosenfeld et al., *op. cit.*; Carol Ezzell, "Genetic Therapy: Just a Nasal Spray Away?" *Science News* 139, no. 16: 246; April 20, 1991; Ronald Kotulak, "Technique May Fight Emphysema," *Chicago Tribune*, April 19, 1991, sec. 1, p. 4; Andrew Skolnick, "Gene Replacement Therapy for Hereditary Emphysema?" *JAMA* 262, no. 18 (1989); November 10, 1989; and Barbara J. Culliton, "Endothelial Cells to the Rescue," *Science* 246 (1989): 750; November 10, 1989.

42. King and Stansfield, *op. cit.*, p. 82; Theodore Friedmann, *Gene Therapy: Fact and Fiction in Biology's New Approaches to Disease* (Cold Spring Harbor, N.Y.: Banbury Center/Cold Spring Harbor Laboratory, 1993), 126; Ezzell, "Genetic Therapy."

43. Ezzell, "Genetic Therapy"; Beverly Merz, "Gene Therapy Enters 'Second Generation,'" *American Medical News*, December 22–29, 1989, 3, 11.

44. P. L. Feigner and G. Rhodes, "Gene Therapeutics," *Nature* 349 (1991): 351–52; January 24, 1991.

45. Ezzell, "Genetic Therapy."

46. Joseph Zabner et al., "Adenovirus-Mediated Gene Transfer Transiently Corrects the Chloride Transport Defect in Nasal Epithelia of Patients with Cystic Fibrosis," *Cell* 75, no. 2: 207–16; October 22, 1993.

47. *Ibid.*

48. *Ibid.*

49. Kathy A. Fackelmann, "Glowing Evidence of Gene Altered Arteries," *Science News* 139, no. 25 (1991): 391; June 22, 1991.

50. Elizabeth G. Nabel et al., "Recombinant Gene Expression in Vivo within Endothelial Cells of the Arterial Wall," *Science* 244 (1989): 1342–44; June 16, 1989. See also Culliton, *op. cit.* (note 41), 246.

51. Jerry E. Bishop, "Michigan Researchers Are Developing Method to Place Genes in Body Tissues," *Wall Street Journal*, September 14, 1990, B4; Elizabeth G. Nabel et al., "Site-Specific Gene Expression in Vivo by Direct Gene Transfer into the Arterial Wall," *Science* 249 (1990): 1285–88; September 14, 1990.

52. Nabel el al., "Site-Specific Gene Expression," 1287.

53. Culliton, *op. cit.* (note 41), 749. See also Leon Jaroff, "Giant Step for Gene Therapy," *Time*, September 24, 1990, 74–76, chart on p. 76.

54. King and Stansfield, *op. cit.* (note 22), 46.

55. Brian E. Henderson et al., "Toward the Primary Prevention of Cancer," *Science* 254 (1991): 1131–38; November 22, 1991.

56. *Ibid.*

57. Natalie Angier, "New Gene Therapy to Fight Cancer Passes First Human Test," *New York Times*, July 18, 1991, B7; Carol Ezzell, "Scientists Seek to Fight Cancer with Cancer," *Science News* 139 (1991): 326; March 25, 1991; and Peter Gorner, "Panel OKs Gene Fight vs. Cancer," *Chicago Tribune*, August 1, 1990, sec. 1, p. 5.

58. Paul T. Golumbek et al., "Treatment of Established Renal Cancer by Tumor Cells Engineered to Secrete Interleukin," *Science* 254 (1991): 713–16; November 1, 1991; David Brown, "Cancer in Mice Is Cured by Gene Therapy," *Washington Post*, November 1, 1991, A22.

59. Ezzell, "Scientists"; Michael Waldholz, "Gene Implants Destroy Cancer in Lab Rodents," *Wall Street Journal*, November 1, 1991, B1.

Some of Rosenberg's peers suggested that this second round of cancer gene therapy experiments was premature and not grounded on solid scientific evidence. Drew Pardoll, whose research Rosenberg himself called "crucial to validating the human studies we are doing," stated that TNF was ineffective in catalyzing an enhanced immune response against cancer in animals (Waldholz, "Gene Implants"). However, even if this first attempt at stimulating patients' immune systems with gene therapy does not work, the idea remains promising and will probably be investigated using genes other than those for TNF.

60. Gary J. Nabel et al., "Immunotherapy of Malignancy by in Vivo Gene Transfer into Tumors," Human Gene Therapy Protocol, submitted October 1991.

61. Robert A. Weinberg, "Tumor Suppressor Genes," *Science* 254 (1991): 1138–46; November 22, 1991.

62. Monica Hollslein et al., "p53 Mutations in Human Cancers," *Science* 253 (1991): 49–53; July 5, 1991.

63. Weinberg, *op. cit.*

64. *Ibid.*, 1145; Ruth Sager, "Tumor Suppressor Genes: The Puzzle and the Promise," *Science* 246 (1989): 1406–12; December 15, 1989.

65. Robert Bookstein et al., "Suppression of Tumorigenicity of Human Prostate Carcinoma Cells by Replacing a Mutated RB Gene," *Science* 247 (1990): 712–15; February 9, 1990; and Beverly Merz, "Use of Anti-Oncogenes Studied," *American Medical News*, February 23, 1990, 8.

66. Arnold J. Levine et al., "The p53 Tumour Suppressor Gene," *Nature* 351 (1991): 453–56; June 6, 1991; Weinberg, *op. cit.*, 1143.

67. Hollstein et al., *op. cit.*

68. Levine et al., *op. cit.*; Weinberg et al., *op. cit.*, 1143.

69. Weinberg, *op. cit.*, 1143; Levine et al., *op. cit.*, 454.

70. Suzanne J. Baker et al., "Suppression of Human Colorectal Carcinoma Cell Growth by Wild-Type p53," *Science* 249 (1990): 912–15; August 24, 1990.

71. Michael Waldholz, "Colon Cancer Growth Is Halted in Tests by Replacing Tumor Suppressor Genes," *Wall Street Journal*, August 24, 1990, B2.

72. Edward H. Oldfield et al., "Gene Therapy for the Treatment of Brain Tumors Using Infra Tumoral Transduction with the Thymidine Kinase Gene and Intravenous Ganciclovir," *Human Gene Therapy* 4, no. 1: 39–69; February 1993.

73. Protocols 9306.044.9306–054.9306–054.9406–077, as listed in the June 1995 Data Management Report, Office of Recombinant DNA Activities, National Institutes of Health.

74. Protocols 9409–084 (Holt/Arteaga) and 9306–052 (Ilan) abstracts.

75. Protocol 9306–048 (Galpin/Casciato), non-technical and scientific abstracts (1993).

76. Philip Greenberg et al., "A Phase 1/11 Study of Cellular Adoptive Immunotherapy Using Genetically Modified CD8 + HIV-Specific T Cells for HIV-Seropositive Patients Undergoing Allogeneic Bone Marrow Transplant," Human Gene Therapy Protocol, submitted October 1991.

77. W. French Anderson, "Gene Therapy for AIDS," *Human Gene Therapy* 5, no. 2: 149–50; February 1994.

78. John Carey, "Gene Therapy: Cells That Carry Messengers of Health," *Business Week,* May 28, 1990, 74; Leon Jaroff, "Giant Step for Gene Therapy," *Time*, September 24, 1990, 76.

79. Anderson, "Gene Therapy for AIDS."

80. *Ibid.*

81. Friedmann, *Gene Therapy: Fact and Fiction* (note 42), 131–132; King and Stansfield, *op. cit.* (note 22), 143, 291, 312.

82. Anderson, "Prospects for Human Gene Therapy" (note 8), 401.

83. King and Stansfield, *op. cit.* (note 22), 143.

84. Theodore Friedmann, "Progress toward Human Gene Therapy," *Science* 244 (1989): 1275; June 16, 1989; David Suzuki and Peter Knudtson, *Genethics: The Clash between the New Genetics and Human Values* (Cambridge: Harvard University Press, 1989), 175.

85. Philippe Leboulch et al., "Mutagenesis of Retroviral Vectors Transducing Human Beta-Globin Gene and Beta-Globin Locus Control Region Derivatives Results in Stable Transmission of an Active Transcriptional Structure," *EMBO Journal* 13, no. 13 (1994): 3065–76; July 1, 1994.

86. Gina Kolata, "First Effort to Treat Muscular Dystrophy," *New York Times*, May 2, 1991, B10; Peter Gorner, "Gene Injections to Fight Muscular Dystrophy," *Chicago Tribune*, April 26, 1991, sec. 1, p. 1; Feigner and Rhodes, *op. cit.* (note 44); Gina Kolata, "Why Gene Therapy Is Considered Scary but Cell Therapy Isn't," *New York Times,* September 16, 1990, E5. See also S. B. England et al., "Very Mild Muscular Dystrophy Associated with the Deletion of 46% of Dystrophin," *Nature* 343 (1990): 180–82; January 11, 1990.

87. England et al., 180.

88. Dr. Law has encountered resistance from some parts of the medical community who believe his human experiments are premature. He has therefore severed his relationship with the Muscular Dystrophy Association and with the University of Tennessee and has established his own research entity, the Cell Therapy Research Foundation, within which he conducts his experiments after institutional review board (IRB) review.

89. Kolata, "First Effort."

90. See, for example, Arthur L. Caplan, H. Tristram Engelhardt Jr., and James J. McCartney, eds., *Concepts of Health and Disease: Interdisciplinary Perspectives* (Reading, Mass.: Addison-Wesley, 1981); and H. Tristram Engelhardt Jr. and Kevin Wm.

Wildes, "Health and Disease: IV. Philosophical Perspectives," in Warren T. Reich, ed., *Encyclopedia of Bioethics,* rev. ed. (New York: Simon & Schuster, 1995), 2:1101–6.

91. Norman Daniels, *Just Health Care* (Cambridge: Cambridge University Press, 1985), 26–32; Christopher Boorse, "Health as a Theoretical Concept," *Philosophy of Science* 44, no. 4: 542–73; December 1977; and Christopher Boorse, "On the Distinction between Disease and Illness," *Philosophy and Public Affairs* 5, no. 1: 49–68; Fall 1975.

92. Two helpful discussions of the just allocation of human gene therapy are Norman Daniels, "The Genome Project, Individuals, and Just Health Care," in Timothy F. Murphy and Marc A. Lappe, eds., *Justice and the Human Genome Project* (Berkeley: University of California Press, 1994), 110–32; and Leonard M. Fleck, "Just Genetics: A Problem Agenda," 133–52.

93. For discussion of this case, see Larry Thompson, "Healy Approves an Unproven Treatment," *Science* 259 (1993): 172; January 8, 1993; and Larry Thompson, "Should Dying Patients Receive Untested Genetic Methods?" *Science* 259 (1993): 452; January 22, 1993.

94. See, for example, Jay Katz, with Alexander Morgan Capron and Eleanor Swift Glass, *Experimentation with Human Beings: The Authority of the Investigator, Subject, Professions, and State in the Human Experimentation Process* (New York: Russell Sage Foundation, 1972).

95. See, for example, Anna C. Mastroianni, Ruth Faden, and Daniel Federman, eds., *Women and Health Research: Ethical and Legal Issues of Including Women in Clinical Studies,* 2 vols. (Washington, D.C.: National Academy Press, 1994).

96. Larry Thompson, "The First Kids with New Genes," *Time,* June 7, 1993, 50–53.

97. Leon Jaroff, "Brave New Babies," *Time,* May 31, 1993, 56–57.

98. Gina Kolata, "Novel Bypass Method: A Dose of New Genes," *New York Times,* December 13, 1994.

99. For early discussion, see Michael Hamilton, ed., *The New Genetics and the Future of Man* (Grand Rapids, Mich.: Eerdmans, 1972).

100. Eliot Marshall, "Gene Therapy's Growing Pains," *Science* 269 (1995): 1050–55; August 25, 1995. See also Gina Kolata, "In the Rush toward Gene Therapy, Some See a High Risk of Failure," *New York Times*, July 25, 1995, p. C3.

101. For further discussion of this point, see Institute of Medicine, Division of Health Promotion and Disease Prevention, *New Vaccine Development: Establishing Priorities,* vol. 1, *Diseases of Importance in the United States* (Washington, D.C.: National Academy Press, 1985), chaps. 3–4.

102. For a summary of studies being conducted in countries other than the United States, see a current issue of *Human Gene Therapy.* For a recent report on gene therapy research in the United Kingdom, see United Kingdom, Health Departments, Gene Therapy Advisory Committee, *First Annual Report: November 1993–December 1994* (London: Department of Health, January 1995).

103. Daniel Glick, "A Genetic Road Map," *Newsweek*, October 2, 1989, 46.

104. *Ibid.*

28

Human Inheritable Genetic Modifications: Assessing Scientific, Ethical, Religious, and Policy Issues

American Association for the Advancement of Science
Mark S. Frankel
Audrey R. Chapman

INTRODUCTION

This report assesses the scientific prospects for inducing controlled inheritable genetic changes in human beings and explores the ethical, religious, and social implications of developing and introducing technologies that would change the genetic inheritance of future generations. The analysis leads to a set of recommendations as to whether and how to proceed. The report is based on the deliberations of a working group of eminent scientists, ethicists, theologians, and policy analysts convened by the American Association for the Advancement of Science (AAAS) and further analysis by project staff.

Rapid breakthroughs in genetic research, spurred by the Human Genome Project, advances in molecular biology, and new reproductive technologies, have advanced our understanding of how we might approach genetic interventions as possible remedies for diseases caused by genetic disorders, particularly for those caused by abnormalities in single genes. Limitations of current medical therapies to treat diseases with a genetic component have led to efforts to develop techniques for treating diseases at the molecular level, by altering a person's cells. To date, most of the research and clinical resources related to gene therapy have been invested in developing techniques for targeting nonreproductive body cells. Somatic gene therapies designed to treat or eliminate disease are intended to affect only the individuals receiving treatment. Very recently, researchers announced credible successes in improving patient health through gene therapy, perhaps signaling that years of research are about to bear fruit.

Recent advances in animal research are also raising the possibility that we will eventually have the technical capacity to modify genes that are transmitted to future generations. This report uses the term inheritable genetic modification (IGM) to refer to any biomedical intervention that can be expected to modify the genome that a person can transfer to his or her offspring. One form of IGM would be to treat the germ or reproductive cells that develop into the egg or sperm of a developing organism and transmit its heritable characteristics. Another form of germ line therapy would be to modify the gametes (sperm and egg cells) or the cells from which they are derived. Still other technologies under development, such as the insertion of artificial chromosomes, would also introduce inheritable genetic changes.

Greater knowledge of genetics is also making it possible to contemplate genetic interventions not only to treat or eliminate diseases but also to "enhance" normal human characteristics beyond what is necessary to sustain or restore good health. Examples would be efforts to improve height or intelligence or to intervene to change certain characteristics, such as the color of one's eyes or hair. Such interventions could be attempted through either somatic modification or IGM.

What reasons do advocates give for developing and applying this technology? In theory, modifying the genes that are transmitted to future generations would have several advantages over somatic cell gene therapy. Inheritable genetic modifications offer the possibility of preventing the inheritance of some genetically based diseases within families rather than repeating somatic therapy generation after generation. Some scientists and ethicists argue that germ line intervention is medically necessary to prevent certain classes of disorders because there are situations where screening and selection procedures will not be applicable, such as when both parents have the same mutation. Because germ line intervention would influence the earliest stage of human development, it also offers the potential for preventing irreversible damage attributable to defective genes before it occurs. Over a long period of time, germ line gene modification could be used to decrease the incidence of certain inherited diseases in the human gene pool currently causing great suffering. By contrast, because somatic cell gene therapy treats only the affected individuals, it could not be used to decrease the incidence of diseases in the same way.

However, there are significant technical obstacles, as yet unresolved, to developing scientific procedures appropriate to inheritable genetic applications. Because these interventions would be transmitted to the progeny of the person treated, there would need to be compelling scientific evidence that these procedures are safe and effective; for those techniques that add foreign material, their stability across generations would need to be determined, based initially on molecular and animal studies, before proceeding with germ line interventions in humans. It is not yet possible to meet these standards. Nor is it possible to predict when we will be able to do so.

IGM also raises profound ethical, theological, and policy issues that need to be thoroughly discussed and evaluated. Efforts to modify genes that are transmitted to future generations have the potential to bring about not only a medical, but also a social revolution, for they offer us the power to mold our children in a variety of novel ways. These techniques could give us extraordinary control over biological properties and personality traits that we currently consider essential to our humanness. Even with the technical ability to proceed, we would still need to determine whether these procedures offer a theologically, socially, and ethically acceptable alternative to other technologies under development to treat genetic diseases. Do we have the wisdom, ethical commitment, and public policies necessary to apply these technologies in a manner that is equitable, just, and respectful of human dignity?

The potential magnitude of these interventions makes it very important to improve societal awareness of the technical possibilities, give careful consideration to the implications of their use, and design a process for sustained public discussion before proceeding. Informed public discussion will require an understanding of the scientific possibilities and risks, as well as the pressing moral concerns this technology raises.

The furor over the possibility of cloning human beings through the application of the somatic cell nuclear transfer technology used to clone the lamb Dolly and subsequently other mammals underscores the importance of undertaking a serious examination of the scientific, ethical, religious, and policy implications of new technologies in advance of scientific breakthroughs. As the media coverage and public reaction to the Roslin Institute's work on mammalian cloning showed, it is far more difficult to have an informed and unemotional public discussion after a scientific discovery is announced than before it becomes a reality.

Scientists and ethicists have called attention to the need for scientific and ethical discussions related to inheritable human genetic interventions for nearly thirty years. As early as 1972, a few scientists warned that prospective somatic cell gene therapy would carry a risk of inadvertently altering germ cells as well as their targeted somatic cells. In 1982, a Presidential Commission declared that "especially close scrutiny is appropriate for any procedure that would create inheritable genetic changes."

To date, however, there has been little sustained public consideration of this topic. While the science is advancing rapidly, our understanding of the ethical, religious, and policy implications has not kept pace. Typically, our society proceeds in a "reactionary mode," scrambling to match our values and policy to scientific developments. But with a scientific advance that raises profound issues related to the possibilities of modifying our genetic futures it is important to plan ahead, to decide whether and how to proceed with its development, and to give direction to this technology through rigorous analysis and public dialogue.

To facilitate public deliberations about IGM, two programs within AAAS—the Scientific Freedom, Responsibility and Law Program and the Program of Dialogue on Science, Ethics, and Religion—coorganized a two-and-a-half-year project assessing scientific, ethical, theological, and policy issues related to inheritable genetic modification (IGM). Our goal was to formulate recommendations as to what, if any, types of applications should be encouraged and what safe-guards should be instituted. Building on a forum on human germ line issues cosponsored by the two programs in September 1997, the project convened a working group of scientists, ethicists, theologians, and policy analysts to develop a series of recommendations. Much of the work was conducted in two subgroups, each of which was broadly multidisciplinary in composition. The first subgroup examined the feasibility of various kinds of human germ line applications, the risks involved, the appropriate scope and limits of germ line research and applications on human subjects, and consent issues. The second subgroup considered the social, ethical, and theological implications of IGM. The working groups met together to formulate findings and craft public policy recommendations. Members of the two working groups are identified in Appendix A.

MAJOR FINDINGS, CONCERNS, AND RECOMMENDATIONS

A majority of the project's working group members endorses the following findings, concerns, and recommendations.

Findings

- The working group concluded that IGM cannot presently be carried out safely and responsibly on humans. Current methods for somatic gene transfer are inefficient and unreliable because they involve addition of DNA to cells rather than correcting or replacing a mutated gene with a normal one. They are inappropriate for human germ line therapy because they cannot be shown to be safe and effective. A requirement for IGM, therefore, is the development of reliable gene correction or replacement techniques.
- With current gene addition technologies, iatrogenic genetic damage could occur as a result of the unintended germ line side effects of somatic cell therapy. These problems seem at least as great as the harmful genetic damage that might arise from intentional germ line transfers. Therefore, attention must also be given to the accompanying side effects of somatic cell therapies already in use or planned.
- The working group identified few scenarios where there was no alternative to IGM for couples to minimize the prospect that their offspring will have a

specific genetic disorder. The further development of somatic cell gene transfer, moreover, will offer more options for treating one's offspring.

- Guided by the theologians—mainline Protestant, Catholic, and Jewish traditions—and ethicists on the working group, the group concluded that religious and ethical evaluations of IGM will depend on the nature of the technology, its impact on human nature, the level of safety and efficacy, and whether IGM is used for therapeutic or enhancement purposes. Ethical considerations related to the social effects of IGM, particularly its implications for social justice, will play a major role in shaping the attitudes of religious communities.

- To date, the private sector has played a prominent role in the funding of somatic cell genetic research, raising questions about the influence of commercial interests on the conduct of researchers and on the scope and direction of the research. Similar questions are likely to surface if IGM research and applications go forward.

Concerns

- The ability of IGM to shape the genetic inheritance of future generations raises major ethical concerns. IGM might change attitudes toward the human person, the nature of human reproduction, and parent-child relationships. IGM could exacerbate prejudice against persons with disabilities. The introduction of IGM in a society with differential access to health care would pose significant justice issues and could introduce new, or magnify existing, inequalities.

- IGM for enhancement purposes is particularly problematic. Enhancement applications designed to produce improvements in human form or function could widen the gap between the "haves" and the "have-nots" to an unprecedented extent. Efforts to improve the inherited genome of persons might commodify human reproduction and foster attempts to have "perfect" children by "correcting" their genomes. Some types of enhancement applications might lead to the imposition of harmful conceptions of normality. The dilemma is that IGM techniques developed for therapeutic purposes are likely to be suitable for enhancement applications as well. Thus, going forward with IGM to treat disease or disability will make it difficult to avoid use of such interventions for enhancement purposes even when this use is considered ethically unacceptable.

Recommendations

- Even in advance of a decision about whether to proceed with IGM as traditionally understood as gene transfer in reproductive cells, a public body should be assigned responsibility to monitor and oversee research and developments in IGM, more broadly conceptualized as any technique aimed at modifying the genes that a person can transmit to his or her offspring. Some interventions that fall within the scope of the working group's definition of IGM are already taking place without the oversight that we believe is necessary.

- It is important to promote extensive public education and discussion to ascertain societal attitudes about proceeding with IGM and to develop a meaningful process for making decisions about the future of this technology. These efforts should be informed by an understanding of the relevant science; involve an extended discussion of the cultural, religious, and ethical concerns associated with IGM; and be as open and inclusive as possible. International consultation on these matters should also be encouraged.

- If a societal decision is made to proceed with IGM, a comprehensive oversight mechanism should be put in place with authority to regulate IGM applications in both the public and private sectors. Such a mechanism would help to promote public safety, develop guidelines for the use of IGM, ensure adequate public participation in policy decisions regarding IGM, and address concerns about commercial influence and conflicts of interest.

- Any protocol for somatic cell transfer in which inheritable modifications are reasonably foreseeable should not proceed without assessing the short- and long-term risks and without proper public oversight.

- Before IGM can proceed, there should be a means in place for assessing the short- and long-term risks and benefits of such interventions. Society must decide how much evidence of safety, efficacy, and moral acceptance will be required before allowing human clinical trials or IGM applications.

- At this time, the investment of public funds in support of the clinical development of technologies for IGM is not warranted. However, basic research should proceed in molecular and cellular biology and in animals that is relevant to the feasibility and effects of germ line modification.

- Human trials of inheritable genetic changes should not be initiated until techniques are developed that meet agreed upon standards for safety and efficacy. In the case of the addition of foreign genetic material, the precise molecular change or the changes in the altered genome should be proven with molecular certainty, probably at the sequence level, to ascertain that no other changes have occurred. Furthermore, the functional effects of the designed alteration should be characterized over multiple generations to preclude slowly developing genetic damage and the emergence of an iatrogenic genetic defect. In the case in which attempts at IGM involve precise correction of the mutant sequence and no addition of foreign material, human trials should not begin before it can be proven at the full genome sequence that only the intended genetic change, limited to only the intended site, has occurred. If it is shown at the full genome sequence level that the sequence of a functionally normal genome has been restored, there will likely be no need for multi-generation evaluation.

- The role of market forces in shaping the future of IGM research and applications should be carefully assessed to ensure that adequate attention is paid to public priorities and sensibilities.

- Existing conflict of interest guidelines governing research should be reviewed and, where appropriate, amended and vigorously enforced to address the increasing role of commercial interests in genetics research. The guidelines should specify when a financial interest in a commercial IGM venture is grounds for precluding an investigator's direct participation in a clinical trial supported by that company. They should require that investigators disclose any financial interests in the research during the informed consent process, and should prohibit researchers with a direct financial interest in a study's outcome from participating in that study's selection of patients, the informed consent process, or the direction of the study.

[handwritten margin note: existing conflict guidelines should be reviewed]

NOTE

Reprinted by permission of the American Association for the Advancement of Science.

29

Germ-Line Genetic Engineering and Moral Diversity: Moral Controversies in a Post-Christian World

H. Tristram Engelhardt Jr.

I. INTRODUCTION: HUMAN NATURE IN THE PLURAL

The prospect of germ-line genetic engineering, the ability to engineer genetic changes that can be passed on to subsequent generations, raises a wide range of moral and public policy questions.[1] One of the most provocative questions is, simply put: Are there moral reasons that can be articulated in general secular terms for accepting human nature as we find it? Or, at least in terms of general secular moral restraints, may we reshape human nature better to meet our own interests, as we define them? This question in turn raises the further question of whether human nature as it now exists has a moral standing akin to sacredness that can be understood in nonreligious terms. This essay will take as a given that it is not possible to show in general secular moral terms that human nature has a sanctity or special moral standing that should guide secular health-care policy.[2] In addition, as this essay shows, it is not possible through appeals to considerations of authorizing consent or beneficence toward others to remedy this failure to establish a sanctity or special moral standing for human nature. Absent a religious or culturally normative understanding of human nature and given the availability of germ-line genetic engineering, there is a plurality of possibilities for refashioning our nature. The unavailability of substantive secular moral constraints on germ-line genetic engineering discloses a secularly licit plurality of possibilities for human nature. The likelihood that we will be able to refashion our human nature reveals how few general secular moral constraints there are to guide us. Paradoxi-

503

cally, the more we are able to reengineer our human nature, the less guidance is available. The plurality of possible conceptions of human well-being that can be pursued through germ-line genetic engineering challenges our self-understanding as humans. Given human freedom, and in the absence of taken-for-granted religious or cultural moral constraints, the likelihood of germ-line genetic engineering opens the possibility of human nature in the plural.

In this essay, the term "human nature" identifies more than merely the incarnation of a mind in a body. "Human nature" is used to identify those particularities of mind and body that mark humans as an identifiable species of rational animals. In this sense of "human nature" and given significant germ-line genetic engineering, numerous human species may once more be identifiable within the genus *Homo*, as was the case in the past (e.g., *Homo neanderthalensis*). They could be identified as distinct not merely through an appeal to an absence of cross-fertile mating, but through reference to substantial differences that would be the result of refashioning human nature in the pursuit of alternative ideals of human well-beings. I acknowledge at the outset that the ways in which humans realize their nature are always determined in part by environment and culture, and presume as well that the genetic material available for determination by environment and culture is primarily afforded in terms that can be acted upon by germ-line genetic engineering. Significant genetic restructuring would then lead to speciation in the sense of (1) creating significantly different human forms and (2) producing infertile mating across types of humans. To talk about changing human nature is to talk about changing our current genetically based constellation of strengths and weaknesses. Of all scientific innovations, germ-line genetic engineering promises to be the most radical and intimate.

In exploring concerns raised by the prospect of germ-line genetic interventions, I do not assume that within the next few decades sufficient strides will have been made in germ-line genetic engineering so as to allow a significant reshaping of the human genome. My assumption is rather that such abilities will be realized over the next centuries. It is only within the last two hundred years that anatomy, pathology, physiology, biochemistry, microbiology, etc., have become basic sciences in the sense of being foundational to the clinical practice of medicine. My assumption is that over the next centuries the genetic sciences and technologies will emerge as cardinal basic sciences in providing a key to the traditional basic sciences. The genetic sciences and technologies will likely be used in effecting major genetic restructurings of both human anatomy and physiology.

We have the luxury of being able to reflect on the moral quandaries we will face, if such advances occur. This essay offers an exploration of one among the issues that will need to be addressed: Are there substantive or only procedural constraints on the refashioning of human nature? As we turn to making ourselves our own objects in the sense of determining how to reshape ourselves through our own technologies, can we articulate any firm moral constraints or guidelines? As we confront the most revolutionary opportunity with regard to our own nature, are we at the same time radically despoiled of firm guideposts? Or, does the very possibility of being able to reshape our nature despoil us of the possibility of securing substantive constraints and helpful moral direction? Can we discover content-full secular moral norms, or are we left with agreement and moral diversity—a likely plurality of outcomes?

II. HUMANITY, HUMANITAS, AND THE NORMATIVELY HUMAN

From Greco-Roman times, in legal codes such as the Institutes of Gaius and the Justinian Code, a notion of the normatively human has directed reflections concerning morality and public policy. When Napoleon remarked regarding Goethe, "Voilà un homme!"[3] he was not identifying species membership, but rather affirming a particular notion of human excellence—Napoleon was using "homme" against a history of celebrating the normatively human. In the Greco-Roman world, many took pains to understand the term *philanthropia* as not merely love of one's fellow man, but rather as an endorsement of those traits that are most excellently human.[4] These concerns became tied to the term *humanitas* and its cognates: Cicero helped popularize these and joined them to the phrase *studio humanitatis*.[5] In the term's various usages in *humanissime* (being very kind), *humanitas* (kindness), *humanitas* (urbanity), *sensus humanitatis* (a feeling of humanity), Cicero is drawing on a view of what distinguishes humans from other animals.

At stake is not merely a description of biological characteristics, but an affirmation of the ways in which humans deport themselves with excellence and grace. Aulus Gellius, in the mid-second century A.D., ties together these various interests in refinement, education, erudition, and the normatively human:

> Those who have spoken Latin and have used the language correctly do not give to the word *humanitas* the meaning which it is commonly thought to have, namely, what the Greeks call "philanthropia," signifying a kind of friendly spirit and good-feeling towards all men without distinction; but they gave to *humanitas* about the force of the Greek *paedeia*; that is, what we call *eruditionem institutionemque in bonas artes*, or "education and training in the liberal arts." Those who earnestly desire and seek after these are most highly humanized. For the pursuit of that kind of knowledge, and the training given by it, have been granted to man alone of all the animals, and for that reason it is termed humanitas, or "humanity."[6]

Though Gellius is manifestly concerned with the cultural and intellectual environment that will educate sensibilities and produce refinement, the Latin *humanitas* suggests that one is drawing the excellences of human refinement out of human nature, so that they can be given full expression.

A background human nature, which can be expressed in certain excellences, is presupposed by the ancients and continues in the humanities to the present. The humanists of the Renaissance understood themselves as cultivating the truly human,[7] as did theorists of the second humanism at the beginning of the nineteenth century, such as Wilhelm von Humboldt (1767–1835) and Friedrich Niethammer (1766–1848).[8] The same can be said regarding theorists of the new humanism, such as Irving Babbitt (1865–1933) and Paul E. Moore (1864–1937), or those of the third humanism, such as Horst Rudiger (1908–) and Werner Jaeger (1888–1961), writing at the end of the nineteenth and the first half of the twentieth centuries. The attempt was again to preserve an accent on those marks of excellence that distinguish humans.[9] A complex set of moral and aesthetic concerns have been expressed around a cluster of concepts associated with that which is nonnatively human.[10]

These reflections concerning the nonnatively human involve a cluster of aesthetic and moral ideals, each of which presupposes particular sensibilities, aesthetic responses, intelligences, and responses to the world. In antiquity, a unity could be given to the nonnatively human by accepting a particular cultural perspective as canonical. Thus, for the Roman world, *humanitas* was unproblematically associated with *romanitas*. *Homo humanus* was *Homo romanus*, as opposed to *Homo barbarus*. Those humans who acted *romaniter* were equivalent to those who acted *humaniter*.[11] These associated understandings were embedded in an intact set of moral practices and traditions. Stoic and Roman reflections on natural law provided a theoretical grounding in human nature of the values associated with the nonnatively human—there was a human nature, and it had value implications.[12] Final grounding and focus were provided by the Judeo-Christian appreciation of a single God creating a single nature. This appreciation was further strengthened in the Christian recognition of the Incarnation as the unique bond between human nature and the nature of God.[13] All of this fortified the expectation that the normatively human could be identified and could guide policy.

III. THE LOSS OF HUMAN NATURE AS A GIVEN FOR MORAL AND POLITICAL THEORY

Against the prospect of germ-line genetic engineering and in a post-Christian world, this expectation is brought into question. The normatively human, insofar as it depends on particularities of inclination, sensibility, and intelligence, is in principle open to being recast through the intrusions of germ-line genetic engineering, as one becomes able to engineer and direct human evolution. Indeed, the mapping of the human genome, the demonstrated feasibility of somatic cell genetic engineering (genetically based changes that cannot be passed on to one's offspring), and the dear possibility of significant germ-line genetic engineering change the relationship of humans to human nature. Human nature, which is still a relatively unalterable given, would become a cardinal point in the human refashioning of the human condition. Whether understood as the gift of God or merely as the deliverance of spontaneous mutations, random selection, genetic drift, cosmic happenstance, and biochemical constraints, human nature has until now been regarded as placing constraints on human freedom. Indeed, it has served as a point of orientation for morality, political theory, and the biomedical sciences. The character of human nature has provided a point of departure for articulating natural law, understanding human sympathies (and their limits), establishing governmental responses to the improvidence of humans, and framing human expectations regarding the roles of government. Human nature as given to us by God or produced for us by evolution has been a secure starting-point not only for arguing what could and should be done, but also for arguing what could not or should not be done. Human nature has been given a more central place than the general character of the universe, for here nature shapes the starting-point of all human undertakings.

Human nature provides bases for expectations regarding aging, suffering, disability, and death. It grounds taken-for-granted, normal, or ordinary expectations regarding human abilities and limitations. It predisposes human beings to accept certain sufferings as natural (e.g., the pain of teething and childbirth) and to view death at certain ages (e.g., in the late nineties) as being acceptable. The likely range of human abilities and limitations is determined by the interaction of humans with the environment and other organisms. Whether a particular environment or set of organisms is threatening, neutral, or congenial depends as much on the character of humans as on the character of the environment or the organisms. The sensitivity of humans to various materials in the environment makes such materials into pollutants. The circumstance that humans can contract tuberculosis, hepatitis, and AIDS makes the organisms associated with these phenomena infectious agents. As one is able to refashion human nature, one will also be able to change the ways in which the environment and other organisms are regarded. One will also change what limitations and abilities are regarded as normal or natural.

This is as much the case with respect to the complex cultural environments in which post-industrial humans live, as with regard to environments understood in more physical terms. Human institutions, laws, and expectations are in great measure framed in terms of expectations regarding human capacities to learn, to know, and to respond effectively. Policies regarding education, police, and welfare are all governed by expectations regarding the usual responses of humans. Such appeals, however, presume that human nature is a given, a source of perennial guidance and constraints. Adaptation is primarily directed toward changing the environment, not toward changing the character of humans. Everything is different if human nature, which supplied enduring content to concrete moral concerns, can itself be changed. Though there will surely be physical, chemical, and biological limits to any self-refashioning of human nature, the particular content of our human nature is now open for significant recasting by genetic engineering. Our human nature, our most intimate place in the cosmos—that in terms of which we have framed much of morality and public policy—is about to become the focus of our manipulating and transforming energies.

IV. FROM SANCTITY TO HAPPENSTANCE: THE LOSS OF ULTIMATE ORIENTATION

There are likely to be substantial cultural constraints on facing the problem of determining whether secular morality can offer firm guidelines or constraints as we move genetically to reengineer our human nature. The difficulties may lie in the remaining/unnoticed influences on secular thought of the cultural force of once widely acknowledged theological commitments.[14] Within the Judeo-Christian religions, there are substantial grounds for hesitations regarding intervening in the human genome. After all once one acknowledges that human nature is created by God, even if created through evolution, the basic design of human nature can be recognized as having a fundamental status through divine endorsement. Indeed, the basic design of human nature can be recognized as having a kind of sanctity or moral standing because as emphasized in the first chapter of Genesis, the good-ness of human nature is guaranteed by God. Moreover, the Christian recognition of the significance of the Incarnation, of God becoming man, includes a Divine affirmation and acceptance of human nature now redeemed from the fall. The nor-matively human can then be recognized in Jesus Christ, through Whom human nature is affirmed by being united to God. The Judeo-Christian heritage discloses a central metaphysical anchor in the reflections on human nature. It indicates where to look for constraints when one turns to refashioning that nature.

Within a secular culture that no longer recognizes humans as creatures of God, much less having a nature taken on by God through the Incarnation, this cosmic centrality of human nature is lost. This circumstance is more profound than the cosmic disorientation engendered by Copernicus, Giordano Bruno and Galileo Galilei, who placed the world in a cosmos in which humans were no longer the center. The post-Darwinian, post-Christian worldview removes human nature from the moral center of things. Human nature becomes the outcome of biological and chemical forces, happenstance and chance, so that its particular characteristics no longer have a claim as moral constraints on human technological powers. There may still be among the secular an agonizing sense that human nature must in some sense be sacred. However, that sense of sanctity cannot coherently be articulated without a source of holiness. A transcendent purpose or grounding for an overrid-ing value is not available, because the secular is immanent. There remains instead a cluster of vague intuitions regarding the impropriety of radically reconstructing human nature. In a post-Christian secular society, those vague intuitions can no longer be placed within the traditions and moral practices that originally gave them their force and substance.

V. SECULAR MORAL CONSTRAINTS: SOME GUIDANCE NO SPECIFIC CONTENT

The cardinal difficulty is that one cannot resolve moral controversies in morally substantive terms (i.e., other than by indicating formal contradictions) without presuming a particular moral vision. In order to know which choices maximize benefits over harms, one must already be able to compare different kinds of harms and benefits (i.e., liberty harms and benefits, equality harms and benefits, etc.). In order to maximize preference satisfaction, one must know how to compare rational versus impassioned preferences, as well as know the correct discount rate for time. In order for an appeal to a hypothetical choice theory to deliver an answer one must have already outfitted the chooser with one among the many possible moral senses or thin theories of the good. That is, if one wishes to resolve a moral controversy in substantive terms, one must already agree on basic moral premises or on an account of moral evidence (which will presuppose a particular guiding moral sense, thin theory of the good, etc.). Substantive, not merely formal, resolutions of moral controversies require substantive guidance so as to be able to choose among alternatives. The attempt rationally to discover moral content either begs the question or involves an infinite regress.

In such circumstances, by sound argument one cannot decide moral controversies about the substantive nature of justice, beneficence, or constraints on refashioning human nature without first considering important substantive issues of moral controversy. However, without agreeing to background moral premises, one can still resolve moral controversies by decision, by agreement of all involved. Such a way of resolving moral controversies will not presuppose a particular moral content or rely on a preponderant consensus. It will instead approach the resolution of controversies on the basis of the authorization of those involved. As a result, contracts and market transactions become the paradigm example of moral controversy resolution and moral authorization. The appeal to agreement becomes the one way of resolving controversies in a morally authoritative way that does not presuppose either a religious revelation or the acceptance of a particular content-full moral view. One finds by default a possibility for the resolution of moral controversies where the authority for the resolution is that of agreement and where the resolution's grammar requires for its possibility using others only with their consent. Appeal to consent is not as such valued, but is integral to a cardinal practice of moral controversy resolution. Those who enter into this practice can understand the general secular morally authoritative resolution of moral controversies without agreement regarding particular content-full moral premises. Those who do not enter will lack any such basis for common morally authoritative action and will in this sense be secular moral out-laws. The appeal to agreement is the disclosure of a secular moral possibility, a kind of transcendental possibility.[15]

As a surrogate for appeals to moral constraints grounded in our character as humans, the appeal to consent or permission offers no content-full guidance. Consent, absent the incorporation of particular notions of the good or the right, involves mere permission: bare concurrence or agreement. To require more from consent than mere agreement in the absence of coercion on the part of those one agrees with is to build into permission particular notions of benefits, harms, and proper conduct, as is the case with the various legal requirements for free and informed need only imagine that genetic engineering will be able to produce children as reliably and safely as is now possible through ordinary feral, in vivo means, so as to meet currently accepted standards of responsible reproduction.

This is not to say that one cannot morally rule out interventions that are malevolent or likely to produce outcomes which the recipients can be anticipated to abhor. Malevolence violates morality without invoking any particular moral content by being directed against morality. To produce children malevolently—that is, to produce children in circumstances that the producers hold to be improper, even if those produced would find them acceptable—is to act contrary to morality as directed to the good. In addition, judgments can be made regarding matters of beneficence, given some substantive agreement regarding the nature of the good. For example, one can envisage that there will be some avoidable outcomes of germ-line genetic engineering, which are easily worse than other outcomes, so that some alternatives can be judged to be preferable. Such comparisons of alternatives can provide guidance insofar as they exhibit a commonality sufficient to allow judgments of better and worse outcomes, preferable and less preferable alternatives.[16]

Novel moral challenges arise from the possibility of substantially different ways of refashioning human nature, which would allow the pursuit of different constellations of goods, so that one would need to compare quite different and mutually exclusive benefits. Given no morally neutral way of choosing among most substantively different visions of the good (i.e., other than the constraints set by permission and nonmalevolence), it is not possible to determine which radical reengineerings of human nature are morally impermissible. Imagine, for example, that one could genetically engineer either gills to allow breathing under water or lungs to live in a very low oxygen environment. Imagine also that one could engineer the ideal sedentary philosopher able to sit for hours reading philosophical tomes and swilling large portions of good port without any adverse health outcomes. Genetic engineering will very likely offer not just better or worse outcomes, but outcomes that realize fundamentally different biological and physiological excellences. It is this possibility of divergent excellences that offers the most novel challenge. Genetic engineering allows the pursuit of different understandings of health and well-being in the absence of substantive guidance regarding what is morally licit.

From the foregoing it does not appear possible to secure secular moral guideposts and constraints on germ-line genetic engineering beyond rather formal ones such as: (1) do not use actual persons without their consent, (2) do not act against future persons in ways one presumes they would not find acceptable, (3) do not act with malevolent intent against future persons, and (4) when commensurable goods are at stake, all else being equal, attempt to maximize the good. Some risky undertakings may violate these conditions so as clearly to be proscribable. One is left with a vast range of possibilities open to be chosen in the absence of general secular moral constraints. Faced with the prospect of being able to refashion human nature, it is not possible to find a content-full, normative, secular understanding of human nature to guide the unparalleled project of refashioning the character of human nature and of shaping evolution by genetic engineering.

Insofar as one seeks to give substantive moral guidance, one asks a question that cannot be answered in the terms in which it is posed. There are no general secular moral restraints on interventions into the human genome that can establish on principle the moral impropriety of refashioning human nature. To arrive at such restraints would require showing (1) that intervening in the human genome is in some sense in principle immoral, or (2) that the use of genetic engineering would always cause more harm than benefit. To show that germ-line genetic interventions are in principle impermissible would require establishing that the current status of the human genome has a sanctity that cannot be established in general secular terms. Secular moral concerns that do not involve moral obligations regarding permission and nonmalevolence collapse into concerns about how most prudently to maximize benefits over harms. Even to show that genetic engineering in general will cause more harm than benefit would require comparing imponderable risks of (1) possible noxious outcomes due to genetic intervention versus (2) possible noxious outcomes due to the unavailability of genetic engineering (which could, e.g., protect humans against new and highly virulent lethal viruses). Such concerns regarding imponderable outcomes at best cancel each other out; particular rankings of benefits and harms must take into account how benefits and harms will likely be perceived by those who come into existence as the result of genetic engineering. All of this depends on particular persons in particular communities with particular human natures.

VI. GENETIC ENGINEERING:
FACING THE POSSIBILITIES

The challenge is to place the enterprise of germ-line genetic engineering within a coherent moral understanding, a set of moral practices, and a moral tradition, such that particular undertakings can be seen to be appropriate and others forbidden. To recast arguments from Alasdair MacIntyre regarding the broken character of Western moral practices (i.e., the view that to resolve moral controversies coherently, one must share intact moral understandings and practices, and such common moral understandings are no longer available in the West), the difficulty is that a framework within which to understand which interventions in human nature are appropriate or inappropriate is no longer at hand, beyond requirements that one act with consent, not be malevolent, and attempt to achieve the good. Since there is no agreement about the nature of the good or the good of human nature, substantive guidance has been lost. Content-full guidance requires intact moral practices and traditions, as MacIntyre has argued.[17] In the absence of such traditions, such as those supplied by an intact religious or cultural perspective, disinclinations to engage in significant germ-line genetic engineering directed toward enhancing particular human abilities or creating new ones will at best appear to be taboos.

Faced with the challenge of justifying general secular moral constraints regarding the refashioning and re-creation of human nature, one encounters numerous understandings of what is morally significant concerning human nature and of which evolutionary goals should be invoked in guiding evolution. Confronted with numerous competing understandings of the normatively natural and of the goals appropriate to humans, one can recast a phrase from MacIntyre and ask *which* human nature, whose evolutionary goals should guide germ-line genetic engineering. As we face the possibility of reconstructing ourselves to meet our own goals, who is the *we* to define the goals? Which goals should be normative? Once the normativity of a theological context is lost, the normative and univocal character of the design of human nature is lost as well. Different moral communities with different framing moral premises and different rules of moral evidence and inference will likely endorse different goals.

VII. DISEASE, HEALTH, AND HUMAN ENHANCEMENT

Because of the deservedly bad reputation that has befallen eugenics in this century because of the moral atrocities of the National Socialists, as well as out of other considerations, there has been an attempt to draw a line between medical interventions to cure diseases and those to enhance function. If one could find firm moral grounds for only treating diseases but never enhancing human functions, then many of the difficult questions regarding how much enhancement is proper could be clearly answered. Interventions aimed at curing disease would not involve the question of the extent to which human nature may or may not appropriately be recast through human germ-line genetic interventions. It is for this reason that somatic cell and germ-line genetic engineering would be relatively unproblematic, if they were directed only toward restoring individuals to age- and sex-appropriate, species-typical levels of species-typical functions.

The difficulty is in understanding why medicine should be restricted to treating only failures to achieve age- and sex-appropriate species-typical levels of species-typical functions, since medicine regularly treats states of affairs that do not satisfy these restraining conditions.[18] Presbyopia, decay of teeth and bones, glucose intolerance after age seventy, benign prostatic hyperplasia in elderly males, and osteoporosis tied to menopause are among the long list of species-typical characteristics and levels of functions to which medicine regularly turns its therapeutic powers. Indeed, the actual practice of medicine does not appear directly concerned with what is species-typical, but rather with what involves unacceptable losses in function, grace, or anatomical form; unacceptable levels of pain or suffering; or what is appreciated as a premature death. Medicine has traditionally been concerned with ameliorating pain and suffering as well as helping individuals achieve the goals they are hindered in realizing, due to the limitations of anatomy, physiology, and psychology.[19]

This complaint-oriented character of medicine is in practice constrained by appeals to usual or ordinary levels of human function, pain, grace, and life expectancy. Such appeals serve to delegitimate complaints, restrain social expectations, or limit the use of scarce resources in terms of a set of usually implicit background understandings of what it is reasonable to seek through medical interventions. Invocations of what is "typical" are used to justify or delegitimate demands for health care. If a ninety-year-old man complains that he has trouble with some shortness of breath after running up three flights of stairs, it is not at all unreasonable to reflect that most men are dead at the age of ninety, and for him to expect the vigor of youth is unreasonable. To want the same well-being as a twenty-year-old is to expect a state of affairs that is not only not species-typical for sex and age, but also not feasible. Once it becomes feasible, the project of allowing individuals to realize species-atypical levels of function for age becomes reasonable. The medical expectations of individuals are placed within the usual and the typical. Notions of disease and illness are freighted with a complex set of subtle evaluations regarding what it is reasonable to expect in aid and treatment from others and what one should accept as one's lot in life.

Notwithstanding contrary trends in the history of medicine, with its culturally conditioned notions of disease and accounts of treatment, considerable energies have been invested in attempting to provide an account of disease and health not dependent on particular, culturally determined understandings. These efforts have in part been expended in the service of establishing an account of medicine that can transcend cultural vagaries. Such efforts have often been combined with a concern to distinguish responses to mere human whim from interventions to treat true somatic diseases.[20] With the advent of concerns with health-care cost containment and the allocation of scarce medical resources, these arguments have been directed to distinguishing between mere desires brought to medicine for fulfillment and true medical needs demanding attention.

The goal has been to find a culture-independent border between those interventions that are truly medical (in that they respond to true diseases) and those that are not truly medical (in that they respond to conditions that are not true diseases, but mere dissatisfactions with elements of the human condition). Were such a border available, one could then direct the energies of physicians and the resources of society toward addressing real medical needs and not face the prospect of a health-care system without firm boundaries regarding what it is committed to treating. A leading strategy has been to identify as diseases those states of affairs that involve deviations from species-typical levels of species-typical functions. The assumption has been that, if one could identify those medical needs which are based on the failure to realize a species-typical level of species-typical function, one would not be hostage to particular cultural understandings of the goals of medicine that might be overexpansive. One could draw a distinction between when medicine is invited to respond to mere human desires and when it is invoked to respond to true human needs.[21] One would also have guidance in the proper use of germ-line genetic engineering. One would find limits to germ-line genetic engineering that have their basis in the true character of diseases and, derivatively, in the proper goals of medicine.

The difficulty lies in the heterogeneous character of the states traditionally addressed therapeutically by medicine. Medicine responds to the absence of (1) age- and sex-appropriate levels of usual human functions; (2) acceptable grace in deportment (e.g., freedom from uncontrollable rhythmic motion); (3) acceptable anatomical form; (4) acceptable freedom from pain, suffering, and vexation; and (5) acceptable freedom from premature death. Any account of appropriateness, acceptability, or prematurity depends on what is usually expected from humans of a particular age and sex. Once one can alter that which is usual, the hope of discovering clear lines evanesces. In addition, once one recognizes the pleomorphic character of human form and function, it becomes difficult to identify particular deportment, form, levels of function, and freedom from suffering as those which are species-typical in the sense of species normative. Moreover, and crucially, insofar as there are diverse ecological niches within which humans now or in the future might wish to thrive, and insofar as there are different and indeed competing notions of human excellence and thriving, there will be numerous competing notions of how humans ought appropriately to understand health or successful adaptation.

An examination of concepts such as disease, health, and successful adaptation discloses a connection between those concepts and various moral and nonmoral values and goals. To be diseased is not to be able, because of biological or psychological hindrances, to do the things one takes to be appropriate, to live an appropriate lifespan, or to have acceptable grace in deportment. Or it is to be subject to pain or suffering considered to be inappropriate. In great measure, in the history of medicine problems have been considered medical insofar as medicine has been able to address them. To treat a problem as medical is to place it within biological and psychological causal accounts, and to expect that medicine can provide prognoses, if not therapeutic benefits. Medicine also brings special role-expectations (i.e., the sick role generally absolves one from being guilty for being ill, though one may be guilty—for having become ill, for having knowingly done things that would harm one's health; it excuses one from certain social duties and may vest one with special welfare rights; the sick role also identifies medical experts and obliges one to seek medical intervention).[22] The social practice of medicine is one directed toward prevention, treatment, cure, and care, not toward punishment, and thus stands apart from other social practices such as the law. Given established human expectations and scarce medical resources, energies are more frequently directed to those hindrances, pains, and limitations that are accepted as inappropriate for men or women of particular ages. But what appears crucial is the feasibility, the costs involved in ameliorating a limitation or burden.

This analysis undermines the possibility of discovering in the results of evolution a morally significant line between diseases and various competing, positive notions of health. Medicine as a social practice is directed toward clusters of complaints that are acknowledged as warrants for medical intervention. These, however, are dependent on particular value expectations. Medically cognizable diseases, illnesses, disabilities, impairments, disorders, and problems are states of affairs that fail to meet particular accepted ideals of function, form, deportment, grace, freedom from pain, and risk of death, and whose treatment is understood within established therapy roles.[23] Once these ideals and expectations change, the goals of medicine change.

VIII. CONCLUSION

Germ-line genetic engineering has the prospect of being the case of scientific innovation par excellence. It may well be able to change the very ways in which humans can understand the task of adapting to their environment and establishing successful social structures. It may well be able to change our concerns regarding what is an environmental threat. It has the prospect of changing the range of social problems to which governments may be moved to attend. But this is only one dimension of germ-line genetic engineering's moral significance. As significant is the possible plurality of human well-beings and ways of fashioning human nature, which it by default discloses. This plurality reveals our inability in general secular moral terms to discover which way of fashioning human nature should determine our future. It therefore presents the possibility that numerous possible future alternatives may be chosen, fracturing mankind into numerous different species of humans. By raising the possibility of substantially changing the character of human responses to the environment and other humans, germ-line genetic engineering discloses how little guidance general secular moral reflection can provide. Any attempt rationally to discover which particular moral content should be guiding or normative either begs the question or involves an infinite regress. As with most significant secular moral and public policy challenges, one finds individuals faced with the choice of creating policy in the absence of general secular, rationally discoverable content-full norms.

NOTES

Reprinted by permission of Cambridge University Press from *Social Philosophy and Policy* 13, no. 2 (1996): 47–62.

    1. A considerable literature has developed regarding the moral and public policy implications of the genome project as well as of the prospects of genetic engineering. See ELSI Bibliography, Ethical, Legal, and Social Implications of the Human Genome Project (Washington, D.C.: U.S. Department of Energy, May 1993); and ELSI Bibliography, 1994 Supplement (Washington, D.C.: U.S. Department of Energy, September 1994).

    2. H. Tristram Engelhardt Jr., "Human Nature Technologically Revisited," *Social Philosophy and Policy*, vol. 8, no. 1 (Autumn 1990): 180–91.

    3. Emil Ludwig, *Napoleon,* trans. Eden and Cedar Paul (New York: Modern Library), 322.

4. See, for example, Werner Jaeger, *Humanism and Theology* (Milwaukee: Marquette University Press, 1943), esp. 87; see also Jaeger, *Paideia: The Idea of Greek Culture*, 3 vols. (Oxford: Oxford University Press, 1945).

5. Franz Beckmann, *Humanitas* (Munster: Aschendorff, 1952).

6. *The Attic Nights of Aulus Gellus*, trans. John C. Roife (Cambridge: Harvard University Press, 1978), 2:457.

7. Paul O. Kristeller, *Renaissance Thought* (New York: Harper & Row, 1961).

8. See, e.g., Friedrich Niethammer, *Der Streit des Philanthropinismus und Humanismus* (Jena: Frommann, 1801).

9. See Norman Foerster, ed., *Humanism and America* (New York: Farrar & Rinehart, 1930); Richard Newald, *Probleme und Gestalten des deutschen Humanismus* (Berlin: Walter de Gruyter, 1963); and Horst Rudiger, *Wesen und Wandlung des Humanismus* (Hamburg: Hoffmann and Campe, 1937).

10. H. Tristram Engelhardt Jr., *Bioethics and Secular Humanism: The Search for a Common Morality* (Philadelphia: Trinity Press International, 1991).

11. Martin Heidegger, "Brief über den Humanismus," in Heidegger, *Wegmarken* (Frankfurt/Main: Vittorio Klostermann, 1976), esp. 319f.

12. Greek, Stoic, and Roman thinkers such as Chrysippus (279–206 B.C.) and Cicero (106–43 B.C.) developed the Greek distinction between positive law *(dikaion nomikon)* and natural law *(dikaion physikon)*. This distinction came to be incorporated into Roman laws such as the Institutes of Gaius (A.D. 161). See, for example, Francis de Zulueta, *The Institutes of Gaius*, 2 vols. (Oxford: Clarendon, 1976).

13. I explore the difficulties of transferring a religious notion of the sanctity of nature into a serviceable secular moral concept in "Human Nature Technologically Revisited" (supra note 2). See also Kurt Bayertz, *Sanctity of Life and Human Dignity* (Dordrecht: Kluwer, 1996).

14. The reader should be given notice: the author is an Orthodox Christian who holds that although reason does not provide canonical moral content, revelation does.

15. This argument is developed more fully in H. Tristram Engelhardt Jr., *The Foundations of Bioethics,* 2d ed. (New York: Oxford University Press, 1996), chaps. 1–4.

16. See Derek Parfit, *Reasons and Persons* (Oxford: Clarendon, 1984), esp. 371–417.

17. See Alasdair MacIntyre, *Whose Justice? Which Rationality?* (Notre Dame, Ind.: University of Notre Dame Press, 1988); MacIntyre, *After Virtue* (Notre Dame, Ind.: University of Notre Dame Press, 1981).

18. See Christopher Boorse, "Health as a Theoretical Concept," *Philosophy of Science* 44 (1977): 542–73.

19. See Engelhardt, *Foundations of Bioethics*, chap. 5.

20. See Edmund D. Pellegrino and David C. Thomasma, *A Philosophical Basis of Medical Practice* (New York: Oxford University Press, 1981); Pellegrino and Thomasma, *For the Patient's Good* (New York: Oxford University Press, 1988); and Leon Kass, "Regarding the End of Medicine and the Pursuit of Health," *Public Interest*, Summer 1975, 11–42.

21. Norman Daniels, *Just Health Care* (New York: Cambridge University Press, 1985).

22. Talcott Parsons, "Definitions of Health and Illness in the Light of American Values and Social Structure," in *Patients, Physicians, and Illness,* ed. E. G. Jaco (Glencoe, Ill.: Free Press, 1958), 165–87.

23. H. Tristram Engelhardt Jr., "Clinical Problems and the Concept of Disease," *Health Disease and Causal Explanations in Medicine,* ed. L. Nordertelt and B. L. Lindhal (Dordrecht: D. Reidel, 1984), 27–41.

VI

HUMAN CLONING AND STEM CELL RESEARCH

CLONING VERSUS TWINS

In a literal sense, cloning means making copies of one thing. We have already discussed generating copies of a piece of DNA with plasmids and bacteria. Human cloning means generating at least one copy of another individual person. In a literal sense, identical twins are genetic copies of each other or clones of each other: they physically look alike and their genetic makeup is identical. This happens when a very early embryo divides and splits into two embryos that become attached to the wall of the uterus and grow into two genetically identical individuals. They are genetically identical because they originate from one embryo containing one set of chromosomes. Since the cloning of Dolly the sheep, however, identical individuals can theoretically exist without being twins. The two cloned individuals are not generated by the splitting of one embryo; rather, a cell from one individual is used to make the second individual. In the simplest terms, the DNA from one cell is put into an unfertilized egg that has had its own DNA removed. Then the egg containing the new chromosomal DNA is tricked into behaving like a newly fertilized embryo and divides to form a fetus and eventually a newborn individual. In both identical twins and human cloning, the genetic information is obtained from only one source—either the splitting of an embryo or one cell of an individual used to generate another individual. Because the genetic information is from the same source, the two individuals are genetically identical. Obviously, the two processes—twinning and cloning an individual—are very different in other ways. Identical twins live their lives in parallel because they were born at the same time. Cloned individuals live their lives in a sequential order to each other. In the case of Dolly, her "mother" had lived her life before a cell was used to clone Dolly. Instead of a sibling relationship between twins, there is a sort of parent/offspring relationship between two cloned individuals because one is born before the other. However, the term "parent/offspring relationship" does not

517

apply to the cloning process because other animals have been cloned from cells obtained from another yet undeveloped embryo. Does this seem confusing? It is difficult to identify parents and children or brothers and sisters in a cloning sense, which is the source of much of the concern about the ethical and social aspects of cloning people.

NATURE VERSUS NURTURE IN CLONING

Social scientists have long discussed nature versus nurture. This is a convenient phrase used to ask a complex question: Are we defined completely by our genetics, by life experiences, or by a metaphysical/spiritual presence? Books have been written on the subject, so we will only discuss certain areas as they apply to cloning. The genetic sequences between two cloned individuals are essentially the same, but not completely. The sequences of DNA at the ends of chromosomes, called telomeres, vary in length between individuals and even cells within the same individual. As a cell ages, the length of the ends of the DNA is shortened. Normally this difference is not noticeable because these sequences do not code for any proteins; however, they do correlate with how many times a cell can divide—in other words, the aging of a cell. Cells that have longer telomeres continue to divide or live longer. Cells that have shorter telomeres eventually stop dividing and die. This raises interesting questions about the telomeres of cloned individuals. Will the cloned individual inherit the shorter telomeres of its parent, particularly if the parent is in advanced age? Or will the length of the telomeres be increased during the cloning process? Some data suggest that the length of the telomeres increases during cloning. At least one aspect of the aging process appears to be reversed.

Clearly, our life experiences in the context of our genetics influence the type of individual we become. We exercise choices in our lives that take us down a certain road. Perhaps a cloned person would choose to take a different road than its donor. Some have said that our genes also "learn" during life's experiences. By taking an extreme situation, we can illustrate this point. If an individual values a tan, she may choose to lie in the sun for long periods of time. Ultraviolet rays can change some of the bonds or chemistry in DNA. The abnormal changes caused by ultraviolet rays are generally fixed by a normal enzyme in the cell that functions to repair altered DNA. However, let us suppose that an individual's gene coding for this special enzyme is slightly different in its sequence so that it is not as efficient as the gene in other people. This slightly mutated enzyme can repair most altered DNA induced by normal exposure to UV light, but it may not be able to correct excessive amounts of DNA alterations caused by large exposures to UV light. Say that the individual who chooses to sunbathe has an enzyme that cannot keep up with the repair process, so the mutations start to accumulate, thus changing the sequence of the DNA slightly from what it originally was. Other cellular events eventually compound this problem of accumulating DNA damage, and cancer develops. Let us suppose this individual chooses to marry an incompatible partner, become addicted to drugs, and gamble away her money with the result that she cannot pay for cancer treatment. Her life then falls apart and she commits suicide from an unhappy life. Some might say that she had a genetic predisposition to commit suicide, but other factors played a much greater role in the final outcome than a slight mutation in a DNA repair enzyme. On the other hand, let's say her twin made different choices resulting in a healthy, happy life. You see the point. We may be genetically predisposed to certain physical features, but our choices certainly have a great influence on how genes might affect our lives. Some would argue that even our decision processes are genetically based, but that level of complexity will not be discussed here, because frankly the answers are not proven for all these issues. The point is that genetically identical people do not turn into absolutely the same individuals and our life's decisions play a profound role in the individuals we become.

METHOD OF CLONING DOLLY THE SHEEP

Much of the process for generating a cloned animal was described in part 4, but we will review the process here. First, unfertilized eggs are obtained from a female animal, a sheep in the case of Dolly (figure 22). The egg cells are at the stage of being ready for fertilization. The chromosomes of the eggs are removed by a scientist who sucks out the DNA with a needle using a microscope. This process is called "enucleation," or removing nuclear material containing the DNA. The eggs no longer have their DNA.

In the case of Dolly, adult cells from another animal served as donor cells and were transferred into the enucleated eggs. The egg cell is much larger in size than the adult cell, so transfer of the donor cell into the enucleated egg is relatively easy to do with the microsurgical equipment. The chromosomal DNA from the adult donor cell is released into the egg using electrical pulses that fuse the outer shell or membrane of the adult cell with the membrane of the egg. In this fusion process the adult cell's chromosomal DNA is released into the egg. As with the embryonic stem cells described in the previous section, the "activation process" is next. Some chemicals and electrical pulses are used to make the egg behave as though it were now fertilized. After the activation step, the egg begins to grow and divide as an embryo. The final step is to put the embryo in a pseudopregnant female animal, an animal that was treated with the appropriate hormones and bred with a vasectomized male animal so that it physiologically behaves as though it were pregnant, even though it is not impregnated with sperm as a prelude to fertilization. Instead, the activated embryos from the procedure are implanted in a pseudopregnant female animal. At this stage in the procedure, if all things were done properly and the cell physiology is working as anticipated, the embryo develops into a fetus and is delivered as a newborn animal.

The generation of Dolly was remarkable because an adult cell—an endothelial mammary gland cell—was used instead of an embryonic stem cell. This was a great surprise because dogma was that adult differentiated cells could not be "reversed" into a stem cell. Experiments similar to Dolly had previously been done using adult cells, but no animals were born. So why did the Dolly experiment work? How could the DNA from an adult cell act as an embryonic stem cell? The solution was relatively simple. When the adult cells were grown in cell culture just before being transferred into the enucleated egg, they were forced to go into a resting stage of the cell cycle when the cells were starved of certain nutrients. This was accomplished through the removal of nearly all the serum normally added to cells in culture. The serum provides many nutrients, so when the serum was removed from the cells, the cells were forced into a resting stage and could not divide and grow. What was not obvious to the scientific community is that this process "reprogrammed" the chromosomal DNA into the beginning stages of differentiation. This likely happened by differentiation proteins becoming detached or removed from the chromosomes. When the reprogrammed chromosomal DNA was released to the enucleated egg, it behaved as an undifferentiated cell, or an embryonic stem cell.

Future research will reveal the complicated details of the process of reprogramming chromosomal DNA, but the above simplistic explanation will help you evaluate the ethical issues of cloning. For example, the Dolly cloning process did not require a mother and a father. Sperm and egg cells never united. Only an adult cell was used. This is the first example of asexual reproduction of a mammal. Grave ethical considerations are associated with the potential cloning of a human being because there would be no family nucleus—no mother and father in the conventional sense. The other ethical consideration is the poor success rate experienced in cloning Dolly and other animals. Hundreds of eggs were needed to have one success. This number could be reduced with improved efficiencies, but many women would have to donate eggs, many women would need to be prepared to receive the cloned eggs, and only a small proportion would deliver a baby. The other ethical issue that has emerged comes from the observation that many embryos do not survive until delivery because of developmental problems, and those that are delivered have subtle or even profound abnormalities.

Moreover, just because one species can be cloned does not mean that another species can be cloned using the same protocols. In fact, there are differences in the protocols for cloning one species compared to another. At present, cloning of primates using adult cells has not been successful. Much experimentation has been required, and success with primates has still not been realized. How much experimentation would be required for human beings? In short, the cloning process is not failsafe, particularly when measured against the stringent standards normally applied in human clinical trials.

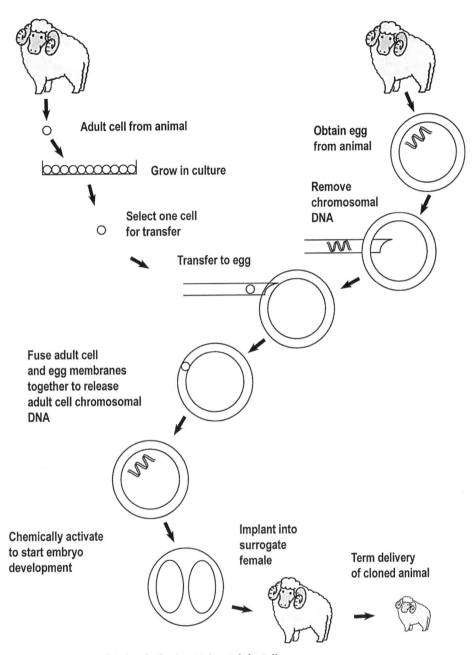

Figure 22.   Steps of Animal Cloning Using Adult Cells

In a strict biological sense, a cloned animal is not an exact copy. A difference between the cloned individual and its pseudoparent is the individual's mitochondrial DNA. There are compartments within the cell called the mitochondria where proteins are produced from DNA within the mitochondria. These proteins are not produced from the regular chromosomal DNA in the nucleus. In normal sexual reproduction, mitochondrial DNA is inherited from only one parent during sexual reproduction, and not both, as is the chromosomal DNA. In the cloning process, mitochondrial DNA could theoretically be obtained from either the adult donor cell or the enucleated egg. Upon inspection, however, it appears that the mitochondrial DNA is inherited solely from the donor egg. Consequently, the nuclear DNA is obtained from the adult donor cell and the mitochondrial DNA is from the enucleated recipient egg. Therefore, cloned animals are not strictly clones because of the presence of egg mitochondrial DNA. It is not known, however, if the different mitochondrial DNA would change the biological or chemical nature of the cloned individual.

CLONING PEOPLE

Technical questions remain about the science of cloning, particularly the cloning of people. First, would it be technically possible to clone a person? The answer may not be known unless much experimentation is done, so are we willing to ethically justify such experimentation? Would there be unacceptable mental or physical defects of the fetus or born individuals? Would the cloned individual be abnormally aged? Could all adult cell types be used for cloning, or is there a limit? What role would "nature versus nurture" play in the characteristics of an individual? What effects would the lack of a conventional mother or father have on the development of a cloned individual? What are the real justifiable advantages to cloning a person? The understanding of the scientific process as described in this section will help the reader better evaluate the ethical issues of cloning.

STEM CELL RESEARCH

Therapeutic Value

The discussion in part 5 of cloning using embryonic stem cells will help the reader understand the issues behind using stem cells for treatment of human diseases. The concept of using stem cells for therapy is based on the ability of stem cells to become different types of cells of the body, such as skin, liver, pancreas, or heart cells. Certain diseases affect the function of organs and tissues in the body. For example, diabetes results from dysfunction of pancreas cells, Parkinson's disease results from abnormal brain function, and hepatitis results from liver dysfunction.

Therefore, stem cells could be nudged chemically to become a desired cell type, which could be transplanted into the tissue of the diseased organ to supply normal functions. Alternatively, the stem cell could be transplanted to the diseased tissue where the surrounding cells would supply the chemical signals for the stem cell to differentiate into the appropriate cell type. When normal function is supplied by transfer of stem cells or by cells derived from stem cells, the symptoms of the disease might be diminished or be reversed. In this regard, it is important to understand that the person may or may not be completely cured from the disease, but a relief from the symptoms could help patients to live with less distress and have a more satisfying life.

Embryonic versus Adult Stem Cells

Stem cells could theoretically be obtained from different sources. As explained in part 5, stem cells can be obtained from early embryos (figure 23). The embryonic stem cells are extracted from the embryo and multiplied in cell culture. At the time they are to be used for therapy, they are chemically nudged to start differentiation into the cells of choice, such as liver, blood, brain, or pancreatic cells, in preparation for transplantation into the patient. The other source of stem cells is a person, not an embryo (figure 24). These stem cells are limited in their ability to become any type of cell and in the numbers of types of cells they could become. At least conceptually, a pancreatic stem cell would be obtained from the pancreas and a liver stem cell from the liver, although the author believes there will be exceptions to this rule. An organ contains many different types of cells, not just one type. The liver stem cell, therefore, would have the ability to become many of those types of liver cells. For many years scientists believed that the bone marrow, where the blood cells are made, was one of the few places in the body that adult stem cells could be obtained.

Adult stem cells have also been identified in other organs, and many believe that most organs in the human body have adult stem cells. Conceptually, stem cells could be obtained from the pancreas to treat pancreatic disease such as diabetes, or skin stem cells could be used to treat burn patients.

Most ethical issues would be avoided by using exclusively adult stem cells for therapy, but research has not progressed to the point to know if adult stem cells will be adequate for therapy. The ethical issues in stem cell research come from the use of embryonic stem cells. Obviously, pro-life and pro-choice arguments have been extended from abortion to the use of embryos for stem cell research, but there are other ethical issues to consider, such as further commercialization of human reproductive biology (e.g., ova [eggs] and embryos having market value). Parents with a diseased child may want to produce more embryos to treat the existing child. Perhaps the greatest ethical concern is to use the human cloning procedure to generate stem cells from the person under treatment for the disease. The greatest problem of tissue or organ transplantation is immunological rejec-

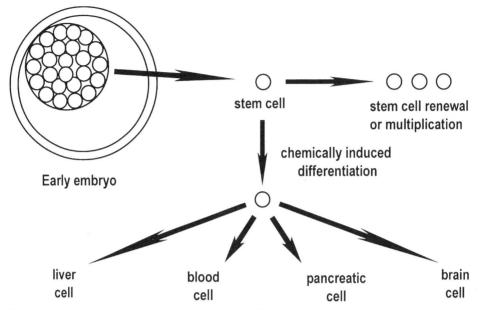

Figure 23. Embryonic Stem Cell Renewal and Differentiation into Useful Therapeutic Cells

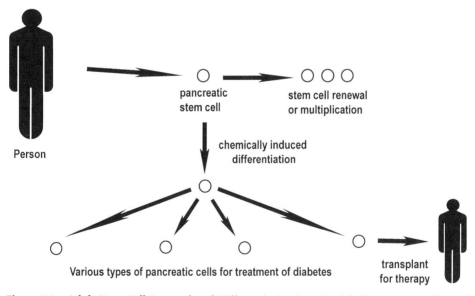

Figure 24. Adult Stem Cell Renewal and Differentiation into Useful Therapeutic Cells

tion. For example, a donor heart and the recipient person are immunologically typed or matched to identify "compatibility" so that there is less chance for rejection and failure of the transplantation. In the same way, stem cells would need to be matched with the recipient patient. Even with compatibility matching, there are risks of rejections. Immunosuppressive drugs are taken to lessen the possibility of rejection. Therefore, the ideal situation would be to use the cloning procedure to generate embryonic stem cells from the patient herself. For example, if the person is being treated for diabetes, skin cells from the patient, which are not diseased, might be used for transfer into an enucleated egg (figure 22). The resulting cloned embryos and the embryonic stem cells derived from the embryos would have been derived from the patient and would be completely compatible for transplantation back into the patient. The embryonic stem cells could then be used for transplantation into the pancreas for treatment of diabetes without having a risk of rejection by the body. Scientifically this is the ideal situation, but the ethical issues surrounding the use of embryos for stem cell research and the use of human cloning would both apply to this scientifically ideal scenario. Which is more ethical: never using embryonic stem cells for treatment or using both embryonic stem cell and cloning technologies to treat debilitating diseases such as spinal cord injury, stroke, Parkinson's disease, or heart disease?

Ethical Issues

As this is being written, a fierce debate is under way about human stem cell research. Stem cells—the beginning cells of all human physiology—can eventually become any cell in the human body. If we could harvest stem cells and promote them to differentiate into organ tissues, we could create liver or pancreatic or brain tissue for transplants. We could potentially save many lives with this treatment as a source of tissue for transplants. As of now this requires harvesting stem cells from human embryos at the earliest stage of development. For those who believe that human life begins at conception, this amounts to destruction of a human life for research purposes, to which they naturally object. Even if we knew that harvesting stem cells would save lives, this would amount to taking one life to save another or even taking one life to save several. Those who believe in the equal value of all lives and those who do not accept utility theory naturally reject this view.

Critics of stem cell research believe that there is a very promising line of research that might provide the same benefits without the use of an embryo. Pluripotent cells are the so-called stem cells of each body tissue. For example, there will be a pluripotent liver or pancreas cell that can and does differentiate into all of the cells of the liver or pancreas. Suppose we could harvest those cells from adult donors; then we could use those cells to grow needed organ tissue without destroying an embryo for research. This is the view developed in the position paper of the Coalition of Americans reprinted below. The coalition opposes the use of embryonic stem cells and rejects all public funding for research that many people deeply regard as immoral. Furthermore, they argue that using these early embryo cells is likely to be highly ineffective when compared to using pluripotent cells. Embryonic stem cells must be differentiated into many stages before they become useful tissues. Pluripotent cells may not need to go through as many stages.

Those who support stem cell research deny a premise about embryonic life that opponents accept. Two different approaches to stem cell research diverge from there. Some supporters, most notably the National Bioethics Advisory Commission, believe that the very earliest embryo, while deserving respect and nuturance, does not have the same moral standing as more fully developed embryos or infants. On this account the early embryo can be used for research purposes. For the national commission, embryos procured from abortion clinics were already aborted anyway so they could be used for stem cell research. Critics regard this as a way to promote abortion and believe that it would encourage abortion for research purposes. The national commission does not support deliberately creating embryos for research but believes that embryos already aborted could be used to harvest stem cells for research.

Notice that this view avoids the question of whether stem cell research will pay off. If it is moral to do the research, then the payoff question is assumed by a pure cost-benefit analysis. The critics argue that prior questions regarding the life of the embryo must be answered first. Supporters of the research are not swayed by that claim.

A report by the American Association for the Advancement of Science (reproduced in chapter 33) rejects the critics' position on the moral standing of the embryo. But it concedes that as a matter of public policy, stem cell harvesting should be separated from the question of abortion. It seems unlikely that embryos would be aborted just to harvest stem cells, but that does not mean we should ignore the position of the critics. Rather, we should try to accommodate the critics as much as possible while still engaging in stem cell research. For the AAAS, the compromise position is to use embryos stored in fertility clinics. In the process of making "test tube babies" embryos are created in the laboratory and then implanted into the mother's womb. Several embryos are created and kept in a storage unit until a successful pregnancy occurs. At that point, the leftover embryos are destroyed. The AAAS group believed that this was a way to create a barrier between stem cell research and abortion. The embryos would be destroyed as the natural outcome of the process of creating new life. As such, they argue, we can stand for the value of life while promoting stem cell research.

The AAAS report assumed the moral acceptability of in vitro fertilization and reimplantation, the classic test tube baby scenario. Since this involves creating embryos that will be destroyed, they argue that from the widespread approval of this procedure that there is a way of promoting stem cell research within a broadly morally acceptable framework.

30

Human Cloning: Report and Recommendations of the National Bioethics Advisory Commission

National Bioethics Advisory Commission

ETHICAL CONSIDERATIONS

The prospect of creating children through somatic cell nuclear transfer has elicited widespread concern, much of it in the form of fears about harms to the children who may be born as a result. There are concerns about possible physical harms from the manipulations of ova, nuclei, and embryos which are parts of the technology, and also about possible psychological harms, such as a diminished sense of individuality and personal autonomy. There are ethical concerns as well about a degradation of the quality of parenting and family life if parents are tempted to seek excessive control over their children's characteristics, to value children according to how well they meet overly detailed parental expectations, and to undermine the acceptance and openness that typify loving families. Virtually all people agree that the current risks of physical harm to children associated with somatic cell nuclear transplantation cloning justify a prohibition at this time on such experimentation. In addition to concerns about specific harms to children, people have frequently expressed fears that a widespread practice of such cloning would undermine important social values, such as opening the door to a form of eugenics or by tempting some to manipulate others as if they were objects instead of persons, and exceeding the moral boundaries inherent in the human condition. Arrayed against these concerns are other important social values, such as protecting personal choice, maintaining privacy and the freedom of scientific inquiry, and encouraging the possible development of new biomedical breakthroughs. As somatic cell nuclear transfer cloning could represent a means of human reproduction for some people, limitations on that choice must be made only when the societal benefits of prohibition clearly outweigh the value of maintaining the private nature of such highly personal decisions. Especially in light of some arguably compelling cases for attempting to create a child through somatic cell nuclear transfer, the ethics of policy making must strike a balance between the values we, as a society, wish to reflect and the freedom of individual choice and any liberties we propose to limit.

One of the key challenges for the Commission has been to understand many of the moral and religious objections to creating human beings using somatic cell nuclear transfer as well as to investigate and articulate the widespread intuitive disapproval of cloning human beings in this manner. This challenge included an initial attempt to examine the plausibility and persuasiveness of these objections and of the counter-arguments or especially compelling and specific cases for deploying this technology. As with the concerns offered in opposition to cloning, those offered in its defense also must be examined for their plausibility and persuasiveness. Religious perspectives were presented in the previous chapter. This chapter focuses on ethical principles not tied to any particular religious tradition, although these broad principles may be incorporated in the teachings of many religions.

The task is made quite difficult by the fact that neither moral philosophers nor religious thinkers can agree on the "best" moral theory; indeed, they often cannot even agree on the practical implications of any single theory. For example, some people base their arguments on an assessment of the particular harms and benefits that would flow to individuals and to society if somatic cell nuclear transfer techniques were to become commonplace. Others express their views by arguing about overarching rights—the child's right to individuality and dignity versus the nucleus donor's right to procreate or the scientist's right to do research. And while moral and even human rights are not necessarily understood as absolute, a choice to violate such rights requires more than a simple balancing of benefits over harms.

While some of the risks and benefits of somatic cell nuclear cloning of human beings are well enough understood to support the conclusion that it should not be permitted at this time, the difficult task of striking the balance among competing rights and interests needs more time for discussion and development. This chapter reviews some of these arguments which may serve as the starting point for a profound and sustained reflection on the significance of creating children through somatic cell nuclear transfer.

The following discussion of issues raised by such cloning begins with an important caveat. Any research or clinical experiment on creating a child in this manner would involve the creation of an embryo. That is, the fusion of a human somatic cell and an egg whose nucleus has been removed would produce a human embryo, with the apparent potential to be implanted in utero and developed to term. Ethical concerns surrounding the issues of embryo research, absent the implantation and carrying to term of an embryo, have recently received extensive analysis and deliberation in our country (National Institutes of Health, 1994). Indeed, as described in Chapter Five, federal funding for human embryo research is severely restricted, although there are few restrictions on human embryo research carried out in the private sector using non-federal funds.

The unique prospect, vividly raised by Dolly, is the creation of a new individual genetically identical to an existing (or previously existing) person—a "delayed" genetic twin. This prospect has been the source of the overwhelming public concern about such cloning. While the creation of embryos for research purposes alone always raises serious ethical questions, the use of somatic cell nuclear transfer to create embryos raises no new issues in this respect. The unique and distinctive ethical issues raised by the use of somatic cell nuclear transfer to create children relate to, for example, serious safety concerns, individuality, family integrity, and treating children as objects. Consequently, the Commission focused its attention on the use of such techniques for the purpose of creating an embryo which would then be implanted in a woman's uterus and brought to term. It also expanded its analysis of this particular issue to encompass activities in both the public and private sectors.

POTENTIAL FOR PHYSICAL HARMS

There is one basis of opposition to somatic cell nuclear transfer cloning on which almost everyone can agree. For reasons outlined in Chapter Two, there is virtually universal concern regarding the current safety of attempting to use this technique in human beings. Even if there were a compelling case in favor of creating a child in this manner, it would have to yield to one fundamental principle of both medical ethics and political philosophy—the injunction, as it is stated in the Hippocratic canon, to "first do no harm." In addition, the avoidance of physical and psychological harm was established as a standard for research in the Nuremberg Code, 1946–49.

At this time, the significant risks to the fetus and physical well-being of a child created by somatic cell nuclear transplantation cloning outweigh arguably beneficial uses of the technique.

It is important to recognize that the technique that produced Dolly the sheep was successful in only 1 of 277 attempts. If attempted in humans, it would pose the risk of hormonal manipulation in the egg donor, multiple miscarriages in the birth mother, and possibly severe developmental abnormalities in any resulting child. Clearly the burden of proof to justify such an experimental and potentially dangerous technique falls on those who would carry out the experiment. Standard practice in biomedical science and clinical care would never allow the use of a medical drug or device on a human being on the basis of such a preliminary study and without much additional animal research. Moreover, when risks are taken with an innovative therapy, the justification lies in the prospect of treating an illness in a patient, whereas, here no patient is at risk until the innovation is employed. Thus, no conscientious physician or Institutional Review Board should approve attempts to use somatic cell nuclear transfer to create a child at this time. For these reasons, prohibitions are warranted on all attempts to produce children through nuclear transfer from a somatic cell at this time.

Even on this point, however, NBAC has noted some difference of opinion. Some argue, for example, that prospective parents are already allowed to conceive, or to carry a conception to term, when there is a significant risk—or even certainty—that the child will suffer from a serious genetic disease. Even when others think such conduct is morally wrong, the parents' right to reproductive freedom takes precedence. Since many of the risks believed to be associated with somatic cell nuclear transfer may be no greater than those associated with genetic disorders, some contend that such cloning should be subject to no more restriction than other forms of reproduction (Brock, 1997).

And, as in any new and experimental clinical procedure, harms cannot be accurately determined until trials are conducted in humans. Law professor John Robertson noted before NBAC on March 13, 1997 that:

> [The] first transfer [into a uterus] of a human [embryo] clone [will occur] before we know whether it will succeed. . . . [Some have argued therefore] that the first transfers are somehow unethical . . . experimentation on the resulting child, because one does not know what is going to happen, and one is . . . possibly leading to a child who could be disabled and have developmental difficulties . . . [But the] child who would result would not have existed but for the procedure at issue, and [if] the intent there is actually to benefit that child by bringing it into being . . . [this] should be classified as experimentation for [the child's] benefit and thus it would fall within recognized exceptions. . . . We have a very different set of rules for experimentation intended to benefit [the experimental subject]. (Robertson, 1997)

But the argument that somatic cell nuclear transfer cloning experiments are "beneficial" to the resulting child rest on the notion that it is a "benefit" to be brought into the world as compared to being left unconceived and unborn. This metaphysical argument, in which one is forced to compare existence with non-existence, is problematic. Not only does it require us to compare something unknowable—non-existence—with something else, it also can lead to absurd conclusions if taken to its logical extreme. For example, it would support the argument that there is no degree of pain and suffering that cannot be inflicted on a child, provided that the alternative is never to have been conceived. Even the originator of this line of analysis rejects this conclusion.

In addition, it is true that the actual risks of physical harm to the child born through somatic cell nuclear transfer cannot be known with certainty unless and until research is conducted on human beings. It is likewise true that if we insisted on absolute guarantees of no risk before we permitted any new medical intervention to be attempted in humans, this would severely hamper if not halt completely the introduction of new therapeutic interventions, including new methods of responding to infertility. The assertion that we should regard attempt at human cloning as "experimentation for [the child's] benefit" is not persuasive.

CLONING AND INDIVIDUALITY

In addition to physical harms, many worry about psychological harms associated with such cloning. One of the forms of psychological harm most frequently mentioned is the possible loss of a sense of uniqueness. Many argue that somatic cell nuclear transfer cloning creates serious issues of identity and individuality and forces us to reconsider how we define ourselves. In his testimony before NBAC March 13, 1997, Gilbert Meilaender commented on the importance of genetic uniqueness not only for individuals but in the eyes of their parents: "Our children begin with a kind of genetic independence of us, their parents. They replicate neither their father nor their mother. That is a reminder of the independence that we must eventually grant them and for which it is our duty to prepare them. To lose even in principle this sense of the child as gift will not be good for children" (Meilaender, 1997). The concept of creating a genetic twin, although separated in time, is one aspect of somatic cell nuclear transfer cloning that most find both troubling and fascinating. The phenomenon of identical twins has intrigued human cultures across the globe, and throughout history (Schwartz, 1996). It is easy to understand why identical twins hold such fascination. Common experience demonstrates how distinctly different twins are, both in personality and in personhood. At the same time, observers cannot help but imbue identical bodies with some expectation that identical persons occupy those bodies, since body and personality remain intertwined in human intuition. With the prospect of somatic cell nuclear transfer cloning comes a scientifically inaccurate but nonetheless instinctive fear of multitudes of identical bodies, each housing personalities that are somehow less than distinct, less unique, and less autonomous than usual.

Is there a moral or human right to a unique identity, and if so would it be violated by this manner of human cloning? For such somatic cell nuclear transfer cloning to violate a right to a unique identity, the relevant sense of identity would have to be genetic identity, that is, a right to a unique unrepeated genome. Even with the same genes, two individuals—for example homozygous twins—are distinct and not identical, so what is intended must be the various properties and characteristics that make each individual qualitatively unique and different than others. Does having the same genome as another person undermine that unique qualitative identity? Along these lines of inquiry some question whether reproduction using somatic cell nuclear transfer would violate what philosopher Hans Jonas called a right to ignorance, or what philosopher Joel Feinberg called a right to an open future, or what Martha Nussbaum called the quality of "separateness" (Jonas, 1974; Feinberg, 1980; Nussbaum, 1990). Jonas argued that human cloning, in which there is a substantial time gap between the beginning of the lives of the earlier and later twin, is fundamentally different from the simultaneous beginning of the lives of homozygous twins that occur in nature. Although contemporaneous twins begin their lives with the same genetic inheritance, they also begin their lives or biographies at the same time, in ignorance of what the twin who shares the same genome will by his or her choices make of his or her life. To whatever extent one's genome determines one's future, each life begins ignorant of what that determination will be, and so remains as free to choose a future as are individuals who do not have a twin. In this line of reasoning, ignorance of the effect of one's genome on one's future is necessary for the spontaneous, free, and authentic construction of a life and self.

A later twin created by cloning, Jonas argues, knows, or at least believes he or she knows, too much about him- or herself. For there is already in the world another person, one's earlier twin, who from the same genetic starting point has made the life choices that are still in the later twin's future. It will seem that one's life has already been lived and played out by another, that one's fate is already determined, and so the later twin will lose the spontaneity of authentically creating and becoming his or her own self. One will lose the sense of human possibility in freely creating one's own future. It is tyrannical, Jonas claims, for the earlier twin to try to determine another's fate in this way.

And even if it is a mistake to believe such crude genetic determinism according to which one's genes determine one's fate, what is important for one's experience of freedom and ability to create a life for oneself is whether one thinks one's future is open and undetermined, and so still to be largely determined by one's own choices. One might try to interpret Jonas's objection so as not to assume either genetic determinism or a belief in it. A later twin might grant that he or she is not destined to follow in his or her earlier twin's footsteps, but that nevertheless the earlier twin's life would always haunt the later twin, standing as an undue influence on the latter's life, and shaping it in ways to which others' lives are not vulnerable.

In a different context, and without applying it to human cloning, Feinberg has argued for a child's right to an open future. This requires that others raising a child not close off the future possibilities that the child would otherwise have by constructing his or her own life. One way this right to an open future would be violated is to deny even a basic education to a child, and another way might be to create the child as a later twin so that he or she will believe its future has already been set by the choices made and the life lived by the earlier twin.

On the other hand, all of these concerns are not only quite speculative, but are directly related to certain specific cultural values. Someone created through the use of somatic cell nuclear transfer techniques may or may not believe that their future is relatively constrained. Indeed, they may believe the opposite. In addition, quite normal parenting usually involves many constraints on a child's behavior that children may resent. Moreover, Feinberg's argument does not apply, if the belief is false and it can be shown to be false.

Thus, a central difficulty in evaluating the implications for somatic cell nuclear transfer cloning of a right either to ignorance or to an open future is whether the right is violated merely because the later twin may be likely to believe that its future is already determined, even if that belief is clearly false and supported only by the crudest genetic determinism. Moreover, what such a twin is likely to believe will depend on the facts that emerge and what scientists and ethicists claim.

CLONING AND THE FAMILY

Among those concerns that are not focused on arguments about harm to the child are a set of worries about use of such cloning as a means of control. There are concerns, for example, about possibly generating large numbers of people whose life choices are limited by their own constrained self-image or by the constraining expectations of others. From this image of less-than-autonomous children comes the fear, however misplaced, of technology creating armies of cloned soldiers, each diminished in his or her physical individuality and thereby diminished in their psychological autonomy. Similarly, this expectation of diminished autonomy underlies the eugenic arguments that have led many to speculate about the possibility of cloning "desirable" or "evil" people, ranging from actors to dictators of various stripes to distinguished religious leaders. Complicating matters even further, this misplaced belief in the ability of genes to fully determine behavior and personality amplifies the image, so that in the end one imagines being able to make armies of either complacent workers, crazed soldiers, brilliant musicians, or beatific saints.

Although such fears are based, as noted in Chapter Two, on gross misunderstandings of human biology and psychology, they are nonetheless fears that have been voiced. In addition, these same concerns also manifest themselves in fears that underlie the characterization of somatic cell nuclear transfer cloning as a form of "making" children rather than "begetting" children. With cloning, the total genetic blueprint of the cloned individual is selected and determined by the human artisans. This, according to Kass,

> would be taking a major step into making man himself simply another one of the man made things. Human nature becomes merely the last part of nature to succumb to the technological project which turns all of nature into raw material at human disposal. . . . As with any product of our making, no matter how excellent, the artificer stands above it, not as an equal but as a superior, transcending it by his will and creative prowess. (Kass, 1997)

For many, this kind of relationship is inconsistent with an ideal of parenting, in which parents embrace not only the similarities between themselves and their children but also the differences, and in which they accept not only the developments they sought to bring about through care and teaching but also the serendipitous developments they never planned for or anticipated (Rothenberg, 1997).

Of course, parents already exercise great control over their offspring, through means as varied as contraception to control of the timing and spacing of births, to genetic screening and use of donor gametes to avoid genetic disorders, to organized medical and educational interventions to guide physical and intellectual development. These interventions exist along a spectrum of control over development. Somatic cell nuclear transfer cloning, some fear, offers the possibility of virtually complete control over one important aspect of a child's development, his or her genome, and it is the completeness of this control, even if only over this partial aspect of human development, that is alarming to many people and invokes images of manufacturing children according to specification. The lack of acceptance this implies for children who fail to develop according to expectations, and the dominance it introduces into the parent-child relationship, are viewed by many as fundamentally at odds with the acceptance, unconditional love, and openness characteristic of good parenting. Meilaender addressed both the mystery of reproduction and fears about it veering toward a means of production in his testimony before NBAC:

> But whatever we say of [other reproductive technologies], surely human cloning would be a new and decisive turn on this road. Far more emphatically a kind of production. Far less a surrender to the mystery of the genetic lottery which is the mystery of the child who replicates neither Father nor Mother but incarnates their union. Far more an understanding of the child as a product of human will. (Meilaender, 1997)

Questions are raised, as well, about the effect such interventions will have on a particular child. Will the child himself or herself feel less independent from the nucleus donor than a child ordinarily would from a parent? Will the knowledge of how one's genetic profile developed in another person at another time leave the child feeling that his character is as predetermined as his eye or hair color? Even if the child feels completely independent of the nucleus donor, will others regard the child as a copy or a successor to that donor? If so, will such expectations on the part of others warp the child's emerging self understanding?

Finally, some critics of such cloning are concerned that the legal or social status of the child arising from nuclear transfer of somatic cells may be uncertain. For some, the disparity between the child's genetic and social identity threatens the stability of the family. Is the child who results from somatic cell nuclear transfer the sibling or the child of its parents? The child or the grandchild of its grandparents? From this perspective the child's psychological and social well-being may be in doubt or even endangered. Ambiguity over parental roles may undermine the child's sense of identity. It may be harder for a child to achieve independence from a parent who is also his or her twin.

At the same time, others are not persuaded by such objections. Children born through assisted reproductive technologies may also have complicated relationships to genetic, gestational, and rearing parents. Skeptics of this point of view note that there is no evidence that confusion over family roles has harmed children born through assisted reproductive technologies, although the subject has not been carefully studied.

POTENTIAL HARMS TO
IMPORTANT SOCIAL VALUES

Those with grave reservations about somatic cell nuclear transfer cloning ask us to imagine a world in which cloning human beings via somatic cell nuclear transfer were permitted and widely practiced. What kind of people, parents, and children would we become in such a world? Opponents fear that such cloning to create children may disrupt the interconnected web of social values, practices, and institutions that support the healthy growth of children. The use of such cloning techniques might encourage the undesirable attitude that children are to be valued according to how closely they meet parental expectations, rather than loved for their own sake. In this way of looking at families and parenting, certain values are at the heart of those relationships, values such as love, nurturing, loyalty, and steadfastness. In contrast, a world in which such cloning was widely practiced would give, the critics claim, implicit approval to vanity, narcissism, and avarice. To these critics, changes that undermine those deeply prized values should be avoided if possible. At a minimum, such undesirable changes should not be fostered by public policies.

On the other hand, others are not persuaded by these objections. First, many social observers point out that if strongly held moral values are in decline, there are likely many complex reasons for this, which would not be addressed by a ban on cloning in this fashion. Furthermore, skeptics argue that people can, and do, adapt in socially redeeming ways to new technologies. In their view, a child born through somatic cell nuclear transfer could be loved and accepted like any other child, and not disrupt important family and kinship relations.

The strength of public reaction, however, reflects a deep concern that somehow many important social values could be harmed in a society where such cloning was widely used. In his testimony before the Commission on March 13, 1997, bioethicist Leon Kass summarized many of the widely held concerns regarding the possibility of cloning human beings via somatic cell nuclear transfer when he noted:

Almost no one sees any compelling reason for human cloning. Almost everyone anticipates its possible misuses and abuses. Many feel oppressed by the sense that there is nothing we can do to prevent it from happening and this makes the prospect seem all the more revolting. Revulsion is surely not an argument. . . . But . . . in crucial cases repugnance is often the emotional bearer of deep wisdom beyond reason's power fully to articulate it. (Kass, 1997)

But some people, however, argue against relying on moral intuition to set public policy. While it is certainly true that repugnance may be the bearer of wisdom, it may also be the bearer of simple and thoughtless prejudice. In her testimony before NBAC on March 14, 1997, bioethicist Ruth Macklin challenged the inclination to take as axiomatic the proposition that to be born as a result of using these techniques is to be harmed or at least to be wronged:

Intuition has never been a reliable epistemological method, especially since people notoriously disagree in their moral intuitions. . . . If objectors to cloning can identify no greater harm than a supposed affront to the dignity of the human species, that is a flimsy basis on which to erect barriers to scientific research and its applications. (Macklin, 1997)

Nevertheless, opponents assert that this new type of cloning tempts human beings to transgress moral boundaries and to grasp for powers that are properly outside human control. Ancient Greek literature and many Biblical interpretations emphasize that human beings occupy a moral position between other forms of life and the divine. In particular, humans should not consider themselves as omnipotent over nature. From this perspective, respecting limits is to respect the appropriate place of humankind in the universe and to ensure that technology is not allowed to push aside critical social and moral commitments. This view need not be tied to a single religious doctrine, a particular view of God, or even a belief in God. However, these objections are often expressed in religious terms. For example, critics talk of how the ability to create children through somatic cell nuclear transfer may tempt us to seek immortality, to usurp the role of God, or to violate divine commands.

On the other hand, some observers do not see this type of cloning as dramatically new or extreme, especially when compared to other assisted reproductive technologies. Robertson notes:

In an important sense cloning is not the most radical thing on the horizon. Much more significant, I think, would be the ability to actually alter or manipulate the genome of offspring. Cloning takes a genome as it is . . . and might replicate it. . . . [T]hat is much less ominous than having an ability to take a given genome and either add or take out a gene which could then lead to a child being born with characteristics other than it would have had with the genome it started with. (Robertson, 1997)

Finally, critics have also raised questions about an inappropriate use of scarce resources. The generation of children through somatic cell nuclear transfer would divert scarce resources, including the skills of researchers and clinicians, from more pressing social and medical needs. These considerations about allocation of resources are particularly pertinent if public funds would be involved. In the words of theologian Nancy Duff:

> When considering research into human cloning we must look at the responsible use of limited resources. . . . [I]t is mandatory to ask whether other research projects will serve a greater number of people than research on human cloning and take the answer to that seriously. (Duff, testimony, 1997)

TREATING PEOPLE AS OBJECTS

Some opponents of somatic cell nuclear cloning fear that the resulting children will be treated as objects rather than as persons. This concern often underlies discussions of whether such cloning amounts to "making" rather than "begetting" children, or whether the child who is created in this manner will be viewed as less than a fully independent moral agent. In sum, will being cloned from the somatic cell of an existing person result in the child being regarded as less of a person whose humanity and dignity would not be fully respected? One reason this discussion can be hard to capture and to articulate is that certain terms, such as "person," are used differently by different people.[3] What is common to these various views, however, is a shared understanding that being a "person" is different from being the manipulated "object" of other people's desires and expectations. Writes legal scholar Margaret Radin,

> The person is a subject, a moral agent, autonomous and self-governing. An object is a non-person, not treated as a self-governing moral agent. . . . [By] "objectification of persons," we mean, roughly, "what Kant would not want us to do."

That is, to objectify a person is to act towards the person without regard for his or her own desires or well-being, as a thing to be valued according to externally imposed standards, and to control the person rather than to engage her or him in a mutually respectful relationship. Objectification, quite simply, is treating the child as an object—a creature less deserving of respect for his or her moral agency. Commodification is sometimes distinguished from objectification and concerns treating persons as commodities, including treating them as a thing that can be exchanged, bought or sold in the marketplace. To those who view the intentional choice by another of one's genetic makeup as a form of manipulation by others, somatic cell nuclear transfer cloning represents a form of objectification or commodification of the child.

Some may deny that objectification is any more a danger in somatic cell nuclear transfer cloning than in current practices such as genetic screening or, in the future perhaps, gene therapy. These procedures aim either to avoid having a child with a particular condition, or to compensate for a genetic abnormality. But to the extent that the technology is used to benefit the child by, for example, allowing early preventive measures with phenylketonuria, no objectification of the child takes place.

When such cloning is undertaken not for any purported benefit of the child himself or herself, but rather to satisfy the vanity of the nucleus donor, or even to serve the need of someone else, such as a dying child in need of a bone marrow donor, then some would argue that it goes yet another step toward diminishing the personhood of the child created in this fashion. The final insult, opponents argue, would come if the child created through somatic cell nuclear transfer is regarded as somehow less than fully equal to the other human beings, due to his or her diminished physical uniqueness and the diminished mystery surrounding some aspects of his or her future, physical development.

EUGENIC CONCERNS

The desire to improve on nature is as old as humankind. It has been played out in agriculture through the breeding of special strains of domesticated animals and plants. With the development of the field of genetics over the past 100 years came the hope that the selection of advantageous inherited characteristics—called eugenics, from the Greek *eugenes* meaning wellborn or noble in heredity—could be as beneficial to humankind as selective breeding in agriculture.

The transfer of directed breeding practices from plants and animals to human beings is inherently problematic, however. To begin, eugenic proposals require that several dubious and offensive assumptions be made. First, that most, if not all, people would mold their reproductive behavior to the eugenic plan; in a country that values reproductive freedom, this outcome would be unlikely absent compulsion. Second, that means exist for deciding which human traits and characteristics would be favored, an enterprise that rests on notions of selective human superiority that have long been linked with racist ideology.

Equally important, the whole enterprise of "improving" humankind by eugenic programs oversimplifies the role of genes in determining human traits and characteristics. Little is known about correlation between genes and the sorts of complex, behavioral characteristics that are associated with successful and rewarding human lives; moreover, what little is known indicates that most such characteristics result from complicated interactions among a number of genes and the environment. While cows can be bred to produce more milk and sheep to have softer fleece, the idea of breeding humans to be superior would belong in the realm of science fiction even if one could conceive how to establish the metric of superiority, something that turns not only on the values and prejudices of those who construct the metric but also on the sort of a world they predict these specially bred persons would face.

Nonetheless, at the beginning of this century eugenic ideas were championed by scientific and political leaders and were very popular with the American public. It was not until they were practiced in such a grotesque fashion in Nazi Germany that their danger became apparent. Despite this sordid history and the very real limitations in what genetic selection could be expected to yield, the lure of "improvement" remains very real in the minds of some people. In some ways, creating people through somatic cell nuclear transfer offers eugenicists a much more powerful tool than any before. In selective breeding programs, such as the "germinal choice" method urged by the geneticist H. J. Muller a generation ago (Kevles, 1995), the outcome depended on the usual "genetic lottery" that occurs each time a sperm fertilizes an egg, fusing their individual genetic heritages into a new individual. Cloning, by contrast, would allow the selection of a desired genetic prototype which would be replicated in each of the "offspring," at least on the level of the genetic material in the cell nucleus.

It might be enough to object to the institution of a program of human eugenic cloning—even a voluntary program—that it would rest on false scientific premises and hence be wasteful and misguided. But that argument might not be sufficient to deter those people who want to push the genetic traits of a population in a particular direction. While acknowledging that a particular set of genes can be expressed in a variety of ways and therefore that cloning (or any other form of eugenic selection) does not guarantee a particular phenotypic manifestation of the genes, they might still argue that certain genes provide a better starting point for the next generation than other genes.

The answer to any who would propose to exploit the science of cloning in this way is that the moral problems with a program of human eugenics go far beyond practical objections of infeasibility. Some objections are those that have already been discussed in connection with the possible desire of individuals to use somatic cell nuclear transfer, that the creation of a child under such circumstances could result in the child being objectified, could seriously undermine the value that ought to attach to each individual as an end in themselves, and could foster inappropriate efforts to control the course of the child's life according to expectations based on the life of the person who was cloned.

In addition to such objections are those that arise specifically because what is at issue in eugenics is more than just an individual act, it is a collective program. Individual acts may be undertaken for singular and often unknown or even unknowable reasons, whereas a eugenics program would propagate dogma about the sorts of people who are desirable and those who are dispensable. That is a path that humanity has tread before, to its everlasting shame. And it is a path to whose return the science of cloning should never be allowed to give even the slightest support.

ARGUMENTS FOR MAINTAINING PERSONAL AUTONOMY AND FREEDOM OF INQUIRY

Arrayed against these concerns about the societal effects of cloning human beings via somatic cell nuclear transfer are arguments for maintaining individual choice over whether to use the technology. These arguments are made on five separate grounds: first, that there is a general presumption in favor of individual liberty; second, that certain actions, such as human reproduction, are particularly personal and should remain free of constraint; third, as a society we ought not limit the freedom of scientific inquiry; fourth, that there are some reasons to create a child through somatic cell nuclear transfer so compelling they should transcend objections to the practice even if it should otherwise be prohibited; and finally, that many of the objections to the use of this technique are largely speculative and unproven.

PRESUMPTIONS IN FAVOR OF PERSONAL AUTONOMY

The presumption in favor of individual liberty stems from a consensus within the United States that one of the most important values we share is a commitment to personal autonomy. In part, this commitment is maintained because of the widespread fear that one's own personal choices might be constrained if subject to collective decision making. To the extent that making a personal choice is a form of personal satisfaction, then the means to maximize our collective satisfaction is to make as many personal choices available as possible (Posner, 1992). In addition, personal autonomy is considered valuable in and of itself, since it is viewed by many as the deepest expression of one's individuality and personality, i.e., the deepest expression of one's self. Thus, commentators have argued that a commitment to individual liberty requires that individuals be left free to create children using somatic cell nuclear transfer if they so choose and if their doing so does not cause significant harms to others (Robertson, 1997; Macklin, 1997).

But such liberty is too broad in scope to be an uncontroversial moral right (Mill, 1859; Rhodes, 1995). As many others have pointed out, granting such untethered primacy to autonomy can ignore the possibility of competing values that are held as dear in some or all circumstances. Thus, principles of equality, virtue, nonmaleficence, and benevolence may compete for primacy with the principle of autonomy. In her March 13, 1997, testimony before NBAC, theologian Lisa Cahill asserted that

> an excessive focus on [autonomy] can prevent us from seeing why other values as well are socially important and protectable and why certain freely chosen practices can still be wrong even if they do not result in immediate or quantifiable harm or

direct infringement on the options of other free agents. . . . A narrow focus on autonomy to freely choose personally preferred goals undermines our ability to talk together about what would go to make up a good society and what we can do concretely to move towards one.

Indeed, some analysts, such as legal scholar Mary Ann Glendon (1991) and sociologist Amitai Etzioni (1990), have argued that the rhetoric of rights and personal autonomy has obscured the correlative values of responsibility, duty, and restraint. And, indeed, while personal autonomy is upheld rhetorically as an ideal, it is often also constrained on behalf of the common good, even in the absence of harm to others, both in personal and public life. This still leaves open, however, the question of when, in particular, other values ought to trump the value of personal liberty.

In their book *Democracy and Disagreement* (1996) political theorists Amy Gutmann and Dennis Thompson set forth some guidelines for when moral arguments ought to be allowed to constrain personal liberty. Among them are: (1) a convincing argument that a particular action is wrong, independent of whatever specific harms it might cause, because it violates, for example, natural law, social convention, or fundamental social values; (2) that the wrong is serious enough to warrant public attention and is otherwise eligible for public regulation; and (3) that regulation or prohibition will not cause more harm than the action that opponents seek to prohibit.

FREEDOM OF REPRODUCTIVE CHOICE

While the discussion of social values, above, might satisfy the first two conditions set down by Gutmann and Thompson, the third condition requires more attention in this case. To determine whether prohibition of somatic cell nuclear transfer cloning would cause more harm than it prevents, one must examine the particular kind of choices that would be constrained. Certain actions, it is argued, deserve special protection from collective decision making, and human reproduction is often cited as an example. Reproduction is an intensely personal phenomenon, most often commencing in the intimacy of coitus, and always resulting in the creation of a parental relationship that redefines one's place in the world. Without reproduction, one remains a child and perhaps a sibling. With reproduction or with its social equivalent, adoption, one becomes a parent, taking on responsibilities for another that necessarily require abandoning some of the personal freedoms enjoyed before. When and how to take on such responsibilities and to change one's life course is necessarily one of the most personal and significant decisions imaginable.

It could be argued that somatic cell nuclear transfer cloning is not covered by the right to reproductive freedom, because whereas assisted reproductive technologies covered by that right are remedies for inabilities to reproduce sexually, somatic cell nuclear transfer cloning is an entirely new means of reproduction; indeed, its critics see it as radically new and as more a means of the mere "manufacturing of humans" than of reproduction. Its asexual nature, for example, leads some to view it as distinctly different from reproduction, which they view as inherently collaborative and sexual. This led one commentator to note that:

> It would be possible for female lineages to proceed without any male contribution at all and it would be possible for one woman to create her own child using her own ovum and DNA. . . . So the child who is truly the child of a single parent would be a genuine revolution in human history and her or his advent should be viewed with immense caution. (Cahill, 1997)

On the other hand, while somatic cell nuclear transfer cloning is a different means of reproduction than sexual reproduction, it is nonetheless a means that can serve individuals' interest in reproducing. If it is not covered by the moral right to reproductive freedom, some argue, that must be not because it is a new means of reproducing, but instead because it has other objectionable moral features, such as eroding human dignity or uniqueness.

Assuming for the sake of discussion that somatic cell nuclear transfer cloning is a form of reproduction, the question remains whether reproductive freedom ought to protect its use. Reproductive freedom includes not only the familiar right to choose not to reproduce, for example by means of contraception, but also the right to reproduce. It is commonly understood to include the use of various artificial reproductive technologies, such as *in vitro* fertilization, and sperm or egg donation. But the case for permitting the use of a particular means of reproduction is strongest when that means is necessary for particular individuals to be able to procreate at all.

It is possible that somatic cell nuclear transfer cloning could be the only technique for individuals to create a genetically related child, but in other cases different means of procreating would also be possible. When individuals have alternative means of procreating, cloning might be chosen because it replicates a particular individual's genome. The reproductive interest in question then is not simply reproduction itself, but a more specific interest in choosing what kind of children to have.

However, the more a reproductive choice is not simply the determination of one's own life but the determination of the nature of another, as in the case of cloning via somatic cell nuclear transfer, the more the interests of that other person—that is, the resulting child—should carry moral weight in decisions that determine its nature (Annas, 1994). In addition to the parents and child, reproduction is also a communal phenomenon. It thrusts a new person into the world, and the whole community has obligations for this new member's well-being.

Thus, the decision to reproduce is rife with consequence both to the new person brought into being and to those who will live and interact with that new person. Naturally, this invites communal commentary on the wisdom of when and how this person is brought into being. And while constitutional law has viewed certain aspects of reproductive choice as fundamental rights, discourse is not so constrained. Thus, one is free to argue, as a matter of ethics, that reproductive choices ought to be made in light of communal values, even while accepting that there are administrative and political reasons for avoiding efforts to embody these moral judgements in the form of laws, whose enforcement would intrude the state into the private realm of family life and conjugal relations to an unacceptable degree.

FREEDOM OF SCIENTIFIC INQUIRY

Another argument made against prohibiting efforts to attempt to create a child through somatic cell nuclear transfer focuses on the need to encourage research and scientific advances. There is no doubt that the freedom of the ethical and responsible pursuit of knowledge has been an enduring American value, supported by scientists and non-scientists alike. Historically. scientific inquiry has been protected and even encouraged because of the great social benefit the public recognizes in maintaining the "sanctity of knowledge and the value of intellectual freedom. But the importance we attach to free scientific inquiry does not mean the pursuit of science without moral constraints. International statements about the ethics of research with human subjects, such as the Nuremberg Code and the Declaration of Helsinki, make it abundantly clear that science, however valuable, must, as scientists and non-scientists agree, observe important moral boundaries. Scientific research, for example, must not endanger community safety or the rights or interests of its human subjects. Likewise, it must not inflict unnecessary suffering on animals.

Thus, both the federal government and the states already regulate the researcher's methods in order to protect the rights of research subjects and community safety. Research may be restricted, for example, to protect the subject's autonomy by requiring informed consent, and by reviewing the choice of who should serve as research subjects against principles of justice. Thus, for example, if the government can show that restrictions on cloning and cloning technology are sufficiently important to the general well-being of individuals or society, such restrictions are likely to be upheld as legitimate, constitutional governmental actions, even if scientists were held to have a First Amendment right of scientific inquiry (Robertson, 1977).

Therefore, even if scientific inquiry were found to be a constitutionally protected activity, the government could regulate to protect against compelling harms, such as the current physical risks posed by the prospective use of somatic cell nuclear transfer techniques to create children. The freedom to pursue knowledge is distinguishable from the right to choose the method for achieving that knowledge, since the method itself may permissibly be regulated. Although the government may not prohibit research in an attempt to prevent the development of new knowledge, it may and should restrict or prohibit the means used by researchers if they involve sufficient harm to others (Robertson, 1977). Ultimately, researchers themselves are responsible for maintaining ethical and scientific standards and must strive to integrate the two in their work.

CONSIDERATION OF EXCEPTIONAL CASES

Even as a matter of ethics, rather than of law, it is quite possible to argue against a wholesale condemnation of somatic cell nuclear transfer cloning of human beings. Some circumstances have been identified in which the choice to create a child in this manner would be understandable, or even, as some have argued, desirable. Consider the following examples:

> A couple wishes to have children, but both adults are earners of a lethal recessive gene. Rather than risk the one in four chance of conceiving a child who will suffer a short and painful existence, the couple considers the alternatives: to forgo rearing children; to adopt; to use prenatal diagnosis and selective abortion; to use donor gametes free of the recessive trait; or to use the cells of one of the adults and attempt to clone a child. To avoid donor gametes and selective abortion, while maintaining a genetic tie to their child, they opt for cloning.

> A family is in a terrible accident. The father is killed, and the only child, an infant, is dying. The mother decides to use some cells from the dying infant in an attempt to use somatic cell nuclear transfer to create a new child. It is the only way she can raise a child who is the biological offspring of her late husband.

> The parents of a terminally ill child are told that only a bone marrow transplant can save the child's life. With no other donor available, the parents attempt to clone a human being from the cells of the dying child. If successful, the new child will be a perfect match for bone marrow transplant, and can be used as a donor without significant risk or discomfort. The net result: two healthy children, loved by their parents, who happen to be identical twins of different ages.

In each of these examples, the impulse to attempt such cloning can be understood. In the first example, the possible complications caused by having a child who is genetically identical to one of the parents is weighed against the value of avoiding selective abortion or of keeping the marital relationship free of the ghost of an anonymous sperm or egg donor. In the second, the psychological complexities of bearing a "replacement" child are weighed against the grief of losing not only a husband but also the possibility of a child who will grow up as a physical reminder of that love. While some may argue that neither case is compelling, because infertility and grief are part of human existence, the intensely personal nature of that infertility or grief argues for an equally personal decision about how to respond. The third case makes what is probably the strongest possible case for cloning a human being, as it demonstrates how this technology could be used for lifesaving purposes. Indeed, the tragedy of allowing the sick child to die because of a moral or political objection to such cloning overall merely points up the difficulty of making policy in this area.

Some would argue that what is more important in these scenarios is how the resulting child will be viewed. Macklin argues that:

> The ethics of these situations must be judged by the way in which the parents nurture and rear the resulting child and whether they bestow the same love and affection on a child brought into existence by a technique of assisted reproduction as they would on a child born in the usual way. (Macklin, 1997)

It may be that a policy which prohibited the creation of children though somatic cell nuclear transfer cloning would ban a handful of scenarios for which some people feel sympathy. Nonetheless, it may be necessary to forbid the practice overall in order to protect other crucial societal values.

MORAL REASONING AND PUBLIC POLICIES

> It is certainly possible that there may be no substantial benefits to society that would result if human cloning were to become a reality. Yet this would constitute a good argument for prohibition only if considerable harms are a likely consequence. We need a realistic portrait, not a recitation of worst case science fiction scenarios before we may conclude that the harms of allowing cloning to proceed in a research context and even beyond are so great that even with regulations and oversight consummate evil will result. (Ruth Macklin, 1997)

> We should proceed with research into human cloning only if compelling arguments can be made for its potential benefits. (Nancy Duff, 1997)

Some citizens may be persuaded that the harms and wrongs described in this chapter are ethically compelling and might be decisive reasons never to permit cloning via somatic cell nuclear transfer. Others may be less certain about the significance of the objections, and unwilling to conclude that somatic cell nuclear transfer cloning would be ethically impermissible, if and when the risks could be shown to be minimal. This range of views is reflected in the testimony, letters, and commissioned papers reviewed by NBAC, and is also characteristic of the commissioners themselves.

NBAC was asked to consider whether public policy should permit, regulate, or prohibit the creation of children through somatic cell nuclear transfer. The formation of public policy in an area as sensitive as procreation requires careful thought and measured deliberation. In the United States, governmental policies that prohibit or regulate human actions require justification because of a general presumption against governmental interference in individual activities. This presumption can be rebutted under various circumstances for a variety of reasons. Many critics of cloning via somatic cell nuclear transfer are concerned, however, that this initial presumption of no interference with individual actions will lead to unwise policies.

Some considerations carry more weight in the public policy arena than they do in the formation of individual judgments. In setting public policy, for example, pragmatic and procedural considerations often, quite appropriately, carry greater weight than in deciding private choices. One reason for this is that the burden of enforcing public policies must be considered. For example, it is extremely intrusive to monitor reproductive decisions by individuals and couples. It may be impractical to have a policy that allows some cases of somatic cell nuclear transfer to create a child, while prohibiting others, even though we make such judgments privately about individual actions. Furthermore, trying to distinguish acceptable from unacceptable reasons will be difficult. People might be led to misrepresent their true reasons in order to fit whatever is deemed "acceptable."

Moreover, the reasoning used to evaluate the desirability of proposed public policies regarding the creation of children through somatic cell nuclear transfer differs somewhat from the reasoning employed in making private decisions. When individuals make judgments they may rely on many sources of wisdom and knowledge, including their religious faith and moral intuitions. People will use their understanding of morality to decide what they should and should not do, as well as to judge the actions of others.

Those engaged in moral discourse about public policy, however, must move beyond such personal considerations, however deeply felt, and develop coherent arguments that will persuade many others to accept a particular point of view. As a result, it is useful to formulate moral convictions in ways that most people can understand and reflect upon. In a pluralistic society, there is no easy way to determine when and which governmental interventions are warranted. No algorithm clearly indicates whether the arguments for governmental intervention in a particular situation are stronger than the arguments against such interventions. Instead, we must engage in moral discourse, debate, and argument in a process of public deliberation. Although closure must be reached, and decisions made, even if there is no consensus, our society has only just begun to reflect seriously on the possibility of creating children through somatic cell nuclear transfer. It may be premature to come to closure on some issues because so little time has been devoted to the issue.

Thus, the ethics of making policy, as opposed to the ethics of cloning itself, require us to return to the guidelines set forth by scholars such as Gutmann and Thompson: are the moral concerns sufficiently strong to justify prohibition or regulation? If so, is the price we pay in the form of constraints on personal liberty or the abridgement of legally protected rights acceptable? Can individual cases be treated as exceptions? Or will making exceptions create more problems, in the form of intrusive inquiries into people's motives, for example, such that making the exceptions causes more harm than good? It is difficult to answer these questions with certainty.

CONCLUSIONS

In summary, the Commission reached several conclusions in considering the appropriateness of public policies regarding the creation of children through somatic cell nuclear transfer. First and foremost, creating children in this manner is unethical at this time because available scientific evidence indicates that such techniques are not safe at this time. Even if concerns about safety are resolved, however, significant concerns remain about the negative impact of the use of such a technology on both individuals and society. Public opinion on this issue may remain divided. Some people believe that cloning through somatic cell nuclear transfer will always be unethical because it undermines important social values and will always risk causing psychological or other harms to the resulting child. In addition, although the Commission acknowledged that there are cases for which the use of such cloning might be considered desirable by some people, overall these cases were insufficiently compelling to justify proceeding with the use of such techniques. Finally, the Commission was not persuaded by objections to a prohibition against such cloning which were based, in part, on the expectation that its use is unlikely to be widespread and, in part, on the belief that many of the assumed harms are purely speculative.

Finally, many scenarios of creating children through somatic cell nuclear transfer are based on the serious misconception that selecting a child's genetic makeup is equivalent to selecting the child's traits or accomplishments. A benefit of more widespread discussion of such cloning would be a clearer recognition that a person's traits and achievements depend heavily on education, training, and the social environment, as well as on genes. Should this type of cloning proceed, however, any children born as a result of this technique should be treated as having the same rights and moral status as any other human being.

Clearly, there is a need for further public deliberation on the serious moral concerns raised by the prospect of cloning human beings. As the Commission proceeded in its review, the members learned from listening to the public and to each other. Many important issues remain unresolved, such as the nature and scope of our moral interest in the freedom to make procreative choices, and whether that freedom should encompass creating a child through somatic cell nuclear transfer cloning. The Commission believes that it is essential to try to understand the diverse reactions to such cloning and the ethical arguments for and against various policies regarding its use. This report is only the beginning of a public process to assess the impact of this new technology.

NOTE

Reprinted by permission of the National Bioethics Advisory Commission.

REFERENCES

Annas, G. J. "Regulatory models for human embryo cloning: The free market, professional guidelines, and government restrictions." *Kennedy Institute of Ethics Journal* 4, no. 3 (1994): 235–49.

Brock, D. "Cloning human beings: An assessment of the ethical issues pro and con." Paper prepared for NBAC, 1997.

Brock, D. W. "The non-identity problem and genetic harm." *Bioethics* 9 (1995): 269–75.

Cahill, L. Testimony presented to the National Bioethics Advisory Commission, March 13, 1997.

Chadwick, R. F. "Cloning." *Philosophy* 57 (1982): 201–9.

Coleman, F. "Playing God or playing scientist: A constitutional analysis of laws banning embryological procedures." 27 *Pacific Law Journal* 17 (1996): 1331.

Duff, N. "Theological Reflections on Human Cloning." Testimony presented to the National Bioethics Advisory Commission, March 13, 1997.

Etzioni, A. *The Moral Dimension.* New York: Free Press, 1990.

Feinberg, J. "The child's right to an open future." In *Whose Child? Children's Rights, Parental Authority, and State Power*, edited by W. Aiken and H. LaFollette. Totowa, N.J.: Rowman & Littlefield, 1980.

Glendon, M. A. *Rights Talk.* New York: Free Press, 1991.

Gutmann, A., and D. Thompson. *Democracy and Disagreement.* Cambridge: Belknap, 1996.

Jonas, H. *Philosophical Essays: From Ancient Creed to Technological Man.* Englewood Cliffs, N.J.: Prentice-Hall, 1974.

Kass, L. "Why we should ban the cloning of human beings." Testimony presented to the National Bioethics Advisory Commission, March 13, 1997.

Kevles, D. J. *In the Name of Eugenics.* Cambridge: Harvard University Press, 1995.

Macklin, R. "Why we should regulate—but not ban—the cloning of human beings." Testimony presented to the National Bioethics Advisory Commission, March 14, 1997.

———. "Splitting embryos on the slippery slope: Ethics and public policy." *Kennedy Institute of Ethics Journal* 4 (1994): 209–26.

Meilaender, G. "Remarks on Human Cloning to the National Bioethics Advisory Commission." Testimony presented to the National Bioethics Advisory Commission, March 13, 1997.

Mill, J. S. *On Liberty.* Indianapolis: Bobbs-Merrill, 1859.

National Institutes of Health. *Report of the Human Embryo Research Panel.* Bethesda, Md.: National Institutes of Health, 1994.

Nussbaum, M. C. "Aristotelian Social Democracy." In *Liberalism and the Good*, edited by R. Bruce Douglass et al., 217–26. 1990.

Parfit, D. *Reasons and Persons.* Oxford: Oxford University Press, 1984.

Posner, R. *Sex and Reason.* Cambridge: Harvard University Press, 1992.

Radin, M. "Reflections on Objectification." *Southern California Law Review* 341 (November 1991): 65.

———. "The Colin Ruagh Thomas O'Fallon Memorial Lecture on Personhood." *Oregon Law Review* 423 (Summer 1995): 74.

Rhodes, R. "Clones, harms, and rights." *Cambridge Quarterly of Healthcare Ethics* 4 (1995): 285–90.

Robertson, J. A. "A ban on cloning and cloning research is unjustified." Testimony presented to the National Bioethics Advisory Commission, March 14, 1997.

———. "The question of human cloning." *Hastings Center Report* 24 (1994): 6–14.

———. "The Scientist's Right to Research: A Constitutional Analysis." *Southern California Law Review* 1203 (1977): 51.

Rothenberg, K. Testimony before the Senate Committee on Labor and Human Resources, March 12, 1997.

Schwartz, H. *The Culture of Copy.* New York: Zone, 1996.

31

The Wisdom of Repugnance

Leon R. Kass

Our habit of delighting in news of scientific and technological breakthroughs has been sorely challenged by the birth announcement of a sheep named Dolly. Though Dolly shares with previous sheep the "softest clothing, woolly, bright." William Blake's question, "Little Lamb, who made thee?" has for her a radically different answer. Dolly was, quite literally, made. She is the work not of nature or nature's God but of man, an Englishman, Ian Wilmut, and his fellow scientists. What's more, Dolly came into being not only asexually—ironically, just like "He [who] calls Himself a Lamb"—but also as the genetically identical copy (and the perfect incarnation of the form or blueprint) of a mature ewe, of whom she is a clone. This long-awaited yet not quite expected success in cloning a mammal raised immediately the prospect—and the specter—of cloning human beings: "I am child and Thou a lamb," despite our differences, have always been equal candidates for creative making only now, by means of cloning we may both spring from the hand of man playing at being God.

After an initial flurry of expert comment and public consternation, with opinion polls showing overwhelming opposition to cloning human beings, President Clinton ordered a ban on all federal support for human cloning research (even though none was being supported) and charged the National Bioethics Advisory Commission to report in ninety days on the ethics of human cloning research. The commission (an eighteen-member panel, evenly balanced between scientists and nonscientists, appointed by the president and reporting to the National Science and Technology Council) invited testimony from scientists, religious thinkers and bioethicists, as well as from the general public. It is now deliberating about what it should recommend, both as a matter of ethics and as a matter of public policy.

Congress is awaiting the commission's report, and is poised to act. Bills to prohibit the use of federal funds for human cloning research have been introduced in

551

the House of Representatives and the Senate; and another bill, in the House, would make it illegal "for any person to use a human somatic cell for the process of producing a human clone." A fateful decision is at hand. To clone or not to clone a human being is no longer an academic question.

TAKING CLONING SERIOUSLY, THEN AND NOW

Cloning first came to public attention roughly thirty years ago following the successful asexual production, in England, of a clutch of tadpole clones by the technique of nuclear transplantation. The individual largely responsible for bringing the prospect and promise of human cloning to public notice was Joshua Lederberg, a Nobel Laureate geneticist and a man of large vision. In 1966, Lederberg wrote a remarkable article in *The American Naturalist* detailing the eugenic advantages of human cloning and other forms of genetic engineering, and the following year he devoted a column in *The Washington Post,* where he wrote regularly on science and society, to the prospect of human cloning. He suggested that cloning could help us overcome the unpredictable variety that still rules human reproduction, and allow us to benefit from perpetuating superior genetic endowments. These writings sparked a small public debate in which I became a participant. At the time a young researcher in molecular biology at the National Institutes of Health (NIH), I wrote a reply to the *Post*, arguing against Lederberg's amoral treatment of this morally weighty subject and insisting on the urgency of confronting a series of questions and objections, culminating in the suggestion that "the programmed reproduction of man will, in fact, dehumanize him."

Over the last 30 years, it has become harder, not easier, to discern the true meaning of human cloning. We have in some sense been softened up to the idea—through movies, cartoons, jokes and intermittent commentary in the mass media, some serious, most lighthearted. We have become accustomed to new practices in human reproduction: not just in vitro fertilization, but also embryo manipulation, embryo donation and surrogate pregnancy. Animal biotechnology has yielded transgenic animals and a burgeoning science of genetic engineering, easily and soon to be transferable to humans.

Even more important, changes in the broader culture make it now vastly more difficult to express a common and respectful understanding of sexuality, procreation, nascent life, family, and the meaning of motherhood, fatherhood and the links between the generations. Twenty-five years ago, abortion was still largely illegal and thought to be immoral, the sexual revolution (made possible by the extramarital use of the pill) was still in its infancy, and few had yet heard about the reproductive rights of single women, homosexual men and lesbians. (Never mind shameless memoirs about one's own incest.) Then one could argue, without embarrassment, that the new technologies of human reproduction—babies without sex—and their confounding of normal kin relations—who's the mother: the egg donor, the surrogate who carries and delivers or the one who rears?—would "undermine the justification and support that biological parenthood gives to the monogamous marriage. Today, defenders of stable, monogamous marriage risk charges of giving offense to those adults who are living in "new family forms" or to those children who, even without the benefit of assisted reproduction, have acquired either three or four parents or one or none at all. Today, one must even apologize for voicing opinions that twenty-five years ago were nearly universally regarded as the core of our culture's wisdom on these matters. In a world whose once-given natural boundaries are blurred by technological change and whose moral boundaries are seemingly up for grabs, it is much more difficult to make persuasive the still compelling case against cloning human beings. As Raskolnikov put it, "man gets used to everything—the beast!"

Indeed, perhaps the most depressing feature of the discussions that immediately followed the news about Dolly was their ironical tone, their genial cynicism, their moral fatigue: "AN UDDER WAY OF MAKING LAMBS" (Nature), "WHO WILL CASH IN ON BREAK-THROUGH IN CLONING?" (*The Wall Street Journal*), "is CLONING BAAAAAAAAD?" (*The Chicago Tribune*). Gone from the scene are the wise and courageous voices of Theodosius Dobzhansky (genetics), Hans Jonas (philosophy), and Paul Ramsey (theology) who, only twenty-five years ago, all made powerful moral arguments against ever cloning a human being. We are now too sophisticated for such argumentation; we wouldn't be caught in public with a strong moral stance, never mind an absolutist one. We are all, or almost all, post-modernists now.

Cloning turns out to be the perfect embodiment of the ruling opinions of our new age. Thanks to the sexual revolution, we are able to deny in practice, and increasingly in thought, the inherent procreative teleology of sexuality itself. But, if sex has no intrinsic connection to generating babies, babies need have no necessary connection to sex. Thanks to feminism and the gay rights movement, we are increasingly encouraged to treat the natural heterosexual difference and its preeminence as a matter of "cultural construction." But if male and female are not normatively complementary and generatively significant, babies need not come from male and female complementarity. Thanks to the prominence and the acceptability of divorce and out-of-wedlock births, stable, monogamous marriage as the ideal home for procreation is no longer the agreed-upon cultural norm. For this new dispensation, the clone is the ideal emblem: the ultimate "single-parent child."

Thanks to our belief that all children should be *wanted* children (the more high-minded principle we use to justify contraception and abortion), sooner or later only those children who fulfill our wants will be fully acceptable. Through cloning, we can work our wants and wills on the very identity of our children, exercising control as never before. Thanks to modern notions of individualism and the rate of cultural change, we see ourselves not as linked to ancestors and defined by traditions, but as projects for our own self-creation, not only as self-made men but also man-made selves; and self-cloning is simply an extension of such rootless and narcissistic self-re-creation.

Unwilling to acknowledge our debt to the past and unwilling to embrace the uncertainties and the limitations of the future, we have a false relation to both: cloning personifies our desire fully to control the future, while being subject to no controls ourselves. Enchanted and enslaved by the glamour of technology, we have lost our awe and wonder before the deep mysteries of nature and of life. We cheerfully take our own beginnings in our hands and, like the last man, we blink.

Part of the blame for our complacency lies, sadly, with the field of bioethics itself, and its claim to expertise in these moral matters. Bioethics was founded by people who understood that the new biology touched and threatened the deepest matters of our humanity: bodily integrity, identity and individuality, lineage and kinship, freedom and self-command, eros and aspiration, and the relations and strivings of body and soul. With its capture by analytic philosophy, however, and its inevitable routinization and professionalization, the field has by and large come to content itself with analyzing moral arguments, reacting to new technological developments and taking on emerging issues of public policy, all performed with a naive faith that the evils we fear can all be avoided by compassion, regulation and a respect for autonomy. Bioethics has made some major contributions in the protection of human subjects and in other areas where personal freedom is threatened; but its practitioners, with few exceptions, have turned the big human questions into pretty thin gruel.

One reason for this is that the piecemeal formation of public policy tends to grind down large questions of morals into small questions of procedure. Many of the country's leading bioethicists have served on national commissions or state task forces and advisory boards, where, understandably, they have found utilitarianism to be the only ethical vocabulary acceptable to all participants in discussing issues of law, regulation and public policy. As many of these commissions have been either officially under the aegis of NIH or the Health and Human Services Department, or otherwise dominated by powerful voices for scientific progress, the ethicists have for the most part been content, after some "values clarification" and wringing of hands, to pronounce their blessings upon the inevitable. Indeed, it is the bioethicists, not the scientists, who are now the most articulate defenders of human cloning: the two witnesses testifying before the National Bioethics Advisory Commission in favor of cloning human beings were bioethicists, eager to rebut what they regard as the irrational concerns of those of us in opposition. We have come to expect from the "experts" an accommodationist ethic that will rubber-stamp all technical innovation, in the mistaken belief that all other goods must bow down before the gods of better health and scientific advance.

If we are to correct our moral myopia, we must first of all persuade ourselves not to be complacent about what is at issue here. Human cloning, though it is in some respects continuous with previous reproductive technologies, also represents something radically new, in itself and in its easily foreseeable consequences. The stakes are very high indeed. I exaggerate, but in the direction of the truth, when I insist that we are faced with having to decide nothing less than whether human procreation is going to remain human, whether children are going to be made rather than begotten, whether it is a good thing, humanly speaking, to say yes in principle to the road which leads (at best) to the dehumanized rationality of Brave New World. This is not business as usual, to be fretted about for a while but finally to be given our seal of approval. We must rise to the occasion and make our judgments as if the future of our humanity hangs in the balance. For so it does.

THE STATE OF THE ART

If we should not underestimate the significance of human cloning, neither should we exaggerate its imminence or misunderstand just what is involved. The procedure is conceptually simple. The nucleus of a mature but unfertilized egg is removed and replaced with a nucleus obtained from a specialized cell of an adult (or fetal) organism (in Dolly's case, the donor nucleus came from mammary gland epithelium). Since almost all the hereditary material of a cell is contained within its nucleus, the renucleated egg and the individual into which this egg develops are genetically identical to the organism that was the source of the transferred nucleus. An unlimited number of genetically identical individuals—clones—could be produced by nuclear transfer. In principle, any person, male or female, newborn or adult, could be cloned, and in any quantity. With laboratory cultivation and storage of tissues, cells outliving their sources make it possible even to clone the dead.

The technical stumbling block, overcome by Wilmut and his colleagues, was to find a means of reprogramming the state of the DNA in the donor cells, reversing its differentiated expression and restoring its full totipotency so that it could again direct the entire process of producing a mature organism. Now that this problem has been solved, we should expect a rush to develop cloning for other animals, especially livestock, in order to propagate in perpetuity the champion meat or milk producers. Though exactly how soon someone will succeed in cloning a human being is anybody's guess, Wilmut's technique, almost certainly applicable to humans, makes *attempting* the feat an imminent possibility.

Yet some cautions are in order and some possible misconceptions need correcting. For a start, cloning is not Xeroxing. As has been reassuringly reiterated, the clone of Mel Gibson, though his genetic double, would enter the world hairless, toothless and peeing in his diapers, just like any other human infant. Moreover, the success rate, at least at first, will probably not be very high: the British transferred 277 adult nuclei into enucleated sheep eggs, and implanted twenty-nine clonal embryos, but they achieved the birth of only one live lamb clone. For this reason, among others, it is unlikely that, at least for now, the practice would be very popular, and there is no immediate worry of mass-scale production of multicopies. The need of repeated surgery to obtain eggs and, more crucially, of numerous borrowed wombs for implantation will surely limit use, as will the expense; besides, almost everyone who is able will doubtless prefer nature's sexier way of conceiving.

Still, for the tens of thousands of people already sustaining over 200 assisted-reproduction clinics in the United States and already availing themselves of in vitro fertilization, intracyto-plasmic sperm injection and other techniques of assisted reproduction, cloning would be an option with virtually no added fuss (especially when the success rate improves). Should commercial interests develop in "nucleus-banking" as they have in sperm-banking; should famous athletes or other celebrities decide to market their DNA the way they now market their autographs and just about everything else; should techniques of embryo and germline genetic testing and manipulation arrive as anticipated, increasing the use of laboratory assistance in order to obtain "better" babies—should all this come to pass, then cloning, if it is permitted, could become more than a marginal practice simply on the basis of free reproductive choice, even without any social encouragement to upgrade the gene pool or to replicate superior types. Moreover, if laboratory research on human cloning proceeds, even without any intention to produce cloned humans, the existence of cloned human embryos in the laboratory, created to begin with only for research purposes, would surely pave the way for later baby-making implantations.

In anticipation of human cloning, apologists and proponents have already made clear possible uses of the perfected technology, ranging from the sentimental and compassionate to the grandiose. They include: providing a child for an infertile couple; "replacing" a beloved spouse or child who is dying or has died; avoiding the risk of genetic disease; permitting reproduction for homosexual men and lesbians who want nothing sexual to do with the opposite sex; securing a genetically identical source of organs or tissues perfectly suitable for transplantation; getting a child with a genotype of one's own choosing, not excluding oneself; replicating individuals of great genius, talent or beauty—having a child who really could "be like Mike"; and creating large sets of genetically identical humans suitable for research on, for instance, the question of nature versus nurture, or for special missions in peace and war (not excluding espionage), in which using identical humans would be an advantage. Most people who envision the cloning of human beings, of course, want none of these scenarios. That they cannot say why is not surprising. What is surprising and welcome, is that, in our cynical age, they are saying anything at all.

THE WISDOM OF REPUGNANCE

"Offensive." "Grotesque." "Revolting." "Repugnant." "Repulsive." These are the words most commonly heard regarding the prospect of human cloning. Such reactions come both from the man or woman in the street and from the intellectuals, from believers and atheists, from humanists and scientists. Even Dolly's creator has said he "would find it offensive" to clone a human being.

People are repelled by many aspects of human cloning. They recoil from the prospect of mass production of human beings, with large clones of look-alikes, compromised in their individuality; the idea of father-son or mother-daughter twins; the bizarre prospects of a woman giving birth to and rearing a genetic copy of herself, her spouse or even her deceased father or mother; the grotesqueness of conceiving a child as an exact replacement for another who has died; the utilitarian creation of embryonic genetic duplicates of oneself, to be frozen away or created when necessary, in case of need for homologous tissues or organs for transplantation; the narcissism of those who would clone themselves and the arrogance of others who think they know who deserves to be cloned or which genotype any child-to-be should be thrilled to receive; the Frankensteinian hubris to create human life and increasingly to control its destiny; man playing God. Almost no one finds any of the suggested reasons for human cloning compelling; almost everyone anticipates its possible misuses and abuses. Moreover, many people feel oppressed by the sense that there is probably nothing we can do to prevent it from happening. This makes the prospect all the more revolting.

Revulsion is not an argument; and some of yesterday's repugnances are today calmly accepted—though one must add, not always for the better. In crucial cases, however, repugnance is the emotional expression of deep wisdom, beyond reason's power fully to articulate it. Can anyone really give an argument fully adequate to the horror which is father-daughter incest (even with consent), or having sex with animals, or mutilating a corpse, or eating human flesh, or even just (just!) raping or murdering another human being? Would anybody's failure to give full rational justification for his or her revulsion at these practices make that revulsion ethically suspect? Not at all. On the contrary, we are suspicious of those who think that they can rationalize away our horror, say, by trying to explain the enormity of incest with arguments only about the genetic risks of inbreeding.

The repugnance at human cloning belongs in this category. We are repelled by the prospect of cloning human beings not because of the strangeness or novelty of the undertaking, but because we intuit and feel immediately and without argument the violation of things that we rightfully hold dear. Repugnance, here as elsewhere, revolts against the excesses of human willfulness, warning us not to transgress what is unspeakably profound. Indeed, in this age in which everything is held to be permissible so long as it is freely done, in which our given human nature no longer commands respect, in which our bodies are regarded as mere instruments of our autonomous rational wills, repugnance may be the only voice left that speaks up to defend the central core of our humanity. Shallow are the souls that have forgotten how to shudder.

The goods protected by repugnance are generally overlooked by our customary ways of approaching all new biomedical technologies. The way we evaluate cloning ethically will in fact be shaped by how we characterize it descriptively, by the context into which we place it, and by the perspective from which we view it. The first task for ethics is proper description. And here is where our failure begins.

Typically, cloning is discussed in one or more of three familiar contexts, which one might call the technological, the liberal and the meliorist. Under the first, cloning will be seen as an extension of existing techniques for assisting reproduction and determining the genetic makeup of children. Like them, cloning is to be regarded as a neutral technique, with no inherent meaning or goodness, but subject to multiple uses, some good, some bad. The morality of cloning thus depends absolutely on the goodness or badness of the motives and intentions of the cloners: as one bioethicist defender of cloning puts it, "the ethics must be judged [only] by the way the parents nurture and rear their resulting child and whether they bestow the same love and affection on a child brought into existence by a technique of assisted reproduction as they would on a child born in the usual way."

The liberal (or libertarian or liberationist) perspective sets cloning in the context of rights, freedoms and personal empowerment. Cloning is just a new option for exercising an individual's right to reproduce or to have the kind of child that he or she wants. Alternatively, cloning enhances our liberation (especially women's liberation) from the confines of nature, the vagaries of chance, or the necessity for sexual mating. Indeed, it liberates women from the need for men altogether, for the process requires only eggs, nuclei and (for the time being) uteri—plus, of course, a healthy dose of our allegedly "masculine" manipulative science that likes to do all these things to mother nature and nature's mothers. For those who hold this outlook, the only moral restraints on cloning are adequately informed consent and the avoidance of bodily harm. If no one is cloned without her consent, and if the clonant is not physically damaged, then the liberal conditions for licit, hence moral, conduct are met. Worries that go beyond violating the will or maiming the body are dismissed as "symbolic"—which is to say, unreal.

The meliorist perspective embraces valetudinarians and also eugenicists. The latter were formerly more vocal in these discussions, but they are now generally happy to see their goals advanced under the less threatening banners of freedom and technological growth. These people see in cloning a new prospect for improving human beings—minimally, by ensuring the perpetuation of healthy individuals by avoiding the risks of genetic disease inherent in the lottery of sex, and maximally, by producing "optimum babies," preserving outstanding genetic material, and (with the help of soon-to-come techniques for precise genetic engineering) enhancing inborn human capacities on many fronts. Here the morality of cloning as a means is justified solely by the excellence of the end, that is, by the outstanding traits or individuals cloned—beauty, or brawn, or brains.

These three approaches, all quintessentially American and all perfectly fine in their places, are sorely wanting as approaches to human procreation. It is, to say the least, grossly distorting to view the wondrous mysteries of birth, renewal and individuality, and the deep meaning of parent-child relations, largely through the lens of our reductive science and its potent technologies. Similarly, considering reproduction (and the intimate relations of family life!) primarily under the political-legal, adversarial and individualistic notion of rights can only undermine the private yet fundamentally social, cooperative and duty-laden character of child-bearing/child-rearing and their bond to the covenant of marriage. Seeking to escape entirely from nature (in order to satisfy a natural desire or a natural right to reproduce!) is self-contradictory in theory and self-alienating in practice. For we are erotic beings only because we are embodied beings, and not merely intellects and wills unfortunately imprisoned in our bodies. And, though health and fitness are clearly great goods, there is something deeply disquieting in looking on our prospective children as artful products perfectible by genetic engineering, increasingly held to our willfully imposed designs, specifications and margins of tolerable error.

The technical, liberal and meliorist approaches all ignore the deeper anthropological, social and, indeed, ontological meanings of bringing forth new life. To this more fitting and profound point of view, cloning shows itself to be a major alteration, indeed, a major violation, of our given nature as embodied, gendered and engendering beings—and of the social relations built on this natural ground. Once this perspective is recognized, the ethical judgment on cloning can no longer be reduced to a matter of motives and intentions, rights and freedoms, benefits and harms, or even means and ends. It must be regarded primarily as a matter of meaning: Is cloning a fulfillment of human begetting and belonging? Or is cloning rather, as I contend, their pollution and perversion? To pollution and perversion, the fitting response can only be horror and revulsion; and conversely generalized horror and revulsion are prima facie evidence of foulness and violation. The burden of moral argument must fall entirely on those who want to declare the widespread repugnances of humankind to be mere timidity or superstition.

Yet repugnance need not stand naked before the bar of reason. The wisdom of our horror at human cloning can be partially articulated, even if this is finally one of those instances about which the heart has its reasons that reason cannot entirely know.

THE PROFUNDITY OF SEX

To see cloning in its proper context, we must begin not, as I did before, with laboratory technique, but with the anthropology—natural and social—of sexual reproduction.

Sexual reproduction—by which I mean the generation of new life from (exactly) two complementary elements, one female, one male, (usually) through coitus—is established (if that is the right term) not by human decision, culture or tradition, but by nature; it is the natural way of all mammalian reproduction. By nature, each child has two complementary biological progenitors. Each child thus stems from and unites exactly two lineages. In natural generation, moreover, the precise genetic constitution of the resulting offspring is determined by a combination of nature and chance, not by human design: each human child shares the common natural human species genotype, each child is genetically (equally) kin to each (both) parent(s), yet each child is also genetically unique.

These biological truths about our origins foretell deep truths about our identity and about our human condition altogether. Every one of us is at once equally human, equally enmeshed in a particular familial nexus of origin, and equally individuated in our trajectory from birth to death—and, if all goes well, equally capable (despite our mortality) of participating, with a complementary other, in the very same renewal of such human possibility through procreation. Though less momentous than our common humanity, our genetic individuality is not humanly trivial. It shows itself forth in our distinctive appearance through which we are everywhere recognized; it is revealed in our "signature" marks of fingerprints and our self-recognizing immune system; it symbolizes and foreshadows exactly the unique, never-to-be-repeated character of each human life.

Human societies virtually everywhere have structured child-rearing responsibilities and systems of identity and relationship on the bases of these deep natural facts of begetting. The mysterious yet ubiquitous "love of one's own" is everywhere culturally exploited, to make sure that children are not just produced but well cared for and to create for everyone clear ties of meaning, belonging and obligation. But it is wrong to treat such naturally rooted social practices as mere cultural constructs (like left- or right-driving, or like burying or cremating the dead) that we can alter with little human cost. What would kinship be without its clear natural grounding? And what would identity be without kinship? We must resist those who have begun to refer to sexual reproduction as the "traditional method of reproduction," who would have us regard as merely traditional, and by implication arbitrary, what is in truth not only natural but most certainly profound.

Asexual reproduction, which produces "single-parent" offspring, is a radical departure from the natural human way, confounding all normal understandings of father, mother, sibling, grandparent, etc., and all moral relations tied thereto. It becomes even more of a radical departure when the resulting offspring is a clone derived not from an embryo, but from a mature adult to whom the clone would be an identical twin; and when the process occurs not by natural accident (as in natural twinning), but by deliberate human design and manipulation; and when the child's (or children's) genetic constitution is preselected by the parent(s) (or scientists). Accordingly, as we will see, cloning is vulnerable to three kinds of concerns and objections, related to these three points: cloning threatens confusion of identity and individuality, even in small-scale cloning; cloning represents a giant step (though not the first one) toward transforming procreation into manufacture, that is, toward the increasing depersonalization of the process of generation and, increasingly, toward the "production" of human children as artifacts, products of human will and design (what others have called the problem of "commodification" of new life); and cloning—like other forms of eugenic engineering of the next generation—represents a form of despotism of the cloners over the cloned, and thus (even in benevolent cases) represents a blatant violation of the inner meaning of parent-child relations, of what it means to have a child, of what it means to say "yes" to our own demise and "replacement."

Before turning to these specific ethical objections, let me test my claim of the profundity of the natural way by taking up a challenge recently posed by a friend. What if the given natural human way of reproduction were asexual, and we now had to deal with a new technological innovation—artificially induced sexual dimorphism and the fusing of complementary gametes—whose inventors argued that sexual reproduction promised all sorts of advantages, including hybrid vigor and the creation of greatly increased individuality? Would one then be forced to defend natural asexuality because it was natural? Could one claim that it earned deep human meaning?

The response to this challenge broaches the ontological meaning of sexual reproduction. For it is impossible, I submit, for there to have been human life—or even higher forms of animal life—in the absence of sexuality and sexual reproduction. We find asexual reproduction only in the lowest forms of life: bacteria, algae, fungi, some lower invertebrates. Sexuality brings with it a new and enriched relationship to the world. Only sexual animals can seek and find complementary others with whom to pursue a goal that transcends their own existence. For a sexual being, the world is no longer an indifferent and largely homogeneous otherness, in part edible, in part dangerous. It also contains some very special and related and complementary beings, of the same kind but of opposite sex, toward whom one reaches out with special interest and intensity. In higher birds and mammals, the outward gaze keeps a lookout not only for food and predators, but also for prospective mates; the beholding of the many splendored world is suffused with desire for union, the animal antecedent of human eros and the germ of sociality. Not by accident is the human animal both the sexiest animal—whose females do not go into heat but are receptive throughout the estrous cycle and whose males must therefore have greater sexual appetite and energy in order to reproduce successfully—and also the most aspiring, the most social, the most open and the most intelligent animal.

The soul-elevating power of sexuality is, at bottom, rooted in its strange connection to mortality, which it simultaneously accepts and tries to overcome. Asexual reproduction may be seen as a continuation of the activity of self-preservation. When one organism buds or divides to become two, the original being is (doubly) preserved, and nothing dies. Sexuality, by contrast, means perishability and serves replacement; the two that come together to generate one soon will die. Sexual desire, in human beings as in animals, thus serves an end that is partly hidden from, and finally at odds with, the self-serving individual. Whether we know it or not, when we are sexually active we are voting with our genitalia for our own demise. The salmon swimming upstream to spawn and die tell the universal story: sex is bound up with death, to which it holds a partial answer in procreation.

The salmon and the other animals evince this truth blindly. Only the human being can understand what it means. As we learn so powerfully from the story of the Garden of Eden, our humanization is coincident with sexual self-consciousness, with the recognition of our sexual nakedness and all that it implies: shame at our needy incompleteness, unruly self-division and finitude; awe before the eternal; hope in the self-transcending possibilities of children and a relationship to the divine. In the sexually self-conscious animal, sexual desire can become eros, lust can become love. Sexual desire humanly regarded is thus sublimated into erotic longing for wholeness, completion and immortality, which drives us knowingly into the embrace and its generative fruit—as well as into all the higher human possibilities of deed, speech and song.

Through children, a good common to both husband and wife, male and female achieve some genuine unification (beyond the mere sexual "union," which fails to do so). The two become one through sharing generous (not needy) love for this third being as good. Flesh of their flesh, the child is the parents' own commingled being externalized, and given a separate and persisting existence. Unification is enhanced also by their commingled work of rearing. Providing an opening to the future beyond the grave, carrying not only our seed but also our names, our ways and our hopes that they will surpass us in goodness and happiness, children are a testament to the possibility of transcendence. Gender duality and sexual desire, which first draws our love upward and outside of ourselves, finally provide for the partial overcoming of the confinement and limitation of perishable embodiment altogether.

Human procreation, in sum, is not simply an activity of our rational wills. It is a more complete activity precisely because it engages us bodily, erotically and spiritually, as well as rationally. There is wisdom in the mystery of nature that has joined the pleasure of sex, the inarticulate longing for union, the communication of the loving embrace and the deep-seated and only partly articulate desire for children in the very activity by which we continue the chain of human existence and participate in the renewal of human possibility. Whether or not we know it, the severing of procreation from sex, love and intimacy is inherently dehumanizing, no matter how good the product.

We are now ready for the more specific objections to cloning.

THE PERVERSITIES OF CLONING

First, an important if formal objection: any attempt to clone a human being would constitute an unethical experiment upon the resulting child-to-be. As the animal experiments (frog and sheep) indicate, there are grave risks of mishaps and deformities. Moreover, because of what cloning means, one cannot presume a future cloned child's consent to be a clone, even a healthy one. Thus, ethically speaking, we cannot even get to know whether or not human cloning is feasible.

I understand, of course, the philosophical difficulty of trying to compare a life with defects against nonexistence. Several bioethicists, proud of their philosophical cleverness, use this conundrum to embarrass claims that one can injure a child in its conception, precisely because it is only thanks to that complained-of conception that the child is alive to complain. But common sense tells us that we have no reason to fear such philosophisms. For we surely know that people can harm and even maim children in the very act of conceiving them, say, by paternal transmission of the AIDS virus, maternal transmission of heroin dependence or, arguably, even by bringing them into being as bastards or with no capacity or willingness to look after them properly. And we believe that to do this intentionally, or even negligently, is inexcusable and clearly unethical.

Cloning creates serious issues of identity and individuality. The cloned person may experience concerns about his distinctive identity not only because he will be in genotype and appearance identical to another human being, but, in this case, because he may also be twin to the person who is his "father" or "mother"—if one can still call them that. What would be the psychic burdens of being the "child" or "parent" of your twin? The cloned individual, moreover, will be saddled with a genotype that has already lived. He will not be fully a surprise to the world. People are likely always to compare his performances in life with that of his alter ego. True, his nurture and his circumstance in life will be different; genotype is not exactly destiny. Still, one must also expect parental and other efforts to shape this new life after the original—or at least to view the child with the original version always firmly in mind. Why else did they clone from the star basketball player, mathematician and beauty queen—or even dear old dad—in the first place?

Since the birth of Dolly, there has been a fair amount of doublespeak on this matter of genetic identity. Experts have rushed in to reassure the public that the clone would in no way be the same person, or have any confusions about his or her identity: as previously noted, they are pleased to point out that the clone of Mel Gibson would not be Mel Gibson. Fair enough. But one is short-changing the truth by emphasizing the additional importance of the intrauterine environment, rearing and social setting: genotype obviously matters plenty. That, after all, is the only reason to clone, whether human beings or sheep. The odds that clones of Wilt Chamberlain will play in the NBA are, I submit, infinitely greater than they are for clones of Robert Reich.

Curiously, this conclusion is supported, inadvertently, by the one ethical sticking point insisted on by friends of cloning: no cloning without the donor's consent. Though an orthodox liberal objection, it is in fact quite puzzling when it comes from people (such as Ruth Macklin) who also insist that genotype is not identity or individuality, and who deny that a child could reasonably complain about being made a genetic copy. If the clone of Mel Gibson would not be Mel Gibson, why should Mel Gibson have grounds to object that someone had been made his clone? We already allow researchers to use blood and tissue samples for research purposes of no benefit to their sources: my falling hair, my expectorations, my urine and even my biopsied tissues are "not me" and not mine. Courts have held that the profit gained from uses to which scientists put my discarded tissues do not legally belong to me. Why, then, no cloning without consent—including, I assume, no cloning from the body or someone who just died? What harm is done the donor, if genotype is "not me"? Truth to tell, the only powerful justification for objecting is that genotype really does have something to do with identity, and everybody knows it. If not, on what basis could Michael Jordan object that someone cloned "him," say, from cells taken from a "lost" scraped-off piece of his skin? The insistence on donor consent unwittingly reveals the problem of identity in all cloning.

Genetic distinctiveness not only symbolizes the uniqueness of each human life and the independence of its parents that each human child rightfully attains. It can also be an important support for living a worthy and dignified life. Such arguments apply with great force to any large-scale replication of human individuals. But they are sufficient, in my view, to rebut even the first attempts to clone a human being. One must never forget that these are human beings upon whom our eugenic or merely playful fantasies are to be enacted.

Troubled psychic identity (distinctiveness), based on all-too-evident generic identity (sameness), will be made much worse by the utter confusion of social identity and kinship ties. For, as already noted, cloning radically confounds lineage and social relations, for "offspring" as for "parents." As bioethicist James Nelson has pointed out, a female child cloned from her "mother" might develop a desire for a relationship to her "father," and might understandably seek out the father of her "mother" who is after all also her biological twin sister. Would "grandpa," who thought his paternal duties concluded, be pleased to discover that the clonant looked to him for paternal attention and support?

Social identity and social ties of relationship and responsibility are widely connected to, and supported by, biological kinship. Social taboos on incest (and adultery) everywhere serve to keep clear who is related to whom (and especially which child belongs to which parents), as well as to avoid confounding the social identity of parent-and-child (or brother-and-sister) with the social identity of lovers, spouses and co-parents. True, social identity is altered by adoption (but as a matter of the best interest of already living children: we do not deliberately produce children for adoption). True, artificial insemination and in vitro fertilization with donor sperm, or whole embryo donation, are in some way forms of "prenatal adoption"—a not altogether unproblematic practice. Even here, though, there is in each case (as in all sexual reproduction) a known male source of sperm and a known single female source of egg—a genetic father and a genetic mother— should anyone care to know (as adopted children often do) who is genetically related to whom.

In the case of cloning, however, there is but one "parent." The usually sad situation of the "single-parent child" is here deliberately planned, and with a vengeance. In the case of self-cloning, the "offspring" is, in addition, one's twin; and so the dreaded result of incest—to be parent to one's sibling—is here brought about deliberately, albeit without any act of coitus. Moreover, all other relationships will be confounded. What will father, grandfather, aunt, cousin, sister mean? Who will bear what ties and what burdens? What sort of social identity will someone have with one whole side—"father's" or "mother's"—necessarily excluded? It is no answer to say that our society, with its high incidence of divorce, remarriage, adoption, extramarital child-bearing and the rest, already confounds lineage and confuses kinship and responsibility for children (and everyone else), unless one also wants to argue that this is, for children, a preferable state of affairs.

Human cloning would also represent a giant step toward turning begetting into making, procreation into manufacture (literally, something "hand-made"), a process already begun with in vitro fertilization and genetic testing of embryos. With cloning, not only is the process in hand, but the total genetic blueprint of the cloned individual is selected and determined by the human artisans. To be sure, subsequent development will take place according to natural processes; and the resulting children will still be recognizably human. But we here would be taking a major step into making man himself simply another one of the man-made things. Human nature becomes merely the last part of nature to succumb to the technological project, which turns all of nature into raw material at human disposal, to be homogenized by our rationalized technique according to the subjective prejudices of the day.

How does begetting differ from making? In natural procreation, human beings come together, complementarily male and female, to give existence to another being who is formed, exactly as we were, *by what we are*: living, hence perishable, hence aspiringly erotic, human beings. In clonal reproduction, by contrast, and in the more advanced forms of manufacture to which it leads, we give existence to a being not by what we are but by what we intend and design. As with any product of our making, no matter how excellent, the artificer stands above it, not as an equal but as a superior, transcending it by his will and creative prowess. Scientists who clone animals make it perfectly clear that they are engaged in instrumental making; the animals are, from the start, designed as means to serve rational human purposes. In human cloning, scientists and prospective "parents" would be adopting the same technocratic mentality to human children: human children would be their artifacts.

Such an arrangement is profoundly dehumanizing, no matter how good the product. Mass-scale cloning of the same individual makes the point vividly; but the violation of human equality, freedom and dignity is present even in a single planned clone. And procreation dehumanized into manufacture is further degraded by commodification, a virtually inescapable result of allowing baby-making to proceed under the banner of commerce. Genetic and reproductive biotechnology companies are already growth industries, but they will go into commercial orbit once the Human Genome Project nears completion. Supply will create enormous demand. Even before the capacity for human cloning arrives, established companies will have invested in the harvesting of eggs from ovaries obtained at autopsy or through ovarian surgery, practiced embryonic genetic alteration, and initiated the stockpiling of prospective donor tissues. Through the rental of surrogate-womb services, and through the buying and selling of tissues and embryos, priced according to the merit of the donor, the commodification of nascent human life will be unstoppable.

Finally, and perhaps most important, the practice of human cloning by nuclear transfer—like other anticipated forms of genetic engineering of the next generation—would enshrine and aggravate a profound and mischievous misunderstanding of the meaning of having children and of the parent-child relationship. When a couple now chooses to procreate, the partners are saying yes to the emergence of new life in its novelty, saying yes not only to having a child but also, tacitly, to having whatever child this child turns out to be. In accepting our finitude and opening ourselves to our replacement, we are tacitly contesting the limits of our control. In this ubiquitous way of nature, embracing the future by procreating means precisely that we are relinquishing our grip, in the very activity of taking up our own share in what we hope will be the immortality of human life and the human species. This means that *our* children are not our children: they are not our property, not our possessions. Neither are they supposed to live our lives for us, or anyone else's life but their own. To be sure, we seek to guide them on their way, imparting to them not just life but nurturing, love, and a way of life; to be sure, they bear our hopes that they will live fine and flourishing lives, enabling us in small measure to transcend our own limitations. Still, their genetic distinctiveness and independence are the natural foreshadowing of the deep truth that they have their own and never-before-enacted life to live. They are sprung from a past, but they take an uncharted course into the future.

Much harm is already done by parents who try to live vicariously through their children. Children are sometimes compelled to fulfill the broken dreams of unhappy parents: John Doe Jr. or the III is under the burden of having to live up to his forebear's name. Still, if most parents have hopes for their children, cloning parents will have expectations. In cloning, such over-bearing parents take at the start a decisive step which contradicts the entire meaning of the open and forward-looking nature of parent-child relations. The child is given a genotype that has already lived, with full expectation that this blueprint of a past life ought to be controlling of the life that is to come. Cloning is inherently despotic, for it seeks to make one's children (or someone else's children) after one's own image (or an image of one's choosing) and their future according to one's will. In some cases, the despotism may be mild and benevolent. In other cases, it will be mischievous and downright tyrannical. But despotism—the control of another through one's will—it inevitably will be.

MEETING SOME OBJECTIONS

The defenders of cloning, of course, are not wittingly friends of despotism. Indeed, they regard themselves mainly as friends of freedom: the freedom of individuals to reproduce, the freedom of scientists and inventors to discover and devise and to foster "progress" in genetic knowledge and technique. They want large-scale cloning only for animals, but they wish to preserve cloning as a human option for exercising our "right to reproduce"—our right to have children, and children with "desirable genes." As law professor John Robertson points out, under our "right to reproduce" we already practice early forms of unnatural, artificial and extramarital reproduction, and we already practice early forms of eugenic choice. For this reason, he argues, cloning is no big deal.

We have here a perfect example of the logic of the slippery slope, and the slippery way in which it already works in this area. Only a few years ago, slippery slope arguments were used to oppose artificial insemination and in vitro fertilization using unrelated sperm donors. Principles used to justify these practices, it was said, will be used to justify more artificial and more eugenic practices, including cloning. Not so, the defenders retorted, since we can make the necessary distinctions. And now, without even a gesture at making the necessary distinctions, the continuity of practice is held by itself to be justificatory.

The principle of reproductive freedom as currently enunciated by the proponents of cloning logically embraces the ethical acceptability of sliding down the entire rest of the slope—to producing children ectogenetically from sperm to term (should it become feasible) and to producing children whose entire genetic makeup will be the product of parental eugenic planning and choice. If reproductive freedom means the right to have a child of one's own choosing, by whatever means, it knows and accepts no limits.

But far from being legitimated by a "right to reproduce," the emergence of techniques of assisted reproduction and genetic engineering should compel us to reconsider the meaning and limits of such a putative right. In truth, a "right to reproduce" has always been a peculiar and problematic notion. Rights generally belong to individuals, but this is a right which (before cloning) no one can exercise alone. Does the right then inhere only in couples? Only in married couples? Is it a (woman's) right to carry or deliver or a right (of one or more parents) to nurture and rear? Is it a right to have your own biological child? Is it a right only to attempt reproduction, or a right also to succeed? Is it a right to acquire the baby of one's choice?

The assertion of a negative "right to reproduce" certainly makes sense when it claims protection against state interference with procreative liberty, say, through a program of compulsory sterilization. But surely it cannot be the basis of a tort claim against nature, to be made good by technology, should free efforts at natural procreation fail. Some insist that the right to reproduce embraces also the right against state interference with the free use of all technological means to obtain a child. Yet such a position cannot be sustained: for reasons having to do with the means employed, any community may rightfully prohibit surrogate pregnancy, or polygamy, or the sale of babies to infertile couples, without violating anyone's basic human "right to reproduce." When the exercise of a previously innocuous freedom now involves or impinges on troublesome practices that the original freedom never was intended to reach, the general presumption of liberty needs to be reconsidered.

We do indeed already practice negative eugenic selection, through genetic screening and prenatal diagnosis. Yet our practices are governed by a norm of health. We seek to prevent the birth of children who suffer from known (serious) genetic diseases. When and if gene therapy becomes possible, such diseases could then be treated, in utero or even before implantation—I have no ethical objection in principle to such a practice (though I have some practical worries), precisely because it serves the medical goal of healing existing individuals. But therapy, to be therapy, implies not only an existing "patient." It also implies a norm of health. In this respect, even germline gene "therapy," though practiced not on a human being but on egg and sperm, is less radical than cloning, which is in no way therapeutic. But once one blurs the distinction between health promotion and genetic enhancement, between so-called negative and positive eugenics, one opens the door to all future eugenic designs. "To make sure that a child will be healthy and have good chances in life": this is Robertson's principle, and owing to its latter clause it is an utterly elastic principle, with no boundaries. Being over eight feet tall will likely produce some very good chances in life, and so will having the looks of Marilyn Monroe, and so will a genius-level intelligence.

Proponents want us to believe that there are legitimate uses of cloning that can be distinguished from illegitimate uses, but by their own principles no such limits can be found. (Nor could any such limits be enforced in practice.) Reproductive freedom, as they understand it, is governed solely by the subjective wishes of the parents-to-be (plus the avoidance of bodily harm to the child). The sentimentally appealing case of the childless married couple is, on these grounds, indistinguishable from the case of an individual (married or not) who would like to clone someone famous or talented, living or dead. Further, the principle here endorsed justifies not only cloning but, indeed, all future artificial attempts to create (manufacture) "perfect" babies.

A concrete example will show how, in practice no less than in principle, the so-called innocent case will merge with, or even turn into, the more troubling ones. In practice, the eager parents-to-be will necessarily be subject to the tyranny of expertise. Consider an infertile married couple, she lacking eggs or he lacking sperm, that wants a child of their (genetic) own, and propose to clone either husband or wife. The scientist-physician (who is also co-owner of the cloning company) points out the likely difficulties—a cloned child is not really their (genetic) child, but the child of only one of them; this imbalance may produce strains on the marriage; the child might suffer identity confusion; there is a risk of perpetuating the cause of sterility; and so on—and he also points out the advantages of choosing a donor nucleus. Far better than a child of their own would be a child of their own choosing. Touting his own expertise in selecting healthy and talented donors, the doctor presents the couple with his latest catalog containing the pictures, the health records and the accomplishments of his stable of cloning donors, samples of whose tissues are in his deep freeze. Why not, dearly beloved, a more perfect baby?

The "perfect baby," of course, is the project not of the infertility doctors, but of the eugenic scientists and their supporters. For them, the paramount right is not the so-called right to reproduce but what biologist Bentley Glass called, a quarter of a century ago, "the right of every child to be born with a sound physical and mental constitution—based on a sound genotype . . . the inalienable right to a sound heritage." But to secure this right, and to achieve the requisite quality control over new human life, human conception and gestation will need to be brought fully into the bright light of the laboratory, beneath which it can be fertilized, nourished, pruned, weeded, watched, inspected, prodded, pinched, cajoled, injected, tested, rated, graded, approved, stamped, wrapped, sealed and delivered. There is no other way to produce the perfect baby.

Yet we are urged by proponents of cloning to forget about the science fiction scenarios of laboratory manufacture and multiple-copied clones, and to focus only on the homely cases of infertile couples exercising their reproductive rights. But why, if the single cases are so innocent, should multiplying their performance be so off-putting? (Similarly, why do others object to people making money off this practice, if the practice itself is perfectly acceptable?) When we follow the sound ethical principle of universalizing our choice—"would it be right if everyone cloned a Wilt Chamberlain (with his consent, of course)? Would it be right if everyone decided to practice asexual reproduction?"—we discover what is wrong with these seemingly innocent cases. The so-called science fiction cases make vivid the meaning of what looks to us, mistakenly, to be benign.

Though I recognize certain continuities between cloning and, say, in vitro fertilization, I believe that cloning differs in essential and important ways. Yet those who disagree should be reminded that the "continuity" argument cuts both ways. Sometimes we establish bad precedents, and discover that they were bad only when we follow their inexorable logic to places we never meant to go. Can the defenders of cloning show us today how, on their principles, we will be able to see producing babies ("perfect babies") entirely in the laboratory or exercising full control over their genotypes (including so-called enhancement) as ethically different, in any essential way, from present forms of assisted reproduction? Or are they willing to admit, despite their attachment to the principle of continuity, that the complete obliteration of "mother" or "father," the complete depersonalization of procreation, the complete manufacture of human beings and the complete genetic control of one generation over the next would be ethically problematic and essentially different from current forms of assisted reproduction?

If so, where and how will they draw the line, and why? I draw it at cloning for all the reasons given.

BAN THE CLONING OF HUMANS

What, then, should we do? We should declare that human cloning is unethical in itself and dangerous in its likely consequences. In so doing, we shall have the backing of the overwhelming majority of our fellow Americans, and of the human race, and (I believe) of most practicing scientists. Next, we should do all that we can to prevent the cloning of human beings. We should do this by means of an international legal ban if possible, and by a unilateral national ban, at a minimum. Scientists may secretly undertake to violate such a law, but they will be deterred by not being able to stand up proudly to claim the credit for their technological bravado and success. Such a ban on clonal baby-making, moreover, will not harm the progress of basic genetic science and technology. On the contrary, it will reassure the public that scientists are happy to proceed without violating the deep ethical norms and intuitions of the human community.

This still leaves the vexed question about laboratory research using early embryonic human clones, specially created only for such research purposes, with no intention to implant them into a uterus. There is no question that such research holds great promise for gaining fundamental knowledge about normal (and abnormal) differentiation, and for developing tissue lines for transplantation that might be used, say, in treating leukemia or in repairing brain or spinal cord injuries—to mention just a few of the conceivable benefits. Still, unrestricted clonal embryo research will surely make the production of living human clones much more likely. Once the genies put the cloned embryos into the bottles, who can strictly control where they go (especially in the absence of legal prohibitions against implanting them to produce a child)? . . .

The president's call for a moratorium on human cloning has given us an important opportunity. In a truly unprecedented way, we can strike a blow for the human control of the technological project, for wisdom, prudence and human dignity. The prospect of human cloning, so repulsive to contemplate, is the occasion for deciding whether we shall be slaves of unregulated progress, and ultimately its artifacts, or whether we shall remain free human beings who guide our technique toward the enhancement of human dignity. If we are to seize the occasion, we must, as the late Paul Ramsey wrote,

> raise the ethical questions with a serious and not a frivolous conscience. A man of frivolous conscience announces that there are ethical quandaries ahead that we must urgently consider before the future catches up with us. By this he often means that we need to devise a new ethic that will provide the rationalization for doing for the future what men are bound to do because of new actions and interventions science

will have made possible. In contrast a man of serious conscience means to say in raising urgent ethical questions that there may be some things that men should never do. The good things that men do can be made complete only by the things they refuse to do.

NOTE

Reprinted by permission of the author from *The New Republic*, June 2, 1997.

32

Genetic Encores: The Ethics of Human Cloning

Robert Wachbroit

The successful cloning of an adult sheep, announced in Scotland this past February, is one of the most dramatic recent examples of a scientific discovery becoming a public issue. During the last few months, various commentators—scientists and theologians, physicians and legal experts, talk-radio hosts and editorial writers—have been busily responding to the news, some calming fears, other raising alarms about the prospect of cloning a human being. At the request of the President, the National Bioethics Advisory Commission (NBAC) held hearings and prepared a report on the religious, ethical, and legal issues surrounding human cloning. While declining to call for a permanent ban on the practice, the Commission recommended a moratorium on efforts to clone human beings, and emphasized the importance of further public deliberation on the subject.

An interesting tension is at work in the NBAC report. Commission members were well aware of "the widespread public discomfort, even revulsion, about cloning human beings." Perhaps recalling the images of Dolly the ewe that were featured on the covers of national news magazines, they noted that "the impact of these most recent developments on our national psyche has been quite remarkable." Accordingly, they felt that one of their tasks was to articulate, as fully and sympathetically as possible, the range of concerns that the prospect of human cloning had elicited.

Yet it seems clear that some of these concerns, at least, are based on false beliefs about genetic influence and the nature of the individuals that would be produced through cloning. Consider, for instance, the fear that a clone would not be an "individual" but merely a "carbon copy" of someone else—an automaton of the sort familiar from science fiction. As many scientists have pointed out, a clone would not in fact be an identical *copy*, but more like a delayed identical *twin*. And just as identical twins are two separate people—biologically, psychologically, morally and legally, though not genetically—so, too, a clone would be a separate person from her non-contemporaneous twin. To think otherwise is to embrace a belief in genetic determinism—the view that genes determine everything about us, and that environmental factors or the random events in human development are insignificant.

The overwhelming scientific consensus is that genetic determinism is false. In coming to understand the ways in which genes operate, biologists have also become aware of the myriad ways in which the environment affects their "expression." The genetic contribution to the simplest physical traits, such as height and hair color, is significantly mediated by environmental factors (and possibly by stochastic events as well). And the genetic contribution to the traits we value most deeply, from intelligence to compassion, is conceded by even the most enthusiastic genetic researchers to be limited and indirect.

It is difficult to gauge the extent to which "repugnance" toward cloning generally rests on a belief in genetic determinism. Hoping to account for the fact that people "instinctively recoil" from the prospect of cloning, James Q. Wilson wrote, "There is a natural sentiment that is offended by the mental picture of identical babies being produced in some biological factory." Which raises the question: once people learn that this picture is mere science fiction, does the offense that cloning presents to "natural sentiment" attenuate, or even disappear? Jean Bethke Elshtain cited the nightmare scenarios of "the man and woman on the street," who imagine a future populated by "a veritable army of Hitlers, ruthless and remorseless bigots who kept reproducing themselves until they had finished what the historic Hitler failed to do: annihilate us." What happens, though, to the "pity and terror" evoked by the topic of cloning when such scenarios are deprived (as they deserve to be) of all credibility?

Richard Lewontin has argued that the critics' fears—or at least, those fears that merit consideration in formulating public policy—dissolve once genetic determinism is refuted. He criticizes the NBAC report for excessive deference to opponents of human cloning, and calls for greater public education on the scientific issues. (The Commission in fact makes the same recommendation, but Lewontin seems unimpressed.) Yet even if a public education campaign succeeded in eliminating the most egregious misconceptions about genetic influence, that wouldn't settle the matter. People might continue to express concerns about the interests and rights of human clones, about the social and moral consequences of the cloning process, and about the possible motivations for creating children in this way.

INTERESTS AND RIGHTS

One set of ethical concerns about human clones involves the risks and uncertainties associated with the current state of cloning technology. This technology has not yet been tested with human subjects, and scientists cannot rule out the possibility of mutation or other biological damage. Accordingly, the NBAC report concluded that "at this time, it is morally unacceptable for anyone in the public or private sector, whether in a research or clinical setting, to attempt to create a child using somatic cell nuclear transfer cloning." Such efforts, it said, would pose "unacceptable risks to the fetus and/or potential child."

The ethical issues of greatest importance in the cloning debate, however, do not involve possible failures of cloning technology, but rather the consequences of its success. Assuming that scientists were able to clone human beings without incurring the risks mentioned above, what concerns might there be about the welfare of clones?

Some opponents of cloning believe that such individuals would be wronged in morally significant ways. Many of these wrongs involve the denial of what Joel Feinberg has called "the right to an open future." For example, a child might be constantly compared to the adult from whom he was cloned, and thereby burdened with oppressive expectations. Even worse, the parents might actually limit the child's opportunities for growth and development: a child cloned from a basketball player, for instance, might be denied any educational opportunities that were not in line with a career in basketball. Finally, regardless of his parents' conduct or attitudes, a child might be burdened by the *thought* that he is a copy and not an "original." The child's sense of self-worth or individuality or dignity, so some have argued, would thus be difficult to sustain.

How should we respond to these concerns? On the one hand, the existence of a right to an open future has a strong intuitive appeal. We are troubled by parents who radically constrict their children's possibilities for growth and development. Obviously, we would condemn a cloning parent for crushing a child with oppressive expectations, just as we might condemn fundamentalist parents for utterly isolating their children from the modern world, or the parents of twins for inflicting matching wardrobes and rhyming names. But this is not enough to sustain an objection to cloning itself. Unless the claim is that cloned parents cannot help but be oppressive, we would have cause to say they had wronged their children only because of their subsequent, and avoidable, sins of bad parenting—not because they had chosen to create the child in the first place. (The possible reasons for making this choice will be discussed below.)

We must also remember that children are often born in the midst of all sorts of hopes and expectations; the idea that there is a special burden associated with the thought "There is someone who is genetically just like me" is necessarily speculative. Moreover, given the falsity of genetic determinism, any conclusions a child might draw from observing the person from whom he was cloned would be uncertain at best. His knowledge of his future would differ only in degree from what many children already know once they begin to learn parts of their family's (medical) history. Some of us knew that we would be bald, or to what diseases we might be susceptible. To be sure, the cloned individual might know more about what he or she could become. But because our knowledge of the effect of environment on development is so incomplete, the clone would certainly be in for some surprises.

Finally, even if we were convinced that clones are likely to suffer particular burdens, that would not be enough to show that it is wrong to create them. The child of a poor family can be expected to suffer specific hardships and burdens, but we don't thereby conclude that such children shouldn't be born. Despite the hardships, poor children can experience parental love and many of the joys of being alive: the deprivations of poverty, however painful, are not decisive. More generally, no one's life is entirely free of some difficulties or burdens. In order for these considerations to have decisive weight, we have to be able to say that life doesn't offer any compensating benefits. Concerns expressed about the welfare of human clones do not appear to justify such a bleak assessment. Most such children can be expected to have lives well worth living; many of the imagined harms are no worse than those faced by children acceptably produced by more conventional means. If there is something deeply objectionable about cloning, it is more likely to be found by examining implications of the cloning process itself, or the reasons people might have for availing themselves of it.

CONCERNS ABOUT PROCESS

Human cloning falls conceptually between two other technologies. At one end we have the assisted reproductive technologies, such as in vitro fertilization, whose primary purpose is to enable couples to produce a child with whom they have a biological connection. At the other end we have the emerging technologies of genetic engineering—specifically, gene transplantation technologies—whose primary purpose is to produce a child that has certain traits. Many proponents of cloning see it as part of the first technology: cloning is just another way of providing a couple with a biological child they might otherwise be unable to have. Since this goal and these other technologies are acceptable, cloning should be acceptable as well. On the other hand, many opponents of cloning see it as part of the second technology: even though cloning is a transplantation of an entire nucleus and not of specific genes, it is nevertheless an attempt to produce a child with certain traits. The deep misgivings we may have about the genetic manipulation of offspring should apply to cloning as well.

The debate cannot be resolved, however, simply by determining which technology to assimilate cloning to. For example, some opponents of human cloning see it as continuous with assisted reproductive technologies; but since they find those technologies objectionable as well, the assimilation does not indicate approval. Rather than argue for grouping cloning with one technology or another, I wish to suggest that we can best understand the significance of the cloning process by comparing it with these other technologies, and thus broadening the debate.

To see what can be learned from such a comparative approach, let us consider a central argument that has been made against cloning, that it undermines the structure of the family by making identities and lineages unclear. On the one hand, the relationship between an adult and the child cloned from her could be described as that between a parent and offspring. Indeed, some commentators have called cloning "asexual reproduction," which clearly suggests that cloning is a way of generating descendants. The clone, on this view, has only one biological parent. On the other hand, from the point of view of genetics, the clone is a sibling, so that cloning is more accurately described as "delayed twinning" rather than as asexual reproduction. The clone, on this view, has two biological parents, not one—they are the same parents as those of the person from whom that individual was cloned.

Cloning thus results in ambiguities. Is the clone an offspring or a sibling? Does the clone have one biological parent or two? The moral significance of these ambiguities lies in the fact that in many societies, including our own, lineage identifies responsibilities. Typically, the parent, not the sibling, is responsible for the child. But if no one is unambiguously the parent, so the worry might go, who is responsible for the clone? Insofar as social identity is based on biological ties, won't this identity be blurred or confounded?

Some assisted reproductive technologies have raised similar questions about lineage and identity. An anonymous sperm donor is thought to have no parental obligations towards his biological child. A surrogate mother may be required to relinquish all parental claims to the child she bears. In these cases, the social and legal determination of "who is the parent" may appear to proceed in defiance of profound biological facts, and to subvert attachments that we as a society are ordinarily committed to upholding. Thus, while the *aim* of assisted reproductive technologies is to allow people to produce or raise a child to whom they are biologically connected, such technologies may also involve the creation of social ties that are permitted to override biological ones.

In the case of cloning, however, ambiguous lineages would seem to be less problematic, precisely because no one is being asked to relinquish a claim on a child to whom he or she might otherwise acknowledge a biological connection. What, then, are the critics afraid of? It does not seem plausible that someone would have herself cloned and then hand the child over to her parents, saying, "You take care of her! She's *your* daughter!" Nor is it likely that, if the cloned individual did raise the child, she would suddenly refuse to pay for college on the grounds that this was not a sister's responsibility. Of course, policymakers should address any confusion in the social or legal assignment of responsibility resulting from cloning. But there are reasons to think that this would be *less* difficult than in the case of other reproductive technologies.

Similarly, when we compare cloning with genetic engineering, cloning may prove to be the less troubling of the two technologies. This is true even though the dark futures to which they are often alleged to lead are broadly alike. For example, a recent *Washington Post* article examined fears that the development of genetic enhancement technologies might "create a market in preferred physical traits." The reporter asked, "Might it lead to a society of DNA haves and have-nots, and the creation of a new underclass of people unable to keep up with the genetically fortified Joneses?" Similarly, a member of the National Bioethics Advisory Commission expressed concern that cloning might become "almost a preferred practice," taking its place "on the continuum of providing the best for your child." As a consequence, parents who chose to "play the lottery of old-fashioned reproduction would be considered irresponsible."

Such fears, however, seem more warranted with respect to genetic engineering than to cloning. By offering some people—in all probability, members of the upper classes—the opportunity to acquire desired traits through genetic manipulation, genetic engineering could bring about a biological reinforcement (or accentuation) of existing social divisions. It is hard enough already for disadvantaged children to compete with their more affluent counterparts, given the material resources and intellectual opportunities that are often available only to children of privilege. This unfairness would almost certainly be compounded if genetic manipulation came into the picture. In contrast, cloning does not bring about "improvements" in the genome: it is, rather, a way of *duplicating* the genome— with all its imperfections. It wouldn't enable certain groups of people to keep getting better and better along some valued dimension.

To some critics, admittedly, this difference will not seem terribly important. Theologian Gilbert Meilaender, Jr., objects to cloning on the grounds that children created through this technology would be "designed as a product" rather than "welcomed as a gift." The fact that the design process would be more selective and nuanced in the case of genetic engineering would, from this perspective, have no moral significance. To the extent that this objection reflects a concern about the commodification of human life, we can address it in part when we consider people's reasons for engaging in cloning.

REASONS FOR CLONING

This final area of contention in the cloning debate is as much psychological as it is scientific or philosophical. If human cloning technology were safe and widely available, what use would people make of it? What reasons would they have to engage in cloning?

In its report to the President, the Commission imagined a few situations in which people might avail themselves of cloning. In one scenario, a husband and wife who wish to have children are both carriers of a lethal recessive gene:

> Rather than risk the one in four chance of conceiving a child who will suffer a short and painful existence, the couple considers the alternatives: to forgo rearing children; to adopt; to use prenatal diagnosis and selective abortion; to use donor gametes free

of the recessive trait; or to use the cells of one of the adults and attempt to clone a child. To avoid donor gametes and selective abortion, while maintaining a genetic tie to their child, they opt for cloning.

In another scenario, the parents of a terminally ill child are told that only a bone marrow transplant can save the child's life. "With no other donor available, the parents attempt to clone a human being from the cells of the dying child. If successful, the new child will be a perfect match for bone marrow transplant, and can be used as a donor without significant risk or discomfort. The net result: two healthy children, loved by their parents, who happen to be identical twins of different ages."

The Commission was particularly impressed by the second example. That scenario, said the NBAC report, "makes what is probably the strongest possible case for cloning a human being, as it demonstrates how this technology could be used for lifesaving purposes." Indeed, the report suggests that it would be a "tragedy" to allow "the sick child to die because of a moral or political objection to such cloning." Nevertheless, we should note that many people would be morally uneasy about the use of a minor as a donor, regardless of whether the child were a result of cloning. Even if this unease is justifiably overridden by other concerns, the "transplant scenario" may not present a more compelling case for cloning than that of the infertile couple desperately seeking a biological child.

Most critics, in fact, decline to engage the specifics of such tragic (and presumably rare) situations. Instead, they bolster their case by imagining very different scenarios. Potential users of the technology, they suggest, are narcissists or control freaks—people who will regard their children not as free, original selves but as products intended to meet more or less rigid specifications. Even if such people are not genetic determinists, their recourse to cloning will indicate a desire to exert all possible influence over the "kind" of child they produce.

The critics' alarm at this prospect has in part to do, as we have seen, with concerns about the psychological burdens such a desire would impose on the clone. But it also reflects a broader concern about the values expressed, and promoted, by a society's reproductive policies. Critics argue that a society that enables people to clone themselves thereby endorses the most narcissistic reason for having children—to perpetuate oneself through a genetic encore. The demonstrable falsity of genetic determinism may detract little, if at all, from the strength of this motive. Whether or not clones will have a grievance against their parents for producing them with this motivation, the societal indulgence of that motivation is improper and harmful.

It can be argued, however, that the critics have simply misunderstood the social meaning of a policy that would permit people to clone themselves even in the absence of the heartrending exigencies described in the NBAC report. This country has developed a strong commitment to reproductive autonomy. (This commitment emerged in response to the dismal history of eugenics—the very history that is sometimes invoked to support restrictions on cloning.) With the exception of practices that risk coercion and exploitation—notably baby-selling and commercial surrogacy—we do not interfere with people's freedom to create and acquire children by almost any means, for almost any reason. This policy does not reflect a dogmatic libertarianism. Rather, it recognizes the extraordinary personal importance and private character of reproductive decisions, even those with significant social repercussions.

Our willingness to sustain such a policy also reflects a recognition of the moral complexities of parenting. For example, we know that the motives people have for bringing a child into the world do not necessarily determine the manner in which they raise him. Even when parents start out as narcissists, the experience of child-rearing will sometimes transform their initial impulses, making them caring, respectful, and even self-sacrificing. Seeing their child grow and develop, they learn that she is not merely an extension of themselves. Of course, some parents never make this discovery; others, having done so, never forgive their children for it. The pace and extent of moral development among parents (no less than among children) are infinitely variable. Still, we are justified in saying that those who engage in cloning will not, by virtue of this fact, be immune to the transformative effects of parenthood—even if it is the case (and it won't always be) that they begin with more problematic motives than those of parents who engage in the "genetic lottery."

Moreover, the nature of parental motivation is itself more complex than the critics often allow. Though we can agree that narcissism is a vice not to be encouraged, we lack a clear notion of where pride in one's children ends and narcissism begins. When, for example, is it unseemly to bask in the reflected glory of a child's achievements? Imagine a champion gymnast who takes delight in her daughter's athletic prowess. Now imagine that the child was actually cloned from one of the gymnast's somatic cells. Would we have to revise our moral assessment of her pleasure in her daughter's success? Or suppose a man wanted to be cloned and to give his child opportunities he himself had never enjoyed. And suppose that, rightly or wrongly, the man took the child's success as a measure of his own untapped potential—an indication of the flourishing life he might have had. Is this sentiment blamable? And is it all that different from what many natural parents feel?

CONCLUSION

Until recently, there were few ethical, social, or legal discussions about human cloning via nuclear transplantation, since the scientific consensus was that such a procedure was not biologically possible. With the appearance of Dolly, the situation has changed. But although it now seems more likely that human cloning will become feasible, we may doubt that the practice will come into widespread use.

I suspect it will not, but my reasons will not offer much comfort to the critics of cloning. While the technology for nuclear transplantation advances, other technologies—notably the technology of genetic engineering—will be progressing as well. Human genetic engineering will be applicable to a wide variety of traits; it will be more powerful than cloning, and hence more attractive to more people. It will also, as I have suggested, raise more troubling questions than the prospect of cloning has thus far.

NOTE

Reprinted by permission from Reports of the Institute for Philosophy and Public Policy, University of Maryland.

SOURCES

National Bioethics Advisory Commission, "Cloning Human Beings: Report and Recommendations," June 9, 1997; James Q. Wilson, "The Paradox of Cloning," *Weekly Standard,* May 26, 1997; Jean Bethke Elshtain, "Ewegenics," *New Republic,* March 31, 1997; R. C. Lewontin, "The Confusion over Cloning," *New York Review of Books,* October 23, 1997; Leon Kass, "The Wisdom of Repugnance," *New Republic,* June 2, 1997; Susan Cohen, "What Is a Baby? Inside America's Unresolved Debate about the Ethics of Cloning," *Washington Post Magazine,* October 12, 1997; Rick Weiss, "Genetic Enhancements' Thorny Ethical Traits," *Washington Post,* October 12, 1997.

33

Stem Cell Research and Applications: Findings and Recommendations

American Association for the Advancement of Science and Institute for Civil Society

FINDINGS AND RECOMMENDATIONS
November 1999

- Human stem cell research holds enormous potential for contributing to our understanding of fundamental human biology. Although it is not possible to predict the outcomes from basic research, such studies will offer the real possibility for treatments and ultimately for cures for many diseases for which adequate therapies do not exist.

The benefits to individuals and to society gained by the introduction of new drugs or medical technologies are difficult to estimate. The introductions of antibiotics and vaccines, for example, have dramatically increased life spans and improved the health of people all over the world. Despite these and other advances in the prevention and treatment of human diseases, devastating illnesses such as heart disease, diabetes, cancer, and diseases of the nervous system such as Alzheimer's disease present continuing challenges to the health and well-being of people everywhere. The science leading to the development of techniques for culturing human stem cells could lead to unprecedented treatments and even cures for these and other diseases.

As with all research, our ability even to contemplate the possibilities offered by stem cell–derived therapies is a result of many years of research. The science of stem cells dates to the mid-1960s, and many papers have been published on the isolation and laboratory manipulation of stem cells from animal models. While

583

these models are imperfect, they are accepted in the scientific community as good initial predictors of what occurs in human beings.

There already exists evidence from animal studies that stem cells can be made to differentiate into cells of choice, and that these cells will act properly in their transplanted environment. In human beings, transplants of hematopoietic stem cells (the cells which eventually produce blood) following treatments for cancer, for example, have been done for years now. Further, somewhat cruder experiments (e.g., the transplantation of fetal tissue into the brains of Parkinson's patients) indicate that the expectation that stem cell therapies could provide robust treatments for many human diseases is a reasonable one. It is only through controlled scientific research that the true promise will be understood.

- This research raises ethical and policy concerns, but these are not unique to stem cell research.

Innovative research and new technologies derived from such research almost always raise ethical and policy concerns. In biomedical research, these issues include the ethical conduct of basic and clinical research as well as the equitable distribution of new therapies. These issues are relevant to discussions about stem cell research and its eventual applications; however, they are part of a constellation of ethical and policy concerns associated with all advances in biomedical research. Guidelines or policies for the use of human biological materials have been issued at many levels, from internal review boards to the National Bioethics Advisory Commission, which recently released a detailed report on the use of such materials. Existing policies cover all aspects of research, from the use of cell lines in laboratories to human subjects protections, that will surface in the consideration of stem cell research.

It is essential that there be a public that is educated and informed about the ethical and policy issues raised by stem cell research and its applications. Informed public discussion of these issues should be based on an understanding of the science associated with stem cell research, and it should involve a broad cross-section of society.

It is essential for citizens to participate in a full and informed manner in public policy deliberations about the development and application of new technologies that are likely to have significant social impact. The understanding of the science is particularly important for discussing ethical and policy issues. Ideally, scientists should communicate the results of their research in ways that will be readily understandable to a diverse audience, and participate in public discussions related to stem cell research.

The ethical and policy issues raised by stem cell research are not unique, but this research has received a significant amount of public attention and there is much to gain by open reflection on the implications of this sensitive area of research. Congressional hearings, public meetings by government agencies, and media coverage have pushed stem cell research issues into a spotlight.

There should be continued support for the open manner that has allowed all those interested to observe or participate in these processes and for a sustained dialogue among scientists, policy makers, ethicists, theologians, and the public to consider issues that emerge with the advancement of stem cell research.

- Existing federal regulatory and professional control mechanisms, combined with informed public dialogue, provide a sufficient framework for oversight of human stem cell research.

The appearance of new technology can evoke apprehension and engender uncertainty among segments of the population about its uses. Where these concerns are related to issues having important ethical and social implications, certain levels of oversight are appropriate. But it is important to create new oversight mechanisms or regulatory burdens only when there are compelling reasons for doing so.

Federal funding would automatically trigger a set of oversight mechanisms now in place to ensure that the conduct of biomedical research is consistent with broad social values and legal requirements. While basic laboratory research with personally non-identifiable stem cells does not pose special ethical or oversight challenges, an elaborate system of review is in place for research involving human subjects, ranging from procurement issues to the conduct of clinical trials. The Federal Common Rule governing human subjects research provides for local and federal agency review of research proposals in such circumstances, weighing risks against benefits and requiring involved and voluntary consent. The Food and Drug Administration (FDA) has the authority to regulate the development and use of human stem cells that will be used as biological products, drugs, or medical devices to diagnose, treat or cure a disease or underlying condition. Further, states should adopt the federal government's *Model Program for the Certification of Embryo Laboratories.*

Complementing these regulatory mechanisms are the National Bioethics Advisory Commission (NBAC), which has demonstrated its legitimate claim to respect for its efforts as a national body to promote public input into social policy related to advances in biomedical research, and the Recombinant DNA Advisory Committee (RAC), which currently has a mandate to review the ethical and policy issues associated with gene therapy and could be authorized to change its mission to broaden its purview. These federal bodies should work with interested stakeholders in the conduct of stem cell research—professional organizations, patient disease groups, religious communities, the Congress, funding agencies and private foundations, industry, and others—so that the public can be assured that appropriate safeguards are in place as this research evolves.

Thus, at the present time, no new regulatory mechanisms are needed to ensure responsible social and professional control of stem cell research in the United States.

- Federal funding for stem cell research is necessary in order to promote investment in this promising line of research, to encourage sound public policy, and to foster public confidence in the conduct of such research.

Realizing the potential health benefits of stem cell technology will require a large and sustained investment in research. The federal government is the only realistic source for such an infusion of funds, for those who are challenged daily by serious diseases that could in the future be relieved by promise for sooner rather than later research results that can be transferred from the bench to the bedside. Without the stimulus of public funding, new treatments could be substantially delayed.

The commitment of federal funds also offers a basis for public review, approval, and monitoring through well established oversight mechanisms that will promote the public's interest in ensuring that stem cell research is conducted in a way that is both scientifically rigorous and ethically proper. Additionally, public funding contributes to sound social policy by increasing the probability that the results of stem cell research will reflect broad social priorities that are unlikely to be considered if the research is carried out in the private sector alone.

There are segments of American society that disagree on moral grounds with using public monies to support certain types of stem cell research. However, public policy in a pluralistic society cannot resolve all the differences that arise in national debates on sensitive social issues. In the context of stem cell research, this leads to three practical conclusions. One is a willingness to permit individuals, whether they are researchers or embryo or fetal tissue donors, to act in conformity with their own moral views on these matters. A second is the commitment to public involvement in research support when this research is related to the promotion and protection of public health, including the acquisition of new molecular and cellular insights into basic human developmental biology. A third is respect for opposing views, especially those based on religious grounds, to the extent that this is consistent with the protection and promotion of public health and safety.

- Public and private research on human stem cells derived from all sources (embryonic, fetal, and adult) should be conducted in order to contribute to the rapidly advancing and changing scientific understanding of the potential of human stem cells from these various sources.

There are three primary sources of stem cells, each with different characteristics as to how many different developmental paths they can follow and how much they can contribute to our understanding of a functioning organism. Embryonic stem cells (ES cells), derived from a very early embryo, and embryonic germ cells (EG cells), collected from fetal tissue at a somewhat later stage of development, have particular promise for a wide range of therapeutic applications because, according to our present knowledge, they are capable of giving rise to virtually any cell type. Research on these primordial cells will also provide a unique opportunity to study human cell biology.

Adult stem cells, obtained from mature tissues, differentiate into a narrower range of cell types. As a result, many cells of medical interest cannot currently be obtained from adult-derived stem cells. It is also less feasible to develop large-scale cultures from adult stem cells. However, it is important to note that, at this time, it is only adult human stem cells that are well-enough understood that they can be reliably differentiated into specific tissue types, and that have proceeded to clinical trials.

Because the study of human stem cells is at an early stage of development, it is difficult to predict outcomes and findings at this point in time. As more research takes place, the full developmental potential of different kinds of stem cells will become better understood.

In view of the moral concerns surrounding the uses of embryonic and fetal tissue voiced by a segment of the American population, strengthening federally and privately funded research into alternative sources and/or methods for the derivation of stem cells, including further initiatives on adult stem cells, should be encouraged. Human stem cell research can be conducted in a fully ethical manner, but it is true that the extraction of embryonic stem cells from the inner mass of blastocysts raises ethical questions for those who consider the intentional loss of embryonic life by intentional means to be morally wrong. Likewise, the derivation of embryonic germ cells from the gonadal tissue of aborted fetuses is problematic for those who oppose abortion. In contrast, adult stem cell research is more broadly acceptable to the American population.

- Public funding should be provided for embryonic stem cell and embryonic germ cell research, but not at this time for activities involved in the isolation of embryonic stem cells, about which there remains continuing debate. This approach will allow publicly funded researchers to move more quickly toward discoveries that will lead to alleviating the suffering caused by human disease.

Although the derivation of human stem cells can be done in an ethical manner, there is enough objection to the process of deriving stem cells to consider recommending against its public funding. Further, for the foreseeable future there will be sufficient material isolated by researchers not using public funding that this exclusion will not have a negative impact on research.

There are many individuals who believe that any use of human embryos other than for achieving a pregnancy is unethical, believing that the embryo is a full human being from the earliest moments in the conception process. However, many religious traditions take a "developmental" view of personhood, believing that the early embryo or fetus only gradually becomes a full human being and thus may not be entitled to the same moral protections as it will later; others hold that while the embryo represents human life, that life may be taken for the sake of saving and preserving other lives in the future. The dialogue about these issues is ongoing in the United States, but these concerns need not exclude publicly funded research activities on cell lines that have already been established.

- Embryonic stem cells should be obtained from embryos remaining from infertility procedures after the embryo's progenitors have made a decision that they do not wish to preserve them. This decision should be explicitly renewed prior to securing the progenitors' consent to use the embryos in ES cell research.

The most ethical source of human primordial stem cells is embryos produced for the process of in vitro fertilization whose progenitors have decided not to implant them and have given full and informed consent for the use of these embryos for research purposes. Two appropriate potential sources of donation are embryos with poor quality that makes them inappropriate for transfer and embryos remaining when couples have definitely completed their family and do not wish to donate the excess embryos to others.

Informed consent requires that the woman or couple, with substantial under-standing and without controlling influences, authorize the use of their spare embryos for research purposes. Because assisted reproduction can be a stressful process, informed consent should be secured in two stages. The two-stage process would also maintain a separation between personnel working with the woman or couple who hope to get pregnant and personnel requesting embryos for stem cell research.

At the beginning of the process, personnel working with the woman or couple who hope to become pregnant should ascertain their preferences as to the future of embryos remaining after the assisted reproduction process. These options should include consent for embryo donation to another couple, consent for donation for research, and consent for destruction of the spare embryos. Once a couple has definitely decided that it has completed its family, then the couple should be approached a second time to secure an explicit consent to use the embryos in ES cell research.

- Persons considering donating their excess embryos for research purposes should be afforded the highest standards of protection for the informed con-sent and voluntariness of their decision.

Securing embryos for the purpose of harvesting stem cells must proceed in a careful fashion for several reasons. These are to protect the interests of the gamete donors, to reassure the public that important boundaries are not being overstepped, to enable those who are ethically uncomfortable with elements of this research to participate to the greatest extent possible, and to ensure the highest quality of research and outcomes possible.

Consonant with good research practice, policies on the procurement of embryos should include at least the following points: (1) women should not undergo extra cycles of ovulation and retrieval in order to produce more "spare" embryos in the hope that some of them might eventually be donated for research; (2) analogous with our current practice for organ donation, there should be a solid "wall" between personnel working with the woman or couple who hope to get pregnant, and personnel requesting embryos for stem cell purposes; (3) women and men, as individuals or as couples, should not be paid to produce embryos, nor should they receive reduced fees for their infertility procedures for doing so; and (4) consent of both gamete donors should be obtained.

- Where appropriate, guidelines that can attract professional and public support for conducting stem cell research should be developed.

At present, stem cell research raises no unique ethical or policy issues. As research advances, issues may emerge that challenge acceptable ethical practices and public policy. Hence, there should be opportunities for public reconsideration of the need for guidelines specifically targeted to human stem cell research. Such efforts should be informed by the most current scientific evidence and should occur through a process that encourages broad involvement by all sectors of society.

Almost two decades of experience with the Recombinant DNA Advisory Committee's (RAC) oversight of recombinant DNA research suggest that the RAC could be an effective institutional focal point within the federal government to facilitate the type of public dialogue on stem cell research proposed here, and to coordinate efforts to develop new guidelines, where needed. The RAC has a proven track record of providing an open forum for sorting out complex ethical issues and of defusing conflict. Furthermore, it has acquired a degree of legitimacy among scientists in both the public and private sectors, with its widely accepted Points to Consider in the design and conduct of gene therapy.

- In order to allow persons who hold diverse moral positions on the status of the early embryo to participate in stem cell research to the greatest degree possible without compromising their principles, and also to foster sound science, stem cells (and stem cell lines) should be identified with respect to their original.

Patients and researchers should be able to avoid participating in stem cell use if the cells were derived in a way that they would consider to be unethical. As a matter of good scientific practice, records are routinely maintained on the sources of biological materials. It is of utmost importance that documentation of the original source of the stem cells can be made readily available to researchers and to potential recipients of stem cell therapies.

- Special efforts should be made to promote equitable access to the benefits of stem cell research.

The therapeutic potential for treating and possibly curing many serious diseases constitutes a major rationale for large-scale investments of public and private resources in human stem cell research. To justify funding stem cell research on the basis of its potential benefits, particularly the use of public resources, however, requires some assurance that people in need will have access to the therapies as they become available.

Several factors make it unlikely that there will be equitable access to the benefits of this research. Unlike other Western democracies, the United States does not have a commitment to universal health care. More than 44 million people lack health insurance and therefore do not have reliable access even to basic health care. Others are underinsured. Moreover, if stem cell research were to result in highly technological and expensive therapies, health insurers might be reluctant to fund such treatments.

Overcoming these hurdles and assuring equitable access to the benefits of stem cell research in this country will be a politically and financially challenging task. It is therefore appropriate to begin considering how to do so now in advance of the development of applications. The federal government should consider ways to achieve equitable access to the benefits derived from stem cell research.

- Intellectual property regimes for stem cell research should set conditions that do not restrict basic research or encumber future product development.

The U.S. Patent and Trademark Office (PTO) has already stated that purified and isolated stem cell products and research tools meet the criteria for patentable subject matter. When research is funded by the private sector, as is currently the case with stem cell research, and is patented, it is a private matter whether and under what terms new intellectual property is obtainable for research purposes or development. This is of particular concern because the private sector will not invest resources in potential applications that they consider to lack commercial value, but that may have considerable therapeutic promise.

Given the promise of stem cell research, it is important to encourage the development of broadly beneficial therapeutic products with widespread access. This objective could be achieved in a variety of ways. Government investment in promising areas of research would enable federal agencies and laboratories to hold patents and to exercise them in ways that enhance development and contribute to the dissemination of this stem cell technology. Congress or the PTO should define a strong research exemption that would give third parties access to stem cell products and research tools for research purposes without having to obtain permission from the patent holder.

Another possibility is to require compulsory licensing under limited and clearly defined circumstances.

- The formation of company-based, independent ethics advisory boards should be encouraged in the private sector.

Private sector research has played a crucial part in the advancement of research on stem cells. The leadership exhibited by the company that has sponsored all of the published human embryonic and germ cell research to date in establishing an external Ethics Advisory Board to develop guidelines for the ethical conduct of such research is laudable. While these private sector boards are not a substitute for public oversight and guidance, they can be a positive influence on the way that industry-funded stem cell research proceeds.

The credibility and impact of such ethics advisory boards will be enhanced if they review ethical issues at the start-up phase of the research, have multidisciplinary membership, including representatives from the local community, give minimum, if any, financial compensation for service, and share their own findings and recommendations with other companies. The latter provision could be especially helpful in developing a "case law" in the private sphere that would inform public efforts to develop national guidelines.

NOTE

Reprinted by permission of the American Association for the Advancement of Science.

34

On Human Embryos and Stem Cell Research: An Appeal for Legally and Ethically Responsible Science and Public Policy

Coalition of Americans for Research Ethics

Recent scientific advances in human stem cell research have brought into fresh focus the dignity and status of the human embryo. These advances have prompted a decision by the Department of Health and Human Services (HHS) and the National Institutes of Health (NIH) to fund stem cell research which is dependent upon the destruction of human embryos. Moreover, the National Bioethics Advisory Commission (NBAC) is calling for a modification of the current ban against federally funded human embryo research in order to permit direct federal funding for the destructive harvesting of stem cells from human embryos. These developments require that the legal, ethical, and scientific issues associated with this research be critically addressed and articulated. Our careful consideration of these issues leads to the conclusion that human stem cell research requiring the destruction of human embryos is objectionable on legal, ethical, and scientific grounds. Moreover, destruction of human embryonic life is unnecessary for medical progress, as alternative methods of obtaining human stem cells and of repairing and regenerating human tissue exist and continue to be developed.

HUMAN EMBRYONIC STEM CELL RESEARCH VIOLATES EXISTING LAW AND POLICY

In November 1998, two independent teams of U.S. scientists reported that they had succeeded in isolating and culturing stem cells obtained from human embryos

593

and fetuses. Stem cells are the cells from which all 210 different kinds of tissue in the human body originate. Because many diseases result from the death or dysfunction of a single cell type, scientists believe that the introduction of healthy cells of this type into a patient may restore lost or compromised function. Now that human embryonic stem cells can be isolated and multiplied in the laboratory, some scientists believe that treatments for a variety of diseases such as diabetes, heart disease, Alzheimer's, and Parkinson's may be within reach. While we in no way dispute the fact that the ability to treat or heal suffering persons is a great good, we also recognize that not all methods of achieving a desired good are morally or legally justifiable. If this were not so, the medically accepted and legally required practices of informed consent and of seeking to do no harm to the patient could be ignored whenever some "greater good" seems achievable.

One of the great hallmarks of American law has been its solicitous protection of the lives of individuals, especially the vulnerable. Our nation's traditional protection of human life and human rights derives from an affirmation of the essential dignity of every human being. Likewise, the international structure of human rights law—one of the great achievements of the modern world—is founded on the conviction that when the dignity of one human being is assaulted, all of us are threatened. The duty to protect human life is specifically reflected in the homicide laws of all 50 states. Furthermore, federal law and the laws of many states specifically protect vulnerable human embryos from harmful experimentation. Yet in recently publicized experiments, stem cells have been harvested from human embryos in ways which destroy the embryos.

Despite an existing congressional ban on federally funded human embryo research, the Department of Health and Human Services (HHS) determined on January 15, 1999 that the government may fund human embryonic stem cell research. The stated rationales behind this decision are that stem cells are not embryos (which itself may be a debatable point) and that research using cells obtained by destroying human embryos can be divorced from the destruction itself. However, even NBAC denies this latter claim, as is evident by the following statement in its May 6, 1999 Draft Report on Stem Cell Research:

> Whereas researchers using fetal tissue are not responsible for the death of the fetus, researchers using stem cells derived from embryos will typically be implicated in the destruction of the embryo. This is true whether or not researchers participate in the derivation of embryonic stem cells. As long as embryos are destroyed as part of the research enterprise, researchers using embryonic stem cells (and those who fund them) will be complicit in the death of embryos.

If the flawed rationales of HHS are accepted, federally funded researchers may soon be able to experiment on stem cells obtained by destroying embryonic human beings, so long as the act of destruction does not itself receive federal funds. However, the very language of the existing ban prohibits the use of federal funds to support "research in which a human embryo or embryos are destroyed, discarded, or knowingly subjected to risk of injury or death." Obviously, Congress' intent here was not merely to prohibit the use of federal funds for embryo destruction, but to prohibit the use of such funds for research dependent in any way upon such destruction. Therefore, the opinion of HHS that human embryonic stem cell research may receive federal funding clearly violates both the language of and intention behind the existing law. Congress and the courts should ensure that the law is properly interpreted and enforced to ban federal funding for research which harms, destroys, or is dependent upon the destruction of human embryos.

It is important to recognize also that research involving human embryos outside the womb—such as embryos produced in the laboratory by in vitro fertilization (IVF) or cloning—has never received federal funding. Initially, this was because a federal regulation of 1975 prevented government funding of IVF experiments unless such experiments were deemed acceptable by an Ethics Advisory Board. Following the failure of the first advisory board to reach a consensus on the matter, no administration chose to appoint a new board. After this regulation was rescinded by Congress in 1993, the Human Embryo Research Panel recommended to the National Institutes of Health (NIH) that certain kinds of harmful nontherapeutic experiments using human embryos receive federal funding. However, these recommendations were rejected in part by President Clinton and then rejected in their entirety by Congress.

Further, it is instructive to note that the existing law which permits researchers to use fetal tissue obtained from elective abortions requires that the abortions are performed for reasons which are entirely unrelated to the research objectives. This law thus prohibits HHS from promoting the destruction of human life in the name of medical progress, yet medical progress is precisely the motivation and justification offered for the destruction of human life that occurs when stem cells are obtained from human embryos.

Current law against funding research in which human embryos are harmed and destroyed reflects well-established national and international legal and ethical norms against the misuse of any human being for research purposes. Since 1975, those norms have been applied to unborn children at every stage of development in the womb, and since 1995 they have been applied to the human embryo outside the womb as well. The existing law on human embryonic research is a reflection of universally accepted principles governing experiments on human subjects— principles reflected in the Nuremberg Code, the World Medical Association's Declaration of Helsinki, the United Nations Declaration of Human Rights, and many other statements. Accordingly, members of the human species who cannot give informed consent for research should not be the subjects of an experiment unless they personally may benefit from it or the experiment carries no significant risk of harming them. Only by upholding such research principles do we prevent treating people as things—as mere means to obtaining knowledge or benefits for others.

It may strike some as surprising that legal protection of embryonic human beings co-exists with the U.S. Supreme Court's 1973 legalization of abortion. However, the Supreme Court has never prevented the government from protecting prenatal life outside the abortion context, and public sentiment also seems even more opposed to government funding of embryo experimentation than to the funding of abortion. The laws of a number of states—including Louisiana, Maine, Massachusetts, Michigan, Minnesota, Pennsylvania, Rhode Island, and Utah—specifically protect embryonic human beings outside the womb. Most of these provisions prohibit experiments on embryos outside the womb. We believe that the above legally acknowledged protections against assaults on human dignity must be extended to all human beings—irrespective of gender, race, religion, health, disability, or age. Consequently, the human embryo must not be subject to willful destruction even if the stated motivation is to help others. Therefore, on existing legal grounds alone, research using stem cells derived from the destruction of early human embryos is proscribed.

HUMAN EMBRYONIC STEM CELL RESEARCH IS UNETHICAL

The HHS decision and the recommendations of NBAC to federally fund research involving the destruction of human embryos would be profoundly disturbing even if this research could result in great scientific and medical gain. The prospect of government-sponsored experiments to manipulate and destroy human embryos should make us all lie awake at night. That some individuals would be destroyed in the name of medical science constitutes a threat to us all. Recent statements claiming that human embryonic stem cell research is too promising to be slowed or prohibited underscore the sort of utopianism and hubris that could blind us to the truth of what we are doing and the harm we could cause to ourselves and others. Human embryos are not mere biological tissues or clusters of cells; they are the tiniest of human beings. Thus, we have a moral responsibility not to deliberately harm them.

An international scientific consensus now recognizes that human embryos are biologically human beings beginning at fertilization and acknowledges the physical continuity of human growth and development from the one-cell stage forward. In the 1970s and 1980s, some frog and mouse embryologists referred to the human embryo in its first week or two of development as a "pre-embryo," claiming that it deserved less respect than embryos in later stages of development. However, some embryology textbooks now openly refer to the term "pre-embryo" as a scientifically invalid and "inaccurate" term which has been "discarded," and others which once used the term have quietly dropped it from new editions. Both the Human Embryo Research Panel and the National Bioethics Advisory Commission have also rejected the term, describing the human embryo from its earliest stages as a living organism and a "developing form of human life." The claim that an early human embryo becomes a human being only after 14 days or implantation in the womb is therefore a scientific myth. Finally, the historic and well-respected 1995 Ramsey Colloquium statement on embryo research acknowledges that:

> The [embryo] is human; it will not articulate itself into some other kind of animal. Any being that is human is a human being. If it is objected that, at five days or fifteen days, the embryo does not look like a human being, it must be pointed out that this is precisely what a human being looks like—and what each of us looked like—at five or fifteen days of development.

Therefore, the term "pre-embryo," and all that it implies, is scientifically invalid.

The last century and a half has been marred by numerous atrocities against vulnerable human beings in the name of progress and medical benefit. In the 19th century, vulnerable human beings were bought and sold in the town square as slaves and bred as though they were animals. In this century, the vulnerable were executed mercilessly and subjected to demeaning experimentation at Dachau and Auschwitz. At mid-century, the vulnerable were subjects of our own government's radiation experiments without their knowledge or consent. Likewise, vulnerable African-Americans in Tuskegee, Alabama, were victimized as subjects of a government-sponsored research project to study the effects of syphilis. Currently, we are witness to the gross abuse of mental patients used as subjects in purely experimental research. These experiments were and are driven by a crass utilitarian ethos which results in the creation of a "sub-class" of human beings, allowing the rights of the few to be sacrificed for the sake of potential benefit to the many. These unspeakably cruel and inherently wrong acts against human beings have resulted in the enactment of laws and policies which require the protection of human rights and liberties, including the right to be protected from the tyranny of the quest for scientific progress. The painful lessons of the past should have taught us that human beings must not be conscripted for research without their permission—no matter what the alleged justification—especially when that research means the forfeiture of their health or lives. Even if an individual's death is believed to be otherwise imminent, we still do not have a license to engage in lethal experimentation—just as we may not experiment on death row prisoners or harvest their organs without their consent.

We are aware that a number of Nobel scientists endorse human embryonic stem cell research on the basis that it may offer a great good to those who are suffering. While we acknowledge that the desire to heal people is certainly a laudable goal and understand that many have invested their lives in realizing this goal, we also recognize that we are simply not free to pursue good ends via unethical means. Of all human beings, embryos are the most defenseless against abuse. A policy promoting the use and destruction of human embryos would repeat the failures of the past. The intentional destruction of some human beings for the alleged good of other human beings is wrong. Therefore, on ethical grounds alone, research using stem cells obtained by destroying human embryos is ethically proscribed.

HUMAN EMBRYONIC STEM CELL RESEARCH IS SCIENTIFICALLY QUESTIONABLE

Integral to the decision to use federal funds for research on human embryonic stem cells is the distinction between stem cells and embryos. HHS has stated that federal funds may be used to support human embryonic stem cell research because stem cells are not embryos. A statement issued by the Office of the General Counsel of HHS regarding this decision asserts that "the statutory prohibition on the use of [government] funds . . . for human embryo research would not apply to research utilizing human pluripotent stem cells because such cells are not a human embryo within the statutory definition. [Moreover, because] pluripotent stem cells do not have the capacity to develop into a human being, [they] cannot be considered human embryos consistent with the commonly accepted or scientific understanding of that term."

It is important to note that the materials used in an experiment, as well as the methods of experimentation, are considered to be part of scientific research. When a scientific study is published, the first part of the article details the methods and materials used to conduct the research. Ethical and scientific evaluation of an experiment takes into account both the methods and materials used in the research process. Therefore, the source of stem cells obtained for research is both a scientifically and ethically relevant consideration.

Research on human embryonic stem cells is objectionable due to the fact that such research necessitates the prior destruction of human embryos; however, the HHS's claim that stem cells are not, and cannot develop into, embryos may itself be subject to dispute. Some evidence suggests that stem cells cultured in the laboratory may have a tendency to recongregate and form an aggregate of cells capable of beginning to develop as an embryo. In 1993, Canadian scientists reported that they successfully produced a live-born mouse from a cluster of mouse stem cells. While it is true that these stem cells had to be wrapped in placenta-like cells in order to implant in a female mouse, it seems that at least some doubt has been cast on the claim that a cluster of stem cells is not embryonic in nature. If embryonic stem cells do indeed possess the ability to form or develop as a human embryo (without any process of activation which affects the transformation of the cell into a human embryo), research on such stem cells could itself involve the creation and/or destruction of human life and would thereby certainly fall under the existing ban on federally funded embryo research. It would be irresponsible for the HHS to conduct and condone human embryonic stem cell research without first discerning the status of these cells. Their use in any research in which they could be converted into human embryos should likewise be banned.

METHODS OF REPAIRING AND REGENERATING HUMAN TISSUE EXIST WHICH DO NOT REQUIRE THE DESTRUCTION OF HUMAN EMBRYOS

While proponents of human embryonic stem cell research lobby aggressively for government funding of research requiring the destruction of human embryos, alternative methods for repairing and regenerating human tissue render such an approach unnecessary for medical progress.

For instance, a promising source of more mature stem cells for the treatment of disease is hematopoietic (blood cell–producing) stem cells from bone marrow or even from the placenta or umbilical cord blood in live births. These cells are already widely used in cancer treatment and in research on treating leukemia and other diseases. Recent experiments have indicated that their versatility is even greater than once thought. For example, given the right environment, bone marrow cells can be used to regenerate muscle tissue, opening up a whole new avenue of potential therapies for muscular dystrophies. In April 1999, new advances were announced in isolating mesenchymal cells from bone marrow and directing them to form fat, cartilage, and bone tissue. Experts in stem cell research believe that these cells may allow for tissue replacement in patients suffering from cancer, osteoporosis, dental disease, or injury.

An enormously promising new source of more mature stem cells is fetal bone marrow, a source which is many times more effective than adult bone marrow and umbilical cord blood. It appears that fetal bone marrow cells do not provoke immune reactions to the same degree as adult or even newborn infant cells. This is true whether the unborn child is the donor or the recipient—that is, fetal cells can be used to treat adults, or adult bone marrow cells can be used to treat a child in the womb without the usual risk of harmful immune reactions. Such cells would not need to be derived from fetuses who were intentionally aborted, but could instead be obtained from spontaneously aborted fetuses or stillborn infants.

In 1999, unprecedented advances were also made in isolating and culturing neural stem cells from living human nerve tissue and even from adult cadavers. Such advances render it quite possible that treatment of neural diseases such as Parkinson's and Alzheimer's, as well as spinal cord injuries, will not depend upon destructive embryo research.

Earlier claims that embryonic stem cells are uniquely capable of "self-renewal" and indefinite growth can also now be seen as premature. For example, scientists have isolated an enzyme, telomerase, which may allow human tissues to grow almost indefinitely. Although this enzyme has been linked to the development of cancer, researchers have been able to use it in a controlled way to "immortalize" useful tissue without producing cancerous growths or other harmful side effects. Thus, cultures of non-embryonic stem cells may be induced to grow and develop almost indefinitely for clinical use.

One of the most exciting new advances in stem cell research is the January 1999 announcement that Canadian and Italian researchers succeeded in producing new blood cells from neural stem cells taken from an adult mouse. Until recently, it was believed that adult stem cells were capable of producing only a particular type of cell: for example, a neural stem cell could develop only into cells belonging to the nervous system. Researchers believed that only embryonic stem cells retained the capacity to form all kinds of tissue in the human body. However, if stem cells taken from adult patients can produce cells and tissues capable of functioning within entirely different systems, new brain tissue needed to treat a patient with Parkinson's disease, for example, might be generated from blood stem cells derived from the patient's bone marrow. Conversely, neural stem cells might be used to produce needed blood and bone marrow. Use of a patient's own stem cells would circumvent one of the major obstacles posed by the use of embryonic stem cells—namely, the danger that tissue taken from another individual would be rejected when transplanted into a patient. Thus, in commenting on this finding, the *British Medical Journal* remarked on January 30, 1999, that the use of embryonic stem cells "may soon be eclipsed by the more readily available and less controversial adult stem cells." Given that the function of the adult stem cells was converted without the cells first having to pass through an embryonic stage, the use of such cells would not be subject to the ethical and legal objections raised by the use of human embryonic stem cells. The director of the NIH has pointed out that evidence that adult stem cells can take on different functions has emerged only from studies on mice. However, his own claim that human embryonic stem cell research can produce treatments for diabetes and other diseases is also based solely on experimental success in mice.

One approach to tissue regeneration that does not rely on stem cells at all, but on somatic cell gene therapy, is already in use as an experimental treatment. A gene that controls production of growth factors can be injected directly into a patient's own cells, with the result that new blood vessels will develop. In early trials, this type of therapy saved the legs of patients who would have otherwise undergone amputation. It was reported in January 1999 that the technique has generated new blood vessels in the human heart and improved the condition of 19 out of 20 patients with blocked cardiac blood vessels. Such growth factors are now being explored as a means for growing new organs and tissues of many kinds.

The above recent advances suggest that it is not even necessary to obtain stem cells by destroying human embryos in order to treat disease. A growing number of researchers believe that adult stem cells may soon be used to develop treatments for afflictions such as cancer, immune disorders, orthopedic injuries, congestive heart failure, and degenerative diseases. Such researchers are working to further research on adult, rather than embryonic, stem cells. In light of these promising new scientific advances, we urge Congress to provide federal funding for the development of methods to repair and regenerate human tissue which do not require the destruction of embryonic human life. However, even if such methods do not prove to be as valuable in treating disease as are human embryonic stem cells, use of the latter in the name of medical progress is still neither legally nor ethically justifiable for the reasons stated in this document.

CONCLUSION

We believe that an examination of the legal, ethical, and scientific issues associated with human embryonic stem cell research leads to the conclusion that the use of federal funds to support any such research that necessitates the destruction of human embryos is, and should remain, prohibited by law. Therefore, we call on Congress to (1) maintain the existing ban against harmful federally funded human embryo research and make explicit its application to stem cell research requiring the destruction of human embryos and (2) provide federal funding for the development of alternative treatments which do not require the destruction of human embryonic life. If anything is to be gained from the cruel atrocities committed against human beings in the last century and a half, it is the lesson that the utilitarian devaluation of one group of human beings for the alleged benefit of others is a price we simply cannot afford to pay.

NOTE

Reprinted by permission of Coalition of Americans for Research Ethics.

Suggestions for Further Study

THE BASICS OF GENETICS AND ETHICS

Barnum, Susan, and Carol Barnum. *Biotechnology: An Introduction.* Brooks Cole, 1997.

Drlica, Karl. *Understanding DNA.* John Wiley, 1996.

Frankenna, William. *Ethics.* Prentice-Hall, 1994.

Rachels, James. *The Elements of Moral Philosophy.* 3d ed. McGraw-Hill, 2000.

Rachels, James, ed. *The Right Thing to Do: Basic Readings in Moral Philosophy.* McGraw-Hill, 1998.

Boylan, Michael, and Kevin Brown. *Genetic Engineering: Science and Ethics on the New Frontier.* Prentice-Hall, 2002.

Freestone, David, and Ellen Hay, eds. *The Precautionary Principle and International Law.* Kluwer, 1996.

Morris, Julien, ed. *Rethinking Risk and the Precautionary Principle.* Butterworth, 2000.

Nelson, Gerald, ed. *Genetically Modified Organisms in Agriculture: Economics and Politics.* Academic Press, 2000.

Peters, Ted. *Playing God.* Routledge, 1997.

Raffensberger, Carolyn, and Joel Ticknor, eds. *Protecting Public Health and the Environment: Implementing the Precautionary Principle.* Island, 1999.

Ridley, Matt. *Genome.* Harper, 2000.

Turner, Ronald Cole. *The New Genesis.* Westminster/John Knox, 1993.

Wildavsky, Aaron. *Searching for Safety.* Transaction, 1988.

AGRICULTURAL BIOTECHNOLOGY

Busch, Lawrence, et al. *Plants, Power, and Profits.* Blackwell, 1993.

Committee on Environmental Effects Associated with the Commercialization of Transgenic Plants. National Research Council. *Environmental Effects of Transgenic Plants.* National Academy Press, 2002.

Committee on Genetically Modified Pest Protected Plants. National Research Council. *Genetically Modified Pest Protected Plants: Science and Regulation.* National Academy Press, 2000.

Comstock, Gary. *Vexing Nature.* Kluwer, 2001.

Kloppenburg, Jack. *First the Seed: The Political Economy of Plant Biotechnology, 1492–2000.* Cambridge University Press, 1988.

Krimsky, Sheldon, and Roger Wrubel. *Agricultural Biotechnology.* University of Illinois Press, 1996.

Rissler, Jane, and Margaret Mellon. *The Ecological Risks of Engineered* Crops. MIT Press, 1996.

Thompson, Paul. *Agricultural Ethics: Research Teaching and Public Policy.* Iowa State University Press, 1998.

Thompson, Paul, et al. *Ethics, Public Policy, and Agriculture.* Macmillan, 1994.

Several volumes published by the National Agricultural Biotechnology Council are very useful, containing papers delivered at their annual meeting. Especially noteworthy are:
Agricultural Biotechnology and the Common Good, 1994.
Agricultural Biotechnology: Novel Products and New Partnerships, 1996.
Agricultural Biotechnology and Environmental Quality, 1998.

FOOD BIOTECHNOLOGY

Charles, Dan. *Lords of the Harvest.* Perseus, 2001.

Mather, Robin. *A Garden of Unearthly Delights.* Plume, 1996.

Pence, Greg. Designer Foods. Rowman & Littlefield, 2002.

———. *The Ethics of Food: A Reader for the Twenty-First Century.* Rowman & Littlefield, 2002.

ANIMAL BIOTECHNOLOGY

Cohen, Carl, and Tom Regan. *The Animal Rights Debate.* Rowman & Littlefield, 2001.

Fishman, Jay, ed. *Xenotransplantation.* New York Academy of Sciences, 1998.

Institute of Medicine, National Academy of Sciences. *Xenotransplantation: Science, Ethics, and Public Policy.* National Academy Press, 1996.

Regan, Tom. *The Case for Animal Rights.* University of California Press, 1985.

———. *Defending Animal Rights.* University of Illinois Press, 2001.

Singer, Peter. *Animal Liberation.* 2d ed. Avon, 1991.

Varner, Gary. *In Nature's Interests.* Oxford University Press, 1998.

HUMAN GENETIC SCREENING AND THERAPY

Andrews, Lori. *Future Perfect.* Columbia University Press, 2001.

Buchanan, Allen, et al. *From Chance to Choice: Genetics and Justice.* Cambridge University Press, 2000.

Chadwick, Ruth, et al., eds. *The Ethics of Genetic Screening.* Kluwer, 1999.

Kristol, William, and Eric Cohen, eds. *The Future Is Now: America Confronts the New Genetics.* Rowman & Littlefield, 2002.

Lemaine, Nicholas, and Richard Vie, eds. *Understanding Gene Therapy.* Springer-Verlag, 2000.

Lyon, Jeff. *Altered Fates.* Norton, 1996.

Resnick, David, et al. *Human Germline Gene Therapy.* R. G. Landes, 1999.

Willer, Roger. *Genetic Testing and Screening.* Kirk House, 1998.

CLONING

Kass, Leon, and James Wilson. *The Ethics of Human Cloning.* AEI Press, 1998.

McGee, Glenn, ed. *The Human Cloning Debate.* 2d ed. Berkeley Hills Books, 2000.

Pence, Gregory, ed. *Flesh of My Flesh: The Ethics of Cloning Humans.* Rowman & Littlefield, 2000.

Shostak, Stanley. *Becoming Immortal: Combining Cloning and Stem Cell Therapy.* SUNY Press, 2002.

Silver, Lee. *Remaking Eden: How Genetic Engineering and Cloning Will Transform the Family.* Avon, 1998.

Study Cases

AGRICULTURAL BIOTECHNOLOGY

Case 1

In February 2000 the Cartagena Protocol on Biosafety to the Convention on Biodiversity was signed in Montreal. As part of its language it permitted countries to ban imports of genetically modified plants and animals in the following language:

> Lack of scientific certainty due to insufficient relevant scientific information and knowledge regarding the extent of the potential adverse effects of a living modified organism on the conservation and sustainable use of biological diversity in the party of import, taking also into account risks to human health, shall not prevent that party from taking a decision as appropriate, with regard to the import of that living modified organism intended for direct use as food or feed, or for processing, in order to avoid or minimize such potential adverse effects.

Do you think that this "precautionary principle" is sound? How would you revise it? Would it or should it affect other technologies such as information technologies?

Case 2

The EPA requires review of all field trials of genetically modified organisms. Suppose you were a member of an EPA group reviewing the application of Bactgene, Inc. to test its improved nitrogen-fixing bacteria for wheat crops. The company claims that it has preliminary data to indicate that the gene cannot be passed on to other crops. A spokesman for Greenpeace opposes the experiment with the following argument: You can't do the study. Studying living organisms that cannot be recalled is inherently dangerous to ecosystems and thus to life itself. The study will destroy the inherent balance and value of nature. Since we have enough food, especially wheat, to feed the planet, no emergency exists to override ecosystem integrity or natural value. Another speaker replied that if this principle were accepted, then no vaccines could have been tested on human volunteers because something might go wrong and you would be upsetting the natural balance between human beings and disease.

What would your response be? Does the concept of ecosystem integrity make sense when ecosystems are actually changing so much in nature? How should we consider future generations and their need for food? What about their need for a stable living system?

Case 3

Red rice is a serious weed in rice production fields in North and South America. Red rice is extremely difficult to control because rice and red rice are different cultivars of the same genus and species and will therefore interbreed. The outcrossing frequency is approximately 2 percent, which is considered high by plant breeding standards. This means that in rice fields infested with red rice, traits of rice can be rapidly transferred to red rice.

Rice has been genetically engineered to have a herbicide-resistant gene. Because of outcrossing, it is apparent that if herbicide-resistant rice were grown in red rice–infested fields, the red rice would acquire herbicide resistance. An international donor agency funded the research that engineered the rice herbicide resistance. But after examining the potential consequences to U.S. rice production, it refused to allow the rice to be released for U.S. production.

A scientist at the international research agency learned of herbicide-resistant rice and requested and received from the donor agency herbicide-resistant rice seed for commercial production. The scientist wanted the seed because using the seed would eliminate the chemical load on the environment and reduce production costs.

In Colombia, typical rice culture consists of tilling the soil, irrigating to germinate all seeds, and chemically killing the emerging seedlings and weeds. This process is repeated three times before the seedbed is considered ready to plant.

1. What are the potential harms from using herbicide-resistant rice?
2. What are the potential benefits?
3. Who would be harmed and who would benefit?
4. On balance, would you produce and sell the seed?

Case 4

As minister for development for Parador, you are deeply concerned about the conditions of poverty in rural communities in your country. Agro Incorporated has come to you with a proposition. They have developed a transgenic soybean that is highly productive in precisely the soils prevalent in the farming areas of your country. They propose a massive program to introduce this crop to your farmers. They will educate farmers on the best growing techniques and will supply on long-term loans seed, chemicals, and harvesters. They will also enter into long-term contracts to buy all of the production for their international food grains and animal feed business.

Some experts believe that this program may provide much needed rural development.

Would you encourage this development? What risks do you see and how would you deal with them? Suppose you set up a committee to develop a recommendation. The committee has the following members:

A representative of the Ministry for Development
A professor of agriculture at the University of Parador
An economist
A representative of the peasant cooperative movement
The head of the Parador Large Growers Association
A representative of the Indigenous Peoples Alliance
A representative of the Ministry for Rural Affairs

Case 5

The United Nations Food and Agriculture Organization (FAO) has convened an international committee of which you are a member. The committee is established to make recommendations to the international community about agricultural biotechnology. The committee recognizes three starting points:

1. There is currently enough food for all of the world's population if it were distributed equitably.
2. If the United Nations declarations on the rights of families to decide on the number of their children are adhered to, population is likely to outgrow food supply in the future.
3. There is a long development time for new genetically modified crops and their widespread use.

Should the committee recommend a vigorous program of development and open field testing of new crops for more productive animal feed or greater harvests for human consumption?

What safeguards should be in place for underdeveloped countries and subsistence farmers worldwide? What about organic growers?

Case 6

Well-known biologist and opponent of biotechnology Dr. Mae-Wan Ho has written:

> Biodiversity and species integrity are inextricably linked. Transgenic technology transgresses both species integrity and species boundaries, leading to unexpected systemic effects on the physiology of the transgenic organisms produced, as well as the balanced ecological relationships on which biodiversity depends. . . .
>
> Nature is interconnected in such a way that each and every species maintains its own integrity and that may be the essence of biodiversity. Biodiversity may simply be a state of coherence for the ecological system akin to that which exists for the organism as a whole. (Mae-Wan Ho and Beatrix Tappeser, "Transgenic Transgression of Species and Species Integrity," available at the Web site "Genetic Engineering and Its Dangers" maintained by Ron Epstein at San Francisco State University)

What might your response be?

FOOD BIOTECHNOLOGY

Case 1

Congressperson Dana Webb has been concerned about the safety of genetically modified food for several years. As reports that as much as 70 percent of food in stores contains some genetic modification surfaced, her concern for what she calls "a massive unregulated experiment" intensified. She has now introduced a bill in Congress that will require extensive safety testing in animals and humans of any genetically modified food approved for human consumption. The FDA believes that this is unnecessary unless some evidence exists to suspect a health risk from a particular food. They believe that if a plant or animal has been altered in ways that human beings have already been exposed to, the risks will already be known and no additional testing is necessary. Webb believes that we should adopt the "precautionary principle" as she understands it and pretest all genetically modified food.

Suppose she held a town meeting on the subject and the following persons spoke:

A person from the Biotechnology Industry Organization
A representative of the Center for Food Safety
A representative of the Food Information Center (supported by big producers)
A representative of the Organic Consumers Association
A representative of the largest grocery chain in her state

A citizen activist on food safety
The president of her state's Farm Bureau

Case 2

The United Nations Food and Agriculture Organization has convened a meeting of leading food-producing countries to conduct a dialogue with the underdeveloped world about biotechnology in food production. One of the first speakers was a member of the Social Justice party of a very poor nation who made the following argument: "You are all concerned with technology; but where is the justice and security for my people? Will they be secure in their food supply when all the land is owned by corporations calling themselves "grocer to the world"? What happens when these poor people have to buy an adequate diet that they used to grow because all they grow now is a cash crop to be processed hundreds of miles away? My people were poor by your standards before, but they had land and dignity and they could feed themselves. You want to make them slaves to corporations that will destroy their food supply and take their land."

As assistant to the U.S. delegate to this conference, you have been asked to prepare a response to this argument. You can rebut the position in whole or in part or you can argue that the United States should support this position.

Case 3

In 1997 a researcher based in Scotland came on British television and claimed to have shown that there were serious adverse effects found in rats who had been fed genetically modified potatoes. At the time he did not provide data to support his charges. Eighteen months later he and a colleague submitted a research report to one of the leading scientific journals in the world supposedly documenting these claims. As is customary, the editor of the journal sent the paper out for review by other specialists.

After criticism the paper was revised three times to meet objections of reviewers. Finally the editor faced a choice. Some reviewers thought the paper should be published; others thought that it was fatally flawed and should be rejected. A third group thought that the study was deeply flawed but believed that it should be published anyway. Their view was that given the public unease with genetically modified foods, failure to publish would simply fuel fears of a conspiracy to hide the truth. On the other hand publishing it would lend the credibility of a major scientific journal to unfounded public fears.

If you were the editor, what would you do?

Case 4

Should food made from genetically modified organisms be labeled even if the food itself has no genetically modified materials but genetically modified plants or animals went into producing it? Consider the controversy over Starlink corn (a GM corn). It was supposed to be limited to use as an animal feed due to unresolved health concerns, but it turned up by mistake in taco shells and other corn products. The product was eventually pulled from the market. Suppose it had been strictly reserved for animal feed. Should the resulting meat products contain a label to reflect the fact that at some point this product involved biotechnology in its production process? Persons who do not come into contact with the genetically modified product cannot be physically affected so they cannot be physically harmed.

Should there be labeling? If so, what should the label say? The actual consumed product (e.g., ham) is technically "GMO free." Is this what the label should say?

Case 5

Apple production in the United States is plagued by the presence of root-infecting parasitic nematodes. Nematodes are small worms, like soil-borne invertebrates, that colonize and infect the roots of apple trees. The consequences of this root infection range from tree death to reduced yield. Typically, nematode control measures include use of methyl bromide, a highly toxic, volatile, ozone-depleting chemical that will be banned worldwide in 2005. In addition, applications of water-soluble, groundwater-contaminating nematicides are required throughout the life of the orchard, which is typically thirty years.

To aid in this battle to control nematode multiplication, the agriculture division of Viper Tech Life Sciences of Garden City, New Jersey, has discovered that a snake venom protein is toxic to nematodes. The coding sequence for the modified venom protein is assembled with a promoter to ensure that it only accumulates in the roots of apple trees. Exhaustive analysis has shown that the venom gene construct is not expressed in any plant tissues outside the root. Field evaluation has demonstrated that transgenic apple trees do not require the application of any nematicides. The modified venom protein is rapidly broken down in the soil.

Divide into groups with members of the following organizations. Each group should try to reach a decision on whether this technology should be deployed.

Apple growers
USDA/FDA
Food processors
Greenpeace
Biotechnology Industry Organization
Organic Growers Association

Case 6

The secretary of agriculture is in the process of developing rules for labeling food as "organic." Many letters to the Department of Agriculture expressed concern about genetically modified food in this context. Should the rules forbid all genetically modified plants and animals as sources for human "organic" food? Would this also include new plant or fruit varieties developed by breeders and/or exotic nonnative species in various areas? If the rules should permit this kind of breeding, why not include crops developed from biotechnology? What about the apples in the previous case? They are exactly like naturally grown apples and they can be grown to meet organic standards.

Case 7

The vice president for consumer affairs of the second largest grocery chain in the United States has a problem. She knows that there is a growing niche market of consumers opposed to genetically modified food. These consumers are concerned about many parts of the biotechnology enterprise: (1) the safety of the food they eat, (2) disruption of native agriculture, (3) domination of the world food supply by corporate giants, and (4) environmental issues like pesticides.

She is considering a program of negative labeling. Her company would get information from producers and label all food that does not contain and was not produced with any genetic modification. But she wonders what the label should say and why. Further, she wonders whether this would not cast other food in an unfavorable light for consumers. Since most people do not worry about genetically modified food, she wonders whether labeling would affect her store more negatively than no labeling.

ANIMAL TRANSGENICS

Case 1

We are not yet ready to clone primates, but for research purposes cloning primates like chimpanzees would be very helpful. JFM Corporation is targeting this problem extensively. They know that there will be mistakes. Many of the fertilized blastocysts will fail to implant. Some pregnancies will not carry to term. Finally some few births will be seriously abnormal and will have to be destroyed. The cloned animals will be cared for according to the best standards. They will be used for medical studies such as the development of an AIDS vaccine. After the study the animals will be destroyed.

In order to meet public criticism and ward off later complaints, JFM has decided to discuss its work with a group of knowledgeable individuals of diverse backgrounds. Break into groups and assume the following roles for discussion:

Someone from the Humane Association of the United States
The director of the primate behavior center at a major university

The director of research for a large pharmaceutical company
The director of the Christian Animal Welfare Association
The president of the HIV Action Network
A teacher of ethics at a university
A researcher at the National Institutes of Health

Case 2

Lamb is one of the most widely consumed meats in the world. Muslims and Jews have dietary rules that forbid eating pork, and in arid climates grass is difficult to grow for cattle. But grazing sheep on a large range is economically inefficient. Factory farming of sheep seems to be the solution. Sheep will be raised to the ideal weight in pens in large "farms" of several thousand pens directly attached to a packing plant. Factory farming, however, is subject to a number of objections concerning the welfare of the animals, and facilities can be a target of radicals such as the Animal Liberation Front.

Lamb's End, Inc. is proposing to mitigate these concerns with biotechnology. They propose to develop a transgenic line of sheep that will carry more weight per animal and lack fully developed brains, thus increasing the profitability and drastically reducing the suffering of each animal.

Is this a wise and moral use of biotechnology?

Case 3

Recently a wealthy individual approached researchers at a major university with this request: his beloved dog Missy "was getting on in years" and they wished "to reproduce her or at least create a genetic duplicate" with cloning. The researchers opened up a DNA bank so that other pet owners could have a cloned genetic duplicate of their loved one. It is called Genetic Savings and Clone. The project set out a strong code of ethics, which follows. What do you think of this project and this code? Are items missing? Is this a good use of scientific talent and financial resources?

Code of Ethics for the Missyplicity Project

1. In accordance with federal law, no animals will be intentionally harmed at any point during this project, including the research phase. In addition no animals will be endangered through lack of attention or care, or by being subjected to risky procedures of any sort.
2. The psychological welfare, happiness, and socialization of all dogs involved with the project shall be considered at all times. Every dog shall be guaranteed a minimum of one hour of outdoor playtime, weather permitting, indoors if necessary, with people hired specifically for this task.
3. Regardless of the source through which dogs are obtained for use as egg donors

or surrogate mothers, at the completion of their role on the project all dogs shall be placed in loving homes.

4. The "turnover rate" before an animal involved with the project is placed in a loving home shall be limited to eight months, with less than six being the goal.

5. Every effort will be taken to minimize the waste of viable embryos, which will only be destroyed if implantation is impossible, or if the embryos are flawed and likely to result in deformities.

6. All dogs born as a result of this project shall be treated as pets and placed in loving homes, even if they are not actual clones of Missy. In the unlikely event that a dog is born with deformities or other problems, it will only be euthanized if it is suffering.

7. No transgenic work of any kind shall be performed.

8. Every effort shall be taken to ensure that the technology and procedures developed for the project will be applied in the future in an ethical and a socially positive manner.

9. No data, personnel, or other resources shall be knowingly shared with people or programs seeking to clone human beings.

Case 4

The FDA is in the process of approving the first experiments in xenotransplantation. This experiment will have a very detailed consent form. It should first describe the procedure and then ask the patient to consent to a detailed list of specific knowledge and behavior. Does the patient know that the organ is not a long-term solution and is only a bridge organ? What kind of follow-up or monitoring is required? How will the follow-up be done? Who will need to be followed? Can you drop out? What will be a "penalty" for doing so?

Divide into groups and have each group come up with a list of things that must be in the consent form. After class all persons should take their group's list and anything else they have heard and write a full consent form. Each group should have the following:

A transplant surgeon
A potential patient with liver disease
A community member
A public health official
A civil liberties lawyer
An ethics teacher
A member of the local transplant survivors group

Case 5

The vice president of Restor Corporation, which specializes in designer organs for human transplant, is faced with a difficult issue for his shareholders and for patients. His company is developing transgenic pig organs, altered by the insertion of two human genes, for use as bridge organs in patients until a human organ can be found. The only rationale for the company to continue in this research is expected return to investors from numerous patients who need a transplant and can't yet get one. Unfortunately, the rules established by the government ensure that the first experimental trials of transgenic pig organs be done on patients with no other hope for life. They will be the sickest patients who have been on the waiting lists the longest. They will not be representative of all those needing a transplant. The larger group might one day be a market for Restor. The initial research data are likely to show that almost all of the patients died despite the bridge organ, from underlying causes, or because the organ failed.

How can he get the necessary data that could be useful in average patients using transgenic organs? Won't the data always show that for every individual patient the cost-benefit ratio results in not using the organ?

HUMAN GENETIC SCREENING AND THERAPY

Case 1

In 1994 Myriad Genetics began marketing, with FDA approval, a test for the BRCA1 gene, better known as the "breast cancer gene." According to the company, if a woman had a first-degree relative with breast cancer and tested positive for the gene, she would have an 85 percent chance of developing breast cancer sometime in her life.

Carrie Jones is twenty-five. Her mother just died of breast cancer, as did her aunt. She has been told by her employer's health insurance company that they will pay for the test, but then the results become part of her record. If she changes jobs, the results will go with her. The insurer will also pay for radical double mastectomy so Carrie will never have to worry about breast cancer again. But if nothing is done when the insurance is renewed with the employer, she will have to pay much higher premiums because her risk is so much higher. The knowledge of BRCA1 could be beneficial. She could be more rigorous in her self-examinations. She could see her doctor more frequently and start mammography earlier.

What would you advise her to do?

Case 2

Huntington's disease is a truly awful affliction that kills thousands of Americans every year. It typically appears without warning at ages 35–45 and is invariably fatal. It is an inherited neuromuscular disorder that involves increasing inability to control the body. Starting perhaps with a slight twitch of the hand or slurred speech, it progresses to inability to walk, and eventually to the whole body writhing in spasms. Death results fifteen years after the first symptoms.

For years it was known that the disease was genetic and inherited in a classic dominant/recessive pattern just like cystic fibrosis. In the 1980s a screening test for a genetic marker for the disease was developed. Those with the disease in their family could now know if they will eventually get the disease or pass the gene on to their children.

Do individuals have an obligation to know this information about themselves so they can plan for the future? Do they have an obligation to know and share it with a spouse before marriage and childbearing? Do parents have an obligation to have their children tested? If combined with prenatal testing, should a fetus be aborted?

Case 3

Dr. James Parkinson and his research team at university medical center are the leading researchers on a major study to identify a genetic marker for the susceptibility to Alzheimer's disease. It is widely suspected that Alzheimer's has some genetic component. If this component could be identified, then those with a family history of the disease could be encouraged to be tested. Then these individuals could be given treatments to slow the disease and be candidates for tests of new therapies being developed.

Dr. Parkinson has met with local gerontologists, nursing home directors, and Alzheimer's family support groups to identify potential patients and family members for such a study. Your father has advanced Alzheimer's, and the director of the James Village Care Center has contacted you and your sister to see if your father can participate and to see if you and your sister (ages twenty-eight and thirty-one) would also like to participate. Dr. Parkinson argues that knowledge is power. You can engage in activities now that will moderate and delay the disease. Most of all you can plan your future.

Your sister is not so sure. Since you can't really avoid the disease, why would you want to know? You are at risk every time you drive on the freeway: does it make sense that you would or should know that you will have a fatal accident in ten years? Live life to the fullest every day and don't worry.

Would you agree to participate in the study? Would you let your eight-year-old be tested?

Case 4

Recently the FBI has proposed creating a database of DNA from all persons convicted of a crime, much as is now done with fingerprints. Civil libertarians argue that DNA carries much more information than just a fingerprint and that the information could be used to discriminate against individuals for employment or insurance.

Some have even suggested including in the database those arrested for a crime.

Many rapists or pedophiles are arrested but their victims are too afraid to testify. The database might be very useful in identifying such individuals in the future.

A review committee has been established to make recommendations about these proposals. It has the following members:

A civil liberties lawyer
A victims rights advocate
A local district court judge
A prison reform advocate
A chief of police
A chief scientist of a testing lab
A local prosecutor

Case 5

Denise Brown and her husband are considering participating in a study of germ-line gene therapy. Denise has a family history of retinoblastoma—a tumor that causes blindness in children, usually only in one eye. The genetic basis of most cases of the disease is well-known: deletions or mutations on the q14 band on chromosome 13. With Denise's family history the Browns do not want to take the 40–50 percent chance of having a baby with this disease, and she does not want it to appear in her family tree again. She is willing to participate in the study to see if it can be prevented. She and her husband know the success rate in the study is likely to be low and that failure to carry a pregnancy to term will be high.

Do you think the study should be done? What questions would you ask the researcher? Is the failure rate relevant? Is it acceptable to create embryos deliberately for research purposes when you know that the failure rate of carrying the embryos to term is very high? Is there a difference between this and normal pregnancy where miscarriages are very common?

Case 6

All research involving gene therapy conducted with federal funds must be approved by the recombinant advisory committee of the National Institutes of Health. The committee has just received a research proposal that it is now considering. Dr. Tom Bryson at a major research university medical center is proposing to attempt germ-line modification for couples in which one member has alcoholism or depression in his or her family lineage. The genetic sequence that is partially responsible for alcoholism and depression is well-known by the time this study is proposed. As with any gene therapy study, there will be failures. Some embryos will not carry to term. Perhaps in some children the desired change will not be found.

Unlike diseases with a wholly genetic cause and no effective treatment, these "diseases" are different. There are a number of effective treatments for depression. Millions of Americans are fully functioning adults as a result. Alcoholism may have a genetic component but groups like Alcoholics Anonymous are a living example of the fact that genes do not determine the presence of alcoholism as they do color blindness.

The committee must make a decision about Dr. Bryson's proposal and it has the following members:

A specialist in genetics with the NIH
A psychiatrist
A member of the National Committee to End Abortion
The national vice president of Alcoholics Anonymous
A philosopher with expertise in ethics and genetics
A disability rights lawyer
A layperson who has suffered depression

CLONING AND STEM CELL RESEARCH

Case 1

At University Medical Center all proposed research involving human subjects must be reviewed by a committee composed of experts and laypeople. They must see that the study is well designed, that it is conducted in a manner designed to avoid harm to subjects, and that patients are adequately informed about the risks, benefits, side effects, and follow-up.

Dr. George Sands, a leading researcher in the field, proposes to start experimental human cloning as a treatment for infertility. From a technical point of view the study looks excellent. But one problem has been raised by a lay member of the committee. Most studies in animals show a small number of very abnormal live births. In most cases in all studies the fertilized blastocyst does not result in a pregnancy and a large number of pregnancies do not result in live births. A small number result in live births, some of which are very abnormal and with conditions incompatible with life beyond a few months. Thus the committee member says you can't do the study without deliberate harm to at least some live infants.

Dr. Sands responds that we think it perfectly acceptable for couples to get pregnant when there is a high risk of miscarriage or even when there is a high risk of an abnormal baby (e.g., cystic fibrosis). In vitro fertilization is also done at the university and it results in live embryos being destroyed after the couple has the child they want. So we already allow procedures that we know will result in the deliberate creation and destruction of life.

What stand should the committee take? It has the following members:

A housewife from the community
The medical center lawyer
A teacher of ethics at the center
The chairman of the Department of Internal Medicine
A psychologist in private practice
A biochemist at the center
A professor of pediatrics

Case 2

A national cloning ethics advisory committee has been established by Congress. It has a threefold task. First it must make a recommendation to Congress whether to repeal a law banning all research in human cloning. Second, it must establish the principles that will guide the first human cloning research. These will include principles of informed consent, least harm to the individual, disposal of failed blastocyst and embryos, and finally what is to be done with the small number of highly abnormal late-term embryos or potential live births. Third, the committee must decide whether cloning is to be limited to infertile couples, couples where one partner carries a serious genetic abnormality, or couples who want another child to provide a perfect donor match of something like bone marrow for an older child.

The committee members are the following:

A medical expert
An expert on animal cloning
A representative of the National Sanctity of Life Association
The president of the National Infertility League
The head of the Natural Birth Council
The executive director of the Association of Disabled Americans
The president of Bio-Born, Inc.

Case 4

You are a nationally prominent cloning researcher and a couple has come to you with a desperate plea for help. Their three-year-old has a rare blood disease and is slowly dying. She needs a bone marrow transplant to have any chance of survival. Unfortunately no donor match has been found.

The parents want to use somatic cells from the child to clone another child. Since the cloned child will be an exact genetic match, he or she will be a perfect bone marrow donor. If all goes well, the parents will wind up with two healthy children and their family will be complete.

Having just taken a continuing education course, you have some worries. Will the procedure work? What will be done with the failures? Is the clone being treated inhumanely? Is he or she being used as a means to someone else's welfare? But without cloning the new child with that genetic makeup would never be. Isn't life better than not being?

Case 5

Mr. and Mrs. Smith have come to see Dr. Morris, a world-famous expert on diseases of the eye.

Mrs. Smith: "Dr. Havens, our pediatrician, says that Ronnie has a condition that makes him go blind in one eye. Is there anything we can do?"

Dr. Morris: "I am very sorry but the only thing to do is remove the affected eye."

Mrs. Smith: "Why did this happen? What did I do?"

Dr. Morris: "Nothing. Don't blame yourself. This condition has some genetic component though we don't fully understand it. You did not choose your genes and neither did your baby."

Mrs. Smith: "Can't we do something? What are doctors doing to find a cure?"

Dr. Morris: "Our research is difficult. We need to study more closely how the disease develops in the fetus. We are doing all we can. But abortion clinics usually don't check for this condition in the fetus, and second-trimester abortions are not that common anyway."

Mrs. Smith: "I want something done. Why not clone Ronnie and I will abort right at the time you need for study? I read somewhere that the British government has approved cloning embryos for research purposes, so civilized people must think that it is acceptable. I'm not killing a baby."

What do you think of Mrs. Smith's idea? Is her reasoning sound? What would you say to her?

Index

References to figures are printed in italics.

About the Editors

Richard Sherlock is professor of philosophy at Utah State University, where he has taught for seventeen years. Prior to that, he taught medical ethics at the University of Tennessee Medical School and at McGill University. He was also professor of moral theology at Fordham University in New York. Dr. Sherlock is the author of more than sixty articles and books in fields as diverse as medical ethics (consent, competency, confidentiality, euthanasia, abortion, and the rights of the disabled), families in medical care, early modern philosophy, philosophical theology, and ethical issues in biotechnology. He has books forthcoming in philosophical theology and the theological meaning of genetic change.

John D. Morrey is a professor and research scientist at Utah State University. His work involves drug discovery research for viral infections, genetically engineering laboratory animals as models for human viral infections, genetically engineering dairy farm animals for production of useful proteins in their milk, and some cloning of animals. Hepatitis B and C viruses, HIV, West Nile virus, and mad cow disease agents are some of the infectious agents he has investigated. He has also been involved with teaching courses and workshops on the ethics of biotechnology or the new biology. The class started as a onetime honors course in 1996 in which the ethics of genetic engineering were emphasized. Because of its success, the course has evolved through summer workshops to eventually a depth (general education) course at Utah State University. Before coming to Utah State University, Dr. Morrey was a staff fellow at the National Institutes of Health. He obtained his Ph.D. from Utah State University and an M.S. from Brigham Young University. Other activities include: founder of Physician Services, a clinical laboratory service; founder of Taped Technologies, which produces instructional audiovisual programs in molecular biology; and president and chief executive officer of PanGenics, Inc., which produced genetically engineered animals. He has authored more than forty-five scientific research publications in the areas of virology, drug discovery, genetically engineered animals, and ethics.